OO Cudaback
1952 February

Introduction
to
mechanics
of
solids

EGOR P. POPOV

*Professor
of
Civil Engineering
University
of California
Berkeley*

Introduction to mechanics of solids

PRENTICE-HALL, INC., Englewood Cliffs, *N.J.*

PRENTICE-HALL INTERNATIONAL, INC., *London*
PRENTICE-HALL OF AUSTRALIA PTY. LTD., *Sydney*
PRENTICE-HALL OF CANADA, LTD., *Toronto*
PRENTICE-HALL OF INDIA PRIVATE LTD., *New Delhi*
PRENTICE-HALL OF JAPAN, INC., *Tokyo*

CIVIL ENGINEERING AND
ENGINEERING MECHANICS SERIES
N. M. Newmark and W. J. Hall, Editors

© 1968 by Prentice-Hall, Inc., Englewood Cliffs, New Jersey

All rights reserved. No part of this book may be reproduced in any form or by any means without permission in writing from the publisher.

Library of Congress Catalog Card Number 68-10135

19 18 17 16

Current printing (last digit):

Printed in the United States of America

Preface

This book grew out of an attempted revision of the author's earlier work *Mechanics of Materials* (Prentice-Hall, Inc., 1952). It soon became apparent that because of the changing attitudes toward this subject in a number of our engineering schools and a few new developments, a more rigorous treatment of the subject was necessary to meet current requirements. As a result, the contents of most of the chapters are completely new, and the others have been thoroughly revised.

It is assumed that the reader has had the usual courses in freshman and sophomore mathematics and has completed a course in statics. Topics which are particularly important from these subjects are reviewed and illustrated by examples as they occur.

In spirit this book is intended to provide a basic text for all engineering students at an undergraduate level. However, the material presented is so fundamental that some graduate students may find it useful for review, especially since some of the material is not altogether elementary. Throughout the text the parallel development of the physical and the mathematical aspects of the subject is deliberately pursued. Numerous problems from the various fields of engineering are provided for solution by the student.

Preface

In preparing this text the point of view of the student was constantly kept in mind. This explains, for example, the choice of the sign convention for the shears and moments in beams. The convention here coincides with the one familiar to the student from his earlier courses in mathematics and statics and is the one he will encounter later in elasticity and continuum mechanics. The same sign convention is retained in developing the stress-transformation equations and for use with Mohr's circle. This usage avoids much confusion and provides greater consistency in the treatment of the subject.

Again, to allow the student time both to assimilate new information and gain confidence in solving some of the simpler stress-analysis problems, the chapters on stress and strain transformation are placed in the middle of the book. If one were treating this subject simply as a branch of applied mathematics, the alternative approach of introducing this topic early would be more logical.

Constitutive relations receive considerable attention, and the behavior under applied loads of structural components made of linearly elastic, nonlinearly elastic, plastic, and viscoelastic solids is described in detail. A reasonably complete treatment of yield and fracture criteria for the multiaxial state of stress is presented.

Statically indeterminate situations are introduced early, but drill in the solution of such problems is postponed until the student has gained greater facility in calculating deflections.

Energy methods, including Castigliano's theorems and virtual work, are carefully developed, and applications are illustrated by examples. The column-stability problem is introduced from the viewpoint of the large deflection theory, and an explanation of beam-columns is included.

The book contains more material than can be covered in a one-quarter or a one-semester course. Since sufficient material is provided for a full year of study, the text may stimulate the development of intermediate-level courses in this vital area of engineering. The large scope of the book provides a considerable degree of flexibility in designing specific courses to meet individual requirements. This has two additional advantages. First, the reader will inadvertently acquire a balanced concept of the contents of the subject. Second, the book can serve as a reference after it has fulfilled its purpose as an introductory text.

In most of the chapters the more specialized or complex material can be deleted without destroying the continuity of the text, making it possible to terminate the study of a particular topic at a desired level. Articles which can be readily deleted are identified with an asterisk in the table of contents.

For the first basic course in mechanics of solids it cannot be overemphasized that great care must be taken to keep the contents to a minimum. For example, whereas the discussion of singularity functions certainly belongs in a complete text on the subject, the necessity for studying this topic in the first basic course is questionable. The fascination which students develop in working with this technique and the time it takes to learn it perhaps can be used to better advantage in mastering more thoroughly any one of the more fundamental ideas of mechanics. A completely adequate introductory course can be built around Chapters 3, 4, 8, and 9, with a minimum amount of supporting material from Chapters 2, 5, 6, and 7. Such a course would emphasize the concepts of stress, strain, and their transformation, together with a discussion of the constitutive relations. A brief study of beam deflections for determinate and indeterminate cases in Chapter 11 and an introduction to the buckling phenomenon in Chapter 14 would round out such a course.

Although numerous alternatives are possible in making selections from the text for the first basic course in mechanics of solids, the following is a possible sequence of 28 consecutive assignments for $1\frac{1}{2}$-hour class meetings requiring 2 to 3 hours of outside preparation: (1) Secs. 1-1 through 2-6; (2) Secs. 2-7 through 2-12; (3) Secs. 2-13, 2-14, and 2-15; (4) Secs. 3-1 through 3-7; (5) Secs. 3-8 and 3-9; (6) Secs. 4-1 through 4-8; (7) Secs. 4-9 through 4-16; (8) Secs. 4-17 and 4-18; (9) Secs. 5-1 through 5-6; (10) Secs. 5-7, 5-8, and 5-9; (11) Secs. 6-1 through 6-6; (12) Secs. 6-7, 6-8, and 6-9; (13) Secs. 7-1, 7-2, and 7-3; (14) Secs. 7-4 and 7-5; (15) Secs. 8-1 through 8-3; (16) Secs. 8-4 and 8-5; (17) Secs. 9-1 through 9-6; (18) Secs. 9-7, 9-9, 9-10, 9-11, and 9-14; (19) Secs. 9-16, 9-17, (9-18), and 9-19; (20) Secs. 9-20; (21) Secs. 10-1, 10-2, 10-5, 10-6, and 10-7; (22) Secs. 11-1, 11-2, 11-3, 11-5, and 11-6; (23) Secs. 11-8 and 11-9; (24) Secs. 12-1 and 12-2; (25) Secs. 12-3 and 12-4; (26) Secs. 13-1, 13-2, and 13-3; (27) Secs. 14-1 through 14-5; and (28) Secs. 14-6, 14-7, and 14-8. Two or three problems for solution by the student should accompany each assignment.

The development of this book was strongly influenced by the author's colleagues, his students, and the numerous books on this subject published both here and abroad. The privilege of studying under S. Timoshenko and T. von Karman remains memorable. Special gratitude is due, however, to all of the author's colleagues in the Division of Structural Engineering and Structural Mechanics at the University of California, Berkeley, who over a period of many years profoundly influenced and helped to shape the point of view recorded in this book. Of this group, years of association and critical discussions of the subject with Professors R. W. Clough, H. D. Eberhart, K. S. Pister, and A. C. Scordelis were especially valuable. Their help, particularly that of Professor Scordelis, has also greatly contributed to the variety of problems for solution by the student assembled in this book. The warm encouragement and suggestions by Professor J. M. Raphael were much appreciated. The author is also very pleased to acknowledge the many other present and former members of the Civil Engineering and the Mechanical Engineering staffs at the University of California, Berkeley, who contributed by their discussions or by furnishing some problems for solution or

Preface both: F. Baron, V. V. Bertero, J. Bouwkamp, B. Bresler, C. B. Brown, G. W. Brown, T. Y. Lin, S. J. Medwadowski, C. L. Monismith, J. Penzien, D. Pirtz, M. Polivka, C. W. Radcliffe, J. L. Sackman, C. F. Scheffey, R. A. Seban, C. M. Smith, R. L. Taylor, and E. L. Wilson. The author also thanks the doctoral students in the Division who read portions of the manuscript and offered valuable suggestions. Among these M. Khojasteh-Bakt, Marion Cottrell, Dale Perry, L. Selna, S. Yaghmai, and R. J. Evans were particularly helpful.

Professor Donald Brandt of the Civil Engineering Department of the City College of New York read an early draft of the manuscript and offered a number of significant suggestions for the improvement of the text for which the author is most grateful.

The Prentice-Hall staff has been very cooperative. Nicholas Romanelli, Joseph Di Domenico and James Beggs contributed much to the imaginative format of the book and to the excellence of the art work. Pamela Fischer deserves thanks for her care in preparing the manuscript for the compositor.

The author is deeply indebted to his wife, Irene, for her continual help with the preparation of manuscript.

E. P. POPOV

El Cerrito, California

Contents

1 Introduction   1

1-1 Purpose and Scope   1
1-2 Method of Sections   3
1-3 Basic Approach   5

2 Axial force, shear, and bending moment   8

2-1 Introduction   8
2-2 General Remarks   9

PART A
CALCULATION OF REACTIONS   11

2-3 Diagrammatic Conventions for Supports   11
2-4 Diagrammatic Conventions for Loading   13
2-5 Classification of Beams   15
2-6 Calculation of Beam Reactions   16

PART B
AXIAL-FORCE, SHEAR, AND MOMENT DIAGRAMS: A DIRECT APPROACH   21

2-7   Application of the Method of Sections   21
2-8   Shear in Beams   21
2-9   Axial Force in Beams   23
2-10 Bending Moment in Beams   23
2-11 Shear, Axial-Force, and Bending-Moment Diagrams   25
2-12 Step-by-Step Procedure   31

PART C
SHEAR AND MOMENT DIAGRAMS: A SUMMATION APPROACH 32

- 2-13 Differential Equations of Equilibrium 32
- 2-14 Shear Diagrams by Summation 34
- 2-15 Moment Diagrams by Summation 36
- *2-16 Further Remarks on the Construction of Shear and Moment Diagrams 42
- *2-17 Moment Diagram and the Elastic Curve 46

PART D
SINGULARITY FUNCTIONS 47

- 2-18 Notation for and Integration of Singularity Functions 47
- Problems for Solution 53

3 Stress and axial loads 62

- 3-1 Introduction 62

PART A
STRESS 62

- 3-2 Definition of Stress 62
- 3-3 Stress Tensor 64
- *3-4 Differential Equations of Equilibrium 67

PART B
STRESSES IN AXIALLY LOADED MEMBERS 69

- 3-5 Axial Load; Normal Stress 69
- 3-6 Axial Load; Bearing Stress 72
- 3-7 Average Shearing Stress 72
- 3-8 Allowable Stresses; Factor of Safety 82
- 3-9 Design of Axially Loaded Members and Pins 85

4 Strain, constitutive laws, and axial deformation 93

- 4-1 Introduction 93

PART A
STRAIN 93

- 4-2 Physical Meaning of Strain 93
- 4-3 Mathematical Definition of Strain 95
- *4-4 Strain Tensor 97

PART B
LINEAR STRESS-STRAIN LAWS AND STRAIN ENERGY 99

- *4-5 Hooke's Law for Anisotropic Materials 99
- 4-6 Hooke's Law for Isotropic Materials 100
- 4-7 Poisson's Ratio 102

	4-8	Thermal Strains 104
	4-9	Elastic Strain Energy for Uniaxial Stress 105
	4-10	Elastic Strain Energy for Shearing Stresses 107
*4-11		Strain Energy for Multiaxial States of Stress 108

PART C
CONSTITUTIVE RELATIONS FOR UNIAXIAL STRESSES 109

- 4-12 Stress-Strain Diagrams 109
- 4-13 Further Remarks on Stress-Strain Diagrams 110
- 4-14 Stress-Strain Diagrams During Unloading and Load Reversals 113
- 4-15 Idealized Stress-Strain Diagrams 114
- 4-16 Linear Viscoelastic Materials 116

PART D
DEFORMATION OF AXIALLY LOADED MEMBERS 123

- 4-17 Deflection of Axially Loaded Members 123
- 4-18 Stress Concentrations 130
  Problems for Solution 134

5 Torsion   143

- 5-1 Introduction 143
- 5-2 Application of Method of Sections 144
- 5-3 Basic Assumptions 145
- 5-4 The Torsion Formula 146
- 5-5 Remarks on the Torsion Formula 149
- 5-6 Design of Circular Members in Torsion 152
- 5-7 Angle of Twist of Circular Members 153
- 5-8 Shearing Stresses and Deformations in Circular Shafts in the Inelastic Range 158
- *5-9 Stress Concentrations 162
- *5-10 Twist of Viscoelastic Circular Bars 164
- *5-11 Solid Noncircular Members 166
- *5-12 Thin-Walled Hollow Members 169
  Problems for Solution 172

6 Bending stresses in beams   177

- 6-1 Introduction 177
- 6-2 Some Important Limitations of the Theory 178
- 6-3 The Basic Kinematic Assumption 178
- 6-4 The Elastic Flexure Formula 181
- *6-5 Pure Bending of Beams with Unsymmetrical Section 185
- 6-6 Computation of the Moment of Inertia 186
- *6-7 Remarks on the Flexure Formula 188

Contents

    6-8   Inelastic Bending of Beams   192
*6-9   Stress Concentrations   200
*6-10   Beams of Two Materials   202
*6-11   Curved Beams   208
         Problems for Solution   213

7 Shearing stresses in beams   219

7-1   Introduction   219
7-2   Some Preliminaries   220
7-3   Shear Flow   223
7-4   The Shearing Stress Formula for Beams   229
*7-5   Limitations of the Shearing Stress Formula   238
*7-6   Further Remarks on the Distribution of Shearing Stresses   239
*7-7   Shear Center   242
      Problems for Solution   246

8 Compound stresses   252

8-1   Introduction   252
8-2   Superposition and Its Limitation   253
8-3   Skew Bending   260
8-4   Eccentrically Loaded Members   265
8-5   Superposition of Shearing Stresses   270
*8-6   Stresses in Closely Coiled Helical Springs   272
*8-7   Deflection of Closely Coiled Helical Springs   274
      Problems for Solution   276

9 Transformation of stress and strain; yield and fracture criteria   283

9-1   Introduction   283

PART A
TRANSFORMATION OF STRESS   284

9-2   The Basic Problem   284
9-3   Equations for the Transformation of Plane Stress   287
9-4   Principal Stresses   288
9-5   Maximum Shearing Stresses   290
9-6   An Important Transformation of Stress   294
9-7   Mohr's Circle of Stress   295
*9-8   Construction of Mohr's Circle of Stress   297
*9-9   Mohr's Circle of Stress for the General State of Stress   302

PART B
TRANSFORMATION OF STRAIN 304

- 9-10 General Remarks 304
- 9-11 Equations for the Transformation of Plane Strain 305
- 9-12 Alternative Derivation of Equation 9-13 307
- *9-13 Mohr's Circle of Strain 308
- *9-14 Strain Measurements; Rosettes 311
- *9-15 Additional Linear Relations Between Stress and Strain and Among E, G, and ν 313

PART C
YIELD AND FRACTURE CRITERIA 316

- 9-16 Preliminary Remarks 316
- 9-17 Maximum Shearing Stress Theory 316
- *9-18 Maximum Distortion Energy Theory 319
- 9-19 Maximum Normal Stress Theory 323
- *9-20 Comparison of Theories; Other Theories 324
  Problems for Solution 326

10 Problems in stress analysis 332

- 10-1 Introduction 332

PART A
ANALYSIS OF STRESSES 333

- 10-2 Investigation of Stress at a Point 333
- *10-3 Members in a State of Two-Dimensional Stress 339
- *10-4 The Photoelastic Method of Stress Analysis 340
- 10-5 Thin Shells of Revolution 344
- 10-6 Equilibrium Equations for Thin Shells of Revolution 346
- *10-7 Remarks on Thin-Walled Pressure Vessels 351

PART B
DESIGN OF MEMBERS TO MEET STRENGTH REQUIREMENTS 352

- *10-8 General Remarks 352
- *10-9 Design of Axially Loaded Members 352
- *10-10 Design of Torsion Members 353
- *10-11 Design Criteria for Prismatic Beams 353
- *10-12 Design of Prismatic Beams 356
- *10-13 Design of Nonprismatic Beams 360
- *10-14 Design of Complex Members 362
  Problems for Solution 365

11 Deflection of beams 378

- 11-1 Introduction 378
- 11-2 Strain-Curvature and Moment-Curvature Relations 379

11-3 The Governing Differential Equation for Deflection of Elastic Beams 382
*11-4 An Alternative Derivation of Equation 11-10 384
11-5 Alternative Differential Equations of Elastic Beams 385
11-6 Boundary Conditions 385
*11-7 Deflection of Viscoelastic Beams 387

PART A
DIRECT INTEGRATION METHODS 389

11-8 Solution of Beam Deflection Problems by Direct Integration 389
11-9 Statically Indeterminate Elastic Beam Problems 400
*11-10 Two Additional Singularity Functions 403
*11-11 Remarks on the Elastic Deflection of Beams 406
*11-12 Elastic Deflection of Beams in Skew Bending 408
*11-13 Inelastic Deflection of Beams 408

PART B
MOMENT-AREA METHOD 411

*11-14 Introduction to the Moment-Area Method 411
*11-15 Derivation of the Moment-Area Theorems 412
Problems for Solution 424

12 Statically indeterminate problems 432

12-1 Introduction 432

PART A
ANALYSIS WITH THE AID OF DISPLACEMENT RELATIONS 433

12-2 A General Approach 433
12-3 Stresses Caused by Temperature 441

PART B
ANALYSIS BY THE METHOD OF SUPERPOSITION 444

12-4 Method of Analysis 444
*12-5 Moment-Area Method for Statically Indeterminate Beams 450
*12-6 The Three-Moment Equation 456
*12-7 Special Cases 459

PART C
LIMIT ANALYSIS OF BEAMS 462

*12-8 Elastic-Plastic Bending of Beams 462
*12-9 Concluding Remarks 468
Problems for Solution 469

13 Energy methods   481

    13-1   Introduction   481
    13-2   Elastic Strain Energy   482
    13-3   Displacements by the Energy Method   484
    13-4   Castigliano's Deflection Theorem   488
  *13-5   Reciprocal Theorem   494
  *13-6   Generalization of Castigliano's Theorems   495
  *13-7   Virtual-Work Method for Deflections   496
  *13-8   Virtual-Work Equations for Elastic Systems   498
  *13-9   Statically Indeterminate Problems   504
             Problems for Solution   506

14 Buckling of columns   515

    14-1   Introduction   515
    14-2   Nature of the Beam-Column Problem   517
  *14-3   Differential Equations for Beam-Columns   521
    14-4   Stability of Equilibrium   524
    14-5   Euler Buckling Load for Pin-Ended Columns   526
    14-6   Elastic Buckling of Columns with Different End Restraints   529
    14-7   Limitation of the Elastic Buckling Formulas   530
    14-8   Generalized Euler Buckling-Load Formulas   532
  *14-9   Eccentrically Loaded Columns   534
*14-10   Design of Columns   535
*14-11   Column Formulas for Concentric Loading   537
*14-12   On the Energy Approach for Determining Buckling Loads   541
             Problems for Solution   543

Appendix tables   553

Index   565

Abbreviations and symbols*

ABBREVIATIONS

allow	allowable
av	average
cr	critical
ft	feet
hp	horsepower
I	I beam
in.	inches
k	kips
kip	kilo-pounds (1,000 lb)
ksi	kips per square inch
lb	pounds (from Latin *libra* meaning weight)
max	maximum
min	minimum
NA	neutral axis
psi	pounds per square inch
rpm	revolutions per minute
ult	ultimate
WF	wide flange beam
yp	yield point

GREEK LETTER SYMBOLS

α	(alpha)	coefficient of thermal expansion, angle

* With very few exceptions, the abbreviations and letter symbols conform with those approved by the American Standards Association.

Abbreviations and symbols

β	(beta)	angle, constant
γ	(gamma)	shearing strain, weight per unit volume
Δ	(delta)	deflection, change of a function
ε, ϵ	(epsilon)	linear strain, positive small number
η	(eta)	coefficient of viscosity
θ	(theta)	angle of elastic curve, angle of twist per unit length
κ	(kappa)	curvature
λ	(lambda)	eigenvalue in column buckling problems
ν	(nu)	Poisson's ratio
Π, π	(pi)	total potential, ratio of circumference to its diameter
ρ	(rho)	radius, radius of curvature
σ	(sigma)	normal stress
τ	(tau)	shearing stress
Φ, φ, ϕ	(phi)	area of $M/(EI)$ diagram, total angle of twist, angle
Ω, ω	(omega)	potential energy, angular velocity

ROMAN LETTER SYMBOLS

A	area, area of cross section, constant
A_{fghj}	partial area of beam cross-sectional area
B	constant
b	breadth, width
C	constant
c	distance from neutral axis or from center of twist to extreme fiber
d	diameter, distance, depth
E	modulus of elasticity in tension or compression
F	force
f	flexibility coefficient, internal virtual force
G	modulus of elasticity in shear
g	acceleration of gravity
h	height, depth of beam
I	moment of inertia of cross-sectional area
J	polar moment of inertia of cross-sectional area
J_c	creep compliance
K	stress concentration factor, effective length factor
k	spring constant, constant, bulk modulus
L	length, span length
M	moment, bending moment
m	mass, moment caused by virtual unit load
N	number of revolutions per minute
n	number, ratio of moduli of elasticity
P	force, concentrated load
p	distributed load per unit length, intensity of pressure
Q	first or statical moment of area A_{fghj} around neutral axis

Abbreviations and symbols

q	shear flow
R	reaction, radius
r	radius, radius of gyration
S	elastic section modulus ($S = I/c$)
s	distance along a line or a curve
T	torque, temperature
t	thickness, width, tangential deviation
t_x	torque per unit length
U	strain energy
V	shearing force, volume
W	work, total weight
u, v, w	displacement components
X, Y	body forces, unknowns
Z	plastic section modulus
x, y, z	coordinates of a material point

Introduction 1

1-1. PURPOSE AND SCOPE

In all engineering construction the component parts of a structure must be assigned definite physical sizes. Such parts must be properly proportioned to resist the actual or probable forces that may be imposed upon them. Thus, the walls of a pressure vessel must be of adequate strength to withstand the internal pressure; the floors of a building must be sufficiently strong for their intended purpose; the shaft of a machine must be of adequate size to carry the required torque; a wing of an airplane must safely withstand the aerodynamic loads which may come upon it in flight or landing. Likewise, the parts of a composite structure must be rigid enough so as not to deflect or "sag" excessively when in operation under the imposed loads. A floor of a building may be strong enough but may deflect excessively, which in some instances may cause misalignment of manufacturing equipment or in other cases result in the cracking of a plaster ceiling attached underneath. Finally, a member may be so thin or slender that, upon being subjected to compressive loading, it will collapse through buckling; i.e., the initial configuration of a member may become unstable. Ability to determine the

*Chapter 1
Introduction*

maximum load which a slender column can carry before buckling occurs or determination of the safe level of vacuum which can be maintained by a vessel is of great practical importance.

In engineering practice all the above requirements must be met with the minimum expenditure of a given material. Aside from cost, at times—as in the design of satellites—the feasibility and success of the whole mission may depend on the weight of a package.

In the past, texts treating the problems suggested above were called either *Strength of Materials* or *Mechanics of Materials*. To reflect current usage and the more rigorous treatment of the subject, the new title *Introduction to Mechanics of Solids* has been selected for this book. Alternatively, the subject could be called the mechanics of solid deformable bodies. It must be pointed out, however, that much of the material presented in this text is a technical theory of deformable bodies in contrast to the mathematical theory of elasticity or the theory of perfectly plastic solids. Here, instead of establishing every step rigorously from the mathematical point of view, simplifying assumptions are introduced to make a reasonable solution of basic problems possible. It is interesting that the very important theory of plates and shells is principally an extension of the technical theory discussed in this text. Regardless of the details of the title or the rigor, this subject involves the analytical methods for determining the **strength**, **stiffness** (deformation characteristics), and **stability** of the various load-carrying members.

Mechanics of solids is a fairly old subject. It is generally dated from the work of Galileo in the early part of the seventeenth century. Prior to his investigations into the behavior of solid bodies under loads, constructors followed precedent and empirical rules. Galileo was the first to attempt to explain the behavior of some of the members under load on a rational basis. He studied members in tension and compression, and notably beams used in the construction of hulls of ships for the Italian navy. Of course much progress has been made since that time, but in passing it must be noted that much is owed in the development of this subject to the French investigators, among whom a group of outstanding men such as Coulomb, Poisson, Navier, St. Venant, and Cauchy, who worked at the break of the nineteenth century, has left an indelible impression on this subject. The subject is far from closed, however. The space age continually brings more exacting and broader demands. Mechanics of solids cuts broadly across all branches of the engineering profession with remarkably many applications. Its methods are needed by the designers of submarines; by the civil engineer in the design of bridges and buildings; by the engineers designing spacecraft; by the mining engineer and the architectural engineer, each of whom is interested in structures; by the mechanical and chemical engineers, who rely upon the methods of this subject for the design of machinery and pressure vessels; by metallurgists, who need the fundamental concepts of this subject in order to understand how further to improve existing materials; finally, by the electrical engineer, who needs the methods of this subject because of the importance of the mechanical engineering phases of many portions of electrical equipment. Technical mechanics of solids has characteristic methods of its own. It is a definite discipline; thus it is one of the most fundamental subjects of an engineering curriculum, standing alongside other basic subjects such as fluid mechanics, thermodynamics, and a basic course in electricity.

The behavior of a member subjected to forces depends not only on the fundamental laws of Newtonian mechanics governing the equilibrium of the forces but also on the mechanical characteristics of the materials of which the member is fabricated. The necessary information regarding the latter comes from the laboratory where materials are subjected to the action of accurately known forces and the behavior of test specimens is observed with particular regard to such phenomena as the occurrence of breaks, deformations, etc. Determination of such phenomena is a vital part of the subject, but this branch of the subject is left to other books.* Here the end results of such investigations are of interest, and this course is concerned with the analytical or mathematical part of the subject in contradistinction to experimentation. For the above reasons, it is seen that mechanics of solids is a blended science of experiment and Newtonian postulates of analytical mechanics. From the latter is borrowed the branch of the science called *statics*, a subject with which the reader of this book is presumed to be familiar and on which the subject of this book primarily depends.

This text will be limited to the simpler topics of the subject as it is an introductory one. However, in spite of the relative simplicity of the methods employed here, the resulting techniques are unusually useful as they do apply to a vast number of technically important problems.

This is essentially a problem course, hence the subject matter can be mastered only by solving numerous problems. The number of formulas necessary for the conventional analysis and design of structural and machine members by the methods of mechanics of solids is not very large. However, throughout this study the student must develop an ability to *visualize* the problem at hand and the nature of the quantities being computed. Complete, carefully drawn, diagrammatic sketches of problems to be solved will pay large dividends in a quicker and more complete mastery of the subject.

1-2. METHOD OF SECTIONS

The main problem of mechanics of solids is the investigation of the internal resistance and the deformation of a solid body subjected to

---

* H. E. Davis, G. E. Troxell and C. T. Wiskocil, *Testing and Inspection of Engineering Materials* (2nd ed.) (New York: McGraw-Hill Book Co., 1955). Also, C. W. Richards, *Engineering Materials Science* (Belmont, Calif.: Wadsworth Publishing Co., Inc., 1961).

*Chapter 1
Introduction*

loads. This requires a study of the nature of the forces set up within a body to balance the effect of the externally applied forces. For this purpose, a uniform method of approach is employed. A complete diagrammatic sketch of the member to be investigated is prepared, on which all of the external forces acting on a body are shown at their respective points of application. Such a sketch is called a *free-body diagram*. All forces acting on a body, including the reactive forces caused by the supports and the weight* of the body itself, are considered external forces. Moreover, since a stable body at rest is in equilibrium, the forces acting on it satisfy the equations of static equilibrium. Thus, if the forces acting on a body such as the one in Fig. 1-1(a) satisfy the equations of static equilibrium and are all shown acting on it, the sketch represents a free-body diagram. Next, since a determination of the internal forces caused

---

* Strictly speaking, the weight of the body, or more generally, the inertial forces due to acceleration, etc., are *body forces* and act throughout the body in a manner associated with the units of volume of the body. In most instances, these body forces can be considered external loads.

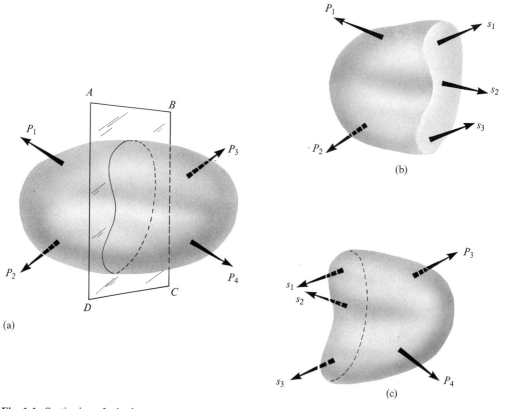

Fig. 1-1. Sectioning of a body.

by the external ones is one of the principal concerns of this subject, an arbitrary section is passed through the body, completely separating it into two parts. The result of such a process may be seen in Figs. 1-1(b) and (c), where an arbitrary plane *ABCD* separates the original solid body of Fig. 1-1(a) into two distinct parts. This process will be referred to as the *method of sections.* Then, if the body as a whole is in equilibrium, any part of it must also be in equilibrium. For such parts of a body, however, some of the forces necessary to maintain equilibrium must act at the cut section. These considerations lead to the following fundamental conclusion: the externally applied forces to one side of an arbitrary cut must be balanced by the internal forces developed at the cut, or briefly, the external forces are balanced by the internal forces. Later it will be seen that the cutting planes will be oriented in a particular direction to fit special requirements. However, the above concept is applied in all problems where internal forces are being investigated.

In discussing the method of sections, it is significant to note that some bodies, although not in static equilibrium, may be in dynamic equilibrium. These problems can be reduced to problems of static equilibrium. First, the acceleration of the part in question is computed, then it is multiplied by the mass of the body, giving a force. The force so computed, if applied to the body in a direction opposite to the acceleration at its mass center, reduces the dynamic problem to one of statics. This is the *d'Alembert principle.* With this point of view, all bodies can be thought of as being instantaneously in a state of static equilibrium. Hence for any body, whether in static or dynamic equilibrium, a free-body diagram can be prepared on which the necessary forces to maintain the body as a whole in equilibrium can be shown. From then on the problem is the same as discussed above.

1-3. BASIC APPROACH

The method of attacking problems in this text follows remarkably uniform lines. At times the procedure is obscured by intermediate steps, but in the final analysis it is always applied. To give an overall view of the subject, a typical procedure is outlined below. A more complete appreciation of many of the listed items will be gained as the subject unfolds. It is suggested that the student periodically review this article after the study of further chapters.

1. From a particular arrangement of structural or machine elements, a single member is isolated. Such a member is indicated on a diagram with all the forces and reactions acting on it. This is the free body of the whole member.

2. The reactions are determined by the application of the equations of statics or of boundary conditions with the appropriate differential equations. In indeterminate problems, statics is supplemented with kinematic considerations.

*Chapter 1
Introduction*

3. At a point where the magnitude of the stress is wanted, a section perpendicular to the axis of the body is passed, and a portion of the body, to either one side of the section or the other, is completely removed.

4. At the section investigated, the system of internal forces necessary to keep the isolated part of the member in equilibrium is determined. In general, this system of forces consists of an axial force, a shear force, a bending moment, and a torque. These quantities are found by treating a part of the member as a free body.

5. With the system of forces at the section properly resolved, the established formulas enable one to determine the stresses at the section considered.

6. If the magnitude of the maximum stress at a section is known, one can provide proper material for such a section; or, conversely, if the physical properties of a material are known, one can select a member of adequate size.

7. In certain other problems, the knowledge of deformation in a member at an arbitrary section caused by the internal forces enables one to predict the deformation of the structure as a whole and hence, if necessary, to design members that do not deflect or sag excessively.

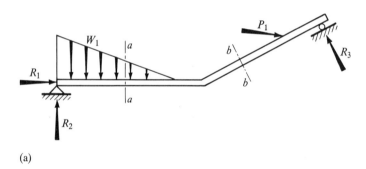

(a)

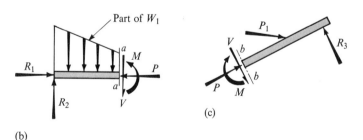

(b)

(c)

P—Axial Force, V—Shear, M—Bending Moment

Fig. 1-2. Free body of a member and two of its parts.

Section 1-3
Basic approach

Figure 1-2 is a schematic illustration of some of the above steps for a two-dimensional problem. Two free bodies isolated from the member by sections *a-a* and *b-b*, respectively, are shown in Figs. 1-2(b) and (c).

An ability to visualize the nature of the quantities computed is essential. All quantities considered in mechanics of solids have a definite physical meaning, many of which can be schematically interpreted. Free-body diagrams help in this immensely.

2 Axial force, shear, and bending moment

2-1. INTRODUCTION

After some preliminary remarks, the subject matter of this chapter is divided into four main parts. Part A is a review of calculation procedures for reactions. Part B discusses a direct approach, based on equations of statics, for determining axial force, shear force, and bending moments at any section of a member. Construction of diagrams for these quantities is then considered. The differential equations for beam equilibrium are derived in Part C, and instructions are given for constructing shear and moment diagrams. Singularity functions for treating discontinuous functions by a method of operational calculus are introduced in Part D. This enables one to treat analytically a broad range of discontinuous loading conditions encountered in practice.

In this chapter attention will be confined to two-dimensional or planar structures and principally to beams since innumerable applications of beams may be found in structures and as elements of machines. The main members supporting floors of buildings are beams, just as an axle of a car is a beam. Many shafts of machinery act simultaneously as torsion members and as beams. With modern materials, the beam is a dominant member of construction.

Beams may be straight or curved, but the major attention of this chapter will be directed toward a study of straight beams. Straight beams occur more frequently in practice; moreover, the system of forces at a section of a straight beam is the same as in a curved one. Hence, if the behavior of a straight beam is understood, little needs to be added regarding curved beams. Further, although in actual installations a straight beam may be vertical, inclined, or horizontal, for convenience the beams discussed here will be shown in a horizontal position. The more general, three-dimensional problems will be encountered in Chapters 8, 10, and 11.

The contents of this chapter may be familiar to some students. Nevertheless, a review of Parts A, B, and C is desirable since a thorough knowledge of this material must be had prior to the study of chapters that follow. The study of Part D may be postponed until later.

2-2. GENERAL REMARKS

For the equilibrium of a solid body, the equations of statics require the fulfillment of the following conditions:

$$\sum F_x = 0 \qquad \sum M_x = 0$$
$$\sum F_y = 0 \qquad \sum M_y = 0 \qquad (2\text{-}1)$$
$$\sum F_z = 0 \qquad \sum M_z = 0$$

The first column states that the sum of all forces acting on a body in any (x, y, z) direction must be zero. The second column notes that the summation of moments of all forces around any axis parallel to any (x, y, z) direction must also be zero for equilibrium. In a planar problem, where all members and forces lie in a single plane such as the x-y plane, relations $\sum F_z = 0$, $\sum M_x = 0$, and $\sum M_y = 0$, although still valid, are trivial. Equation 2-1 should be already familiar to the student.

These equations of statics are directly applicable to deformable solid bodies, using their initial dimensions. The deformations tolerated in engineering structures are usually negligible in comparison with the overall dimensions of structures. Therefore, with the exception of rare cases, for purposes of obtaining the forces in members, the initial undeformed dimensions of members are used in computations.

There are problems where equations of statics are not sufficient to determine the forces acting on the member. For example, the reactions for a straight beam, supported vertically at three points, shown in Fig. 2-1, cannot be determined from statics alone. In this planar problem there are four unknown reaction components, but only three independent equations of statics

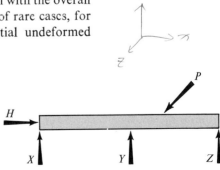

Fig. 2-1. A statically indeterminate beam.

*Chapter 2
Axial force, shear,
and bending moment*

are available. Such problems are termed <u>*externally statically indeterminate*</u>. The consideration of them is postponed until Chapters 11 and 12. For the present, and in the following eight chapters of this text, for the most part only <u>externally statically determinate</u> members will be considered; i.e., all the external reactions acting on such bodies can be determined by Eq. 2-1. There is no dearth of statically determinate problems that are significant practically. Once a better acquaintance with this subject is acquired, extension of procedures to statically indeterminate situations will be relatively simple.

As noted above, reactions in a statically determinate member can be calculated using Eq. 2-1. This is usually the first step in analysis. In the next phase, a section through a body is made to isolate the selected part. By reapplying Eq. 2-1, the force components required for equilibrium of the isolated part of the body are determined. **Throughout this text the force components will be resolved along a right-hand system of Cartesian axes** as shown in Fig. 2-2(a). Usually the x and y axes will be taken in the plane of the paper with the z axis pointing toward the reader. In general, the origin may be taken at any convenient point or moved along the member. However for bars it will usually be located at the left end at the centroid of the cross-sectional area.

The positive sense of the force components on the cut section viewed

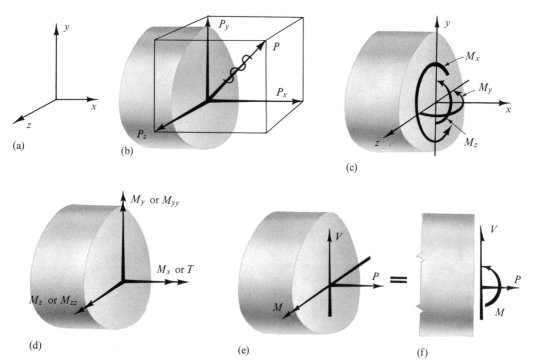

Fig. 2-2. Definition of positive force components at a section of a body.

toward the origin coincides with the positive directions of the coordinate axes, Fig. 2-2(b). The equilibrating forces on the back or the far side of a segment (not shown in the figure) act in the directions opposite to those of the positive coordinate axes. On this basis, at the near section of a member, a force vector P can be expressed as

$$P = P_x i + P_y j + P_z k$$

where P_x, P_y, P_z are the components of the force P, and i, j, k are the unit vectors along the x, y, and z axes, respectively.

The three components of moment which can occur at a section of a member act around the three coordinate axes as shown in Fig. 2-2(c). In this text these quantities will be represented alternatively by double-headed vectors as in Fig. 2-2(d). The sense of these vectors follows the right-hand screw rule.* For emphasis, to show the axis around which a particular bending moment such as M_y or M_z acts, double subscripts will occasionally be employed. Since M_x is the torque which acts on a member, it will be identified by the letter T.

In much the same manner as a force vector, moment vector M can be expressed as

$$M = M_x i + M_y j + M_z k$$

where $M_x \equiv T$, $M_y \equiv M_{yy}$, and $M_z \equiv M_{zz}$ are the components of M along the coordinate axes, as in Fig. 2-2(d).

For planar problems the notation for and the diagrammatic representation of the force components used in this text are shown in Figs. 2-2(e) and (f).

The analysis relating the forces in Fig. 2-2 to the intensity of internal forces (stresses) and deformations which they cause in the members will be given in subsequent chapters. As will be shown, these basic problems are internally statically indeterminate. Again, in general, the equations of statics are not sufficient to solve such problems; deformation assumptions as well as properties of materials must be used.

Section 2-3 Diagrammatic conventions for supports

PART A
CALCULATION OF REACTIONS

2-3. DIAGRAMMATIC CONVENTIONS FOR SUPPORTS

In studying beams it is imperative to adopt diagrammatic conventions for their supports and loadings inasmuch as several kinds of supports and a great variety of loads are possible. A thorough mastery of and adherence to such conventions avoid much confusion and minimize

---

* A right-hand screw advances with the sense of the vector when it is twisted in the sense indicated by the moment couple.

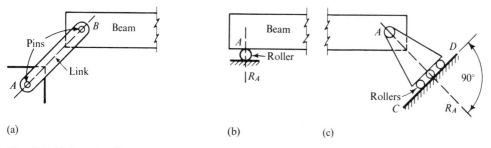

(a) (b) (c)

Fig. 2-3. Link and roller supports. (The only possible line of action of the reaction is shown by the dashed lines.)

the chances of making mistakes. These conventions form the pictorial language of engineers. As stated in the introduction, for convenience the beams will usually be shown in a horizontal position.

Three types of supports are recognized for beams loaded with forces acting in the same plane. These are identified by the kind of resistance they offer to the forces. One support is physically realized by a *roller* or a *link*. It is capable of resisting a force in only one specific line of action. The link in Fig. 2-3(a) can resist a force only in the direction of line *AB*. The roller in Fig. 2-3(b) can resist only a vertical force, and the rollers in Fig. 2-3(c) can resist only a force which acts perpendicularly to the plane *CD*. A reaction of this type corresponds to a single unknown when equations of statics are applied. For inclined reactions the *ratio* between the two components is fixed.

Another support that may be used for a beam is a *pin*. In actual construction such a support is realized by using a detail as in Fig. 2-4(a). In this text such supports will be represented diagrammatically as in Fig. 2-4(b). A pinned support is capable of resisting a force acting in any direction of the plane. Hence, in general, the reaction at such a support may have two components, one in the horizontal and one in the vertical direction. Unlike the ratio applying to the roller or link support, that between the reaction components for the pinned support is not fixed. To determine these two components, two equations of statics must be used.

The third support used for beams is capable of resisting a force in any direction and is also capable of resisting a couple or a moment. Physically such a support is obtained by building a beam into a brick wall,

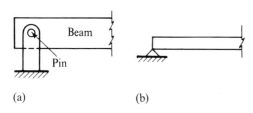

(a) (b)

Fig. 2-4. Pinned support: (a) actual, (b) diagrammatic.

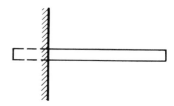

Fig. 2-5. Fixed support.

casting it into concrete, or welding an end of a beam to the main structure. A system of three forces can exist at such a support, two components of force and a moment. Such a support is called a *fixed support*, i.e., the built-in end is fixed or prevented from rotating. The standard convention for indicating it is in Fig. 2-5.

To differentiate fixed supports from the roller and pin supports, which are not capable of resisting moment, the latter two are termed *simple supports*. Figure 2-6 summarizes the foregoing distinctions between the three types of supports, and the kind of resistance offered by each type is shown. Practicing engineers normally assume the supports to be of one

*Section 2-4
Diagrammatic
conventions for
loading*

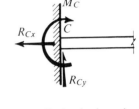

Fig. 2-6. The three common supports.

of the three types by "judgment," although in actual construction supports for beams do not always clearly fall into these classifications. A more refined investigation of this aspect of the problem is beyond the scope of this text.

2-4. DIAGRAMMATIC CONVENTIONS FOR LOADING

Beams are called upon to support a variety of loads. Frequently a force is delivered to the beam through a post, a hanger, or a bolted detail as in Fig. 2-7(a). Such arrangements apply the force over a very limited portion of the beam and are idealized, for the purposes of beam analysis, as *concentrated* forces. These are shown diagrammatically in Fig. 2-7(b).

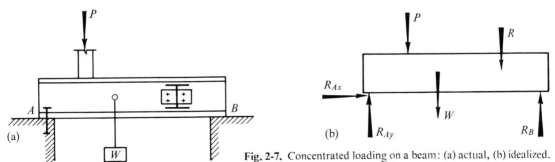

Fig. 2-7. Concentrated loading on a beam: (a) actual, (b) idealized.

13

*Chapter 2
Axial force, shear,
and bending moment*

On the other hand, in many instances the forces are applied over a considerable portion of the beam. For example, in a warehouse goods may be piled up along the length of a beam. Such forces are termed *distributed loads*. Many types occur. Among these, two are particularly important: the *uniformly distributed* loads and the *uniformly varying* loads. The first could easily be an idealization of the warehouse load just mentioned, where the same kind of goods is piled up to the same height along the beam. Likewise the beam itself, if of constant cross-sectional area, is an excellent illustration of the same kind of loading. A realistic situation and a diagrammatic idealization are in Fig. 2-8. This load is usually expressed in pounds per lineal foot (or inch) of the beam, unless specifically noted otherwise, and is abbreviated as lb per ft or lb/ft.

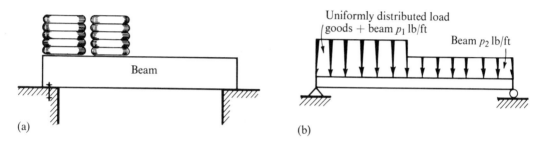

Fig. 2-8. Distributed loading on a beam: (a) actual, (b) idealized.

Uniformly varying loads act on the vertical and inclined walls of a vessel containing liquid. This is illustrated in Fig. 2-9, where it is assumed that the vertical beam is 1-ft wide and γ (lb per cubic foot) is the unit weight of the liquid. For this type of loading, it should be carefully noted that the maximum intensity of the load of p lb per foot is applicable only to an infinitesimal length of the beam. It is twice as large as the average intensity of pressure. Hence the total force exerted by such a loading on a beam is $\frac{1}{2}ph$ lb. Horizontal bottoms of vessels containing liquid are loaded uniformly. Aerodynamic loadings are of a distributed type.

Finally, it is conceivable to load a beam with a concentrated moment applied to the beam essentially at a point. One possible arrangement of many for applying a concentrated moment is in Fig. 2-10(a), and its diagrammatic representation as used in this text is in Fig. 2-10(c).

The necessity for a complete understanding of the foregoing symbolic representation for supports and forces cannot be overemphasized.

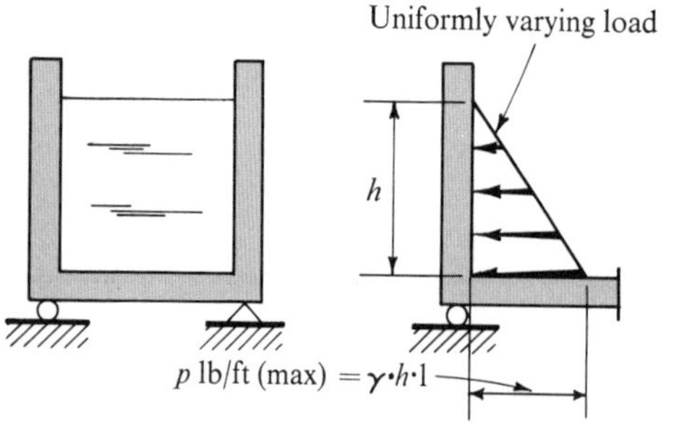

Fig. 2-9. Hydrostatic loading on a vertical wall.

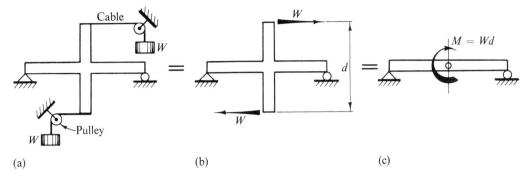

Fig. 2-10. A method of applying a concentrated moment to a beam.

Note particularly the kind of resistance offered by the different kinds of supports and the manner of representing forces at such supports. These notations will be used to construct free-body diagrams for beams.

2-5. CLASSIFICATION OF BEAMS

Beams are classified into several groups, depending primarily on the kind of support used. Thus if the supports are at the ends and are either pins or rollers, the beams are *simply supported* or *simple* beams, Fig. 2-11(a) and (b). The beam becomes a *fixed* beam or *fixed-ended* beam, Fig. 2-11(c), if the ends have fixed supports. Likewise, following the same scheme of nomenclature, the beam shown in Fig. 2-11(d) is a beam fixed at one end and simply supported at the other. Such beams are also called *restrained* beams as an end is "restrained" from rotation. A beam fixed at one end and completely free at the other has a special name, a *cantilever* beam, Fig. 2-11(e). If the beam projects beyond a support, the beam is said to have an *overhang*. Thus the beam shown in Fig. 2-11(f) is an overhanging beam. If intermediate supports are provided for a physically continuous member acting as a beam, Fig. 2-11(g), the beam is termed a *continuous* beam. For all beams the distance L between supports is called a *span*. In a continuous beam there are several spans which may be of varying lengths.

In addition to classifying beams on the basis of supports, descriptive clauses pertaining to the loading are often used. Thus the beam shown in Fig. 2-11(a) is a simple beam with a concentrated load, and the one in Fig. 2-11(b) is a simple beam with a uniformly distributed load. Other beams are similarly described.

Often it is also meaningful to further classify beams into statically determinate and statically indeterminate beams. If the beam, loaded in a plane, is statically determinate, the number of unknown reaction components does not exceed three. These unknowns may always be determined

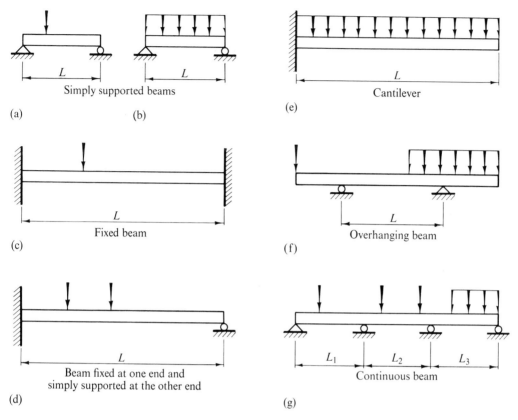

Fig. 2-11. Types of beams.

from the equations of static equilibrium. The next article will briefly review the methods of statics for computing reactions for statically determinate beams. An investigation of statically indeterminate beams will be postponed until Chapters 11 and 12.

2-6. CALCULATION OF BEAM REACTIONS

In much of the subsequent work the analysis of members will begin with the determination of the reactions. When all the forces are applied in one plane, three equations of static equilibrium are available for this purpose. The application of such equations to several beam problems is illustrated below and is intended to serve as a review of this important procedure. In applying the equations of equilibrium, the deformation of beams, being small, can be neglected. For stable beams the small amount of deformation that does take place changes the points of application of the forces imperceptibly.

EXAMPLE 2-1

Find the reactions at the supports for a simple beam loaded as shown in Fig. 2-12(a). Neglect the weight of the beam.

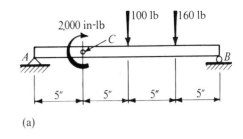

SOLUTION

The loading of the beam is already given in diagrammatic form. The nature of the supports is examined next and the unknown components of these reactions are boldly indicated on the diagram. The beam, with the unknown reaction components and all the applied forces, is redrawn in Fig. 2-12(b) to deliberately emphasize this important step in constructing a free-body diagram. At A, *two* unknown reaction components may exist, since the end is pinned. The reaction at B can only act in a vertical direction since the end is on a roller. The points of application of all forces are carefully noted. After a free-body diagram of the beam is made, the equations of statics are applied to obtain the solution.

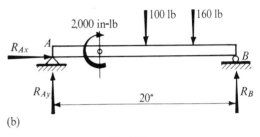

Fig. 2-12

$\sum F_x = 0 \qquad\qquad R_{Ax} = 0$

$\sum M_A = 0 \circlearrowleft +, \qquad 2{,}000 + 100(10) + 160(15) - R_B(20) = 0$

$$R_B = +270 \text{ lb} \uparrow$$

$\sum M_B = 0 \circlearrowleft +, \qquad R_{Ay}(20) + 2{,}000 - 100(10) - 160(5) = 0$

$$R_{Ay} = -10 \text{ lb} \downarrow$$

Check:

$\sum F_y = 0 \uparrow +, \qquad -10 - 100 - 160 + 270 = 0$

Note that $\sum F_x = 0$ uses up one of the three independent equations of statics, thus only two additional reaction components may be determined from statics. If more unknown reaction components or moments exist at the support, the problem becomes statically indeterminate. In Fig. 2-11 the beams shown in parts c, d, and g are statically indeterminate beams as may be proved by examining the number of unknown reaction components (verify this statement).

Note that the concentrated moment applied at C enters only into the expressions for the summation of moments. The positive sign of R_B indicates that the direction of R_B has been correctly assumed in Fig. 2-12(b). The inverse is the case of R_{Ay}, and the vertical reaction at A is downward. Note that a check on the arithmetical work is available if the calculations are made as shown.

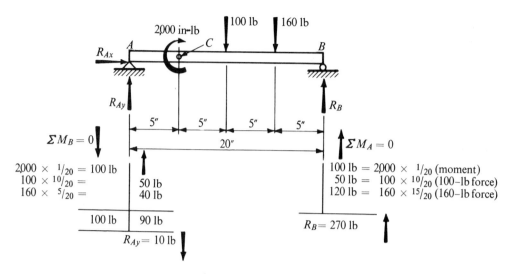

Fig. 2-13

ALTERNATE SOLUTION

In computing reactions some engineers prefer to make calculations in the manner indicated in Fig. 2-13. Fundamentally this involves the use of the same principles. Only the details are different. The reactions for every force are determined one at a time. The total reaction is obtained by summing these reactions. This procedure permits a running check of the computations as they are performed. For every force the sum of its reactions is equal to the force itself. For example, for the 160-lb force, it is easy to see that the upward forces of 40 lb and 120 lb total 160 lb. On the other hand, the concentrated moment at C, being a couple, is resisted by a couple. It causes an *upward* force of 100 lb at the right reaction and a *downward* force of 100 lb at the left reaction.

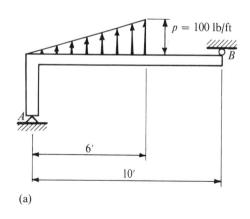
(a)

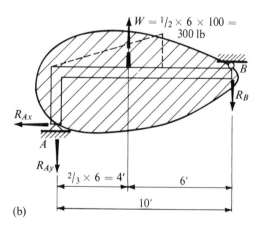
(b)

Fig. 2-14

EXAMPLE 2-2

Find the reactions for the partially loaded beam with a uniformly varying load shown in Fig. 2-14(a). Neglect the weight of the beam.

*Section 2-6
Calculation of
beam reactions*

SOLUTION

An examination of the supporting conditions indicates that there are three unknown reaction components, hence the beam is statically determinate. These and the applied load are shown in Fig. 2-14(b). Note particularly that for computing the reactions the configuration of the member is not important. A crudely shaped outline bearing no resemblance to the actual beam is indicated to emphasize this point. However, this new body is supported at points A and B in the same manner as the original beam.

For calculating the reactions the distributed load is replaced by an equivalent concentrated force. This force is equal to the sum of the distributed forces acting on the beam. It acts through the centroid of the distributed forces. These pertinent quantities are marked on the working sketch, Fig. 2-14(b). After a free-body diagram is prepared the solution follows by applying the equations of static equilibrium.

$\sum F_x = 0$ $R_{Ax} = 0$

$\sum M_A = 0 \circlearrowright +,$ $+300(4) - R_B(10) = 0,$ $R_B = 120 \text{ lb} \downarrow$

$\sum M_B = 0 \circlearrowleft +,$ $-R_{Ay}(10) + 300(6) = 0,$ $R_{Ay} = 180 \text{ lb} \downarrow$

Check:

$\sum F_y = 0 \uparrow +,$ $-180 + 300 - 120 = 0$

EXAMPLE 2-3

Determine the reactions at A and B for the "weightless" beam shown in Fig. 2-15(a). The applied loads are given in kilo-pound or 1,000-lb units called *kips*, which are designated by k.

SOLUTION

A free-body diagram is shown in Fig. 2-15(b). At A there are two unknown reaction components, R_{Ax} and R_{Ay}. At B the reaction R_B acts normal to the supporting plane and constitutes a single unknown. It is expedient to replace this force by the two components R_{By} and R_{Bx}, which in this particular problem are numerically equal. Similarly, it is best to replace the inclined force with the two components shown. These steps reduce the problem to one where all forces are either horizontal or

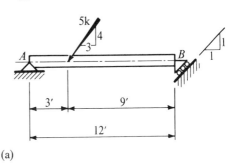

(a)

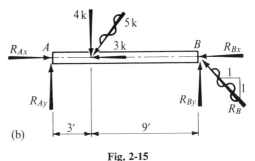

(b)

Fig. 2-15

*Chapter 2
Axial force, shear,
and bending moment*

vertical. This is of great convenience in applying the equations of static equilibrium.

$\sum M_A = 0 \circlearrowright +, \qquad +4(3) - R_{By}(12) = 0, \qquad R_{By} = 1\text{ k} \uparrow = |R_{Bx}|$

$\sum M_B = 0 \circlearrowright +, \qquad +R_{Ay}(12) - 4(9) = 0, \qquad R_{Ay} = 3\text{ k} \uparrow$

$\sum F_x = 0 \rightarrow +, \qquad +R_{Ax} - 3 - 1 = 0, \qquad R_{Ax} = 4\text{ k} \rightarrow$

$R_A = \sqrt{4^2 + 3^2} = 5\text{ k}$

$R_B = \sqrt{1^2 + 1^2} = \sqrt{2}\text{ k}$

Check:

$\sum F_y = 0 \uparrow +, \qquad +3 - 4 + 1 = 0$

Occasionally *hinges** or *pinned joints* are introduced into beams. A hinge is capable of transmitting only horizontal and vertical forces. No moment can be transmitted at a hinged joint. Therefore the point where a hinge occurs is a particularly convenient location for separation of the structure into parts for purposes of computing the reactions. This process is illustrated in Fig. 2-16. Each part of the beam so separated is treated independently. Each hinge provides an extra axis around which moments may be taken to determine reactions. The introduction of a hinge or hinges into a continuous beam in many cases makes the system statically determinate. The introduction of a hinge into a determinate beam results in a beam which is not stable. Note that the reaction at the hinge for one beam acts in an opposite direction on the other beam.

* Another type of connection is shown in Fig. 11-12.

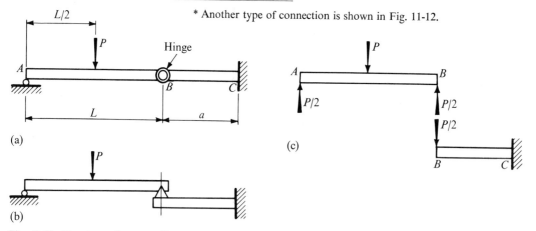

Fig. 2-16. Structures "separated" at hinges to determine the reactions by statics.

PART B
AXIAL-FORCE, SHEAR, AND MOMENT DIAGRAMS: A DIRECT APPROACH

2-7. APPLICATION OF THE METHOD OF SECTIONS

The main objective of this chapter is to establish procedures for determining the forces which exist at a section of a beam. To obtain these forces the method of sections, the basic approach in mechanics of solids, is applied.

The analysis of any beam is begun by preparing a free-body diagram. The reactions are computed next. This is always possible provided the beam is statically determinate. After the reactions are determined, they become known forces, and, in the subsequent steps of the analysis, no distinction need be made between the applied and the reactive forces. Then repeated use is made of the concept that if a whole body is in equilibrium, any part of it is likewise in equilibrium.

For concreteness consider a beam, such as shown in Fig. 2-17(a), with certain concentrated and distributed forces acting on it together with reactions presumed to be known. Any part of this beam to either side of an imaginary cut, as X-X, which is made perpendicular to the axis of the member, can be treated as a free body. Separating this beam at the section X-X, two segments, shown in Figs. 2-17(b) and (c), are obtained. Note particularly that the imaginary section goes through the distributed load and separates it too. Either one of these beam segments is in equilibrium and the conditions of equilibrium require the existence of a system of forces at the cut section of the beam. In general, at a section of a beam, a vertical force, a horizontal force, and a moment are necessary to maintain the part of the beam in equilibrium. These quantities take on special significance in beams and therefore will be discussed separately.

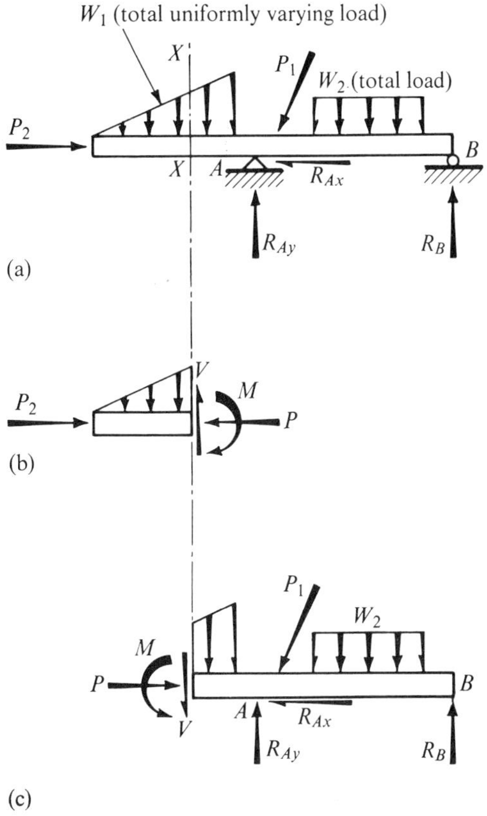

Fig. 2-17. An application of the method of sections to a statically determinate beam.

2-8. SHEAR IN BEAMS

To maintain a segment of a beam such as shown in Fig. 2-17(b) in equilibrium there must be an internal vertical force V at the cut to satisfy

*Chapter 2
Axial force, shear,
and bending moment*

the equation $\Sigma F_y = 0$. This internal force V, acting at right angles to the axis of the beam, is called the *shear* or the *shearing force*. The shear is numerically equal to the algebraic sum of all the vertical components of the external forces acting on the isolated segment, but it is opposite in direction. Given the qualitative data shown in Fig. 2-17(b), V is opposite in direction to the downward load to the left of the section. Similarly, the shear at the same section is also equal numerically, and is opposite in direction, to the sum of all vertical forces to the right of the section, Fig. 2-17(c). The latter sum must, of course, include the vertical reaction components. Whether the right-hand segment or the left is used to determine the shear at a section is immaterial—arithmetical simplicity governs. Shears at any other section may be found similarly.

At this time a significant observation must be made. The same shear shown in Figs. 2-17(b) and (c) at the section X-X is opposite in direction in the two diagrams. For the part of the downward load W_1 to the left of section X-X, the beam at the section provides an upward support to maintain vertical forces in equilibrium. Conversely, the loaded portion of the beam exerts a downward force on the beam as in Fig. 2-17(c). At a section "two directions" of shear must be differentiated, depending upon which segment of the beam is considered. This follows from the familiar action-reaction concept of statics.

The direction of the shear at section X-X would be reversed in both diagrams if the distributed load W_1 were acting upward. Frequently a similar reversal in the direction of shear takes place at one section or

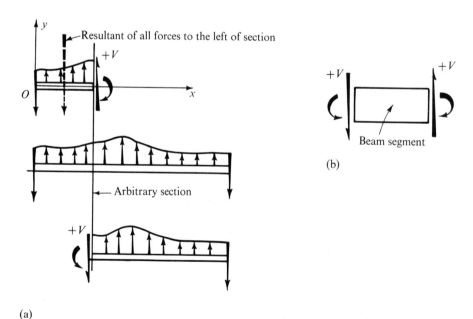

(a)

Fig. 2-18. Definition of positive shear.

another along a beam for reasons which will become apparent later. The adoption of a sign convention is necessary to differentiate between the two possible directions of shear. The definition of positive shear is illustrated in Fig. 2-18(a). An upward internal force acting on the left side of a cut or a downward force acting on the right side of the same cut corresponds to positive shear. Positive shears for an element are shown in Fig. 2-18(b). The shear at section *X-X* of Fig. 2-17 is a positive shear.

Section 2-10 Bending moment in beams

2-9. AXIAL FORCE IN BEAMS

In addition to the shear *V*, a horizontal force such as *P* in Fig. 2-17(b) or (c) may be necessary at a section of a beam to satisfy the conditions of equilibrium. The magnitude and sense of this force follows from a particular solution of the equation $\Sigma F_x = 0$. If the horizontal force *P* acts toward the cut, it is called a *thrust*; if away from the cut, it is termed *axial tension*. In referring to either of these forces the term *axial force* is used. *The line of action of the axial force should always be directed through the centroid of the beam's cross-sectional area.*

Sections along a beam may be examined for the magnitude of the axial force in the above manner. In conformity with the sign convention of Fig. 2-2, a tensile force at a section is positive.

2-10. BENDING MOMENT IN BEAMS

The existence of a shear and an axial force at a section of a beam assures that two of the requirements for equilibrium of a beam segment are met. With these forces the equations $\Sigma F_x = 0$ and $\Sigma F_y = 0$ are satisfied. The remaining condition of static equilibrium for a planar problem is $\Sigma M_z = 0$. This, in general, can be satisfied only by developing a couple or an *internal resisting moment* within the cross-sectional area of the cut to counteract the moment caused by the external forces. The internal resisting moment must act in a direction opposite to the external moment to satisfy the governing equation $\Sigma M_z = 0$. Likewise it follows from the same equation that *the magnitude of the internal resisting moment equals the external moment*. These moments tend to bend a beam in the plane of the loads and are usually referred to as *bending moments*.

The internal bending moment *M* is in Fig. 2-17(b) (page 21). It can be developed only within the cross-sectional area of the beam and is equivalent to a couple. To determine this moment necessary to maintain the equilibrium of a segment, the sum of the moments caused by the forces may be taken around any point in the plane; of course, all forces times their arms must be included in the sum. The internal forces *V* and *P* are no exception. To exclude the moments caused by these forces from the sum, it is usually most convenient in numerical problems to *select the point of intersection of these two internal forces as the point around which the moments are summed*. Both *V* and *P* have arms of zero length at this point, which is located on the centroid of the cross-sectional area of the beam.

Instead of considering the segment to the left of section X-X, the right segment of the beam, Fig. 2-17(c) (page 21), may be used to determine the internal bending moment. As explained above, this internal moment is equal to the external moment of the applied forces (including reactions), providing the summation of moments is made around the centroid of the section at the cut. In Fig. 2-17(b) the resisting moment may be physically interpreted as a pull on the top fibers of the beam and a push on the lower ones. The same interpretation applies to the same moment in Fig. 2-17(c).

If the load W_1 in Fig. 2-17(a) were acting in the opposite direction, the resisting moments in Figs. 2-17(b) and (c) would reverse. This and similar situations require a sign convention for the bending moments. This convention is associated with a definite physical action of the beam. For example, in Figs. 2-17(b) and (c) the internal moments shown pull on the top portion of the beam and compress the lower. This tends to increase the length of the top surface of the beam and to contract the lower surface. A continuous occurrence of such moments along the beam makes the beam deform convex upward, i.e., "shed water." Such bending moments are assigned a negative sign. Conversely, a positive moment is defined as one that produces compression in the top part and tension in the lower part of a beam's cross section. Under such circumstances the beam assumes a shape that "retains water." For example, a simple beam supporting a group of downward forces deflects down as shown in exaggerated form in Fig. 2-19(a), a fact immediately suggested by physical intuition. In such a beam a detailed investigation of bending moments

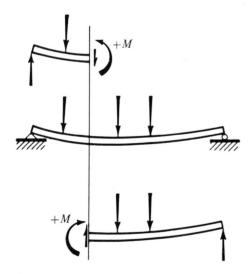

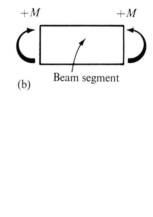

(b) Beam segment

(a)

Fig. 2-19. The definition of a positive bending moment.

along the beam shows that all are positive. The sense of a positive bending moment at a section of a beam is defined in Fig. 2-19(b). This sign convention again agrees with that of Fig. 2-2.

Section 2-11
Shear, axial-force,
and bending-moment
diagrams

2-11. SHEAR, AXIAL-FORCE, AND BENDING-MOMENT DIAGRAMS

By the methods discussed above, the magnitude and sense of shear, axial force, and bending moment may be obtained at any section of a beam. Moreover, with the sign conventions adopted for these quantities, plots of their functions may be made on separate diagrams. On such diagrams, from a base line equal to the length of a beam, ordinates may be laid off equal to the computed quantities. When the plotted points are interconnected by lines, a graphical representation of the function is obtained. Such diagrams, depending on the kind of quantities they depict, are called respectively *the shear diagram*,* *the axial-force diagram, or the bending-moment diagram*. With the aid of such diagrams, the magnitudes and locations of the various quantities become immediately apparent. It is convenient to make these plots directly below the free-body diagram of the beam, using the same horizontal scale for the length of the beam. Draftsmanlike precision in making such diagrams is usually unnecessary as the significant ordinates are generally marked with their respective numerical values.

The axial-force diagrams are not so commonly used as the shear and the bending-moment diagrams because the majority of beams investigated in practice are loaded by forces which act perpendicular to the axis of the beam. For such loadings of a beam, there are no axial forces at any section.

Shear and moment diagrams are exceedingly important. From them a designer sees at a glance the kind of performance that is required from a beam at every section. The procedure discussed above of sectioning a beam and finding the system of forces at the section is fundamental and should be learned well by the student. The following examples illustrate the procedure further.

EXAMPLE 2-4

Construct shear, axial-force, and bending-moment diagrams for the weightless beam shown in Fig. 2-20(a) subjected to the inclined force $P = 5$ kips.

SOLUTION

A free-body diagram of the beam is shown in Fig. 2-20(b). Reactions

---

* In many texts on structural analysis the positive direction of shear is taken opposite to that adopted here. If positive shears as defined in this text are plotted downward, the outlines of the shear diagrams for the two sign conventions become identical.

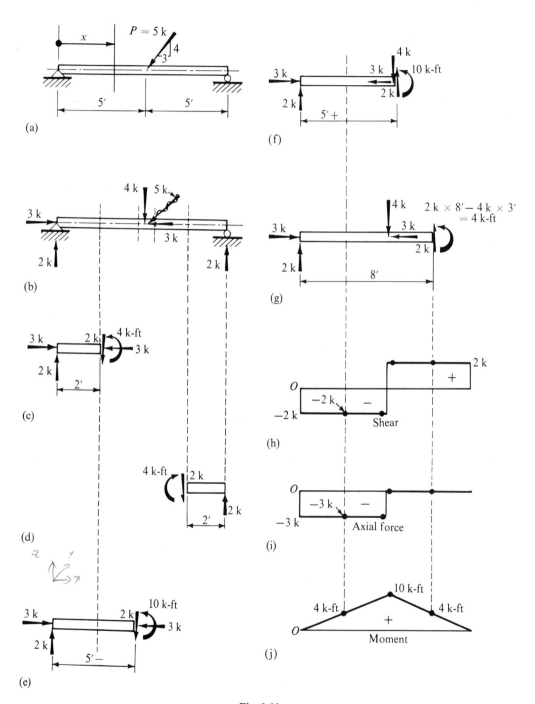

Fig. 2-20

follow from inspection after the applied force is resolved into the two components. Then several sections through the beam are investigated, as shown in Figs. 2-20(c), (d), (e), (f), and (g). In every case the same question is posed: What are the necessary internal forces to keep the segment of the beam in equilibrium? The corresponding quantities are recorded on the respective free-body diagrams of the beam segment. The ordinates for these quantities are indicated by heavy dots in Figs. 2-20(h), (i), and (j), with due attention paid to their sign.

*Section 2-11
Shear, axial-force, and bending-moment diagrams*

Note that the free bodies shown in Figs. 2-20(d) and (g) are alternates as they furnish the same information, and normally both would not be made. Note that a section just to the left of the applied force has one sign of shear, Fig. 2-20(e), but just to the right, Fig. 2-20(f), it has another. This indicates the importance of determining shears on either side of a concentrated force.

In this particular case, after a few individual points have been established on the three diagrams in Figs. 2-20(h), (i), and (j), the behavior of the respective quantities across the whole length of the beam may be reasoned out. Thus, although the segment of the beam shown in Fig. 2-20(c) is 2 ft long, it may vary in length anywhere from zero to just to the left of the applied force, and no change in the shear and the axial force occurs. Hence the ordinates in Figs. 2-20(h) and (i) remain constant for this segment of the beam. On the other hand, the bending moment depends directly on the distance from the supports; hence it varies linearly as shown in Fig. 2-20(j). Similar reasoning applies to the segment shown in Fig. 2-20(d), enabling one to complete the three diagrams on the right-hand side. The use of the free body of Fig. 2-20(g) for completing the diagram to the right of center yields the same result.

Frequently, in addition to or instead of the shear or moment diagrams, analytical expressions for these functions are necessary. For the origin of x at the left end of the beam, the following relations apply:

$V = -2$ k for $0 < x < 5$

$V = +2$ k for $5 < x < 10$

$M = +2x$ k-ft for $0 \leq x \leq 5$

$M = +2x - 4(x-5) = (+20 - 2x)$ k-ft for $5 \leq x \leq 10$

These expressions can be easily established by mentally replacing the distances of 2 ft and 8 ft, respectively, in Figs. 2-20(c) and (g) by an x.

EXAMPLE 2-5

Construct shear and bending-moment diagrams for the beam loaded with the forces shown in Fig. 2-21(a).

SOLUTION

An arbitrary section at a distance x from the left support isolates the beam segment shown in Fig. 2-21(b). This section is applicable for any value of x just to the left of the applied force P. The shear, regardless

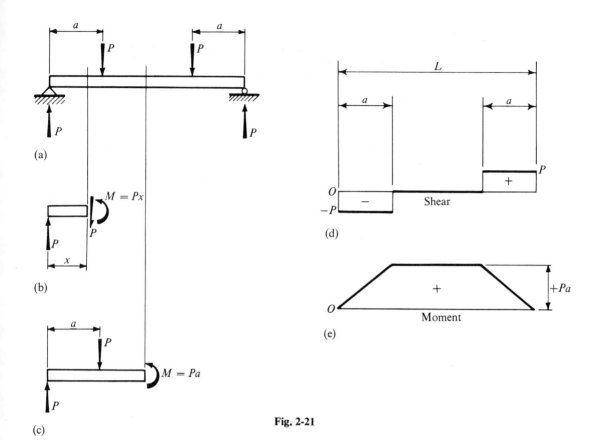

Fig. 2-21

of the distance from the support, remains constant and is $-P$. The bending moment varies linearly from the support, reaching a maximum of $+Pa$.

An arbitrary section applicable anywhere between the two applied forces is shown in Fig. 2-21(c). No shearing force is necessary to maintain the equilibrium of a segment in this part of the beam. Only a constant bending moment of $+Pa$ must be resisted by the beam in this zone. Such a state of bending or flexure is called *pure* bending.

Shear and bending-moment diagrams for this loading condition are shown in Figs. 2-21(d) and (e), respectively. No axial-force diagram is necessary as there is no axial force at any section of the beam.

EXAMPLE 2-6

Plot a shear and a bending-moment diagram for a simple beam with a uniformly distributed load, Fig. 2-22(a).

SOLUTION

The best way of solving this problem is to write down algebraic expressions for the quantities sought. For this purpose an arbitrary section taken at a distance x from the left support is used to isolate the segment shown in Fig. 2-22(b). Since the applied load is continuous, this section is typical and applies to any section along the length of the beam.

Section 2-11
Shear, axial-force, and bending-moment diagrams

The shear V is the negative of the left upward reaction plus the load to the left of the section. The internal bending moment M resists the moment caused by the reaction on the left less the moment caused by the forces to the left of the same section. The summation of moments is performed around an axis at the section. Although it is customary to isolate the left-hand segment, similar expressions may be obtained by considering the right-hand segment of the beam, with due attention paid to sign conventions. The plots of the V and the M functions are shown in Fig. 2-22(c) and (d).

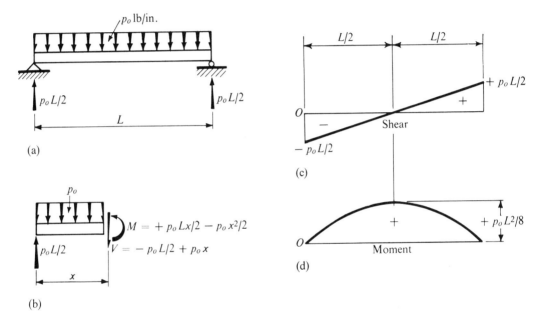

Fig. 2-22

EXAMPLE 2-7

Determine shear, axial-force, and bending-moment diagrams for the cantilever loaded with an inclined force at the end, Fig. 2-23(a).

SOLUTION

The inclined force is replaced by the two components shown in Fig. 2-23(b) and the reaction is determined. The three unknowns at the support follow from the familiar equations of statics.

A segment of the beam is shown in Fig. 2-23(c); from this segment it may be seen that the shearing force and the axial force remain the same regardless of the distance x. On the other hand, the bending moment is a variable quantity. A summation of moments around C gives $(PL - Px)$ acting in the direction shown. This represents a *negative moment*. The moment at the support is likewise a *negative bending moment* as it tends

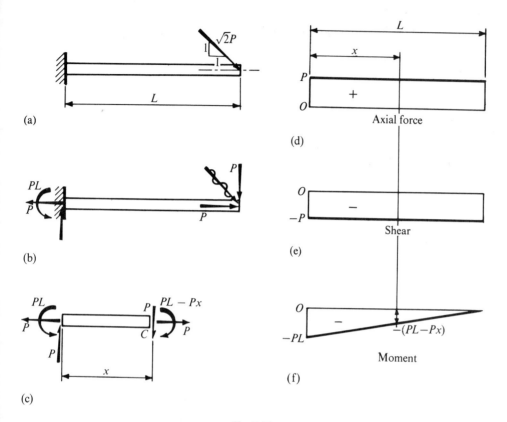

Fig. 2-23

to pull on the upper fibers of the beam. The three diagrams are plotted in Figs. 2-23(d), (e), and (f).

EXAMPLE 2-8

Consider a curved beam whose centroidal axis is bent into a semicircle of 10-in. radius as shown in Fig. 2-24(a). If this member is being pulled by the 1,000-lb forces shown, find the axial force, the shear, and the bending moment at the section $A\text{-}A$, $\alpha = 45°$. The centroidal axis and the applied forces all lie in the same plane.

SOLUTION

There is no essential difference between the method of attack in this problem and that in a straight-beam problem. The body as a whole is examined for conditions of equilibrium. From the conditions of the problem here, such is already the case. A segment of the beam is isolated next. This is shown in Fig. 2-24(b). Section $A\text{-}A$ is taken perpendicular to the axis of the beam. Before determining the quantities wanted at the cut, the applied force P is resolved into components parallel and perpendicular to the cut. These directions are taken respectively as the y

and x axes. This resolution replaces P by the components shown in Fig. 2-24(b). From $\Sigma F_x = 0$, the axial force at the cut is $+707$ lb. From $\Sigma F_y = 0$, the shear is 707 lb in the direction shown. The bending moment at the cut may be determined in several different ways. For example, if $\Sigma M_O = 0$ is used, note that the lines of action of the applied force P and of the shear at the section pass through O. Therefore only the axial force at the centroid of the cut times the radius needs to be considered, and the resisting bending moment is $707(10) = 7{,}070$ in-lb acting in the direction shown. An alternative solution may be obtained by applying $\Sigma M_C = 0$. At C, a point lying on the centroid, the axial force and the shear intersect. The bending moment is then the product of the applied force P and the 7.07-in. arm. In both these methods of determining bending moment, use of the components of the force P is avoided as this is more involved arithmetically.

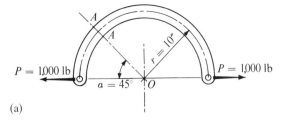

(a)

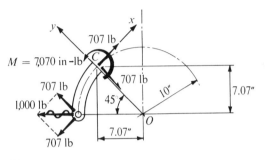

(b)

Fig. 2-24

It is suggested that the student complete this problem in terms of a general angle α. Several interesting observations may be made from such a general solution. The moments at the ends will vanish for $\alpha = 0°$ and $\alpha = 180°$. For $\alpha = 90°$ the shear vanishes and the axial force becomes equal to the applied force P. Likewise the maximum bending moment is associated with $\alpha = 90°$.

2-12. STEP-BY-STEP PROCEDURE

In beam analysis it is exceedingly important to be able to determine the shear, the axial force, and the bending moment at any section. The technique of obtaining these quantities is unusually clear cut and systematic. To lend further emphasis, the steps used in all such problems are summarized. This summary is intended to aid the student in an orderly analysis of problems. Shear memorization of this procedure is discouraged.

1. Make a good sketch of the beam on which all the applied forces are clearly noted and located by dimension lines from the supports.
2. Boldly indicate the unknown reactions (colored pencil may be used to advantage). Rember that a roller support has one unknown, a pinned support has two unknowns, and a fixed support has three unknowns.
3. Replace all the inclined forces (known and unknown) by components acting parallel and perpendicular to the beam axis.*

---

* More ingenuity may be required for curved beams.

4. Apply the equations of statics to obtain the reactions.* A check of the reactions computed in the manner indicated in Examples 2-1, 2-2, and 2-3 is highly desirable.

5. Pass a section at the desired location through the beam perpendicular to its axis. This imaginary section cuts only the beam and isolates the forces which act on the segment.

6. Select a segment to either side of the proposed section and redraw this segment, indicating all external forces acting on it. This must include all the reaction components.

7. Indicate the three possible unknown quantities at the cut section, i.e., show P, V, and M, assuming their directions.

8. Apply the equations of equilibrium to the segment and solve for the quantities P, V, and M.

This procedure enables one to determine the shear, the axial force, and the bending moment at any section of a beam. Signs for these quantities follow from the definitions given earlier. If diagrams for this system of internal forces are wanted, several sections may have to be investigated. Do not fail to determine the abrupt change in shear at concentrated forces and the abrupt change in bending-moment value at points where concentrated moments are introduced. Algebraic expressions for the same quantities sometimes are also necessary.

In the above discussion the construction of shear and moment diagrams was illustrated principally for horizontal members. For inclined members, except for directing the coordinate axes along and perpendicular to the axis of a bar, the procedure is the same. In curved and in spatial structural systems the directions of the axes are along the axes of the member or members. In such cases one of the coordinate axes is taken tangent to the axis of the member—as shown for example in Fig. 2-24. To conform with the diagrammatic scheme used in this text for horizontal beams, the ordinates for bending moment in curved and spatial systems should be plotted on the compression side† of a section.

PART C
SHEAR AND MOMENT DIAGRAMS: A SUMMATION APPROACH

2-13. DIFFERENTIAL EQUATIONS OF EQUILIBRIUM

Instead of the direct approach of cutting a beam and determining shear and moment at a section by statics, an efficient alternative procedure can be used. For this purpose certain fundamental differential relations

---

* This step may be avoided in cantilevers by proceeding from the free end.
† In some texts on structural analysis the opposite scheme is used.

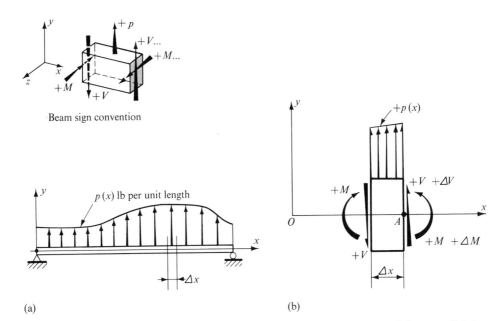

Fig. 2-25. Beam and an element cut out from it by two adjoining sections Δx apart.

must be derived. These can be used for the construction of shear and moment diagrams as well as for the calculation of reactions.

Consider a beam element Δx long, isolated by two adjoining sections taken perpendicular to its axis, Fig. 2-25(a). Such an element is shown as a free body in Fig. 2-25(b). All the forces shown acting on this element have positive sense (for definitions see Figs. 2-2, 2-18, and 2-19). The positive sense of the distributed external force p is taken to coincide with the direction of the positive y axis. As the shear and the moment may each change from one section to the next, note that on the right side of the element these quantities are respectively designated $V + \Delta V$ and $M + \Delta M$.

From the condition for equilibrium of vertical forces, one obtains*

$$\sum F_y = 0 \uparrow +, \quad -V + p\,\Delta x + (V + \Delta V) = 0 \quad \text{or} \quad \Delta V/\Delta x = -p \quad (2\text{-}2)$$

For equilibrium the summation of moments around A also must be zero. So, upon noting that from point A the arm of the distributed force is $\Delta x/2$, one has

$$\sum M_A = 0 \circlearrowleft +, \quad (M + \Delta M) + V\,\Delta x - M - (p\,\Delta x)(\Delta x/2) = 0$$

or

$$\frac{\Delta M}{\Delta x} = -V + \frac{p\,\Delta x}{2} \quad (2\text{-}3)$$

---

* No variation of $p(x)$ within Δx need be considered since, in the limit as $\Delta x \to 0$, the change in p becomes negligibly small. This simplification is not an approximation.

Equations 2-2 and 2-3 in the limit as $\Delta x \to 0$ yield the following two basic differential equations:

$$\lim_{\Delta x \to 0} \frac{\Delta V}{\Delta x} \equiv \frac{dV}{dx} = -p \qquad (2\text{-}4)$$

and

$$\lim_{\Delta x \to 0} \frac{\Delta M}{\Delta x} \equiv \frac{dM}{dx} = -V \qquad (2\text{-}5)$$

By substituting Eq. 2-5 into Eq. 2-4, another useful relation is obtained:

$$\frac{d}{dx}\left(\frac{dM}{dx}\right) = \frac{d^2M}{dx^2} = p \qquad (2\text{-}6)$$

This differential equation can be used for determining reactions of statically determinate beams, whereas Eqs. 2-4 and 2-5 are very convenient for construction of shear and moment diagrams. These applications will be discussed next.

2-14. SHEAR DIAGRAMS BY SUMMATION

By transposing and integrating Eq. 2-4, one obtains the basic relation for the shear V:

$$V(x) = -\int_0^x p\,dx + C_1 \qquad (2\text{-}7)$$

Here it is seen that, except for a possible constant of integration C_1, the shear at a section is the negative of the integral of the vertical forces acting on the beam from the left end to the section in question. Likewise, between any two definite sections of a beam, the change in shear is the negative of all the vertical forces included between these sections. If no force occurs between any two sections, no change in shear takes place. If a concentrated force comes into the summation, a discontinuity or a "jump" in the value of the shear occurs. The continuous summation process remains valid since a concentrated force may be thought of as a distributed force extending for an infinitesimal distance along the beam.

On the basis of the above reasoning, a shear diagram may be established by a simple summation process. For this purpose, the reactions must always be determined first. Then the vertical components of forces and reactions are successively summed from the left end of the beam. The shear at a section is equal to the sum of all vertical forces up to that section and has a sense opposite to the sum of these forces.

When the shear diagram is constructed from the loading diagram by the summation process, the analyst should pose two questions. First, is

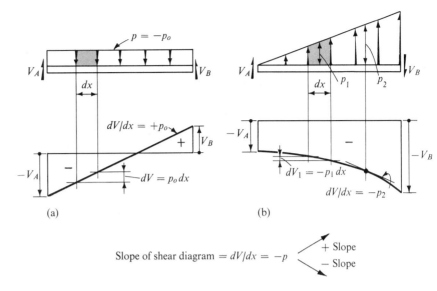

Fig. 2-26. The relation between the loading and the shear diagrams.

the change in shear positive or negative in the particular segment of the beam considered? This depends entirely on whether the forces act up or down. If they act upward, the change is negative. Second, at what rate does the change in shear take place? This question again is answered by the loading diagram since the slope of the shear diagram $dV/dx = -p$. For example, since in Fig. 2-26(a) the load $p = -p_o$ acts downward and is of constant magnitude, the shear curve for this segment is a straight line of constant, positive slope. In Fig. 2-26(b) the variable load acts upward; hence the change in shear from left to right is negative. Since $+p_1 < +p_2$, the slope of the shear diagram becomes more negative to the right. The curve that fits this condition is concave downward.

It is also useful to note that the rate of change of the slope of the shear diagram equals the negative of the rate of change of the load; i.e.,

$$\frac{d}{dx}\left(\frac{dV}{dx}\right) = -\frac{dp}{dx} \quad \text{since} \quad \frac{dV}{dx} = -p \quad \text{(2-8)}$$

On this basis, since in Fig. 2-26(a) the load is uniform, no change in the slope of the shear diagram occurs. In Fig. 2-26(b), since the load increases at a constant rate, the slope of the shear diagram decreases at a constant rate, becoming more negative.

Do not fail to note that a mere systematic consecutive summation of the vertical components of the forces with reversed signs is all that is necessary to obtain the shear diagram. In progressing consecutively from the left end of the beam, the diagram must close at the right end of the beam since just beyond the last vertical force or reaction no shear acts through the beam. The fact that the diagram closes offers an important check on the arithmetical calculations. This check should not be ignored. It permits one to obtain solutions independently with almost complete assurance of being correct. The semigraphical procedure of integration outlined above is very convenient in practical problems. It is the basis for sketching qualitative shear diagrams rapidly.

From the physical point of view, the shear sign convention is not completely consistent. Whenever beams are analyzed, a shear diagram drawn from one side of the beam is opposite in sign to a diagram constructed by looking at the same beam from the other side. The reader should verify this statement on some simple cases, such as a cantilever with a concentrated force at the end and a simply supported beam with a concentrated force in the middle. For design purposes the sign of the shear is usually unimportant.

2-15. MOMENT DIAGRAMS BY SUMMATION

To formulate the summation procedure for establishing moment diagrams, Eq. 2-5 is transposed and integrated. Thus

$$M(x) = -\int_0^x V\,dx + C_2 \tag{2-9}$$

where C_2 is a constant of integration. This equation is completely analogous to Eq. 2-7, developed for the construction of shear diagrams. The term $V\,dx$ (corresponding to $p\,dx$ of the former case) is shown graphically by the shaded areas of the shear diagrams in Fig. 2-27. The sum of these areas between definite sections through a beam corresponds to the above definite integral. If the ends of a beam are on rollers—pin-ended or free— the starting and the terminal moments are zero. If the end is built in, in statically determinate beams the end moment is known from the reaction calculations. If the fixed end of a beam is on the left, this moment with the proper sign is the initial constant of integration C_2.

By proceeding continuously along the beam from the left end and by taking the negative of the sum of the areas of the shear diagram, ordinates for the moment diagram are obtained. This process of deriving the moment diagram from the shear diagram by summation is exactly the same as that employed earlier to go from loading to shear diagrams. The change in moment in a given segment of a beam is equal to the negative of the area of the corresponding shear diagram. The slope of the bending-moment curve is determined by noting the corresponding magnitude and sign of the shear since, according to Eq. 2-5, $dM/dx = -V$. Examples of shear and moment diagrams are in Fig. 2-27, where variable

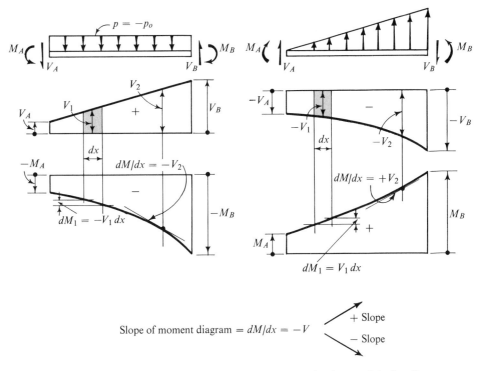

Fig. 2-27. The relation between the shear and the bending-moment diagrams.

shears are seen to cause nonlinear variation of the moment. A constant shear produces a uniform rate of change in the bending moment, resulting in a straight line in the moment diagram. If no shear occurs along a certain portion of a beam, no change in moment takes place.

By a fundamental theorem of calculus, Eq. 2-5 implies that the maximum or minimum moment occurs at a point where the shear is zero as the derivative of M is then zero. This occurs at a point where the shear changes sign.

In a bending-moment diagram obtained by summation, at the right end of the beam, an invaluable check on the work is available. The terminal conditions for the moment must be satisfied. If the end is free or pinned, the computed sum must equal zero. If the end is built in, the end moment computed by summation equals the one calculated initially for the reaction. These are the "boundary conditions" and must always be satisfied.

EXAMPLE 2-9

Construct shear and moment diagrams for the symmetrically loaded beam in Fig. 2-28(a) using the summation procedure.

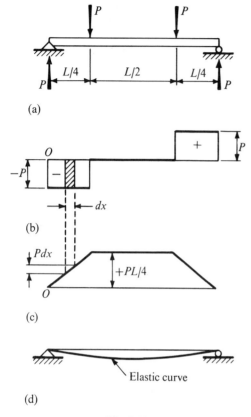

Fig. 2-28

SOLUTION

The reactions are each equal to P. To obtain the shear diagram, Fig. 2-28(b), the summation of forces is started from the left end. The left reaction acts upward so an ordinate on the shear diagram at this force equal to $-P$ must be plotted downward. Since there are no other forces until the quarter point, no change in the magnitude of the shear ordinate can be made until that point. Then a downward force P brings the ordinate back to the base line, and this zero ordinate remains until the next downward force P is reached, where the shear changes to $+P$. At the right end the upward reaction closes the diagram and provides a check on the work. This shear diagram is antisymmetrical.

The moment diagram, Fig. 2-28(c), is obtained by taking the negative of the sum of the areas of the shear diagram. As the beam is simply supported, the moment at both ends is zero. The sum of the negative portion of the shear diagram causes an increase in the moment diagram at a constant rate along the beam until the quarter point is reached, where the moment is $+PL/4$. This moment remains constant in the middle half of the beam. No change in the moment can be made in this zone as there is no corresponding shear area.

Beyond the second force, the moment decreases by $-P\,dx$ in every dx. Hence the moment diagram in this zone has a constant, negative slope. Since the positive and the negative areas of the shear diagram are equal, at the right end the moment is zero. This is as it should be, as the end is on a roller. Thus a check on the work is obtained. This moment diagram is symmetrical.

EXAMPLE 2-10

Construct shear and bending-moment diagrams for the beam loaded as shown in Fig. 2-29(a) using the summation procedure.

SOLUTION

Reactions must be calculated first, and, before proceeding further, the inclined force is resolved into its horizontal and vertical components. The horizontal reaction at A is 30 kips and acts to the right. From $\Sigma M_A = 0$, the vertical reaction at B is found to be 37.5 kips (check this). Similarly, the reaction at A is 27.5 kips. The sum of the vertical reaction components is 65 kips and equals the sum of the vertical forces.

With reactions known, the summation of forces is begun from the left end of the beam to obtain the shear diagram, Fig. 2-29(b). At first, the downward distributed load is large, then it decreases. Hence, the shear diagram in the zone CA at first has a large, positive slope, which gradually decreases resulting in a curved line, which is concave downward. The total downward force from C to A is 15 kips, which is the positive ordinate of the shear diagram, just to the left of the support A. At A, upward reaction of 27.5 kips moves the ordinate of the shear diagram downward to −12.5 kips. This value of the shear applies to a section through the beam just to the right of the support A. The total change in the shear at A is equal to the reaction, but this total does not represent the shear through the beam.

No forces are applied to the beam between A and D, hence there is no change in the value of the shear. At D, the 40 kip downward component of the concentrated force raises the value of the shear to +27.5 kips. Similarly, the value of the shear is lowered to −10 kips at B. Since between E and F the uniformly distributed load acts downward, an increase in shear takes place at a constant rate of 1 kip per foot. Thus at F the shear becomes zero, which serves as the final check.

To construct the moment diagram shown in Fig. 2-29(c) by the summation method, areas of the shear diagram in Fig. 2-29(b) must be continuously summed from the left end and taken with the opposite signs to yield the moment. For the segment CA the shear gradually increases to the right; therefore in the moment diagram a curve concave downward results. The moment at A is equal to the area of the shear diagram for the segment CA with reversed sign. This area is enclosed by a curved line, and it may be determined by integration. This procedure often is tedious, and, instead of using it, the bending moment at A may be obtained from the fundamental definition of a moment at a section. By passing a section through A and isolating the segment CA, the moment at A is found. The remaining areas of the shear diagram in this example are easily determined.

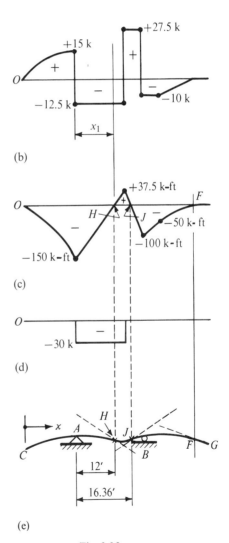

Fig. 2-29

Due attention must be paid to the signs of these areas. It is convenient to arrange the work in tabular form. At the right end of the beam, the customary check is obtained.

$$M_A \ldots -\tfrac{1}{2}(15)2(10) = -150.0 \text{ kip-ft} \qquad \text{(moment around } A\text{)}$$

$$+12.5(15) = +187.5 \qquad (-1) \times \text{(shear area } A \text{ to } D\text{)}$$

$$M_D \ldots \qquad +37.5 \text{ kip-ft}$$

$$-27.5(5) = -137.5 \qquad (-1) \times \text{(shear area } D \text{ to } B\text{)}$$

$$M_B \ldots \qquad -100.0 \text{ kip-ft}$$

$$+10(5) = +50.0 \qquad (-1) \times \text{(shear area } B \text{ to } E\text{)}$$

$$M_E \ldots \qquad -50.0 \text{ kip-ft}$$

$$+\tfrac{1}{2}(10)10 = +50.0 \qquad (-1) \times \text{(shear area } E \text{ to } F\text{)}$$

$$\text{Check:} \ldots M_F = 0.0 \text{ kip-ft}$$

The diagram for the axial force is in Fig. 2-29 (d). The compressive force acts only in the segment AD of the beam.

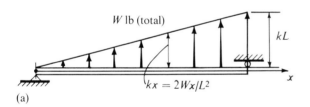

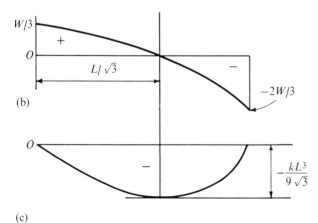

Fig. 2-30

EXAMPLE 2-11

By using Eq. 2-6 subject to the prescribed boundary conditions determine the functions for shear and moment for a simply supported beam loaded as in Fig. 2-30(a). Show the results in shear and moment diagrams. The total applied upward load is W lb.

SOLUTION

Since the load varies uniformly, let the load intensity p at x be kx lb per inch. The total load $W = kL^2/2$. Therefore, $k = 2W/L^2$. Using the constant k just found, one can express the loading function p in terms of the applied load W. On this basis, Eq. 2-6 becomes

$$\frac{d^2M}{dx^2} = p = +kx = +\frac{2W}{L^2}x$$

where the constant k is positive since the applied load acts in an upward direction. Integrating this differential equation twice, one obtains

$$\frac{dM}{dx} = +\frac{kx^2}{2} + C_1 \quad \text{and} \quad M = +\frac{kx^3}{6} + C_1 x + C_2$$

Since $dM/dx = -V$, if the reaction on the left were known, the constant C_1 could be evaluated from the first of the above equations. However, it can be noted directly from the boundary conditions that $M = 0$ at $x = 0$ and at $x = L$, i.e., $M(0) = 0$ and $M(L) = 0$. Therefore since

$$M(0) = 0, \quad C_2 = 0$$

and, similarly, since $M(L) = 0$,

$$kL^3/6 + C_1 L = 0 \quad \text{or} \quad C_1 = -kL^2/6$$

It can be easily verified that except for the sign this value for C_1 is the reaction on the left. Here it was found by solving a boundary-value problem without the use of the conventional procedure used in statics.

After C_1 and C_2 are determined, the expressions for the shear and moment are known:

$$V = -dM/dx = -(kx^2/2) + (kL^2/6)$$

and

$$M = +(kx^3/6) - (kL^2 x/6)$$

These functions are plotted in Figs 2-30(b) and (c). The largest moment occurs at $dM/dx = -V = -kx^2/2 + kL^2/6 = 0$; i.e., at $x_1 = L/\sqrt{3}$. By substituting this value of x_1 into the expression for moment, one finds that the largest moment $M = -kL^3/(9\sqrt{3})$.

The attractive features of the boundary-value approach used in the solution of this problem can be applied so far only in problems where the loading p is a continuous function between the supports. Extension to more general cases is given in Part D of this chapter; statically indeterminate beams are treated in Chapter 11.

Section 2-15
Moment diagrams by summation

2-16. FURTHER REMARKS ON THE CONSTRUCTION OF SHEAR AND MOMENT DIAGRAMS

In the derivation of moment diagrams by summation of shear-diagram areas, no possibility of an external, concentrated moment acting on the infinitesimal element was included. Therefore the derived summation process applies only up to the point of application of an external moment. *At a section just beyond an externally applied moment, a different bending moment is required to maintain the segment of a beam in equilibrium.* For example, in Fig. 2-31 an external clockwise moment M_A is acting on the element of the beam at A. Then, if the internal clockwise moment on the left is M_B, for equilibrium of the element, the resisting counterclockwise moment on the right is $M_B + M_A$. Situations with other sense of moments may be similarly analyzed. At the point of the externally applied moment, a discontinuity or a jump equal to the concentrated moment appears in the moment diagram. Hence, in applying the summation process, due regard must be given the concentrated moments as their effect is not included in the shear-diagram-area summation process. The summation process may be applied up to the point of application of a concentrated moment. At this point a vertical jump equal to the external moment must be made in the diagram. The direction of this vertical jump in the diagram depends upon the sense of the concentrated moment and is best determined with the aid of a sketch analogous to Fig. 2-31. After the discontinuity in the moment diagram is passed, the summation process of the shear-diagram areas may be continued over the remainder of the beam.

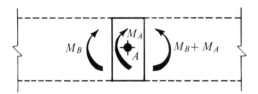

Fig. 2-31. An external concentrated moment acting on an element of a beam.

EXAMPLE 2-12

Construct the bending-moment diagram for a horizontal beam loaded as in Fig. 2-32(a).

SOLUTION

By taking moments about either end of the beam, the vertical reactions are found to be $P/6$. At A the reaction acts down, at C it acts up. From $\Sigma F_x = 0$ it is known that at A a horizontal reaction equal to P acts to the left. The shear diagram is drawn next, Fig. 2-32(b). It has a constant, positive ordinate for the whole length of the beam. After this, by using the summation process, the moment diagram shown in Fig. 2-31(c) is constructed. The moment at the left end of the beam is zero since the support is pinned. The total change in moment from A to B is given by the area of the shear diagram between these sections taken with the reversed sign; it equals $-2Pa/3$. The moment diagram in the zone AB has a constant, negative slope. For further analysis, an element is isolated

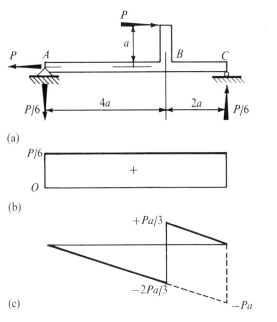

Fig. 2-32

from the beam as shown in Fig. 2-32(d). The moment on the left side of this element is known to be $-2Pa/3$, and the concentrated moment caused by the applied force P about the beam's centroidal axis is Pa. Hence, for equilibrium, on the right side of the element the moment must be $+Pa/3$. At B an upward jump of $+Pa$ is made in the moment diagram, and just to the right of B the ordinate is $+Pa/3$. Beyond B, the summation of the shear diagram area is continued. The area between B and C taken with the reversed sign is equal to $-Pa/3$. This value closes the moment diagram at the right end of the beam, and thus the boundary conditions are satisfied. Note that the inclined lines in the moment diagram are parallel, for if the summation of the shear-diagram area were continued uninterrupted by the concentrated moment, the ordinate on the right would be $-Pa$. Of course, this does not satisfy the boundary condition of the problem.

EXAMPLE 2-13

Construct shear and moment diagrams for the member shown in Fig. 2-33(a). Neglect the weight of the beam.

SOLUTION

In this case, unlike all cases considered so far, definite dimensions are assigned for the depth of the beam. The beam, for simplicity, is assumed to be rectangular in its cross-sectional area, consequently its longitudinal axis lies 3 in. below the top of the beam. Note carefully that this beam is not supported on its axis.

43

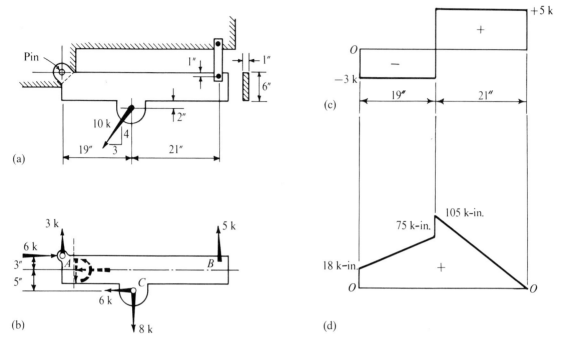

Fig. 2-33

A free-body diagram of the beam with the applied force resolved into components is shown in Fig. 2-33(b). Reactions are computed in the usual manner. Moreover, since the shear diagram is concerned only with the vertical forces, it is easily constructed and is shown in Fig. 2-33(c).

In constructing the moment diagram in Fig. 2-33(d), particular care must be exercised. As was emphasized earlier, the bending moments may always be determined by considering a segment of a beam, and they are most conveniently computed by taking moments of external forces around a point on the centroidal axis of the beam. Thus, by passing a section just to the right of A and considering the left segment, it may be seen that a positive moment of 18 kip-in. is resisted by the beam at this end. Hence the plot of the moment diagram must start with an ordinate of $+18$ kip-in. The other point of the beam where a concentrated moment occurs is C. Here the horizontal component of the applied force induces a clockwise moment of $6(5) = 30$ kip-in. around the neutral axis. Just to the right of C this moment must be resisted by an additional positive moment. This causes a discontinuity in the moment diagram. The summation process of the shear-diagram areas applies for the segments of the beam where no external moments are applied. The necessary calculations are carried out below in tabular form.

M_A . . . . $+6(3) = +\ 18$ kip-in.
$+3(19) = +\ 57$ $(-1) \times$ (shear area A to C)

Moment just to left of $C = +75$ kip-in.

$$\frac{+6(5) = +30}{\text{Moment just to right of } C = +105 \text{ kip-in.}} \quad \text{(external moment at } C\text{)}$$

Check: $$\frac{-5(21) = -105}{M_B = 0} \quad (-1) \times \text{(shear area } C \text{ to } B\text{)}$$

Section 2-16
Further remarks on the construction of shear and moment diagrams

Note that in solving this problem the forces are considered *wherever they actually act on the beam*. The investigation for shear and moments at a section of a beam determines what the beam is actually experiencing. At times this differs from the procedure of determining reactions where the actual framing or configuration of a member is not important.

In engineering practice it is common to find several members rigidly joined to form a structure. Such a structure may be treated by the methods already discussed if it can be separated into statically determinate, individual beams. To illustrate, consider the structure in Fig. 2-34(a). Beginning at point A, the portions of the structure AB, BC, and CD may be successively isolated as free bodies, and the system of forces at each of the cut sections may be determined. The reader should verify these forces, which are shown in Fig. 2-34(b). Thence, shear and moment diagrams may be constructed for each part, using the previously described procedures.

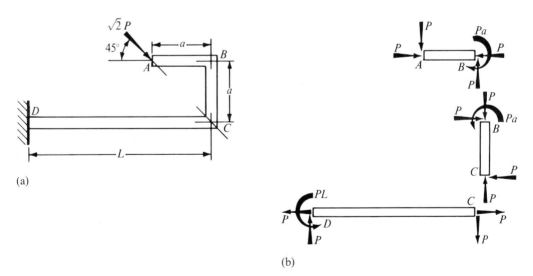

Fig. 2-34. A statically determinate frame separated into individual beams.

2-17. MOMENT DIAGRAM AND THE ELASTIC CURVE

It was stated in Art. 2-10 that a positive moment causes a beam to deform concave upward or to "retain water," and vice versa; hence the shape of the deflected axis of a beam may be definitely established from the sign of the moment diagram. The trace of this axis of a loaded beam in a deflected position is known as the *elastic curve*. It is customary to show the elastic curve on a sketch where the actual small deflections tolerated in practice are greatly *exaggerated*. A sketch of the elastic curve clarifies the physical action of a beam. Moreover, it forms a useful basis for quantitative calculations of beam deflections to be discussed in Chapter 11. Some of the preceding examples for which bending-moment diagrams were constructed will be used to illustrate the physical action of a beam.

An inspection of Fig. 2-28(c) shows that the bending moment throughout the length of the beam is positive. Accordingly, the elastic curve shown in Fig. 2-28(d) is concave up at every point. The ends of the beam are assumed to rest on immovable supports.

In the more complex moment diagram of Fig. 2-29(c), zones of positive and negative moment occur. Corresponding to the zones of negative moment, a definite curvature of the elastic curve that is concave down takes place, Fig. 2-29(e). On the other hand, for the zone *HJ*, where the positive moment occurs, the concavity of the elastic curve is upward. Where curves join, as at *H* and *J*, there are lines which are *tangent* to the two joining curves, since the beam is physically continuous. Also note that the free end *FG* of the beam is tangent to the elastic curve at *F*. There is no curvature in *FG* since the moment is zero in that segment of the beam.

On the elastic curve the point of transition into reverse curvature is called the *point of inflection* or contraflexure. At this point, the moment changes its sign, and the beam is not called upon to resist any moment. This fact often makes these points a desirable place for field connections and so their location is calculated. A procedure for determining points of inflection will be illustrated in Example 2-14, which follows a summary of the above discussion.

The important process of establishing the elastic curve qualitatively may be summarized as follows:

1. Draw a bending-moment diagram.

2. Sketch the elastic curve, corresponding to the signs of moments without reference to the supports, on the moment diagram.

3. If the beam is on two supports, "bodily lift" the curve so drawn and "set it" on the supports; if it is a cantilever, the end of the curve is tangent to the built-in end.

EXAMPLE 2-14

Find the location of the inflection points for the beam analyzed in Example 2-10, Fig. 2-29(a).

SOLUTION

*Section 2-18
Notation for and integration of singularity functions*

By definition, an inflection point corresponds to a point on a beam where the bending moment is zero. Hence, an inflection point may be located by setting up an algebraic expression for the moment in the segment of a beam where such a point is anticipated, and solving this relation equated to zero. By measuring x from the end C of the beam, Fig. 2-29(e), one finds that the bending moment for the segment AD of the beam is $M = -\frac{1}{2}(15)(2)(x-5) + (27.5)(x-15)$. A solution for x is obtained by simplifying and setting this expression equal to zero:

$$M = 12.5x - 337.5 = 0 \qquad x = 27 \text{ ft}$$

Therefore, the inflection point occurring in the segment AD of the beam is $27 - 15 = 12$ ft from the support A.

Similarly, by writing an algebraic expression for the bending moment for the segment DB and setting it equal to zero, the location of the inflection point J is found:

$$M = -\tfrac{1}{2}(15)(2)(x-5) + 27.5(x-15) - 40(x-30) = 0$$

Whence $x = 31.36$ ft, and the distance $AJ = 16.36$ ft.

Often a more convenient method for finding the inflection points consists of utilizing the known relations between the shear and moment diagrams. Thus, since the moment at A is -150 kip-ft, the point of zero moment occurs when the shear-diagram area with the reversed sign from A to H equals this moment, i.e.,

$$-150 + (-1)(-12.5x_1) = 0$$

Hence the distance $AH = 150/12.5 = 12$ ft as before.

Similarly, beginning with a known positive moment of $+37.5$ kip-ft at D, the second inflection point is known to occur when a portion of the shear-diagram area with the reversed sign between D and J reduces this value to zero. Hence, the distance $DJ = 37.5/27.5 = 1.36$ ft, or the distance $AJ = 15 + 1.36 = 16.36$ ft, Fig. 2-29(e), as before.

PART D
SINGULARITY FUNCTIONS*

2-18. NOTATION FOR AND INTEGRATION OF SINGULARITY FUNCTIONS

As was pointed out earlier, analytical expressions for the shear $V(x)$ and the moment $M(x)$ of a given beam may be necessary. If the loading $p(x)$ is a continuous function between the supports, solution of the

---

* This part may be omitted without destroying the continuity of the text. Some readers may find it advantageous to study this material later with Chapter 11.

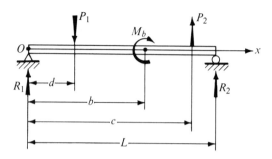

Fig. 2-35. A loaded beam.

differential equation $d^2M/dx^2 = p(x)$ is a convenient approach for determining $V(x)$ and $M(x)$ (see Example 2-11). Now this will be extended to situations in which the loading function is discontinuous. For this purpose the notation of operational calculus will be used. For the function $p(x)$ only polynomials with *positive* integral powers of x, including $0, -1,$ and -2, will be considered. The treatment of other functions is beyond the scope of this text. For the functions considered, however, the method is perfectly general. Further applications of this approach will be given in Chapter 11 for calculating deflections of beams.

Consider a beam loaded as in Fig. 2-35. Since the applied loads are point (concentrated) loads, four distinct regions exist to which different bending moment expressions apply. These are

$$M = R_1 x \qquad \text{when} \qquad 0 \leq x \leq d$$

$$M = R_1 x - P_1(x-d) \qquad \text{when} \qquad d \leq x < b$$

$$M = R_1 x - P_1(x-d) + M_b \qquad \text{when} \qquad b < x \leq c$$

$$M = R_1 x - P_1(x-d) + M_b + P_2(x-c) \qquad \text{when} \qquad c \leq x \leq L$$

All four equations can be written as one, providing we define the following symbolic function:

$$\langle x-a \rangle^n = \begin{cases} 0 & \text{for} \quad 0 < x < a \\ (x-a)^n & \text{for} \quad a < x < \infty \end{cases} \qquad (2\text{-}10)$$

where $n \geq 0 \quad (n = 0, 1, 2, \ldots)$

The expression enclosed by the pointed brackets is nonexistent until x reaches a. For x beyond a, the expression becomes an ordinary binomial. For $n = 0$ and for $x > a$, the function is unity. On this basis, the four separate functions for $M(x)$ given above for the beam of Fig. 2-35 can be combined into one expression which is applicable across the whole span:*

$$M = R_1 \langle x-0 \rangle^1 - P_1 \langle x-d \rangle^1 + M_b \langle x-b \rangle^0 + P_2 \langle x-c \rangle^1$$

---

* This approach was first introduced by A. Clebsch in 1862. O. Heaviside in his *Electromagnetic Theory* initiated and greatly extended the methods of operational calculus. In 1919 W. H. Macaulay specifically suggested the use of special brackets for beam problems. The reader interested in further and/or more rigorous development of this topic should consult texts on mathematics treating Laplace transforms.

Here the values of a are 0, d, b, and c respectively.

To work with this function further, it is convenient to introduce two additional symbolic functions. One is for the concentrated load, treating it as a degenerate case of a distributed load. The other is for the concentrated moment, treating it similarly. Rules for integrating all these functions must be also established. In this discussion the heuristic (nonrigorous) approach will be followed.

A concentrated (point) force may be considered as an enormously strong distributed load acting over a small interval ϵ, Fig. 2-36(a). By treating ϵ as a constant, the following is true

$$\lim_{\epsilon \to 0} \int_{a-\epsilon/2}^{a+\epsilon/2} \frac{P}{\epsilon}\, dx = P \qquad (2\text{-}11)$$

Here it can be noted that P/ϵ has the dimensions of lb/in. and corresponds to the distributed load $p(x)$ in the earlier treatment. Therefore as $\langle x - a \rangle^1 \to 0$, by an analogy of $\langle x - a \rangle^1$ to ϵ, for a concentrated force at $x = a$

$$p = P\langle x - a \rangle_*^{-1} \qquad [\text{lb/in.}] \qquad (2\text{-}12)$$

For p, this expression is dimensionally correct, although $\langle x - a \rangle_*^{-1}$ at

*Section 2-18
Notation for and integration of singularity functions*

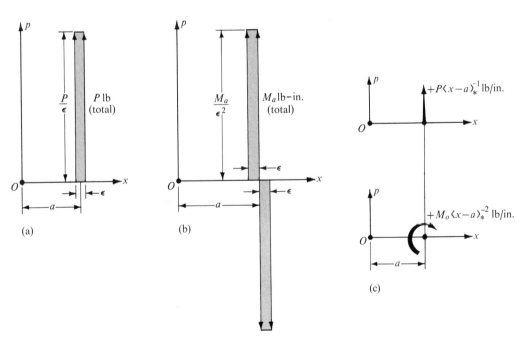

Fig. 2-36. Concentrated force P and moment M_a: (a) and (b) considered as distributed load; (c) symbolic notation for P and M as p.

$x = a$ becomes infinite and by definition is zero everywhere else. Thus, it is a *singular function*. In Eq. 2-12 the asterisk subscript of the bracket is a reminder of the fact that according to Eq. 2-11 the integral of this expression extending over the range ϵ remains bounded and upon integration yields point force itself. Therefore, a special symbolic rule of integration must be adopted:

$$\int_0^x P\langle x - a \rangle_*^{-1} dx = P\langle x - a \rangle^0 \tag{2-13}$$

The coefficient P in the above functions is known as the *strength* of singularity. For P equal to unity, the *unit point load function* $\langle x - a \rangle_*^{-1}$ is also called the *Dirac delta* or the *unit impulse function*.

By analogous reasoning, see Fig. 2-36(b), the loading function p for concentrated moment at $x = a$ is

$$p = M_a \langle x - a \rangle_*^{-2} \quad \text{[lb/in.]} \tag{2-14}$$

This function in being integrated twice defines two symbolic rules of integration. The second integral, except for the exchange of P by M, has already been stated as Eq. 2-13.

$$\int_0^x M_a \langle x - a \rangle_*^{-2} dx = M_a \langle x - a \rangle_*^{-1} \tag{2-15a}$$

$$\int_0^x M_a \langle x - a \rangle_*^{-1} dx = M_a \langle x - a \rangle^0 \tag{2-15b}$$

In Eq. 2-14 the expression is correct dimensionally since p has the units of lb/in. For M_a equal to unity, one obtains the *unit point moment function* $\langle x - a \rangle_*^{-2}$, which is also termed the *doublet* or *dipole*. This function is also singular being infinite at $x = a$ and zero elsewhere. However, after integrating twice a bounded result is obtained. Equations 2-12, 2-14, and 2-15a are symbolic in character. The relation of these equations to the given point loads is clearly evident from Eqs. 2-13 and 2-15a or b.

The integral of binomial functions in pointed brackets for $n \geq 0$ is given by the following rule:

$$\int_0^x \langle x - a \rangle^n dx = \frac{\langle x - a \rangle^{n+1}}{n+1} \quad \text{for} \quad n \geq 0 \tag{2-16}$$

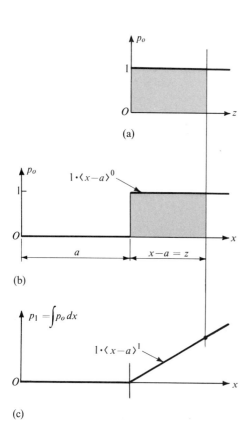

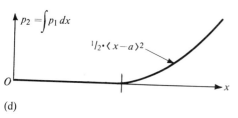

Fig. 2-37. Typical integrations.

This integration process is shown in Fig. 2-37. If a is set equal to zero, one obtains conventional integrals.

EXAMPLE 2-15

Using symbolic functional notation, determine $V(x)$ and $M(x)$ caused by the loading in Fig. 2-38(a).

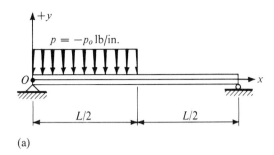

(a)

SOLUTION

To solve this problem Eq. 2-6 can be used. The applied load $p(x)$ acts downward and begins at $x = 0$. Therefore, a term $p = -p_o$, or $p_o\langle x - 0\rangle^0$, which means the same, must exist. This function, however, propagates across the whole span, see Fig. 2-38(b). To terminate the distributed load at $x = L/2$ as required in this problem, another function $+p_o\langle x - L/2\rangle^0$ must be added. The two expressions together represent correctly the applied load.

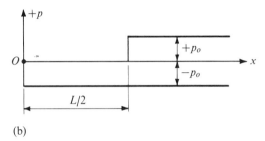

(b)

Fig. 2-38

For this simply supported beam the known boundary conditions are $M(0) = 0$ and $M(L) = 0$. These are used to determine the reactions:

$$\frac{d^2M}{dx^2} = +p = -p_o\langle x - 0\rangle^0 + p_o\langle x - L/2\rangle^0$$

$$\frac{dM}{dx} = -V = -p_o\langle x - 0\rangle^1 + p_o\langle x - L/2\rangle^1 + C_1$$

$$M(x) = -\tfrac{1}{2} p_o\langle x - 0\rangle^2 + \tfrac{1}{2} p_o\langle x - L/2\rangle^2 + C_1 x + C_2$$

$$M(0) = C_2 = 0$$

$$M(L) = -\tfrac{1}{2} p_o L^2 + \tfrac{1}{2} p_o(L/2)^2 + C_1 L = 0$$

hence $\qquad C_1 = +\tfrac{3}{8} p_o L$

and $\quad V(x) = +p_o\langle x - 0\rangle^1 - p_o\langle x - L/2\rangle^1 - \tfrac{3}{8} p_o L$

$$M(x) = -\tfrac{1}{2} p_o\langle x - 0\rangle^2 + \tfrac{1}{2} p_o\langle x - L/2\rangle^2 + \tfrac{3}{8} p_o L x$$

After the solution is obtained, these relations are more easily read by rewriting them in conventional form:

Chapter 2
Axial force, shear,
and bending moment

$$V = -\tfrac{3}{8} p_o L + p_o x$$
$$M = +\tfrac{3}{8} p_o L x - \tfrac{1}{2} p_o x^2$$
when $\quad 0 < x \leq (L/2)$

$$V = -\tfrac{3}{8} p_o L + \tfrac{1}{2} p_o L = +\tfrac{1}{8} p_o L$$
$$M = \tfrac{1}{8} p_o L^2 - \tfrac{1}{8} p_o L x$$
when $\quad (L/2) \leq x < L$

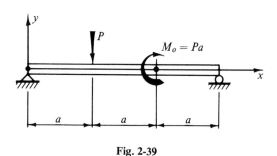

Fig. 2-39

The reactions can be checked by conventional statics. By setting $V = 0$, the location of maximum moment can be found. A plot of these functions is left for the reader to complete.

EXAMPLE 2-16

Find $V(x)$ and $M(x)$ for a beam loaded as in Fig. 2-39. Use singularity functions and treat it as a boundary-value problem.

SOLUTION

By making direct use of Eqs. 2-12 and 2-14 the function $p(x)$ can be written in symbolic form. From the conditions $M(0) = 0$ and $M(L) = 0$, with $L = 3a$, the constants of integration can be found:

$$d^2M/dx^2 = p = -P\langle x - a \rangle_*^{-1} + Pa\langle x - 2a \rangle_*^{-2}$$

$$dM/dx = -V = -P\langle x - a \rangle^0 + Pa\langle x - 2a \rangle_*^{-1} + C_1$$

$$M = -P\langle x - a \rangle^1 + Pa\langle x - 2a \rangle^0 + C_1 x + C_2$$

$$M(0) = C_2 = 0$$

and

$$M(3a) = -2Pa + Pa + 3C_1 a = 0$$

hence $\quad C_1 = +\tfrac{1}{3} P = \tfrac{1}{3} P\langle x - 0 \rangle^0$

and

$$V(x) = -\tfrac{1}{3} P\langle x - 0 \rangle^0 + P\langle x - a \rangle^0 - Pa\langle x - 2a \rangle_*^{-1}$$

$$M(x) = +\tfrac{1}{3} P\langle x - 0 \rangle^1 - P\langle x - a \rangle^1 + Pa\langle x - 2a \rangle^0$$

In the final expression for $V(x)$ the last term has no value if the expression is written in conventional form. Such terms are used only as tracers during the integration process.

It is suggested that the reader check the reactions by conventional statics, write out $V(x)$ and $M(x)$ for the three ranges of the beam within which these functions are continuous, and compare these with a plot of the shear and moment diagrams constructed by the summation procedure.

A suggestion of the way to represent a uniformly varying load, Fig. 2-40(a), acting on a part of a beam is indicated in Fig. 2-40(b). Three separate functions are needed to define the given load completely.

In the above discussion it has been tacitly assumed that the reactions are at the ends of the beams. If such is not the case, the unknown constants C_1 and C_2 must be introduced into Eq. 2-6 as point loads, i.e., as

$$C_1 \langle x - a \rangle_*^{-1} \quad \text{and} \quad C_2 \langle x - b \rangle_*^{-1}$$

This is the condition shown in Fig. 2-40(c). No additional constants of integration are necessary in a solution obtained in this manner.

The advantage of using singularity functions will become especially apparent in subsequent chapters where the solution of statically indeterminate problems is studied. Illustrations of solutions for statically indeterminate beam problems using these functions are in Chapter 11.

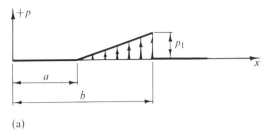
(a)

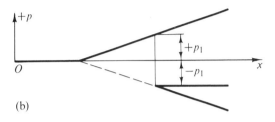
(b)

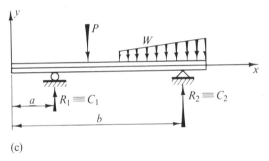
(c)

Fig. 2-40. Illustrations for formulating singularity functions.

PROBLEMS FOR SOLUTION

2-1 and 2-2. For the planar structures shown in the figures find the reactions caused by the applied loads. *Ans. Prob. 2-1.* $R_{By} = 4$ k; *Prob. 2-2.* $R_{Cy} = 9.37$ k.

2-3 through 2-5. For the beams shown in the figures determine the axial force, the shear, and the bending moment midway between the supports caused by the applied loads. *Ans. Prob. 2-5.* $V = 1$ k, $M = -13.5$ k-ft.

2-6 through 2-13. For the planar structures shown in the figures determine the axial force, the shear, and the bending moment at sections a–a. Except for Prob. 2-7, neglect the weight of the members. In every case, draw a free-body of the isolated part of the structure and clearly show on it the sense of the computed quantities. Choose convenient coordinate systems for the presentation of results. *Ans. Prob. 2-8.* $P = 2$ k, $V = -1$ k, $M = -2$ k-ft.; *Prob. 2-9.* $P = -3.43$ k, $V = 1.71$ k, $M = 61.7$ k-in.; *Prob. 2-10.* $P = -14$ k, $V = 2$ k, $M = 5$ k-ft.; *Prob. 2-11.* $P = -30.6$ k, $V = 12.9$ k, $M = 134$ k-ft.; *Prob. 2-12.* $P = -40$ k, $V = 10$ k, $M = -80$ k-ft.; *Prob. 2-13.* $P = -4$ k, $V = -2$ k, $M = -20$ k-in.

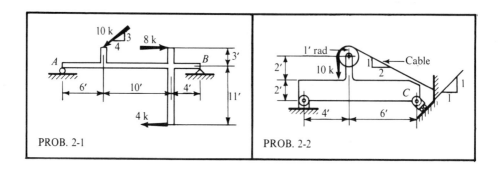

PROB. 2-1

PROB. 2-2

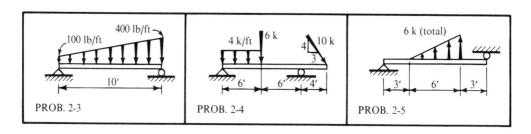

PROB. 2-3

PROB. 2-4

PROB. 2-5

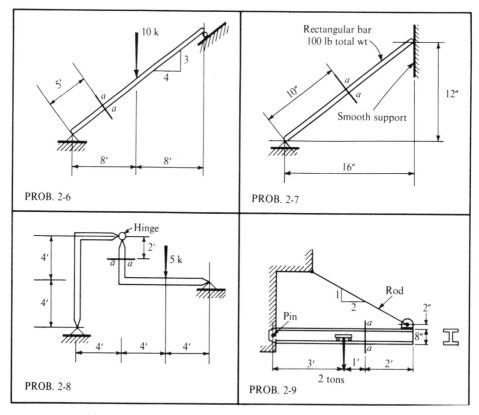

PROB. 2-6

PROB. 2-7

PROB. 2-8

PROB. 2-9

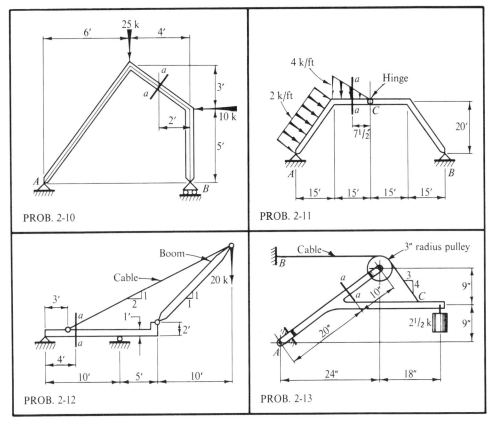

PROB. 2-10

PROB. 2-11

PROB. 2-12

PROB. 2-13

2-14 through 2-19. For the beams loaded as shown in the figures write general expressions for the shear and bending moments for each region over the length of the member. Also plot the corresponding shear and moment diagrams. *Ans.* Maximum moment in parentheses by the figure.

For additional loading conditions see other problems in this chapter.

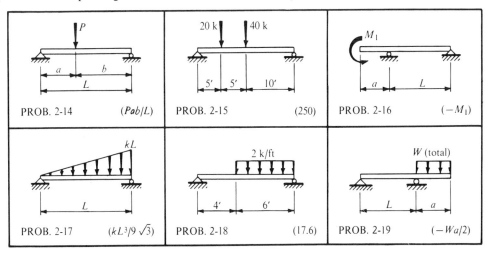

PROB. 2-14   (Pab/L)

PROB. 2-15   (250)

PROB. 2-16   ($-M_1$)

PROB. 2-17   ($kL^3/9\sqrt{3}$)

PROB. 2-18   (17.6)

PROB. 2-19   ($-Wa/2$)

2-20. Write general equations for the internal shear $V(x)$ and bending moment $M(x)$ for the data given in Prob. 2-3.

2-21. Write $V(x)$ and $M(x)$ for each region for the beam load as in Prob. 2-4.

2-22. Same as above for data of Prob. 2-5.

2-23. Establish general algebraic equations for the internal axial force, shear, and bending moment for the curved bar of Example 2-8 loaded as in Fig. 2-24. Plot the results on a polar diagram.

2-24. A rectangular bar bent into a semicircle is built in at one end and is subjected to an internal radial pressure of p lb per unit length (see figure). Write the general expressions for $P(\theta)$, $V(\theta)$, and $M(\theta)$, and plot the results on a polar diagram. Show positive directions assumed for P, V, and M on a free-body diagram.

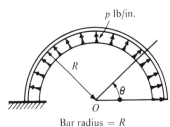

Bar radius $= R$

PROB. 2-24

2-25. A planar frame having the dimensions shown in the figure is subjected to a horizontal load $P = 12.5$ kips. Write the general expressions for P, V, and M for each part of the structure, using appropriate coordinates. Also plot the moment diagram for the whole structure on the compression side of the members. *Ans.* $M_{max} = 35$ k-ft.

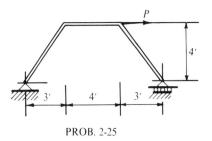

PROB. 2-25

2-26. A bar is made in the shape of a right angle as shown in the figure and is built in at one of its ends. (a) Write the general expressions for V, M, and T (torque) caused by the application of a force F normal to the plane of the bent bar. Plot the results. (b) If in addition to the applied force F the weight of the bar p lb per unit length is also to be considered, what system of internal force components develops at the built-in end?

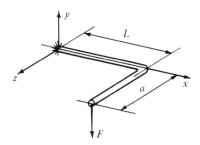

PROB. 2-26

2-27. A welded pipe assembly has three right-angle turns as shown in the figure. (a) Write the general expressions for the internal force components P_x, P_y, P_z, M_x, M_y, and M_z for each part of the assembly caused by $F_x = 100$ lb and $F_z = 50$ lb. (b) Plot the results found in (a). In doing this do not determine the resultant of the bending moments at each section, but rather show the variation in the moments as it occurs in the

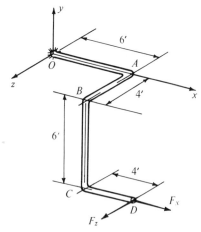

PROB. 2-27

horizontal and vertical planes. (c) If in addition to the applied forces F_x and F_z it is known that the pipe weighs 10 lb per foot and must be considered in the analysis, what is the system of the internal force components at the built-in end?

2-28. A motor drives a shaft with two pulleys as shown in the figure. The belt tensions on the two 10-in.-diameter pulleys have been determined. (a) Plot a moment diagram caused by the vertical force components acting on the shaft, i.e., plot the moment diagram for the xy plane. (b) Plot a moment diagram caused by the horizontal force components, i.e., for the xz plane. (c) Plot the torque diagram. (Note that from $\Sigma M_x = 0$ and the information on the belt tensions the input torque T is known. Also see Fig. 10-24.)

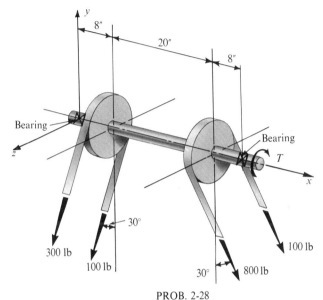

PROB. 2-28

2-29. A circular ring with three hinges in it at A, B, and C is subjected to the loading shown in the figure. Write mathematical expressions for $P(\theta)$, $V(\theta)$ and $M(\theta)$ for the region $0 < \theta <$

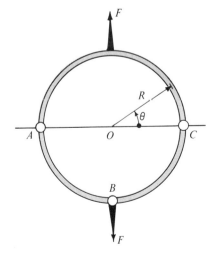

PROB. 2-29

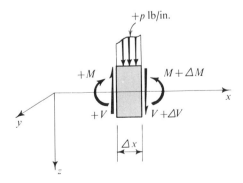

PROB. 2-30

$\pi/2$. Show positive directions assumed for P, V, and M on a free-body diagram.

2-30. If the positive senses of p, V, and M are defined as shown in the figure, derive the relations equivalent to Eqs. 2-4, 2-5, and 2-6. (See Fig. 11-2(b).)

2-31 through 2-53. For the beams loaded in one plane as shown in the figures, neglecting the weight of the members, solve as directed:
  A. Without formal computations, sketch shear and moment diagrams directly below a diagram of the given loaded member.
  B. Same as **A**, and, in addition, show the shape of the elastic curve.
  C. Plot shear, moment, and, wherever significant, axial force diagrams for the main horizontal members. Determine all critical ordinates.
  D. Same as **C**, and, in addition, determine the points of inflection and show the shape of the elastic curve.
  Ans. All shear and moment diagrams must close. The largest moment is given in parentheses by the figures in the units of the problem.

  For additional loading conditions see other problems in this chapter.

2-54 through 2-56. The moment diagrams for beams supported at A and B are as shown in the figures. How are these beams loaded? All curved lines represent parabolas, i.e., plots of equations of the second degree. (*Hint:* construction of shear diagrams aids the

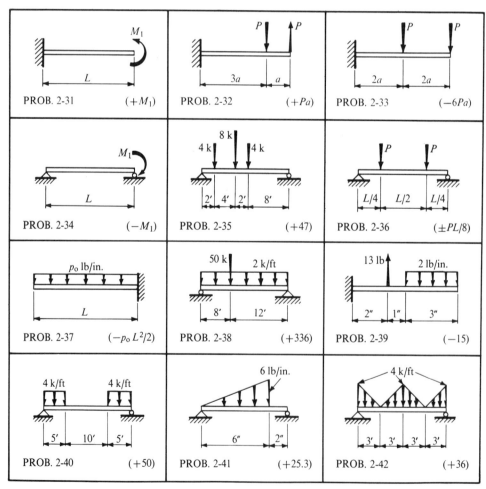

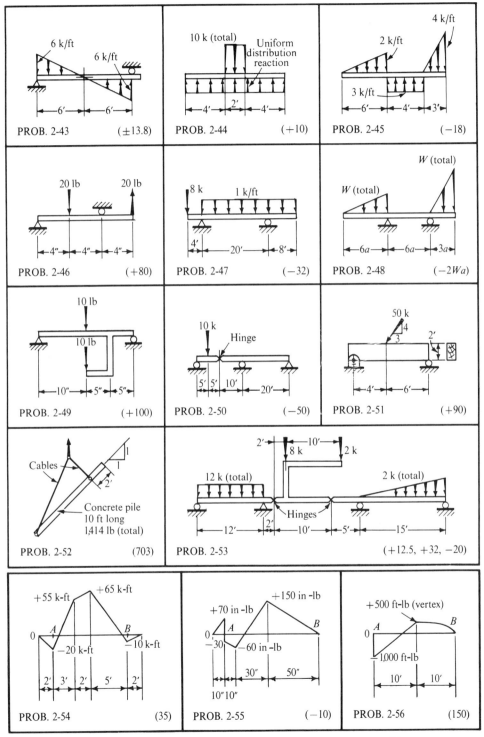

solution.) *Ans.* Reaction at *A* in parentheses by the figure.

2-57. A truck is standing on a raft; it weighs 7½ tons loaded. Assume that 0.1 of the total load is carried by each of the front wheels, and 0.4 by each of the rear wheels. Assume that the two main longitudinal beams of the raft are 6 ft apart, i.e., each beam carries one-half of the truck. Also assume that each of the groups of pontoons provide reactions which may be treated as being uniformly distributed. Plot shear and moment diagrams for each beam for the truck in the position shown. Indicate the critical values using foot-pound units. *Ans.* +18,000 ft-lb (maximum).

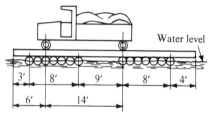

PROB. 2-57

2-58. A small narrow barge is loaded as shown in the figure. Plot shear and moment diagrams for the applied loading. *Ans.* -10^k (maximum), $+50$ k-ft (maximum).

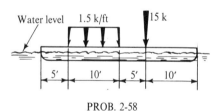

PROB. 2-58

2-59. A 12-in. O.D. steel pipe weighing 50 lb per foot is held by means of rigidly attached yokes in an inclined position as shown in the figure. Plot shear and moment diagrams for this pipe giving the values of all critical ordinates. Joints *A*, *B*, and *C* are pinned.

2-60. The load distribution for a small, single-engine airplane in flight may be idealized as shown in the figure. In this diagram the vector *A* represents the weight of

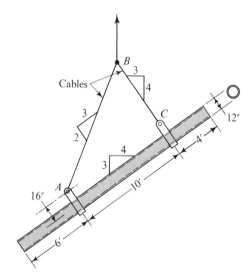

PROB. 2-59

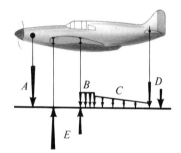

PROB. 2-60

the engine, *B* the uniformly distributed cabin weight, *C* the weight of the aft fuselage, and *D* the forces from the tail control surfaces. The upward forces *E* are developed by the two longerons from the wings. Using this data construct plausible, qualitative shear and moment diagrams for the fuselage.

2-61 through 2-63. For the beams loaded as shown in the figures, using Eq. 2-6, (a) find $V(x)$ and $M(x)$. Check reactions by conventional statics. (b) Plot the shear and moment diagrams.

2-64 through 2-72. For the beams loaded as shown in the figures, using singularity functions and Eq. 2-6, (a) find $V(x)$ and $M(x)$. Check reactions by conventional statics. (b) Plot the shear and moment diagrams.

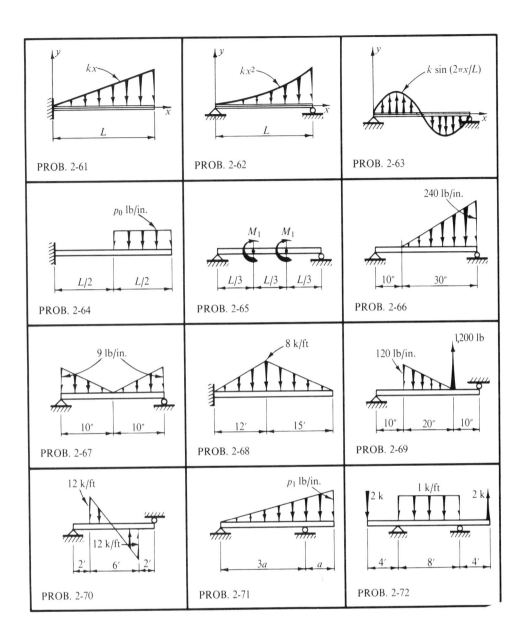

3 Stress and axial loads

3-1. INTRODUCTION

It was pointed out in Chapter 1 that the nature of forces set up within a body to balance the effect of the externally applied forces is a part of the main problem in the mechanics of solids. Specialized procedures for applying the method of sections in order to determine the system of force components at a cut through a beam were treated in Chapter 2. In this chapter the method of sections will be carried further in order to isolate an infinitesimal element and to define the concept of stress. In Part A, the general case of stress is considered; in Part B, procedures are outlined for determining stresses in axially loaded rods. A few examples of shearing-stress calculation are also considered, and a definition of the safety factor is given.

PART A STRESS

3-2. DEFINITION OF STRESS

In general, the internal forces acting on infinitesimal areas of a cut are of varying magnitudes and directions, as was shown earlier in Figs. 1-1(b)

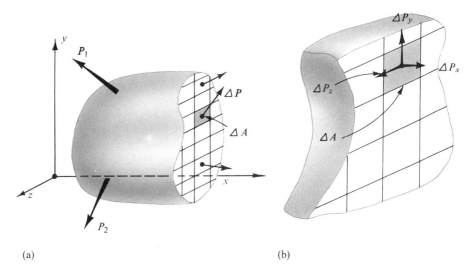

Fig. 3-1. Sectioned body: (a) free body with some internal forces, (b) enlarged view with components of $\Delta \mathbf{P}$.

and (c) and as is again shown in Fig. 3-1(a). These forces are vectorial in nature and they maintain the externally applied forces in equilibrium. In the mechanics of solids it is particularly significant to determine the intensity of these forces on the various portions of the cut as resistance to deformation and to forces depends on these intensities. In general, they vary from point to point and are inclined with respect to the plane of the cut. It is customary to resolve these intensities perpendicular and parallel to the section investigated. As an example, the components of a force vector $\Delta \mathbf{P}$ acting on an area ΔA are shown in Fig. 3-1(b). In this particular diagram, the cut through the body is perpendicular to the x axis, and the directions of ΔP_x and of the normal to ΔA coincide.

Since the components of the intensity of force per unit area—i.e., of *stress*—hold true only at a point, the mathematical definition* of stress is

$$\tau_{xx} = \lim_{\Delta A \to 0} \frac{\Delta P_x}{\Delta A}, \qquad \tau_{xy} = \lim_{\Delta A \to 0} \frac{\Delta P_y}{\Delta A}, \qquad \text{and} \qquad \tau_{xz} = \lim_{\Delta A \to 0} \frac{\Delta P_z}{\Delta A}$$

where, in all three cases, the first subscript of τ (tau) indicates that the plane perpendicular to the x axis is considered, and the second designates the direction of the stress component. In the next article all possible combinations of subscripts for stress will be discussed.

The intensity of the force perpendicular to or normal to the section is called the *normal stress* at a point. It is customary to refer to normal stresses

---

* As $\Delta A \to 0$, some question from the atomic point of view exists in defining stress in this manner. However, a homogeneous model for nonhomogeneous matter appears to have worked well.

that cause traction or tension on the surface of a cut as *tensile stresses*. On the other hand, those that are pushing against the cut are *compressive stresses*. In this book normal stresses will usually be designated by the letter σ (sigma) instead of by a double subscript on τ. A single subscript then suffices to designate the direction of the axis. The other components of the intensity of force act parallel to the plane of the elementary area. These components are called *shearing stresses*. Shearing stresses will be always designated by τ.

The reader should form a clear mental picture of the stresses called normal and those called shearing. To repeat, normal stresses result from force components perpendicular to the plane of the cut, and shearing stresses result from components parallel to the plane of the cut.

It is seen from the above definitions of normal and shearing stresses that since they represent the intensity of force on an area, stresses are measured in units of force divided by units of area. In the English system, the usual units for stress are pounds per square inch, abbreviated in this text as *psi*. In many cases it will be found convenient to use as a unit of force the coined word *kip*, meaning kilo-pound or 1,000 lb. The stress in kips per square inch is abbreviated *ksi*.

It should be noted that stresses multiplied by the respective areas on which they act give forces, and it is the sum of these forces at an imaginary cut that keeps a body in equilibrium.

3-3. STRESS TENSOR

If, in addition to the cutting plane implied in the free body of Fig. 3-1, another plane an infinitesimal distance away and parallel to the first were passed through the body, an elementary slice of the body would be isolated. Then, if an additional two pairs of planes were passed normal to the first pair, a cube of infinitesimal dimensions would be isolated from the body. Such a cube is shown in Fig. 3-2. All stresses acting on this cube are identified on the diagram. As noted earlier, the first subscripts on the τ's associate the stress with a plane perpendicular to a given axis; the second designate the direction of the stress. On the *near faces* of the cube, i.e., on the faces away from the origin, the directions of stress are positive if they coincide with the positive directions of the axes. On the faces of the cube toward the origin, from the action-reaction equilibrium concept, positive stresses act in the direction opposite to the positive directions of the axes. (Note that for normal stresses, by changing the symbol for stress from τ to σ, a single subscript on σ suffices to define this

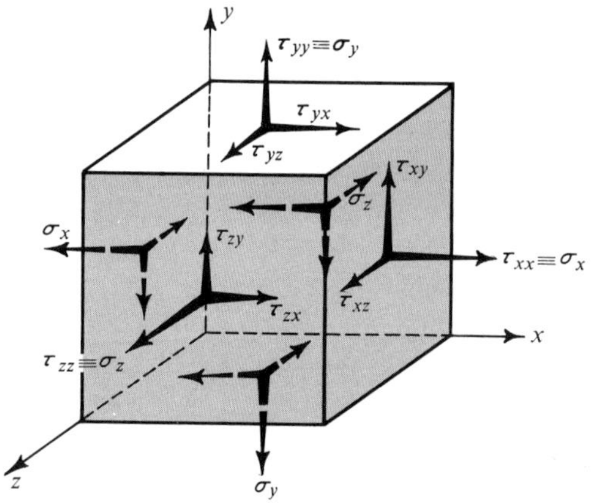

Fig. 3-2. The most general state of stress acting on an element. All stresses have positive sense.

quantity without ambiguity.) The designations for stresses in Fig. 3-2 are widely used in the mathematical theories of elasticity and plasticity. The sign convention here agrees with the one introduced earlier in Fig. 2-2.

An examination of the stress symbols in Fig. 3-2 shows that there are three normal stresses $\tau_{xx} \equiv \sigma_x$, $\tau_{yy} \equiv \sigma_y$, $\tau_{zz} \equiv \sigma_z$, and six shearing stresses τ_{xy}, τ_{yx}, τ_{yz}, τ_{zy}, τ_{zx}, τ_{xz}. By contrast, a force vector P has only three components P_x, P_y, and P_z. These can be written in an orderly manner as a column vector:

$$\begin{pmatrix} P_x \\ P_y \\ P_z \end{pmatrix}$$

Analogously, the stress components can be assembled as follows:

$$\begin{pmatrix} \tau_{xx} & \tau_{xy} & \tau_{xz} \\ \tau_{yx} & \tau_{yy} & \tau_{yz} \\ \tau_{zx} & \tau_{zy} & \tau_{zz} \end{pmatrix} \equiv \begin{pmatrix} \sigma_x & \tau_{xy} & \tau_{xz} \\ \tau_{yx} & \sigma_y & \tau_{yz} \\ \tau_{zx} & \tau_{zy} & \sigma_z \end{pmatrix} \quad \textbf{(3-1)}$$

This is a matrix representation of the *stress tensor*. It is a second-rank tensor requiring two indices to identify its elements or components. A vector is a first-rank tensor, and a scalar is a zero-rank tensor. Sometimes, for brevity, a stress tensor is written in indicial notation as τ_{ij}, where it is understood that i and j can assume designations x, y, and z as noted in Eq. 3-1.

Next, it will be shown that the stress tensor is symmetric, i.e., $\tau_{ij} = \tau_{ji}$. This follows directly from the equilibrium requirements for an element. For this purpose, let the dimensions of the infinitesimal element be dx, dy, and dz, and sum the moments of forces about an axis such as the z axis in Fig. 3-2. Neglecting the infinitesimals of higher order* this process is equivalent to taking the moment about the z axis in Fig. 3-3(a) or, in its two-dimensional representation, in Fig. 3-3(b). Thus

$$M_C = 0 \circlearrowleft +, \quad +(\tau_{yx})(dx\,dz)(dy) - (\tau_{xy})(dy\,dz)(dx) = 0$$

where the expressions in parentheses correspond respectively to stress, area, and moment arm. Simplifying,

$$\tau_{yx} = \tau_{xy} \quad \textbf{(3-2)}$$

---

* The possibility of an infinitesimal change in stress from one face of the cube to another and the possibility of the presence of body (inertial) forces exist. By first considering an element $(\Delta x)(\Delta y)(\Delta z)$ and proceeding to the limit, it can be shown rigorously that these quantities are of higher order and therefore negligible.

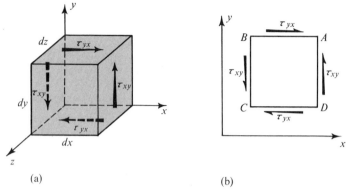

(a) (b)

Fig. 3-3. An element of a body in pure shear.

Similarly it can be shown that $\tau_{xz} = \tau_{zx}$ and $\tau_{yz} = \tau_{zy}$. Hence the subscripts for the shearing stresses are commutative, i.e., their order may be interchanged, and the stress tensor is symmetric.

The implication of Eq. 3-2 is very important. The fact that subscripts are commutative signifies that shearing stresses on mutually perpendicular planes of an infinitesimal element are numerically equal. Moreover, it is possible to have an element in equilibrium only when *shearing stresses occur on four sides of an element simultaneously.* That is, in any body where shearing stresses exist, two pairs of such stresses act on mutually perpendicular planes. Hence $\Sigma M_z = 0$ is not satisfied by a

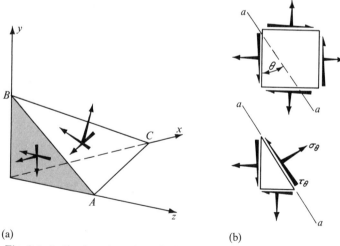

(a) (b)

Fig. 3-4. Inclined sections through element: (a) three-dimensional case, (b) two-dimensional case.

single pair of shearing stresses. On diagrams, as in Fig. 3-3(b), the arrowheads of the shearing stresses must meet at diametrically opposite corners of an element to satisfy the equilibrium conditions.

In subsequent situations more than two pairs of shearing stresses will seldom act on an element simultaneously. Hence the subscripts used above to identify the planes and the directions of the shearing stresses become superfluous. In such cases shearing stresses will be designated by τ without any subscripts. However, one must remember that shearing stresses always occur in two pairs.

It should be noted that the conventional system of axes may not yield the most significant information about the state of stress at a point. In some cases stresses must be examined on inclined planes such as the plane *ABC* in Fig. 3-4(a). This process is termed the *transformation of stress* from one set of axes to another. The two-dimensional counterpart, shown in Fig. 3-4(b), will be studied in detail in Chapter 9.

Using the procedures of stress transformation, some of which will be discussed later, for a particular set of coordinates, one in general can always diagonalize the stress tensor to read

$$\begin{pmatrix} \sigma_1 & 0 & 0 \\ 0 & \sigma_2 & 0 \\ 0 & 0 & \sigma_3 \end{pmatrix} \quad \text{and} \quad \begin{pmatrix} \sigma_1 & 0 & 0 \\ 0 & \sigma_2 & 0 \\ 0 & 0 & 0 \end{pmatrix}$$

for a two-dimensional case of plane stress where $\sigma_3 = 0$. Note the absence of shearing stresses. For the three-dimensional case, the stresses are said to be *triaxial* since three stresses are necessary to describe the state of stress completely. For two-dimensional cases the stresses are *biaxial*. Plane stress occurs in thin sheets stressed in two different directions. For axially loaded members, which are discussed in the next part of this chapter, only one element of the stress tensor survives; such a state of stress is referred to as *uniaxial*. In Chapters 9 and 10 an inverse problem will be discussed: how this one term can be resolved to yield four elements of a stress tensor.

3-4. DIFFERENTIAL EQUATIONS OF EQUILIBRIUM

An infinitesimal element of a body must be in equilibrium. For a two-dimensional case the system of stresses acting on an infinitesimal element $(dx)(dy)(1)$ is shown in Fig. 3-5. In this problem, the element is assumed to be 1 in. thick in the direction perpendicular to the plane of the paper. Note that the possibility of an increment in stresses from one face of the element to another is accounted for. For example, since the rate of change of σ_x in the x direction is $\partial \sigma_x / \partial x$ and a step of dx is made, the

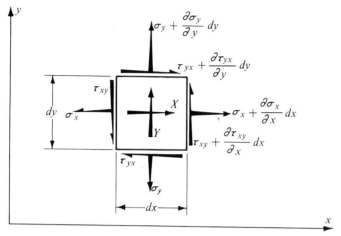

Fig. 3-5. Infinitesimal element with stresses and body forces acting on it.

increment is $(\partial \sigma_x/\partial x)\, dx$. The partial derivative notation has to be used to differentiate between the directions.

The inertial or body forces, such as those caused by the weight or the magnetic effect, are designated X and Y and are associated with the unit volume of the material. With these notations

$$\sum F_x = 0 \rightarrow +,$$

$$\left(\sigma_x + \frac{\partial \sigma_x}{\partial x}\, dx\right)(dy \times 1) - \sigma_x(dy \times 1)$$

$$+ \left(\tau_{yx} + \frac{\partial \tau_{yx}}{\partial y}\, dy\right)(dx \times 1) - \tau_{yx}(dx \times 1) + X(dx\, dy \times 1) = 0$$

Simplifying and recalling that $\tau_{xy} = \tau_{yx}$ holds true, one obtains the basic equilibrium equation for the x direction. This equation, together with an analogous one for the y direction, reads

$$\left. \begin{array}{l} \dfrac{\partial \sigma_x}{\partial x} + \dfrac{\partial \tau_{xy}}{\partial y} + X = 0 \\[6pt] \dfrac{\partial \tau_{yx}}{\partial x} + \dfrac{\partial \sigma_y}{\partial y} + Y = 0 \end{array} \right\} \text{Static equilibrium} \quad (3\text{-}3)$$

The moment equilibrium of the element requiring $\sum M_z = 0$ is assured by having $\tau_{xy} = \tau_{yx}$.

It can be shown that for the three-dimensional case, a typical equation from a set of three is

$$\frac{\partial \sigma_x}{\partial x} + \frac{\partial \tau_{xy}}{\partial y} + \frac{\partial \tau_{xz}}{\partial z} + X = 0 \qquad (3\text{-}4)$$

Note that in deriving the above equations mechanical properties of the material have not been used. This means that these equations are applicable whether a material is elastic, plastic, or viscoelastic. Also it is very important to note that there are not enough equations of equilibrium to solve for the unknown stresses. In the two-dimensional case, in the two parts of Eq. 3-3 there are three unknown stresses σ_x, σ_y, and τ_{xy}. For the three-dimensional case there are six stresses, but only three equations. Thus all problems in stress analysis are internally statically intractable or indeterminate. In technical mechanics of solids such as that presented in this text this indeterminacy is eliminated by introducing appropriate assumptions, which is equivalent to having additional equations.

PART B
STRESSES IN AXIALLY LOADED MEMBERS

3-5. AXIAL LOAD; NORMAL STRESS

In many practical situations, if the direction of the imaginary plane cutting a member is judiciously selected, the stresses that act on the cut will be found both particularly significant and simple to determine. One such important case occurs in a straight axially loaded rod in tension, provided a plane is passed perpendicular to the axis of the rod.* The tensile stress acting on such a cut is the maximum stress as any cut not perpendicular to the axis of the rod provides a larger surface for resisting the applied force. The maximum stress is the most significant one as it tends to cause the failure of the material.†

To obtain an algebraic expression for this maximum stress, consider the case illustrated in Fig. 3-6(a). If the rod is assumed weightless, two equal and opposite forces P are necessary, one at each end, to maintain equilibrium. Then, since the body as a whole is in equilibrium, any part of it is also in equilibrium. A part of the rod to either side of the cut b-b is in equilibrium. At the cut, where the cross-sectional area of the rod is A, a force equivalent to P as shown in Figs. 3-6(b) and (c) must be developed. Whereupon, from the definition of stress, the normal stress, or the stress

---

* Some materials exhibit a greater relative strength to normal stresses than to shearing stresses. For such materials failure takes place on an oblique plane. This is discussed in Chapter 10.
† Immediately following this article some readers may wish to study Art. 10-2 and Example 10-1, where the stresses on inclinded planes are considered. This can be either preceded or followed by the study of Part A of Chapter 9.

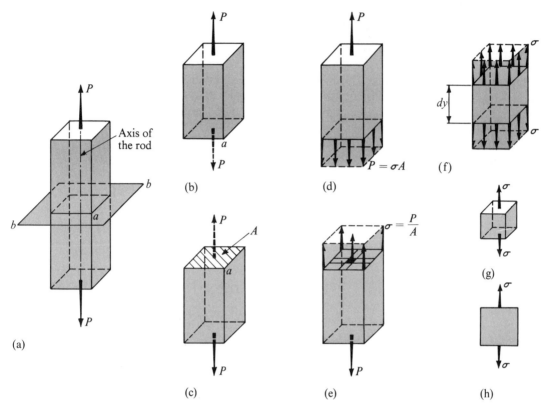

Fig. 3-6. Successive steps in the analysis of a body for stress.

which acts perpendicularly to the cut, is

$$\sigma = \frac{P}{A} \quad \text{or} \quad \frac{\text{force}}{\text{area}} \quad \left[\frac{\text{lb}}{\text{in.}^2}\right] \tag{3-5}$$

This normal stress is uniformly distributed over the cross-sectional area A.* The nature of the quantity computed by Eq. 3-5 may be seen graphically in Figs. 3-6(d) and (e). In general, the force P is a resultant of a number of forces to one side of the cut or another.

If an additional cut parallel to the plane b-b in Fig. 3-6(a) were made, the isolated section of the rod could be represented as in Fig. 3-6(f), and upon further "cutting," an infinitesimal cube as in Fig. 3-6(g) results. The only kind of stresses that appear here are the normal stresses on the two surfaces of the cube. Such a state of stress on an element is referred to as *uniaxial stress*. In practice, isometric views of a cube as shown in

---

* Equation 3-5 strictly applies only if the cross-sectional area is constant along the rod. For a discussion of situations where an abrupt discontinuity in the cross-sectional area occurs, see Art. 4-18.

Fig. 3-6(g) are seldom employed; the diagrams are simplified to look like those of Fig. 3-6(h). Nevertheless, the student must never lose sight of the three-dimensional aspect of the problem at hand.

*Section 3-5
Axial load;
normal stress*

At a cut the system of tensile stresses computed by Eq. 3-5 provides an equilibrant to the externally applied force. When these normal stresses are multiplied by the corresponding infinitesimal areas and then summed over the whole area of a cut, the summation is equal to the applied force P. Thus the system of stresses is statically equivalent to the force P. Moreover, the resultant of this sum must act through the centroid of a section. Conversely, to have uniform stress distribution in a rod, the applied axial force must act through the centroid of the cross-sectional area investigated.

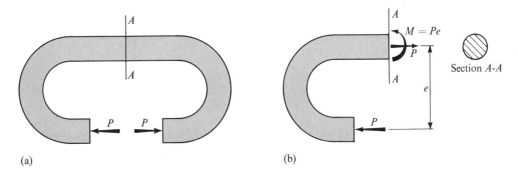

Fig. 3-7. A member with a nonuniform stress distribution at Section A-A.

For example, in the machine part shown in Fig. 3-7(a) the stresses cannot be obtained from Eq. 3-5 alone. Here, at a cut such as A-A, a statically equivalent system of forces developed within the material must consist not only of the force P but also of a bending moment M that maintains the externally applied force in equilibrium. This causes nonuniform stress distribution in the member, which will be treated in Chapter 8.

In accepting Eq. 3-5, it must be kept in mind that the materials' behavior is idealized. Each and every particle of a body is assumed to contribute equally to the resistance of the force. A perfect homogeneity of the material is implied by such an assumption. Real materials such as metals consist of a great many grains, and wood is fibrous. In real materials some particles will contribute more to the resistance of a force than others. Stresses as shown in Figs. 3-6(d) and (e) actually do not exist. The diagram of true stress distribution varies in each particular case and is a highly irregular, jagged affair. However, on the average, or statistically speaking, computations based on Eq. 3-5 are correct, and hence the computed stress represents a highly significant quantity.

Similar reasoning applies to compression members. The maximum normal or compressive stress may again be obtained by passing a section

*Chapter 3
Stress and
axial loads*

perpendicular to the axis of a member and applying Eq. 3-5. The stress so obtained will be of uniform intensity as long as the resultant of the applied forces coincides with the centroid of the area at the cut. However, one must exercise additional care when compression members are investigated. These may be so slender that they may not behave in the fashion considered. For example, an ordinary yardstick under a rather small axial compression force has a tendency to buckle sidewise and collapse. The consideration of such instability of compression members is deferred until Chapter 14. Equation 3-5 is applicable only to axially loaded compression members that are rather chunky, i.e., to short blocks. As will be shown in Chapter 14, a block whose least dimension is approximately one-tenth of its length may usually be considered a short block. For example, a 2-in.-by-4-in. wooden piece may be 20 in. long and still be considered a short block.

3-6. AXIAL LOAD; BEARING STRESS

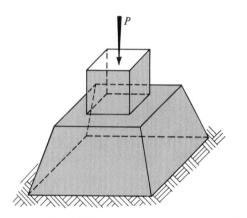

Fig. 3-8. Bearing stresses occur between the block and pier.

Situations often arise where one body is supported by another. If the resultant of the applied forces coincides with the centroid of the contact area between the two bodies, the intensity of force, or the stress, between the two bodies can again be determined from Eq. 3-5. It is customary to refer to this normal stress as a *bearing stress*. Figure 3-8, where a short block bears on a concrete pier and the latter bears on the soil, illustrates such a stress. The bearing stresses are obtained by dividing the applied force P by the corresponding area of contact.

3-7. AVERAGE SHEARING STRESS

Another situation frequently arising in practice is shown in Figs. 3-9(a), (c), and (e). In all of these cases the forces are transmitted from one part of a body to the other by causing stresses in the plane parallel to the applied force. To obtain stresses in such instances, cutting planes such as *A-A* are selected and free-body diagrams* as in Figs. 3-9(b) (d), and (f) are used. The forces are transmitted through the respective cut areas. Hence, assuming that the stresses which act in the plane of these cuts are uniformly distributed, one obtains a relation for stress

---

* A small unbalance in moment equal to Pe exists in the first two cases shown in Fig. 3-9, but, being small, it is commonly ignored.

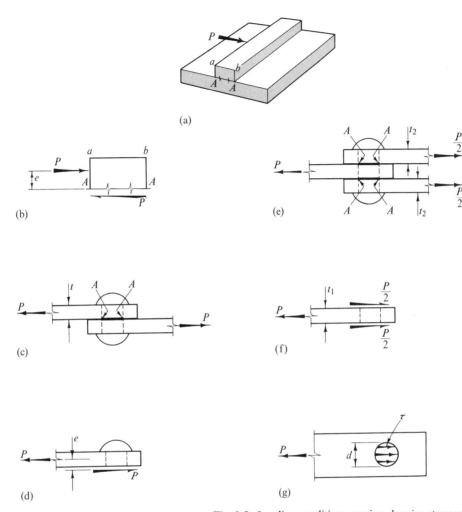

Fig. 3-9. Loading conditions causing shearing stresses.

$$\tau = \frac{P}{A} \quad \text{or} \quad \frac{\textbf{force}}{\textbf{area}} \quad \left[\frac{\text{lb}}{\text{in.}^2}\right] \quad (3\text{-}6)$$

where τ by definition is the shearing stress, P is the total force acting across and parallel to the cut, and A is the cross-sectional area of the cut member. For reasons to be discussed later, unlike normal stress, the shearing stress given by Eq. 3-6 is only approximately true. For the cases shown, the shearing stresses actually are distributed in a nonuniform fashion across the area of the cut. Thus the quantity given by Eq. 3-6 represents an average shearing stress.

The shearing stress as computed by Eq. 3-6 is shown diagrammatically in Fig. 3-9(g). Note that for the case shown in Fig. 3-9(e) there

Chapter 3
Stress and
axial loads

are two planes of the rivet which resist the force. Such a rivet or a bolt is referred to as being in *double shear*.

In cases such as those in Figs. 3-9(c) and (e), as the force P is applied, a highly irregular pressure develops between the bolt and the plates. The average nominal intensity of this pressure is obtained by dividing the force transmitted by the projected area of the bolt onto the plate. This is referred to as the *bearing stress*. The bearing stress in Fig. 3-9(c) is $\sigma_b = P/(td)$, where t is the thickness of the plate and d is the diameter of the rivet. For the case in Fig. 3-9(e) the bearing stresses for the middle plate and the outer plates are $\sigma_1 = P/(t_1 d)$ and $\sigma_2 = P/(2t_2 d)$, respectively.

EXAMPLE 3-1

The beam BE in Fig. 3-10(a) is used for hoisting machinery. It is anchored by two bolts at B, and at C it rests on a parapet wall. The essential details are given in the figure. Note that the bolts are threaded as shown in Fig. 3-10(d) with $d = 0.620$ in. at the root of the threads. If this arrangement is used to lift equipment of 1 ton (2,000 lb), determine the stress in the bolts BD and the bearing stress at C. Assume that the weight of the beam is negligible in comparison with the loads handled.

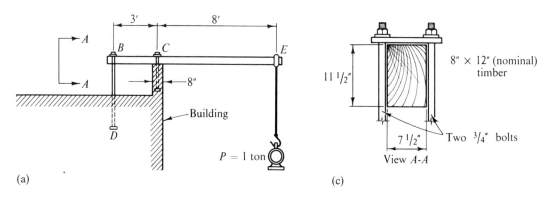

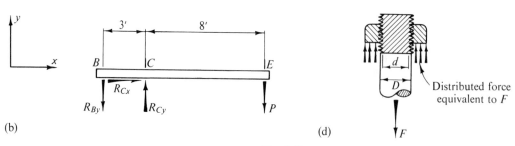

Fig. 3-10

SOLUTION

*Section 3-7
Average shearing stress*

To solve this problem, the actual situation is idealized, and a free-body diagram is made on which all known and unknown forces are indicated. This is shown in Fig. 3-10(b). The vertical reactions at B and C are unknown. They are indicated respectively as R_{By} and R_{Cy}, where the first subscript identifies the location and the second the line of action of the unknown force. As the long bolts BD are not effective in resisting the horizontal force, only an unknown horizontal reaction at C is assumed and marked as R_{Cx}. The applied known force P is shown in its proper location. After a free-body diagram is prepared, the equations of statics are applied and solved for the unknown forces.

$\sum F_x = 0 \qquad R_{Cx} = 0$

$\sum M_B = 0 \circlearrowright +, \quad +2{,}000(8+3) - R_{Cy}(3) = 0, \quad R_{Cy} = 7{,}333 \text{ lb} \uparrow$

$\sum M_C = 0 \circlearrowright +, \quad +2{,}000(8) - R_{By}(3) = 0, \qquad R_{By} = 5{,}333 \text{ lb} \downarrow$

Check: $\sum F_y = 0 \uparrow +, \quad -5{,}333 + 7{,}333 - 2{,}000 = 0$

These steps complete and check the work of determining the forces. The various areas of the material that resist these forces are determined next and Eq. 3-5 is applied.

Cross-sectional area of one $\tfrac{3}{4}$-in. bolt: $A = \pi(0.75/2)^2 = 0.442 \text{ in.}^2$ This is not the minimum area of a bolt; threads reduce it.

The cross-sectional area of one $\tfrac{3}{4}$-in. bolt at the root of the threads is

$$A_{net} = \pi(0.620/2)^2 = 0.302 \text{ in.}^2$$

Maximum normal tensile stress* in each of the two bolts BD:

$$\sigma_{max} = \frac{R_{By}}{2A} = \frac{5{,}333}{2(0.302)} = 8{,}800 \text{ psi}$$

Tensile stress in the shank of the bolts BD:

$$\sigma = \frac{5{,}333}{2(0.442)} = 6{,}000 \text{ psi}$$

Contact area at C:

$$A = 7.5(8) = 60 \text{ in.}^2$$

Bearing stress at C:

$$\sigma_b = \frac{R_{Cy}}{A} = \frac{7{,}333}{60} = 122 \text{ psi}$$

The calculated stress for the bolt shank can be represented in the manner of Eq. 3-1 as

---

* See also discussion on stress concentrations, Art. 4-18.

Chapter 3
Stress and
axial loads

$$\begin{pmatrix} 0 & 0 & 0 \\ 0 & +6,000 & 0 \\ 0 & 0 & 0 \end{pmatrix} \text{psi}$$

where it is arbitrarily assumed that the y axis is in the direction of the applied load. In ordinary problems the complete result is implied but is seldom written down in such detail.

EXAMPLE 3-2

The concrete pier shown in Fig. 3-11(a) is loaded at the top with a uniformly distributed load of 600 lb per square foot. Investigate the state of stress at a level of 4 ft above the base. Concrete weighs approximately 150 lb per cubic foot.

SOLUTION

In this problem the weight of the structure itself is appreciable and must be included in the calculations.

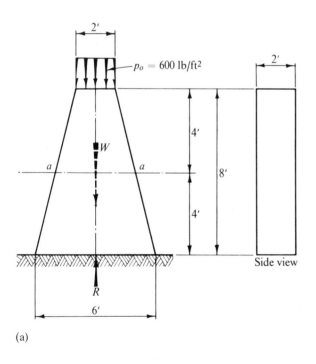
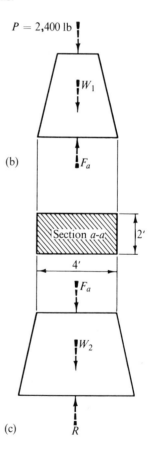

Fig. 3-11

Weight of the whole pier:

$$W = (2 + 6)2(8)150/2 = 9{,}600 \text{ lb}$$

Total applied force:

$$P = 600(2)2 = 2{,}400 \text{ lb}$$

From $\Sigma F_y = 0$, reaction at base:

$$R = W + P = 12{,}000 \text{ lb}$$

These forces are shown in the diagrams schematically as concentrated forces acting through their respective centroids. Then, to determine the stress at the desired level, the body is cut into two separate parts. A free-body diagram for either part is sufficient to solve the problem. For comparison the problem is solved both ways.

Using upper part of the pier as a free body, Fig. 3-11(b), weight of the pier above the cut:

$$W_1 = (2 + 4)2(4)150/2 = 3{,}600 \text{ lb}$$

From $\Sigma F_y = 0$, force at the cut: $F_a = P + W_1 = 6{,}000$ lb. Hence, using Eq. 3-5, the normal stress at the level a-a is

$$\sigma_a = \frac{F_a}{A} = \frac{6{,}000}{2(4)} = 750 \text{ lb per square foot} \quad \text{or} \quad \frac{750}{144} = 5.2 \text{ psi}$$

This stress is compressive as F_a acts on the cut.

Using lower part of the pier as a free body, Fig. 3-11(c), weight of the pier below the cut:

$$W_2 = (4 + 6)2(4)150/2 = 6{,}000 \text{ lb}$$

From $\Sigma F_y = 0$, force at the cut:

$$F_a = R - W_2 = 6{,}000 \text{ lb}$$

The remainder of the problem is the same as before. The pier considered here has a vertical axis of symmetry, making the application of Eq. 3-5 possible.*

EXAMPLE 3-3

A bracket of negligible weight shown in Fig. 3-12(a) is loaded with a force P of 3 kips. For interconnection purposes the bar ends are clevised (forked). Pertinent dimensions are shown in the figure. Find the normal stresses in the members AB and BC and the bearing and shearing stresses for the pin C. All pins are 0.375 in. in diameter.

---

* Strictly speaking the solution obtained is not exact as the sides of the pier are sloping. If the included angle between these sides is large, this solution is altogether inadequate. For further details see S. Timoshenko and J. N. Goodier, *Theory of Elasticity* (2nd ed.) (New York: McGraw-Hill Book Company, 1951, p. 96).

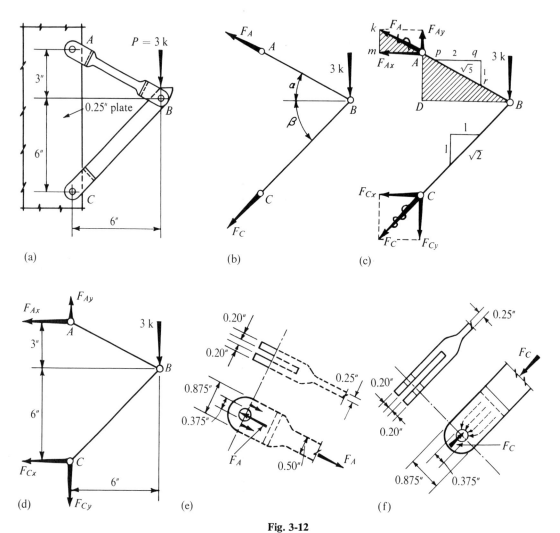

Fig. 3-12

SOLUTION

First an idealized free-body diagram consisting of the two bars pinned at the ends is prepared, Fig. 3-12(b). As there are no intermediate forces acting on the bars and the applied force acts through the joint at B, the forces in the bars are directed along the lines AB and BC, and the bars AB and BC are loaded axially. The magnitudes of the forces are unknown and are labeled F_A and F_C in the diagram.* These forces may be determined graphically by completing a triangle of forces F_A, F_B, and P. These forces may also be found analytically from two simultaneous

---

* In frameworks it is convenient to assume all unknown forces are tensile. A negative answer in the solution then indicates that the bar is in compression.

equations $\Sigma F_y = 0$ and $\Sigma F_x = 0$, written in terms of the unknowns F_A and F_C, a known force P, and two known angles α and β. Both these procedures are possible. However, in this course usually it will be found advantageous to proceed in a different way. Instead of treating forces F_A and F_C directly, their components are used; and instead of $\Sigma F = 0$, $\Sigma M = 0$ becomes the main tool.

*Section 3-7
Average shearing stress*

Any force may be resolved into components. For example, F_A may be resolved into F_{Ax} and F_{Ay} as in Fig. 3-12(c). Conversely, if any one of the components of a directed force is known, the force itself may be determined. This follows from similarity of dimension and force triangles. In Fig. 3-12(c) the triangles Akm and BAD are similar triangles (both are shaded in the diagram). Hence, if F_{Ax} is known,

$$F_A = (AB/DB)F_{Ax}$$

Similarly, $F_{Ay} = (AD/DB)F_{Ax}$. However, note further that AB/DB or AD/DB are ratios, hence relative dimensions of members may be used. Such relative dimensions are shown by a little triangle on the member AB and again on BC. In the problem at hand

$$F_A = (\sqrt{5}/2)F_{Ax} \quad \text{and} \quad F_{Ay} = F_{Ax}/2$$

Adopting the above procedure of resolving forces, the revised free-body diagram, Fig. 3-12(d), is prepared. Two components of force are necessary at the pin joints. After the forces are determined by statics, Eq. 3-5 is applied several times, thinking in terms of a free body of an individual member:

$$\Sigma M_C = 0 \circlearrowleft +, \quad +F_{Ax}(3+6) - 3(6) = 0, \quad F_{Ax} = +2 \text{ kips}$$

$$F_{Ay} = F_{Ax}/2 = 2/2 = 1 \text{ kip},$$

$$F_A = 2(\sqrt{5}/2) = +2.23 \text{ kips}$$

$$\Sigma M_A = 0 \circlearrowright +, \quad +3(6) + F_{Cx}(9) = 0,$$

$$F_{Cx} = -2 \text{ kips} \quad (\text{compression})$$

$$F_{Cy} = F_{Cx} = -2 \text{ kips},$$

$$F_C = \sqrt{2}(-2) = -2.83 \text{ kips}$$

Check: $\Sigma F_x = 0, \quad F_{Ax} + F_{Cx} = 2 - 2 = 0$

$\Sigma F_y = 0, \quad F_{Ay} - F_{Cy} - P = 1 - (-2) - 3 = 0$

Stress in main bar AB:

$$\sigma_{AB} = \frac{F_A}{A} = \frac{2.23}{(0.25)(0.50)} = 17.8 \text{ ksi} \quad (\text{tension})$$

Stress in clevis of bar AB, Fig. 3-12(e):

$$(\sigma_{AB})_{\text{clevis}} = \frac{F_A}{A_{\text{net}}} = \frac{2.23}{2(0.20)(0.875 - 0.375)} = 11.2 \text{ ksi} \quad \text{(tension)}$$

Stress in main bar BC:

$$\sigma_{BC} = \frac{F_C}{A} = \frac{2.83}{(0.875)(0.25)} = 12.9 \text{ ksi} \quad \text{(compression)}$$

In the compression member the net section at the clevis need not be investigated; see Fig. 3-12(f) for the transfer of forces. The bearing stress at the pin is more critical. Bearing between pin C and clevis:

$$\sigma_b = \frac{F_C}{A_{\text{bearing}}} = \frac{2.83}{(0.375)(0.20)2} = 18.8 \text{ ksi}$$

Bearing between the pin C and the main plate:

$$\sigma_b = \frac{F_C}{A} = \frac{2.83}{(0.375)(0.25)} = 30.1 \text{ ksi}$$

Double shear in the pin C:

$$\tau = \frac{F_C}{A} = \frac{2.83}{2\pi(0.375/2)^2} = 12.9 \text{ ksi*}$$

For a complete analysis of this bracket, other pins should be investigated. However, it may be seen by inspection that the other pins in this case are stressed the same amount as computed above, or less.

The advantages of the method used in the above example for finding forces in members should now be apparent. It can also be applied with success in a problem such as the one shown in Fig. 3-13. The force F_A transmitted by the curved member AB acts through points A and B since the forces applied at A and B must be collinear. By resolving this force at A', the same procedure may be followed. Wavy lines through F_A and F_C indicate that these forces are replaced by the two components shown. Alternatively, the force F_A may be resolved at A, and since $F_{Ay} = (x/y)F_{Ax}$, the application of $\Sigma M_C = 0$ yields F_{Ax}.

In frames where the applied forces do not act through a joint, proceed as above as far as possible. Then isolate an individual member, and using its free-body diagram, complete the determination of forces.

---

* Considering the pin in a two-dimensional state of stress $\tau_{xy} = \tau_{yx}$, the tensor representation of the results becomes $\begin{pmatrix} 0 & 12.9 \\ 12.9 & 0 \end{pmatrix}$ ksi

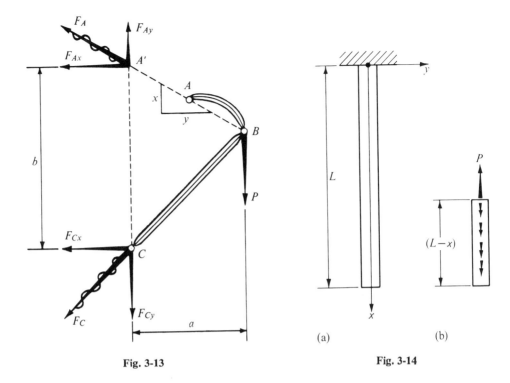

Fig. 3-13

Fig. 3-14

If inclined forces are acting on the structure, resolve them into convenient components.

EXAMPLE 3-4

A 1-in.^{2} rod L in. long is suspended vertically as shown in Fig. 3-14(a). The unit weight of the material is γ. Determine the normal stress in this rod using differential equations of equilibrium.

SOLUTION

With the axes shown in the figure, $\tau_{xy} = 0$, and only the first part of Eq. 3-3 has relevance. The body force $X = \gamma$. By virtue of the boundary condition at the free end of the rod $\sigma_x(L) = 0$. On this basis, setting up a differential equation, intergating it, and determining the constant of integration from the boundary conditions, one has

$$\frac{d\sigma_x}{dx} + \gamma = 0 \quad \text{and} \quad \sigma_x = -\gamma x + C_1$$

$$\sigma_x(L) = -\gamma L + C_1 = 0 \quad \text{and} \quad \sigma_x = (L - x)\gamma$$

This result can be easily checked by cutting the rod $(L - x)$ above the free end, Fig. 3-14(b), and applying Eq. 3-5. Only very few problems can be analyzed using Eq. 3-3 alone. In more general problems deformations must be considered simultaneously in the analysis.

81

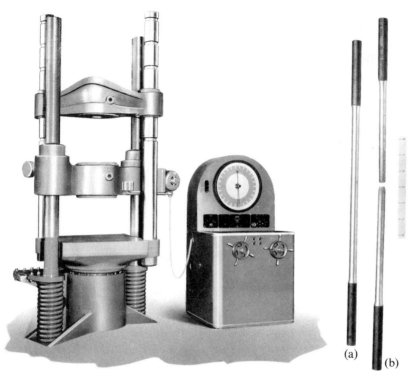

Fig. 3-15. Universal testing machine. (Courtesy Baldwin-Lima-Hamilton Corp.) **Fig. 3-16.** A typical tension-test specimen of mild steel: (a) before fracture, (b) after fracture.

3-8. ALLOWABLE STRESSES; FACTOR OF SAFETY

The determination of stresses would be altogether meaningless were it not for the fact that physical testing of materials in a laboratory provides information regarding a resistance a material has to stress. In a laboratory, specimens of known material, manufacturing process, and heat treatment are carefully prepared to desired dimensions. Then these specimens are subjected to successively increasing known forces. In the usual test, a round rod is subjected to tension and the specimen is loaded until it finally ruptures. The force necessary to cause rupture is called the *ultimate load.* By dividing this ultimate load by the original cross-sectional area of the specimen, the *ultimate strength* (stress) of a material is obtained. Figure 3-15 shows a testing machine used for this purpose. Figure 3-16 is a photograph of a tension-test specimen. The tensile test is used most widely. However, compression, bending, torsion, and shearing tests are also employed. Table 1 of the Appendix gives ultimate strengths and other physical properties for a few materials.

For the design of members the stress level called the *allowable stress* is set considerably lower than the ultimate strength found in the "static" test mentioned above. This is necessary for several reasons. The exact magnitudes of the forces that may act upon the designed structure are seldom accurately known. Materials are not entirely uniform. Some of the materials stretch unpermissible amounts prior to an actual break, so to hold down deformations, stresses must be kept low.* Some materials seriously corrode. Some materials flow plastically under a sustained load, a phenomenon called *creep*. With a lapse of time this may cause large deformations that cannot be tolerated.

For applications where a force comes on and off the structure a number of times, the materials cannot withstand the ultimate stress of a static test. In such cases the ultimate strength depends on the number of times the force is applied as the material works at a particular stress level. Figure 3-17 shows the results of tests† on a number of the same kind of specimens at different stresses. Experimental points indicate the number of cycles required to break the specimen at a particular stress when

Section 3-8
Allowable stresses; factor of safety

---

* See Chapter 4 for more details.
† J. L. Zambrow and M. G. Fontana, "Mechanical Properties, including Fatigue, of Aircraft Alloys at Very Low Temperatures," *Transactions of the American Society for Metals*, **41** (1949), 498.

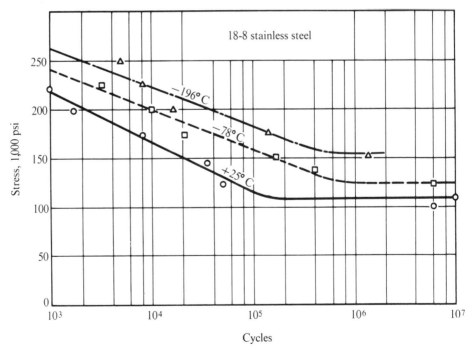

Fig. 3-17. Fatigue strength of 18-8 stainless steel at various temperatures (reciprocating-beam test).

applying a fluctuating load. Such tests are called *fatigue tests* and the corresponding curves are termed *S-N* (stress-number) diagrams. As may be seen from Fig. 3-17, at smaller stresses the material can withstand an ever-increasing number of cycles of load application. For some materials, notably steels, the *S-N* curve for low stresses becomes essentially horizontal. This means that at a low stress an infinitely large number of reversals of stress may take place before the material fractures. The stress at which this occurs is called the *endurance limit* of the material. This limit, being dependent on stress, is measured in pounds per square inch.

Some care must be exercised in interpreting the *S-N* diagrams, particularly with regard to the range of the applied stress. In some tests complete reversal (tension to compression) of stress is made, in others the applied load is varied in a different manner, such as tension to no load and back to tension. The major part of fatigue testing is done on specimens in bending.

In some cases another item deserves attention. As materials are manufactured they are often rolled, peened, and hammered. In castings materials cool unevenly. These processes set up high internal stresses, which are called *residual stresses*. In cases treated in this text the materials are assumed to be initially entirely free of such stresses.

The aforementioned facts, coupled with the impossibility of determining stresses accurately in complicated structures and machines, necessitate a substantial reduction of stress compared to the ultimate strength of a material in a static test. For example, ordinary steel will withstand an ultimate stress in tension of 60,000 psi and more. However, it deforms rather suddenly and severely at the stress level of about 40,000 psi, and it is customary in the United States to use an allowable stress of around 25,000 psi for structural work. This allowable stress is even further reduced to about 10,000 psi for parts that are subjected to alternating loads because of the fatigue characteristics of the material. Fatigue properties of materials are of utmost importance in mechanical equipment. Many failures in machine parts can be traced to disregard of this important consideration. See Art. (4-18.)

Large companies, as well as city, state, and federal authorities, prescribe or recommend allowable stresses for different materials, depending on the application.* Often such stresses are called the allowable *fiber*† stresses.

Since according to Eq. 3-5 stress times area is equal to a force, the allowable and ultimate stresses may be converted into the allowable and ultimate forces or "loads" which a member may resist. Also, a significant

---

* For example, see the building construction code of any large city.
† The adjective *fiber* in the above sense is used for two reasons. Many original experiments were made on wood, which is fibrous in character. Also, in several derivations that follow, the concept of a continuous filament or fiber in a member is a convenient concept for visualizing its action.

ratio may be formed:

$$\frac{\text{ultimate load for a member}}{\text{allowable load for a member}}$$

This ratio is called a *factor of safety* and must always be greater than unity. Although not commonly used, perhaps a better term for this ratio is *factor of ignorance*.

This factor is identical to the ratio of ultimate to allowable stress for tension members. For more complexly stressed members, the former definition is implied, although the ratio of stresses is actually used. As will become apparent from subsequent reading, the two are not synonymous since stresses seldom vary linearly with the applied load.

In the aircraft industry the term *factor of safety* is replaced by another defined as

$$\frac{\text{ultimate load}}{\text{design load}} - 1$$

and is known as the *margin of safety*. In normal usage this also reverts to

$$\frac{\text{ultimate stress}}{\text{maximum stress caused by the design load}} - 1$$

3-9. DESIGN OF AXIALLY LOADED MEMBERS AND PINS

The design of members for axial forces is rather simple. From Eq. 3-5 the required area of a member is

$$A = P/\sigma_{\text{allow}} \tag{3-7}$$

In all statically determinate problems the axial force P is determined directly from equations of equilibrium and the intended use of the material sets the allowable stress. For tension members, the area A so computed is the required net cross-sectional area of a member. For short compression blocks, Eq. 3-7 is also applicable; however, for slender members, do not attempt to use the above equation prior to study of the chapter on columns.

The simplicity of Eq. 3-7 is unrelated to its importance. A large number of problems requiring its use occur in practice. The following problems illustrate some applications of Eq. 3-7 as well as provide additional review in statics.

EXAMPLE 3-5

Reduce the weight of bar *AB* in Example 3-3 by using a better material, chrome-vanadium steel. The ultimate strength of this steel is approximately 120,000 psi. Use a factor of safety of $2\frac{1}{2}$.

Chapter 3
Stress and
axial loads

SOLUTION

$\sigma_{\text{allow}} = 120/2.5 = 48$ ksi. From Example 3-3 the force in the bar AB: $F_A = +2.23$ kips. Required area: $A_{\text{net}} = 2.23/48 = 0.0464$ in.^{2} Adopt: 0.20-in.-by-0.25-in.-bar. This provides an area of $(0.20)(0.25) = 0.050$ in.^{2}, which is slightly in excess of the required area. Many other proportions of the bar are possible.

With the cross-sectional area selected, the actual or working stress is somewhat below the allowable stress: $\sigma_{\text{actual}} = 2.23/(0.050) = 44.6$ ksi. The actual factor of safety is $120/(44.6) = 2.69$, and the actual margin of safety is 1.69.

In a complete redesign, clevis and pins should also be reviewed and, if possible, decreased in dimensions.

EXAMPLE 3-6

Select members FC and CB in the truss of Fig. 3-18(a) to carry an inclined force P of 150 kips. Set the allowable tensile stress at 20,000 psi.

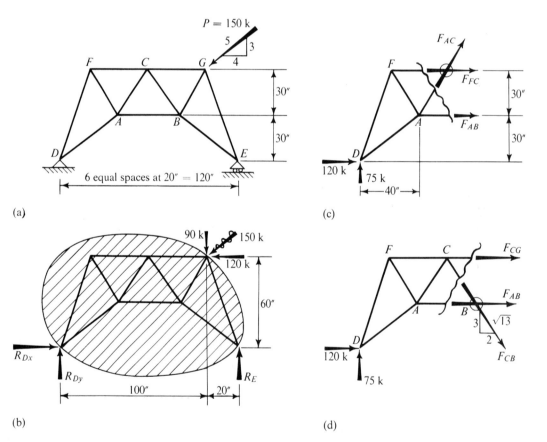

Fig. 3-18

SOLUTION

Section 3-9
Design of axially loaded members and pins

If all members of the truss were to be designed, forces in all members would have to be found. In practice this is often done by constructing a Maxwell-Cremona* diagram or analyzing the truss by the method of joints. However, if only a few members are to be designed or checked, the method of sections is quicker.

It is generally understood that a planar truss such as shown in the figure is stable in the direction perpendicular to the plane of the paper. Practically this is accomplished by introducing braces at right angles to the plane of the truss. In this example the design of compression members is avoided as this will be treated in the chapter on columns.

To determine the forces in the members to be designed, the reactions for the whole structure are computed first. This is done by completely disregarding the interior framing. Only reaction and force components definitely located at their points of application are indicated on a free-body diagram of the whole structure, Fig. 3-18(b). After the reactions are determined, free-body diagrams of a part of the structure are used to determine the forces in the members considered, Figs. 3-18(c) and (d).

Using free body in Fig. 3-18(b):

$\sum F_x = 0$ $R_{Dx} - 120 = 0$, $R_{Dx} = 120$ kips

$\sum M_E = 0 \circlearrowleft +$, $+R_{Dy}(120) - 90(20) - 120(60) = 0$,

$R_{Dy} = 75$ kips

$\sum M_D = 0 \circlearrowright +$, $R_E(120) + 120(60) - 90(100) = 0$,

$R_E = 15$ kips

Check: $\sum F_y = 0$, $+75 - 90 + 15 = 0$

Using free body in Fig. 3-18(c):

$\sum M_A = 0 \circlearrowleft +$, $+F_{FC}(30) + 75(40) - 120(30) = 0$,

$F_{FC} = +20$ kips

$A_{FC} = F_{FC}/\sigma_{\text{allow}} = 1$ in.^{2} (use ½-in.-by-2-in. bar)

Using free body in Fig. 3-18(d):

$\sum F_y = 0$, $-(F_{CB})_y + 75 = 0$, $(F_{CB})_y = +75$ kips

$F_{CB} = \sqrt{13}(F_{CB})_y/3 = +90.2$ kips

$A_{CB} = F_{CB}/\sigma_{\text{allow}} = 90.2/(20) = 4.51$ in.^{2} (use two bars 1⅛ in. by 2 in.)

---

* For example, see H. Sutherland and H. L. Bowman, *Structural Theory* (4th ed.) (New York; John Wiley & Sons, Inc., 1950), p. 47.

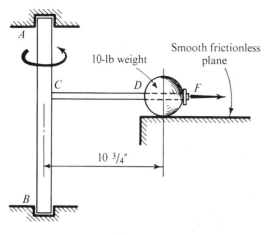

Fig. 3-19

EXAMPLE 3-7

Consider the idealized dynamic system shown in Fig. 3-19. The shaft AB rotates at a constant angular velocity of 600 rpm. A light rod CD is attached to this shaft at point C and at the end of this rod a 10-lb weight is fastened. In describing a complete circle the weight at D spins on a "frictionless" plane. Select the size of the rod CD so that the stress in it will not exceed 10,000 psi. In calculations neglect the weight of the rod.

SOLUTION

The acceleration of gravity g is 32.2 ft per second per second or 32.2(12) = 386 in. per second per second. The angular velocity ω is $600(2\pi)/60 = 20\pi$ radians per second.* For the given motion the body W is accelerated toward the center of rotation with an acceleration of $\omega^2 R$, where R is the distance CD. By multiplying this acceleration a by the mass m of the body, force F is obtained. This force acts in the opposite direction to that of the acceleration (d'Alembert principle), see Fig. 3-19.

$$F = ma = \frac{W}{g}\omega^2 R = \frac{10}{386}(20\pi)^2(10.75) = 1{,}100 \text{ lb}$$

$$A_{\text{net}} = \frac{F}{\sigma_{\text{allow}}} = \frac{1{,}100}{10{,}000} = 0.11 \text{ in.}^2$$

A $\tfrac{3}{8}$-in. round rod provides the required cross-sectional area. The additional pull at C caused by the mass of the rod, not considered above, is

$$F_1 = \int_0^R (m_1\, dr)\omega^2 r$$

where m_1 is the mass of the rod per inch of length and $(m_1\, dr)$ is its infinitesimal mass at a variable distance r from the vertical rod AB. The total pull at C caused by the rod and the weight W at the end is $F + F_1$.

PROBLEMS FOR SOLUTION

3-1. By analogy to Eq. 3-4, another equilibrium equation for a three-dimensional element in Cartesian coordinates may be written as

$$\frac{\partial \tau_{yx}}{\partial x} + \frac{\partial \sigma_y}{\partial y} + \frac{\partial \tau_{yz}}{\partial z} + Y = 0$$

(a) Write out the third equilibrium equation

---

* 2π radians correspond to one complete revolution of the shaft; the 60 in the denominator converts revolutions per minute to revolutions per second.

with Z as the body force. (b) With the aid of a diagram of an elementary cube on which all pertinent stresses are shown, verify Eq. 3-4.

3-2. Show that the differential equations of equilibrium for a two-dimensional plane stress problem in polar coordinates are

$$\frac{\partial \sigma_r}{\partial r} + \left(\frac{1}{r}\right)\frac{\partial \tau_{r\theta}}{\partial \theta} + \frac{\sigma_r - \sigma_\theta}{r} = 0$$

$$\left(\frac{1}{r}\right)\frac{\partial \sigma_\theta}{\partial \theta} + \frac{\partial \tau_{r\theta}}{\partial r} + \frac{2\tau_{r\theta}}{r} = 0$$

The symbols are defined in the figure. Body forces are neglected in this formulation.

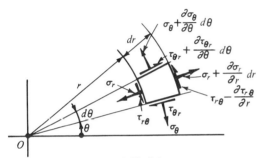

PROB. 3-2

3-3. A rod of variable cross-section built in at one end is subjected to three axial forces as shown in the figure. Find the maximum normal stress. *Ans.* 22.5 ksi.

PROB. 3-3

3-4. Determine the bearing stresses caused by the applied force at A, B, and C for the structure shown in the figure.

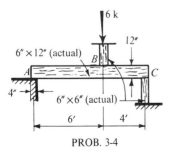

PROB. 3-4

3-5. A lever mechanism used to lift panels of a portable army bridge is shown in the figure. Calculate the shearing stress in pin A caused by a load of 500 lb.

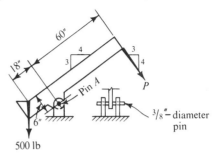

PROB. 3-5

3-6. Calculate the shearing stress in pin A of the bulldozer if the total forces acting on the blade are as shown in the figure. Note that there is a $1\frac{1}{2}$-in.-diameter pin on each side of the bulldozer. Each pin is in single shear.

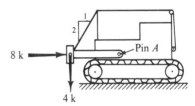

PROB. 3-6

3-7. A 6-in.-by-6-in. (actual size) wooden post delivers a force of 12 kips to a concrete footing, as shown in Fig. 3-8. (a) Find the bearing stress of the wood on the concrete. (b) If the allowable pressure on the soil is 1 ton per square foot, determine the required dimensions in plan view of a square footing. Neglect the weight of the footing. *Ans.* 333 psi, 29.5 in.

3-8. For the structure shown in the figure, calculate the size of the bolt and the area of

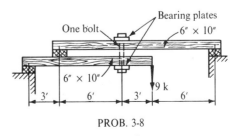

PROB. 3-8

89

the bearing plates required if the allowable stresses are 18,000 psi in tension and 500 psi in bearing. Neglect the weight of the beams.

3-9. Two 10-in.-wide by $\frac{3}{4}$-in.-thick plates are joined together by means of two cover plates each $\frac{1}{2}$ in. thick as shown in the figure. Eight $\frac{7}{8}$-in. bolts are used on each side of the joint in tight-fitting holes. If this joint is subjected to a tensile force $P = 160$ kips, find (a) the shearing stress in the bolts, (b) the tensile stresses in the main plate at sections 1-1, 2-2, 3-3, and 4-4, (c) the maximum tensile stress in the cover plates. Assume that each rivet transmits one-eighth of the applied force. Draw free-body diagrams for the isolated elements in each part of the analysis. (*Note:* High-strength bolts develop very large frictional forces between the assembled elements. The suggested analysis is reasonable only if ordinary bolts are used.)

as shown in the figure. The total weight supported by each of the two hangers is 15 k. Determine the shearing stresses in the 1-in.-diameter pins at points A and B due to the weight of the tank. Neglect the weight of the hangers and assume that contact between the tank and the hangers is frictionless. *Ans.* 5.97 ksi.

3-11. Find the stress in the mast of the derrick shown in the figure. All members are in the same vertical plane and are joined by pins. The mast is made from an 8 in. standard steel pipe weighing 28.55 lb per foot. (See Appendix Table 8.) Neglect the weight of the members. *Ans.* -446 psi.

3-12. It is desired to check the capacity of the derrick structure shown in the figure. All members are made of steel and have the same cross-sectional area of 8 in.² Determine the

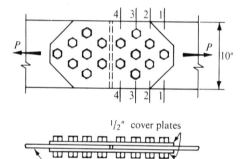

PROB. 3-9

3-10. A 6-ft-diameter cylindrical tank is to be supported at each end by a hanger arranged

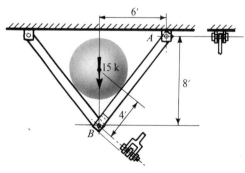

PROB. 3-10

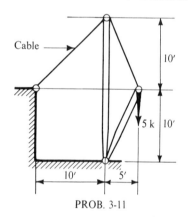

PROB. 3-11

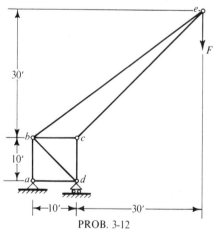

PROB. 3-12

maximum allowable load F if the allowable stresses are 20,000 psi in tension and 15,000 psi in compression. All joints are pinned. *Ans.* 21.2 k.

3-13. A signboard 15 ft by 20 ft in area is supported by two frames as shown in the figure. All members are actually 2 in. by 4 in. in cross section. Calculate the stress in each member due to a horizontal wind load on the sign of 20 lb per square foot. Assume that all joints are connected by pins and that one-quarter of the total wind force acts at B and at C. Neglect the possibility of buckling of the compression members. Neglect the weight of the structure.

3-14. Two high-strength rods of different sizes are attached at A and C and support a load W at B as shown in the figure. What load W can be supported? The ultimate strength of the rods is 160 ksi, and the factor of safety is to be 4. The rod AB has $A = 0.20$ in.^{2}; the rod BC has $A = 0.10$ in.^{2}

3-15. What is the required diameter of the pin B for the bell-crank mechanism shown in the figure if an applied force of 12 kips at A is resisted by a force P at C? The allowable shearing stress is 15,000 psi. *Ans.* 0.60 in.

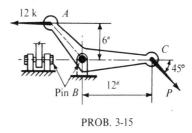

PROB. 3-15

3-16. Find the required cross-sectional areas for all tension members in Example 3-6. The allowable stress is 20 ksi.

3-17. A tower used for a high line is shown in the figure. If it is subjected to a horizontal force of 120 kips and the allowable stresses are 15 ksi in compression and 20 ksi in tension, what is the required cross-sectional area of each member? All members are pin-connected.

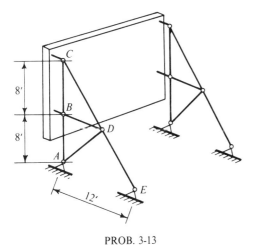

PROB. 3-13

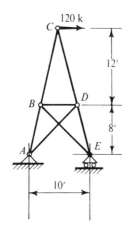

PROB. 3-17

PROB. 3-14

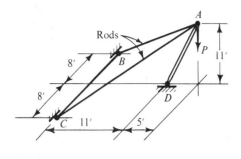

PROB. 3-18

3-21.* A pin-connected frame for supporting a force P is shown in the figure. The stress σ in both members AB and BC is to be the same. Determine the angle α necessary to achieve the minimum weight of construction. Members AB and BC have a constant cross section. Ans. $\cos^2 \alpha = 1/3$ or $\alpha \approx 55°$.

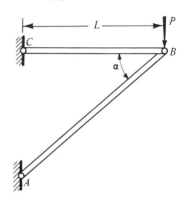

3-18. To support a load, $P = 20$ tons, determine the necessary diameter for the rods AB and AC for the tripod shown in the figure. Neglect the weight of the structure and assume that the joints are pin-connected. No allowance need be made for threads. The allowable tensile stress is 18 ksi.

3-19. A weight W is suspended from a ceiling by three wires of equal size. If the common point of suspension is taken as the origin, the coordinates, measured in feet, of the three points where the wires are attached to the ceiling are (2, −3, 6), (3, 2, 6), and (−3, 0, 6). If the highest stressed wire can carry 50 lb, what may be the weight W?

PROB. 3-21

3-22. Three equal weights W are attached to a thin rod at equal intervals a. Another piece of the rod of length a attaches the first weight to a vertical shaft in a manner analogous to that shown in Fig. 3-19. If this assembly is rotated around the vertical axis with an angular velocity ω, what are the stresses in the three segments of the rod? Neglect the weight of the rod and assume negligible friction between the weights and the horizontal plane on which they move.

3-23. A bar of constant cross-sectional area A is rotated around one of its ends in a horizontal plane with a constant angular velocity ω. The unit weight of the material is γ. Determine the variation of the stress σ along the bar.

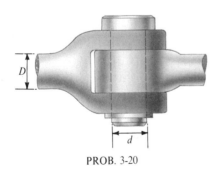

PROB. 3-20

3-20. A joint for transmitting a tensile force is to be made by means of a pin as shown in the figure. If the diameter of the rods being connected is D, what should the diameter d of the pin be? Assume that the allowable shearing stress in the pin is one-half the maximum tensile stress in the rods. (In Art. 9-17 it will be shown that this ratio for the allowable stresses is an excellent assumption for many materials.)

3-24. (a) Show that the body force in a thin circular disc of mass m per unit volume of material and rotating with a constant angular velocity ω is $m\omega^2 r$. (b) Add the body force found in (a) to the first equation of Prob. 3-2, and show that the equilibrium equation for stress distribution symmetrical about an axis is

$$\frac{d(r\sigma_r)}{dr} - \sigma_\theta + m\omega^2 r^2 = 0$$

---

* V. I. Feodosiev, *Strength of Materials* (3rd ed.) (Moscow: Nauka, 1964), p. 39.

Strain, constitutive laws, and axial deformation 4

4-1. INTRODUCTION

The analysis of the deformation of a solid body parallels in importance the analysis of stress and will be the primary objective of this chapter. This requires the precise definition of strain, which is the intensity of deformation. This is done in Part A. The linear relationship between stress and strain in the form of the generalized Hooke's law and the strain energy of an element are considered in Part B. A number of possible relationships between stress and strain for the uniaxial state of stress, including plastic behavior and time-dependent effects, are discussed in Part C. In Part D, some of the developed procedures are applied to axially loaded members. Lastly, some additional limitations that must be imposed on Eqs. 3-5 and 3-7 are pointed out.

PART A STRAIN

4-2. PHYSICAL MEANING OF STRAIN

A solid body subjected to a change of temperature or to an external load deforms. For example, while a specimen is being subjected to an increasing force

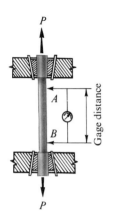

Fig. 4-1. Diagram of a tension specimen in a testing machine.

P as shown in Fig. 4-1, a change in length of the specimen occurs between any two points, such as A and B. Initially, two such points can be selected an arbitrary distance apart. Thus, depending on the test, either a 2-in. or an 8-in. distance is commonly used. This initial distance between the two points is called a *gage distance*. In an experiment the change in the length of this distance is noted. With the same load and a longer gage distance, a larger deformation is observed, or vice versa. Therefore it is more fundamental to refer to the observed elongation per unit of length of the gage, i.e., to its intensity of deformation.

If l_0 is the original gage length and l is the observed length under load, the elongation $\Delta l = l - l_0$. The elongation per unit of length ε (epsilon) is

$$\varepsilon = \int_{l_0}^{l} \frac{dl}{l_0} = \frac{\Delta l}{l_0} \tag{4-1}$$

This elongation per unit of length is termed *linear strain*. It is a dimensionless quantity, but it is customary to refer to it as having the dimensions of inches per inch. Sometimes it is given in per cent. The quantity ε is a very small one. In most engineering applications of the type considered in this text it is of the order of magnitude of 0.1 per cent. This is very small indeed.*

In addition to the linear strain described above, a body in general may also be strained linearly in two additional directions. In analytical treatment the three directions are usually taken orthogonally to each other and are identified by the subscripts x, y, and z. Finally, in general,

---

* Where strains are large, as, for example, in metal forming, it is necessary to introduce the so-called *natural strain*. It is related to the definition given by Eq. 4-1. One simply has to replace l_0 under the integral by l and to integrate:

$$\bar{\varepsilon} = \int_{l_0}^{l} \frac{dl}{l} = \ln \frac{l}{l_0} = \ln(1 + \varepsilon)$$

For small strains both definitions nearly coincide.

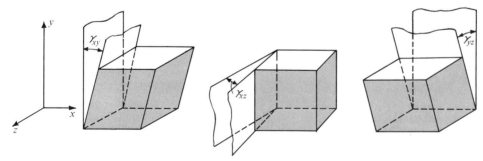

Fig. 4-2. Possible shearing deformations of an element.

a body also may be deformed as in Fig. 4-2. Such deformations cause a change in the initial right angles between imaginary lines in a body, and this angle change defines the *shearing strain*. More precise mathematical definitions of the above quantities will be discussed next.

Section 4-3
Mathematical
definition of strain

4-3. MATHEMATICAL DEFINITION OF STRAIN

Since strains generally vary from point to point, the definitions of strain must relate to an infinitesimal element. With this in mind, consider a linear strain taking place in one direction as in Fig. 4-3(a). Some points

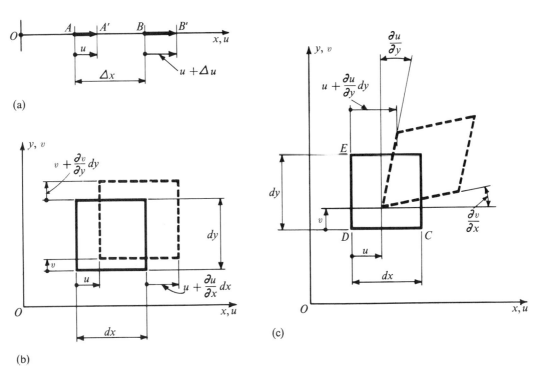

Fig. 4-3. Strained elements in initial and final positions.

like A and B move to A' and B' respectively. During straining, Point A experiences a displacement u. The displacement of Point B is $u + \Delta u$ since in addition to the rigid body displacement u, common to the whole element Δx, a stretch Δu takes place within the element. On this basis, the definition of the linear strain is

$$\varepsilon = \lim_{\Delta x \to 0} \frac{\Delta u}{\Delta x} = \frac{du}{dx} \qquad (4\text{-}2)$$

95

Chapter 4
Strain, constitutive laws, and axial deformation

If a body is strained in orthogonal directions, as shown for a two-dimensional case in Fig. 4-3(b), subscripts must be attached to ε to differentiate between the directions of the strains. For the same reason it is also necessary to change the ordinary derivatives to partial ones. Therefore, if, at a point of a body, u, v, and w are the three displacement components occurring, respectively, in the x, y, and z directions of the coordinate axes, the basic definitions of linear strain become

$$\varepsilon_x = \frac{\partial u}{\partial x}, \qquad \varepsilon_y = \frac{\partial v}{\partial y}, \qquad \varepsilon_z = \frac{\partial w}{\partial z} \tag{4-3}$$

Note that the double subscripts, analogously to those of stress, can be used for linear strain. Thus

$$\varepsilon_x \equiv \varepsilon_{xx}, \qquad \varepsilon_y \equiv \varepsilon_{yy}, \qquad \varepsilon_z \equiv \varepsilon_{zz} \tag{4-4}$$

where one of the subscripts designates the direction of the linear element, and the other the direction of the displacement. Positive signs apply to elongations.

In addition to linear strains, an element can also experience a shearing strain as shown for example in the x-y plane in Fig. 4-3(c). This inclines the sides of the deformed element in relation to the x and the y axes. Since v is the displacement in the y direction, as one moves in the x direction $\partial v/\partial x$ is the slope of the initially horizontal side of the infinitesimal element. Similarly, the vertical side tilts through an angle $\partial u/\partial y$. On this basis, the initially right angle CDE is reduced by the amount $(\partial v/\partial x) + (\partial u/\partial y)$. Therefore, for small angle changes, the definition of the shearing strain associated with the xy coordinates is

$$\gamma_{xy} = \gamma_{yx} = \frac{\partial v}{\partial x} + \frac{\partial u}{\partial y} \tag{4-5}$$

To arrive at this expression it is assumed that tangents of small angles are equal to the angles themselves in radian measure. Positive sign for the shearing strain applies when the element is deformed as in Fig. 4-3(c). (This deformation corresponds to the positive directions of the shearing stresses, see Fig. 3-3.)

The definitions for the shearing strains for the xz and yz planes are similar to Eq. 4-5:

$$\gamma_{xz} = \gamma_{zx} = \frac{\partial w}{\partial x} + \frac{\partial u}{\partial z}, \qquad \gamma_{yz} = \gamma_{zy} = \frac{\partial w}{\partial y} + \frac{\partial v}{\partial z} \tag{4-6}$$

Note that in Eqs. 4-5 and 4-6 the subscripts on γ can be permuted. This is permissible since no meaningful distinction can be made between the two sequences of each alternative subscript.

In examining Eqs. 4-3, 4-5, and 4-6, note that these six strain-displacement equations depend only on three displacements u, v, and w. Therefore, the equations cannot be independent. Three independent equations can be developed showing the interrelationships among ε_{xx}, ε_{yy}, ε_{zz}, γ_{xy}, γ_{yz}, and γ_{zx}. The number of such equations reduces to one for a two-dimensional case. The derivation and the application of these equations, known as the *equations of compatibility*, are given in texts on the theory of elasticity.

4-4. STRAIN TENSOR

The linear and the shearing strains defined in the preceding article together express the strain tensor, which is highly analogous to the stress tensor already discussed. It is necessary, however, to modify the relations for the shearing strains in order to have a tensor, an entity which must obey certain laws of transformation.* Thus, the physically attractive

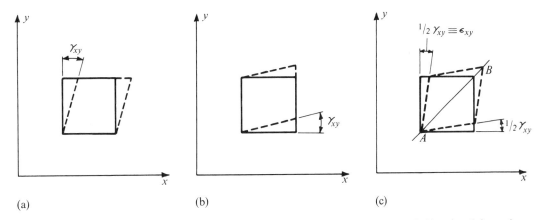

Fig. 4-4. Shearing deformations.

definition of the shearing strain as the change in angle γ is not acceptable when the shearing strain is a component of a tensor. This heuristically may be attributed to the following. In Fig. 4-4(a), positive γ_{xy} is measured from the vertical direction. The same positive γ_{xy} is measured from the horizontal direction in Fig. 4-4(b). In Fig. 4-4(c) the same amount of shearing deformation is shown to consist of two $\gamma_{xy}/2$'s. The deformed elements in Figs. 4-4(a) and (b) can be obtained by rotating the element in Fig. 4-4(c) as a rigid body through an angle of $\gamma_{xy}/2$. The scheme shown in Fig. 4-4(c) is the correct one for defining the shearing-strain component as an element of a tensor. Since in this definition the element is not

---

* Rigorous discussion of this question is beyond the scope of this text. A better appreciation of it will develop, however, after the study of Chapter 9, where strain transformation for a two-dimensional case is considered.

rotated as a rigid body, the strain is said to be *pure* or *irrotational*. Following this approach, one redefines the shearing strains as

$$\varepsilon_{xy} = \varepsilon_{yx} = \frac{\gamma_{xy}}{2} = \frac{\gamma_{yx}}{2}$$

$$\varepsilon_{yz} = \varepsilon_{zy} = \frac{\gamma_{yz}}{2} = \frac{\gamma_{zy}}{2} \qquad (4\text{-}7)$$

$$\varepsilon_{zx} = \varepsilon_{xz} = \frac{\gamma_{zx}}{2} = \frac{\gamma_{xz}}{2}$$

From these equations the strain tensor in matrix representation can be assembled as follows:

$$\begin{pmatrix} \varepsilon_x & \frac{\gamma_{xy}}{2} & \frac{\gamma_{xz}}{2} \\ \frac{\gamma_{yx}}{2} & \varepsilon_y & \frac{\gamma_{yz}}{2} \\ \frac{\gamma_{zx}}{2} & \frac{\gamma_{zy}}{2} & \varepsilon_z \end{pmatrix} = \begin{pmatrix} \varepsilon_{xx} & \varepsilon_{xy} & \varepsilon_{xz} \\ \varepsilon_{yx} & \varepsilon_{yy} & \varepsilon_{yz} \\ \varepsilon_{zx} & \varepsilon_{zy} & \varepsilon_{zz} \end{pmatrix} \qquad (4\text{-}8)$$

The strain tensor is symmetric. Mathematically the notation employed in the last expression is particularly attractive and has wide acceptance in continuum mechanics (elasticity, plasticity, rheology, etc.). Just as for the stress tensor, using indicial notation one can write ε_{ij} for the strain tensor.

Analogously to the stress tensor, the strain tensor can be diagonalized, having only ε_1, ε_2, and ε_3 as the surviving components. For a two-dimensional problem $\varepsilon_3 = 0$; and one has the case of *plane strain*. The tensor for this situation is

$$\begin{pmatrix} \varepsilon_{xx} & \varepsilon_{xy} & 0 \\ \varepsilon_{yx} & \varepsilon_{yy} & 0 \\ 0 & 0 & 0 \end{pmatrix} \quad \text{or} \quad \begin{pmatrix} \varepsilon_1 & 0 & 0 \\ 0 & \varepsilon_2 & 0 \\ 0 & 0 & 0 \end{pmatrix} \quad \text{or} \quad \begin{pmatrix} \varepsilon_1 & 0 \\ 0 & \varepsilon_2 \end{pmatrix} \qquad (4\text{-}9)$$

The transformation of strain suggested by Eq. 4-9 will be considered in Chapter 9.

The reader should note that in discussing the concept of strain, the mechanical properties of the material were not involved. The equations are applicable whatever the mechanical behavior of the material. However, only small strains are defined by the presented equations. Also note that strains give only the relative displacement of points; rigid body displacements do not affect the strains.

PART B
LINEAR STRESS-STRAIN LAWS AND STRAIN ENERGY*

4-5. HOOKE'S LAW FOR ANISOTROPIC MATERIALS

As was pointed out in Chapter 3, there are in general six possible components of stress, and, as was shown in the preceding article, there are also six components of strain. The linear relationship between stress and strain is the simplest relationship between these quantities. For example, one can state $\tau = C\varepsilon$, or, conversely, $\varepsilon = A\tau$, where C and A are elastic constants, and A is the reciprocal of C. Experimental justification for the use of these constants and a more precise definition of the term *elastic* are discussed in Part C of this chapter.

The linear relationship of forces and deformations, or of stresses and strains as suggested above, has become known as *Hooke's law*.† Since there are several components of stress and strain, in formulating the general Hooke's law, use is made of the *principle of superposition*, which asserts that the resultant stress or strain in a system subjected to several forces is the algebraic sum of their effects when applied separately. This is true if each strain is directly and linearly related to the stress causing it and if the strains due to one stress component cause no abnormally large effect on another stress. Fortunately, the strains in most engineering structures are small, which permits the application of the principle of superposition. On this basis, relating each of the six strains to each of the six stress components, the linear stress-strain relations become

$$\varepsilon_{xx} = \varepsilon_x = A_{11}\tau_{xx} + A_{12}\tau_{yy} + A_{13}\tau_{zz} + A_{14}\tau_{xy} + A_{15}\tau_{yz} + A_{16}\tau_{zx}$$

$$\varepsilon_{yy} = \varepsilon_y = A_{21}\tau_{xx} + A_{22}\tau_{yy} + A_{23}\tau_{zz} + A_{24}\tau_{xy} + A_{25}\tau_{yz} + A_{26}\tau_{zx}$$

$$\varepsilon_{zz} = \varepsilon_z = A_{31}\tau_{xx} + A_{32}\tau_{yy} + A_{33}\tau_{zz} + A_{34}\tau_{xy} + A_{35}\tau_{yz} + A_{36}\tau_{zx}$$

$$\varepsilon_{xy} = \gamma_{xy}/2 = A_{41}\tau_{xx} + A_{42}\tau_{yy} + A_{43}\tau_{zz} + A_{44}\tau_{xy} + A_{45}\tau_{yz} + A_{46}\tau_{zx}$$

$$\varepsilon_{yz} = \gamma_{yz}/2 = A_{51}\tau_{xx} + A_{52}\tau_{yy} + A_{53}\tau_{zz} + A_{54}\tau_{xy} + A_{55}\tau_{yz} + A_{56}\tau_{zx}$$

$$\varepsilon_{zx} = \gamma_{zx}/2 = A_{61}\tau_{xx} + A_{62}\tau_{yy} + A_{63}\tau_{zz} + A_{64}\tau_{xy} + A_{65}\tau_{yz} + A_{66}\tau_{zx}$$

(4-10)

---

* Some readers may find it more convenient to study Arts. 4-12 and 4-13 before studying this part.
† Actually Robert Hooke, an English scientist, worked with springs and not with rods. In 1676 he announced an anagram "c e i i i n o s s s t t u v," which in Latin is *Ut Tensio sic Vis* (the force varies as the stretch).

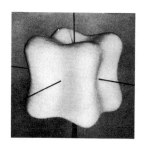

Fig. 4-5. Variation of mechanical properties of a gold crystal with directions. (After E. Schmidt and W. Boas, *Kristallplastizität, Struktur und Eigenschaften der Materie*, Bd XVII (Berlin: Springer, 1935), p. 201.)

These equations appear to have 36 possible constants $A_{11}, A_{12}, \ldots, A_{66}$. However, through energy considerations it can be shown* that the number of independent constants is 21. These are symmetrical to either side of the main diagonal, i.e., $A_{12} = A_{21}$, etc., or, in general, $A_{ij} = A_{ji}$. All must be determined experimentally. It is assumed that the material is *homogeneous*, i.e., that it has the same properties everywhere.†

Hooke's law in the most general form, given by Eq. 4-10, is applicable to homogeneous anisotropic materials such as single crystals. These materials possess different mechanical properties in different directions with reference to their crystallographic planes; see Fig. 4-5. Note the interesting strain response to stress, given by Eq. 4-10. For example, the linear strain ε_{xx} is caused not only by the normal stresses but also by the shearing stresses. This equation asserts that even if shearing stresses are applied, a hypothetical linear strain ε_{xx} occurs. In real materials this effect is usually very small, and for this reason the general form of Hooke's law given by Eq. 4-10 is very seldom used.

Wood has decidedly different properties in the longitudinal, radial, and transverse directions, i.e., in the three orthogonal directions. Such properties are said to be *orthotropic*. For these materials Eq. 4-10 simplifies; it can be shown‡ that only nine independent constants survive. In this form Hooke's law is employed in the study of the behavior of several manufactured materials such as filament-reinforced plastics, fillers in sandwich construction, rolled metals,§ etc. Such studies are beyond the scope of this text. Instead, further widely accepted simplifications in the law will be examined in the next article.

4-6. HOOKE'S LAW FOR ISOTROPIC MATERIALS

For homogeneous isotropic materials, i.e., materials having the same properties in all directions, Eq. 4-10 greatly simplifies. It can be shown‖ that for such a condition $A_{11} = A_{22} = A_{33}$, $A_{12} = A_{13} = A_{23}$,

---

* I. S. Sokolnikoff, *Mathematical Theory of Elasticity* (New York: McGraw-Hill Book Company, 1956), p. 61.
† Since for physical reasons Eq. 4-10 is nonsingular, it can be written in an inverse form; i.e., each of the six stress components can be linearly related to the six strain components. For example

$$\tau_{xx} = C_{11}\varepsilon_{xx} + C_{12}\varepsilon_{yy} + C_{13}\varepsilon_{zz} + C_{14}\varepsilon_{xy} + C_{15}\varepsilon_{yz} + C_{16}\varepsilon_{zx}$$

‡ Sokolnikoff, *Mathematical Theory of Elasticity*, p. 61.
§ Rolling operations produce preferential orientation of crystalline grains in some materials.
‖ Sokolnikoff, *Mathematical Theory of Elasticity*, p. 62. Also, Y. C. Fung, *Foundations of Solid Mechanics* (Englewood Cliffs, N.J.: Prentice-Hall, Inc., 1965), p. 128.

$A_{44} = A_{55} = A_{66}$, and as before $A_{12} = A_{21}$, $A_{13} = A_{31}$, $A_{23} = A_{32}$. All other constants vanish. These simplifications yield the so-called *generalized Hooke's law*, in the sense that it is a generalization of the initially suggested relation for the uniaxial stress condition.

**Section 4-6
Hooke's law for isotropic materials**

The generalized Hooke's law for isotropic material can be written keeping the above remarks in mind. Usual engineering notation will be employed requiring the following changes: $A_{11} = 1/E$, $A_{12} = -\nu/E$, and $A_{44} = 1/(2G)$. Then

$$\varepsilon_x = \frac{\sigma_x}{E} - \nu\frac{\sigma_y}{E} - \nu\frac{\sigma_z}{E}$$

$$\varepsilon_y = -\nu\frac{\sigma_x}{E} + \frac{\sigma_y}{E} - \nu\frac{\sigma_z}{E}$$

$$\varepsilon_z = -\nu\frac{\sigma_x}{E} - \nu\frac{\sigma_y}{E} + \frac{\sigma_z}{E} \tag{4-11}$$

$$\gamma_{xy} = \tau_{xy}/G$$

$$\gamma_{yz} = \tau_{yz}/G$$

$$\gamma_{zx} = \tau_{zx}/G$$

In these equations the constant E is called the *modulus of elasticity*, the elastic modulus, or *Young's modulus*.* For uniaxial stress, when all stresses but one, normal stress, are zero, E is a constant of proportionality relating this normal stress to its linear strain. For example, $\sigma_x = E\varepsilon_x$. Graphically E is the slope of a line on a stress-strain diagram, see Fig. 4-6. Since ε is dimensionless, E has the units of stress. In the English system of units it is usually measured in pounds per square inch. The theory discussed in this text applies only to small deformations. For such cases ε is a small quantity, but E is a very large one. The approximate values of E are tabulated for a few materials in Table 1 of the Appendix. For most steels, E is between 29×10^6 and 30×10^6 psi. The constant of proportionality G is called the *shearing modulus of elasticity* or the modulus of rigidity. The dimensions of G are the same as those of E. The constant ν (nu) is called *Poisson's ratio*.† An understanding of its meaning requires some further comments, which are made in the next article.

Equation 4-11 as stated above implies that there are three elastic constants E, ν, and G. However, as will be shown in Art. 9-15, for isotropic material a relationship exists among the three constants. Therefore, for isotropic materials there are only two elastic constants. The connecting

---

* Young's modulus is named in honor of Thomas Young, an English scientist. His *Lectures on Natural Philosophy*, published in 1807, contains a definition of the modulus of elasticity.

† Named after S. D. Poisson, a French scientist who formulated this concept in 1828.

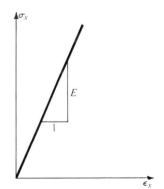

Fig. 4-6. Linear relation between uniaxial stress and linear strain.

Chapter 4
Strain, constitutive laws, and axial deformation

equation is

$$G = \frac{E}{2(1+\nu)} \tag{4-12}$$

4-7. POISSON'S RATIO

From experiments it is known that in addition to the deformation of materials in the direction of the applied normal stress, another remarkable property can be observed in all solid materials, namely, that at right angles to the applied stress, a certain amount of lateral (transverse) expansion or contraction takes place. This phenomenon is illustrated in Figs. 4-7(a) and (b), where the deformations are greatly exaggerated. For clarity, this

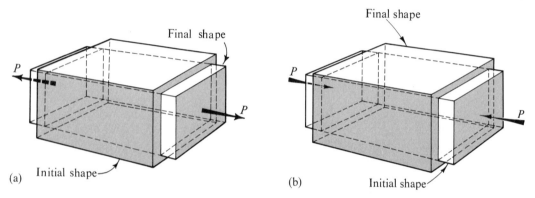

Fig. 4-7. Lateral contraction and expansion of solid bodies subjected to axial forces (Poisson effect).

may be restated: If a solid body is subjected to axial tension, it contracts laterally; on the other hand, if it is compressed, the material "squashes out" sidewise. With this in mind, directions of lateral deformations are easily determined, depending on the sense of the applied normal stress.

For a general theory, it is necessary to refer to these lateral deformations on the basis of deformation per unit of length of the transverse dimension, i.e., on the basis of lateral linear strains. In Eq. 4-11 a provision for the mathematical description of this phenomenon is included. For example, if only $\sigma_x \neq 0$, $\varepsilon_x = \sigma_x/E$ and $\varepsilon_y = \varepsilon_z = -\nu\sigma_x/E$. The negative sign accounts for contraction of the material in the y and z directions as it is being stretched in the x direction. The ratio of the absolute value of the strain in the lateral direction to the strain in the axial direction is Poisson's ratio, i.e.,

$$\nu = -\frac{\varepsilon_y}{\varepsilon_x} = -\frac{\varepsilon_z}{\varepsilon_x} = \frac{\text{lateral strain}}{\text{axial strain}} \tag{4-13}$$

Note that this definition applies only for strains caused by uniaxial stress. As stated in Eq. 4-11, superposition is applicable to situations of multiaxial stress; the separate strain effects caused by each stress are summed in the required direction. For example, each of the tensile stresses in the y and z directions cause a negative strain in the x direction as a result of Poisson's effect.

From experiments it is known that the value of ν fluctuates for different materials over a relatively narrow range. Generally it is in the neighborhood of 0.25 to 0.35. In extreme cases values as low as 0.1 (some concretes) and as high as 0.5 (rubber) occur. The latter value is the largest possible for isotropic materials. It is normally attained by materials during plastic flow and signifies constancy of volume.* The Poisson effect exhibited by materials causes no additional stresses other than those considered earlier unless the transverse deformation is inhibited or prevented.

*Section 4-7
Poisson's ratio*

EXAMPLE 4-1

Consider a carefully conducted experiment in which an aluminum bar of $2\frac{1}{4}$ in. diameter is stressed in a testing machine as in Fig. 4-8. At a certain instant the applied force P is 32 kips while the measured elongation of the rod is 0.00938 in. in a 12-in. gage length and the dimension of the diameter is decreased by 0.000585 in. Calculate the two physical constants ν and E of the material.

SOLUTION

Transverse strain:

$$\varepsilon_t = \frac{\Delta_t}{D} = \frac{-0.000585}{(2.25)} = -0.000260 \text{ in. per in.}$$

where Δ_t is the total change in dimension of the diameter of the rod.

Axial strain:

$$\varepsilon_a = \frac{\Delta}{L} = \frac{0.00938}{12} = 0.000782 \text{ in. per in.}$$

Poisson's ratio:

$$\nu = -\frac{\varepsilon_t}{\varepsilon_a} = \frac{0.000260}{0.000782} = 0.333$$

As the area of the rod $A = \pi(2.25/2)^2 = 3.976$ in.^{2}, from Eq. 3-5,

$$\sigma = \frac{P}{A} = \frac{32,000}{3.976} = 8,030 \text{ psi}$$

and

$$E = \frac{\sigma}{\varepsilon} = \frac{8,030}{0.000782} = 10.3 \times 10^6 \text{ psi}$$

Fig. 4-8

---

* A. Nadai, *Theory of Flow and Fracture of Solids*, Vol. 1 (New York: McGraw-Hill Book Company, 1950).

In practice, when a study of physical quantities such as E and ν is being made, it is best to work with the corresponding stress-strain diagram to be sure that the quantities determined are associated with the linear range of the behavior of the material. Note that since the deformations are very small, it makes no difference whether the initial or the final lengths are used in computing strains.

EXAMPLE 4-2

A 2-in. cube of steel is subjected to a uniform pressure of 30,000 psi acting on all faces. Determine the change in dimension between two parallel faces of the cube. Let $E = 30 \times 10^6$ psi and $\nu = \frac{1}{4}$.

SOLUTION

Using Eq. 4-11 to determine strain, noting that pressure is a compressive stress, and recognizing that the strains remain constant within the interval considered,

$$\varepsilon_x = \frac{(-30{,}000)}{(30)10^6} - \frac{(-30{,}000)}{4(30)10^6} - \frac{(-30{,}000)}{4(30)10^6}$$

$$= -(5)10^{-4} \text{ in. per in.}$$

$$\Delta_x = u = \varepsilon_x L_x = -(5)10^{-4} \times 2 = -10^{-3} \text{ in.} \quad \text{(contraction)}$$

where L_x is the length of the cube in the x direction. In this case, by virtue of symmetry, $u = v = w$.

4-8. THERMAL STRAINS

Beside stresses, changes in temperature can also cause deformation of materials. For homogeneous isotropic materials, a change in temperature of δT degrees causes uniform linear strain in every direction. Expressed as an equation, the *thermal strains* are

$$\varepsilon_x = \varepsilon_y = \varepsilon_z = \alpha \, \delta T \tag{4-14}$$

where α is the coefficient of linear thermal expansion for a particular material. It is experimentally determined. Over a moderate range of temperature change, α remains reasonably constant. In the English system of units it is given in inches per inch per degree Fahrenheit. Typical values of α for a few materials are given in Table 1 of the Appendix.

For isotropic materials, a change in temperature causes no shearing strains, i.e., $\gamma_{xy} = \gamma_{yz} = \gamma_{zx} = 0$.

The linear thermal strain for small strains is directly additive to the linear strains due to stress. On this basis a typical modification of Eq. 4-11 to include thermal strain is

$$\varepsilon_x = \frac{1}{E}\sigma_x - \frac{\nu}{E}\sigma_y - \frac{\nu}{E}\sigma_z + \alpha \, \delta T \tag{4-15}$$

An increase in temperature δT is taken positive.

4-9. ELASTIC STRAIN ENERGY FOR UNIAXIAL STRESS

In mechanics, energy is defined as the capacity to do work, and work is the product of a force and the distance in the direction the force moves. In solid, deformable bodies, stresses multiplied by their respective areas are forces, and deformations are distances. The product of these two quantities is the *internal work* done in a body by externally applied forces. This internal work is stored in a body as the *internal elastic energy of deformation* or *the elastic strain energy*. Methods of computing this internal energy will be discussed next.

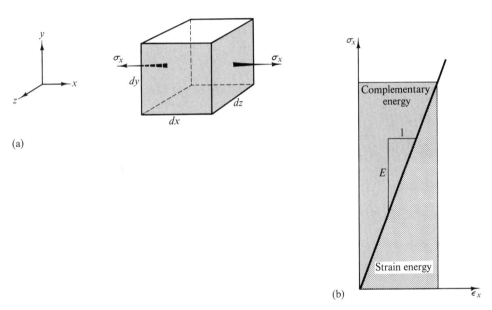

Fig. 4-9. (a) An element in tension and (b) a stress-strain diagram.

Consider an infinitesimal element, such as shown in Fig. 4-9(a), subjected to a normal stress σ_x. The force acting on the right or the left face of this element is $\sigma_x \, dy \, dz$, where $dy \, dz$ is an infinitesimal area of the element. Because of this force, the element elongates an amount $\varepsilon_x \, dx$, where ε_x is strain in the x direction. If the element is made of a linearly elastic material, stress is proportional to strain, Fig. 4-9(b). Therefore, if the element is initially free of stress, the force which finally acts on the element increases linearly from zero until it attains its full value. The average force acting on the element while deformation is taking place is $\sigma_x \, dy \, dz/2$. This average force multiplied by the distance through which it acts is the work done on the element. For a perfectly elastic body no energy is dissipated, and the work done on the element is stored as recoverable internal strain energy. Thus, the internal elastic strain energy U for an infinitesimal element subjected to uniaxial stress is

$$dU = \underbrace{\underbrace{\tfrac{1}{2}\,\sigma_x\,dy\,dz}_{\text{average force}} \times \underbrace{\varepsilon_x\,dx}_{\text{distance}}}_{\text{work}} = \tfrac{1}{2}\,\sigma_x\varepsilon_x\,dx\,dy\,dz = \tfrac{1}{2}\,\sigma_x\varepsilon_x\,dV \quad (4\text{-}16)$$

where dV is the volume of the element.

By recasting Eq. 4-16, one obtains the strain energy stored in an elastic body per unit volume of the material, or its *strain-energy density* U_o. Thus

$$\frac{dU}{dV} = U_o = \frac{\sigma_x \varepsilon_x}{2} \quad (4\text{-}17)$$

This expression may be graphically interpreted as an area under the inclined line on the stress-strain diagram, Fig. 4-9(b). The corresponding area enclosed by the inclined line and the vertical axis is called the *complementary energy*. For linearly elastic materials the two areas are equal. Expressions analogous to Eq. 4-17 apply to the normal stresses σ_y and σ_z and to the corresponding linear strains ε_y and ε_z.

EXAMPLE 4-3

Two bars made of linearly elastic material whose proportions are shown in Fig. 4-10 are to absorb the same amount of energy delivered by axial forces. Compare the stresses in the two bars caused by the same input of energy.

SOLUTION

The bar shown in Fig. 4-10(a) is of uniform cross-sectional area, therefore the normal stress in it σ_1 is constant throughout. By using Eq. 4-17, adapting Eq. 4-11 for the uniaxial case ($\sigma_x = E\varepsilon_x$), and integrating over the volume V of the bar, one obtains the total energy for the bar:

$$U_1 = \int_V \frac{\sigma_1^2}{2E}\,dV = \frac{\sigma_1^2}{2E}\int_V dV = \frac{\sigma_1^2}{2E}(AL)$$

where A is the cross-sectional area of the bar, and L is its length.

The bar shown in Fig. 4-10(b) is of variable cross section. Therefore, if the stress σ_2 acts in the lower part of the bar, the stress in the upper part is $\sigma_2/2$. Again using Eq. 4-11 for the uniaxial case and integrating over the volume of the bar, one finds that the total energy which this bar will absorb in terms of the stress σ_2 is

$$U_2 = \int_V \frac{\sigma^2}{2E}\,dV = \frac{\sigma_2^2}{2E}\int_{\text{lower part}} dV + \frac{(\sigma_2/2)^2}{2E}\int_{\text{upper part}} dV$$

$$= \frac{\sigma_2^2}{2E}\left(\frac{AL}{4}\right) + \frac{(\sigma_2/2)^2}{2E}\left(2A\,\frac{3L}{4}\right) = \frac{\sigma_2^2}{2E}(\tfrac{5}{8}\,AL)$$

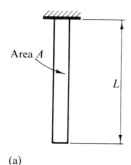
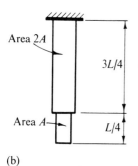

Fig. 4-10

If both bars are to absorb the same amount of energy, $U_1 = U_2$ and

$$\frac{\sigma_1^2}{2E}(AL) = \frac{\sigma_2^2}{2E}(\tfrac{5}{8}\,AL) \quad \text{or} \quad \sigma_2 = 1.265\,\sigma_1$$

The enlargement of the cross-sectional area over a part of the bar in the second case is actually detrimental. For the same energy load, the stress in the "reinforced" bar is 26.5 per cent higher than that in the first bar. This situation is not found in the design of members for static loads.*

4-10. ELASTIC STRAIN ENERGY FOR SHEARING STRESSES

An expression for the elastic strain energy for an infinitesimal element in pure shear may be established in a manner analogous to that for one in uniaxial stress. Thus consider an element in a state of shear as

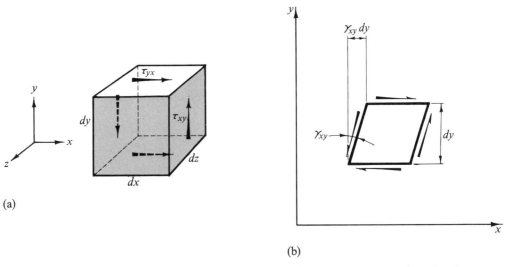

Fig. 4-11. An element for deriving the expression of strain energy due to shearing stresses.

shown in Fig. 4-11(a). The deformed shape of this element is shown in Fig. 4-11(b), where it is assumed that the bottom plane of the element is fixed in position.† As this element is deformed, the force on the top plane reaches a final value of $\tau_{xy}\,dx\,dz$. The total displacement of this force for small deformation of the element is $\gamma_{xy}\,dy$, Fig. 4-11(b). Therefore, since the external work done on the element is equal to the internal, recoverable,

---

* Stresses at the transition of cross-sectional areas are discussed at the end of this chapter.
† This assumption does not make the expression less general.

elastic strain energy,

$$dU_\text{shear} = \underbrace{\tfrac{1}{2}\,\tau_{xy}\,dx\,dz}_{\text{average force}} \times \underbrace{\gamma_{xy}\,dy}_{\text{distance}} = \tfrac{1}{2}\,\tau_{xy}\gamma_{xy}\,dx\,dy\,dz = \tfrac{1}{2}\,\tau_{xy}\gamma_{xy}\,dV \quad (4\text{-}18)$$

where dV is the volume of the infinitesimal element.

By recasting Eq. 4-18, the strain-energy density for shear becomes

$$\left(\frac{dU}{dV}\right)_\text{shear} = \frac{\tau_{xy}\gamma_{xy}}{2} \quad (4\text{-}19)$$

Analogous expressions apply for the shearing stresses τ_{yz}, τ_{zx} with the corresponding shearing strains γ_{yz} and γ_{zx}.

4-11. STRAIN ENERGY FOR MULTIAXIAL STATES OF STRESS

The strain-energy expressions for a three-dimensional state of stress follow directly by superposition of the energies of each stress component. The strain-energy density for the most general case is

$$dU/dV = U_o = \tfrac{1}{2}\sigma_x\varepsilon_x + \tfrac{1}{2}\sigma_y\varepsilon_y + \tfrac{1}{2}\sigma_z\varepsilon_z$$
$$+ \tfrac{1}{2}\tau_{xy}\gamma_{xy} + \tfrac{1}{2}\tau_{yz}\gamma_{yz} + \tfrac{1}{2}\tau_{zx}\gamma_{zx} \quad (4\text{-}20)$$

Upon substituting into this equation the relations for strains as given by Eq. 4-11 and after some algebraic manipulations, one obtains

$$U_o = \frac{1}{2E}(\sigma_x^2 + \sigma_y^2 + \sigma_z^2) - \frac{\nu}{E}(\sigma_x\sigma_y + \sigma_y\sigma_z + \sigma_z\sigma_x)$$
$$+ \frac{1}{2G}(\tau_{xy}^2 + \tau_{yz}^2 + \tau_{zx}^2) \quad (4\text{-}21)$$

as the expression for the elastic strain energy per unit volume for isotropic materials. For situations where there are no shearing stresses, the last term in the equation vanishes. For the case of plane stress with $\sigma_z = 0$ and $\tau_{xz} = \tau_{yz} = 0$, Eq. 4-21 again greatly simplifies.

An equation for U_o analogous to Eq. 4-21 can be established in terms of strains rather than stresses. This is most easily done by recasting the generalized Hooke's law equations to give stresses in terms of strains. In general, for a stressed elastic body the total strain energy is obtained by integration over its volume:

$$U = \iiint U_o\,dx\,dy\,dz \quad (4\text{-}22)$$

Equations 4-21 and 4-22 are very important. The first plays a key role in establishing the laws of plasticity; the second is widely used in stress analysis by energy methods. In this text these items are discussed principally in Chapters 9 and 13.

Section 4-12
Stress-strain
diagrams

PART C
CONSTITUTIVE RELATIONS FOR UNIAXIAL STRESSES

4-12. STRESS-STRAIN DIAGRAMS

In the mechanics of solids the behavior of real materials under load is of primary importance. Experiments, mainly tension and compression tests, provide the basic information on this behavior. In them, the overall, macroscopic behavior of specimens is used to formulate empirical or phenomenological laws. Such formulations are referred to as the *constitutive laws* or the *constitutive relations*. Books on materials science* attempt to provide the reasons for the observed behavior.

It should be apparent from the previous discussion that for general purposes it is more fundamental to report the strain of a rod in tension or compression than to report the elongation of its gage. Similarly, stress is a more significant parameter than force since the effect on a material of an applied force P depends primarily on the cross-sectional area of the member. As a consequence, in the experimental study of the mechanical properties of materials, it is customary to plot diagrams of the relationship between stress and strain in a particular test. Such diagrams for most practical purposes are assumed to be independent of the size of the specimen and of its gage length. In these diagrams, it is customary to use the ordinate scale for stress and the abscissa for strain.

Experimentally determined stress-strain diagrams differ widely for different materials. Even for the same material they differ depending on the temperature at which the test was conducted, the speed of the test, and a number of other variables. However, speaking broadly, two types of diagrams result from experiments at constant temperatures on materials which do not exhibit time dependence. One type, characteristic of mild steel and a few other materials, is shown in Fig. 4-12(a). The other types, typical of many materials, are shown in Fig. 4-12(b). Such diverse materials as tool steel, concrete, and copper have the general shapes of the

---

* See, for example, Z. D. Jastrzebski, *Nature and Properties of Engineering Materials* (New York: John Wiley & Sons, Inc., 1959); L. H. Van Vlack, *Elements of Materials Science* (2nd ed.) (Reading, Mass.: Addison-Wesley Publishing Co., Inc., 1964); J. Wulff, ed., *The Structure and Properties of Materials*, Vols. I and III (New York: John Wiley & Sons, Inc., 1965).

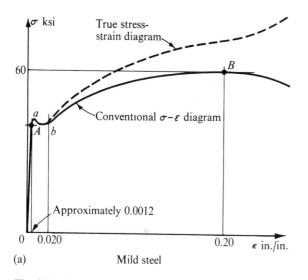
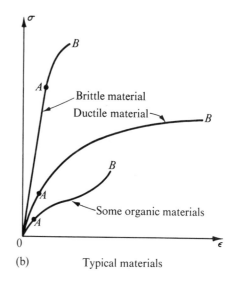

Fig. 4-12. Stress-strain diagrams.

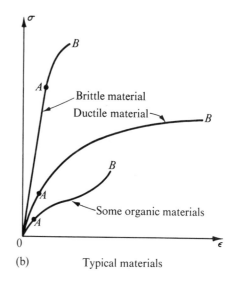

Fig. 4-13. Typical contraction of a specimen of mild steel in tension near the breaking point.

two upper curves, although the extreme values of strain that these materials can withstand differ drastically. The "steepness" of these curves varies considerably. Numerically speaking, each material has its own curve. The terminal point on a stress-strain diagram represents the complete failure (rupture) of a specimen. Materials capable of withstanding large strains are referred to as *ductile materials*. The converse applies to *brittle materials*.

Stresses are usually computed on the basis of the original area of a specimen; such stresses are often referred to as *conventional* or *engineering* stresses. On the other hand it is known that some transverse contraction or expansion of a material always takes place. For mild steel, especially near a breaking point, this effect, referred to as *necking*, is particularly pronounced, see Fig. 4-13. Brittle materials do not exhibit it at usual temperatures, although they too contract transversely a little in a tension test and expand in a compression test. Dividing the applied force by the corresponding actual area of a specimen at the same instant gives the so-called *true stress*. A plot of true stress vs. strain is called a *true stress-strain diagram*, see Fig. 4-12(a).

4-13. FURTHER REMARKS ON STRESS-STRAIN DIAGRAMS

Several important items should be observed in connection with stress-strain diagrams. One of the most important pertains to the somewhat vaguely defined point A, see Fig. 4-12. It lies on a straight line which starts from the origin and closely follows the path of the stress-strain curve. Point *A* is termed the *proportional limit* of the material. The slope of the line from 0 to *A* is the elastic modulus *E*. Physically *E* represents the stiffness of the material to an imposed load.

For all real materials for at least some distance from the origin, to a sufficient degree of accuracy the experimental values of stress vs. strain lie essentially on a straight line. This holds true almost without reservation for glass. It is true for mild steel up to some point, such as *A* in Fig. 4-12(a). It holds nearly true close to the failure point for many high-grade alloy steels. On the other hand, the straight part of the curve hardly exists for concrete, annealed copper, or cast iron. Nevertheless, up to some point such as *A*, Fig. 4-12(b), the relationship between stress and strain may be said to be linear for all materials. This sweeping idealization and generalization is the basis of Hooke's law. Therefore, Hooke's law applies only up to the proportional limit of the material. This is highly significant as in much of the subsequent treatment the derived formulas are based on this law. Clearly then, such formulas are limited in many cases to the behavior of the material in the lower range of stress.

The highest points in the diagrams (*B* in Figs. 4-12(a) and (b)) correspond to the ultimate strength of a material. Stress associated with the remarkably long plateau *ab* in Fig. 4-12(a) is termed the *yield point* of a material. As will be brought out later, this remarkable property of mild steel and other ductile materials is significant in stress analysis. For the present, note that at an essentially constant stress, strains 15 to 20 times those that take place up to the proportional limit occur during yielding. At the yield point a large amount of deformation takes place at a constant stress. The yielding phenomenon is absent in brittle materials.

A study of stress-strain diagrams shows that the yield point is so near the proportional limit that for most purposes the two may be taken as one. However, it is much easier to locate the former. For materials which do not possess a well-defined yield point, one is actually "invented" by the use of the so-called *offset method*. This is illustrated in Fig. 4-14, where a line offset an arbitrary amount of 0.2 per cent of strain is drawn parallel to the straight-line portion of the initial stress-strain diagram. Point *C* is then taken as the *yield point* of the material as 0.2 per cent offset.

Finally, the technical definition of the *elasticity* of a material should be given. In such usage it means that a material is able to regain completely its original dimensions upon removal of the applied forces, i.e., the body completely recovers its original shape. Thus elastic behavior implies the absence of any permanent deformation. Some elastic materials

*Section 4-13
Further remarks on stress-strain diagrams*

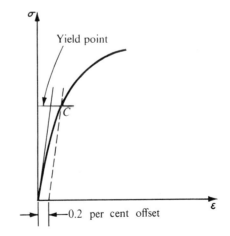

Fig. 4-14. Offset method of determining the yield point of a material.

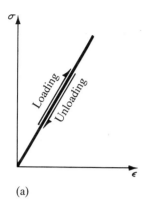
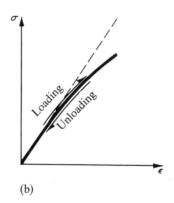

Fig. 4-15. Elastic materials: (a) linear, (b) nonlinear.

exhibit an essentially linear relationship between stress and strain, Fig. 4-15(a); such materials are called *linearly elastic materials*. Some other elastic materials, Fig. 4-15(b), exhibit curvature in their stress-strain diagrams. Such materials are called *nonlinearly elastic materials*. The stress at which permanent deformation or set in the material occurs is the *elastic limit* of the material. For linearly elastic materials, the elastic limit corresponds to the proportional limit.

For the majority of materials, stress-strain diagrams obtained for short compression blocks are reasonably close to those found in tension. However, for some materials the diagrams differ drastically, depending on the sense of the applied force. For example, cast iron and concrete are very weak in tension but not in compression.

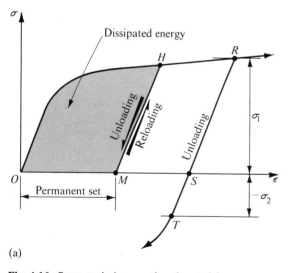
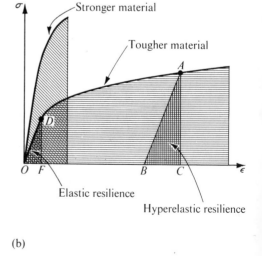

Fig. 4-16. Some typical properties of materials.

4-14. STRESS-STRAIN DIAGRAMS DURING UNLOADING AND LOAD REVERSALS

Inelastic and plastic materials exhibit important phenomena if loading is not monotonically increased. During an unloading process (characterized by a line such as HM in Fig. 4-16(a)), the response is essentially linearly elastic with the elastic modulus of the original material, although a permanent strain or set is observed. On reloading, the material again behaves linearly elastically and can again reach point H. Beyond H, if the material is loaded further, it generates the continuation of the original curve. Upon unloading at R, the material again follows essentially a straight line to S at a no-load condition, and then on to T if loaded in the opposite direction. Note that the absolute ordinate of T is smaller than that of R. This typical effect was first observed by Bauschinger* and bears his name.

According to Eq. 4-17, for linearly elastic material in a uniaxial state of stress, the strain energy per unit volume is $\sigma_x \varepsilon_x / 2$. Alternatively, according to Eq. 4-21, $U_o = \sigma_x^2 / 2E$. Substituting the value of the stress at the proportional limit into this equation gives an index of the ability of the material to store or absorb energy without permanent deformation. The quantity so found is termed the *modulus of resilience* and is used to select materials for applications where energy must be absorbed by members. For example, steel with a proportional limit of 30,000 psi and an E of 30×10^6 psi has a modulus of resilience of $\sigma^2/(2E) = (30,000)^2/2(30)10^6 = 15$ in-lb per cubic inch, whereas a good grade of Douglas Fir, having a proportional limit of 6,450 psi and an E of 1,920,000 psi. has a modulus of resilience of $(6,450)^2/2(1,920,000) = 10.8$ in-lb per cubic inch.

By reasoning analogous to the above, the area under a complete stress-strain diagram, Fig. 4-16(b), gives a measure of the ability of a material to resist an energy load up to fracture and is called its *toughness*. The larger the total area under the stress-strain diagram, the tougher the material. In the inelastic range, only a small part of the energy absorbed by a material is recoverable. Most of the energy is dissipated in permanently deforming the material. The energy which may be recovered when a specimen has been stressed to a point such as A or D in Fig. 4-16(b) is represented by the triangles ABC and DOF, respectively. The line AB of the triangle ABC is parallel to the elastic line OD. The areas enclosed by the respective triangles, as shown in the figure, represent the *elastic* and the *hyperelastic resilience* of the material.

When a material is loaded cyclically into the inelastic range, the dissipated energy per cycle is given by the area enclosed by the non-

---

* Johann Bauschinger (1833–93) was a professor of mechanics at the Polytechnical Institute of Munich, Germany.

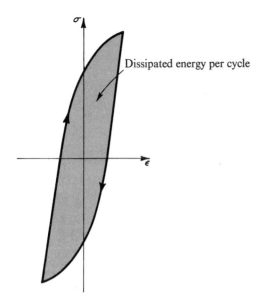

Fig. 4-17. Hysteresis loop for an inelastic material.

coincident lines of the stress-strain diagram, Fig. 4-17. The loop formed is called the *hysteresis loop*. Usually there is a tendency for a small hysteresis loop to develop along a line such as *HM* in Fig. 4-16(a).

4-15. IDEALIZED STRESS-STRAIN DIAGRAMS

For analytical treatment of material behavior it is convenient to idealize the experimentally determined stress-strain diagrams. A group of widely used diagrams for the uniaxial state of stress is shown in Fig. 4-18. In Fig. 4-18(a) the relationship for linearly elastic material is restated. As was pointed out earlier, this is the basis for Hooke's law. Very few materials behave in this manner to their ultimate strength. However, for nearly every material this relationship holds true for small deformations. Because of the simplicity of Hooke's law, it

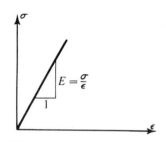
(a) Linearly elastic material

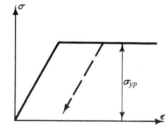
(b) Elastic, perfectly plastic material

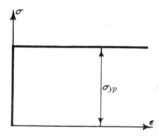
(c) Rigid, perfectly plastic material

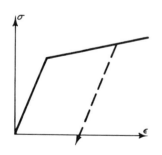
(d) Elastic, linearly hardening material

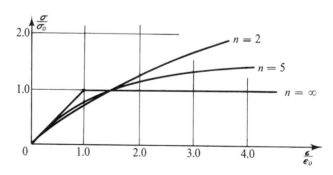
(e) Ramberg–Osgood diagrams

Fig. 4-18. Idealized stress-strain diagrams.

is a common practice to approximate moderately nonlinear response with it.

Some materials, after exhibiting an elastic response, deform very large amounts at a practically constant stress. Mild steel and nylon are notable examples. An unlimited amount of deformation or yielding at constant stress defines an ideally or perfectly plastic material. A material exhibiting a linearly elastic response followed by a perfectly plastic one is shown in Fig. 4-18(b). If the elastic range is negligible in comparison with the plastic one, the idealization of a rigid, perfectly plastic material is employed, Fig. 4-18(c).

Stress-strain diagrams for many materials can be approximated by bilinear diagrams as in Fig. 4-18(d). In such idealizations it is common to refer to the first stage as the *elastic* range, and to the second as the *strain* or *stress-hardening* range. The more general term *linear-hardening* is shown on the diagram.

A convenient equation capable of representing a wide range of stress-strain curves has been developed by Ramberg and Osgood.* This equation† is

$$\frac{\varepsilon}{\varepsilon_o} = \frac{\sigma}{\sigma_o} + \tfrac{3}{7}\left(\frac{\sigma}{\sigma_o}\right)^n \tag{4-23}$$

where ε_o, σ_o, and n are the characteristic constants for a material. The constants ε_o and σ_o correspond to the yield point, which, for all cases other than that of ideal plasticity, is found by the offset method (see Fig. 4-14). The exponent n determines the shape of the curve. Note that Eq. 4-23 is written in dimensionless form, a convenient scheme in general analysis. One of the important advantages of Eq. 4-23 is the fact that it is a continuous mathematical function. For example, an *instantaneous* or *tangent modulus* E_t defined as

$$E_t = \frac{d\sigma}{d\varepsilon} \tag{4-24}$$

can be readily and uniquely determined.

Shearing stress shearing strain diagrams can be also obtained from experiments with the various materials and are susceptible to idealizations analogous to those in Fig. 4-18. As will become clear after the study of Chapter 5, such experiments are best performed on thin-walled tubes subjected to controlled torque. Except for Hooke's law, generalizations

---

* W. Ramberg and W. R. Osgood, *Description of Stress-Strain Curves by Three Parameters*, National Advisory Committee on Aeronautics, TN 902, 1943.
† The coefficient $\tfrac{3}{7}$ is chosen somewhat arbitrarily; different values have been used in some investigations.

of stress-strain relations for biaxial and triaxial states of stress and strain are very difficult and complex problems. Much remains unsettled in this challenging area of mechanics. Additional information on yielding and fracture of materials in complex states of stress will be presented in Chapter 9.

4-16. LINEARLY VISCOELASTIC MATERIALS

In the preceding discussion of stress-strain relations, it is tacitly assumed that the materials are inviscid, i.e., they exhibit no time-dependent flow or creep phenomena. However, asphalt pavements, solid propellant in rocket motors, high-polymer plastics, and concrete, as well as machine elements at elevated temperatures, gradually deform under stress, and such deformations are usually not fully recoverable. A few elementary notions of this problem are considered below for the uniaxial state of stress. A more complete investigation is the concern of rheology.*

For elastic materials stress is said to be a function of strain only. On the other hand, for viscous materials stress depends not only on strain but also on the rate at which the strain is applied. This may be clarified by examining the conceptual models in Fig. 4-19. For the linearly elastic spring the stress is proportional to the strain. For an element with a viscous liquid in the dashpot, the higher the strain rate, the higher the stress necessary to maintain the motion of the applied force. For brevity, the strain rate—the derivative of the strain with respect to time—will be designated by an ε with a dot over it.

In the above terms, for an elastic material $\sigma = \sigma(\varepsilon)$; but for a viscoelastic material, since the stress is a function of both the strain and the strain rate, $\sigma = \sigma(\varepsilon, \dot{\varepsilon})$. The simplest relation among these quantities can be stated as

$$\sigma = E\varepsilon + \eta\dot{\varepsilon} \qquad (4\text{-}25)$$

where the constant η (eta) is the coefficient of viscosity. The last term linearly relates the stress to the strain rate, as shown in Fig. 4-19(b). If this term is zero, one obtains an ordinary Hooke's law. The material behavior described by Eq. 4-25 is associated with the names of Voigt and Kelvin,† who first used it in the analysis of viscoelastic materials. For this reason the idealized material of Eq. 4-25 is referred to as the *Voigt-Kelvin solid*.

Although certainly not fundamental, it is convenient to introduce a conceptual model to clarify the meaning of Eq. 4-25. Such a model is

---

* See, for example, F. R. Eirich, ed., *Rheology* (New York: Academic Press, Inc., 1956).

† W. Voigt (1850–1919), a theoretical physicist, taught at Göttingen University, Germany. Lord Kelvin (William Thomson) (1824–1907) was a British physicist.

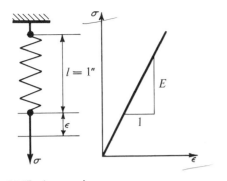

(a) Hookean spring

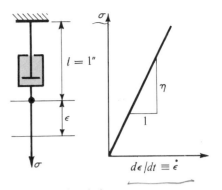

(b) Newtonian dashpot

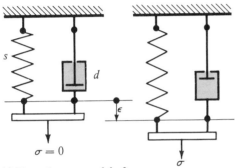

(c) Two-element model of a Voigt–Kelvin solid

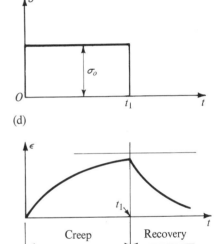

(d)

(e) Creep curve

Fig. 4-19. Models of material response.

obtained by placing a spring and a dashpot in parallel as in Fig. 4-19(c). As the stress σ is applied, the same strain is induced in the spring and in the dashpot, i.e., $\varepsilon_d = \varepsilon_s = \varepsilon$, where the subscripts designate the dashpot (d) and the spring (s), respectively. The total stress (force) σ is the sum of the stress σ_d and σ_s, i.e., $\sigma = \sigma_d + \sigma_s$. By using Hooke's law and a linear stress-strain rate relationship for the Newtonian liquid, one obtains Eq. 4-25, which may be written as

$$\dot{\varepsilon} + (E/\eta)\varepsilon = \sigma/\eta \qquad (4\text{-}25a)$$

EXAMPLE 4-4

Determine the creep of a Voigt–Kelvin solid subjected to a constant stress σ_o. Initially the model is unstrained.

SOLUTION

Noting that the stress $\sigma = \sigma_o$ is constant, one can show that the homogeneous and the particular solutions of Eq. 4-25(a) give

$$\varepsilon = Ae^{-(E/\eta)t} + \sigma_o/E$$

where A is a constant which can be found from the condition that $\varepsilon(0) = 0 = A + \sigma_o/E$, i.e., $A = -\sigma_o/E$. Therefore,

$$\varepsilon = (1 - e^{-(E/\eta)t})(\sigma_o/E)$$

As time increases, the strain asymptotically approaches the maximum strain associated with the elastic spring until, finally, all the applied stress is carried by the spring, and the dashpot becomes inactive. If the stress is removed at an earlier time as in Fig. 4-19(d), an asymptotic recovery of the strain takes place, Fig. 4-19(e).

The above solution shows that the Voigt-Kelvin material exhibits a delayed elastic response; for this reason it is termed an *anelastic* material. Its behavior can be likened to that of an elastic sponge filled with a viscous fluid in which ultimately all the applied load is carried by the elastic core. Based on experimental evidence, it is known that such behavior is not very typical of most materials. Another linear combination of stress, stress rate, and strain rate can be formulated, which is more representative.

By endowing a body with an instantaneous elastic response together with a time-dependent displacement, one can obtain a reasonable approximation of the behavior of many viscoelastic materials. The simplest model having such properties can be visualized as a combination in series of a linear spring and a linear dashpot as in Fig. 4-20. Material of this type is called a *Maxwell solid*.* In the model in Fig. 4-20(a), if a stress σ is applied, the stress (force) through the dashpot (d) is the same as that through the spring (s), i.e., $\sigma_d = \sigma_s = \sigma$. However, since each element of the model contributes to the total strain, $\varepsilon = \varepsilon_s + \varepsilon_d$, where the subscripts as before designate the spring (s) and the dashpot (d), respectively. The strain relation must be differentiated with respect to time since for viscous materials only the connection between the stress and the strain rate is known. On the other hand, for elastic materials with E constant, on differentiating Hooke's law with respect to time, one has $\dot{\varepsilon} = \dot{\sigma}/E$. Then upon adding the strain rates for the two elements and simplifying, one obtains the basic differential equation for the response of the Maxwell solid:

$$\dot{\varepsilon} = \dot{\varepsilon}_s + \dot{\varepsilon}_d = \dot{\sigma}/E + \sigma/\eta \quad \text{or} \quad \dot{\sigma} + (E/\eta)\sigma = E\dot{\varepsilon} \quad \text{(4-26)}$$

---

* James Clerk Maxwell (1831–79), a renowned British physicist, made a number of important contributions to the mechanics of solids.

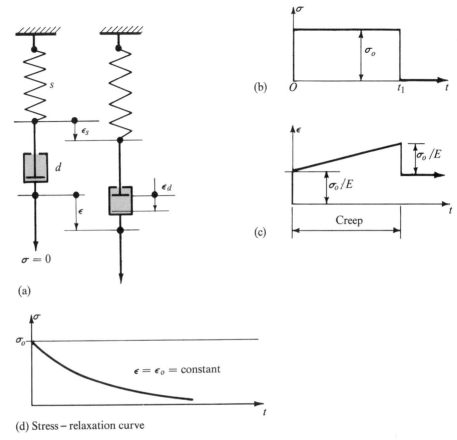

Fig. 4-20. Two-element model of a Maxwell solid.

For the Maxwell solid in pure shear an analogous expression applies:

$$\dot{\gamma} = \dot{\tau}/G + \tau/\bar{\eta} \qquad (4\text{-}26a)$$

where, as before, dots over the quantities represent their derivatives with respect to time, and $\bar{\eta}$ is the coefficient of viscosity in shear.

EXAMPLE 4-5

A Maxwell solid is subjected to a step loading as shown in Fig. 4-20(b); i.e., a constant stress σ_o acts during a time interval $0 < t < t_1$. Determine the strain response.

SOLUTION

In this case the applied stress σ does not change with time, hence $\dot{\sigma} = 0$. At time $t = 0$ an instantaneous elastic strain $\varepsilon_o = \sigma_o/E$ occurs, which is the initial constant of integration. Upon release of the stress, this strain is completely recovered. On this basis, using Eq. 4-26,

119

$$\frac{d\varepsilon}{dt} = \frac{\sigma_o}{\eta} \quad \text{or} \quad \varepsilon = \frac{\sigma_o}{\eta}t + C_1 \quad \text{and} \quad \varepsilon = \frac{\sigma_o}{E} + \frac{\sigma_o}{\eta}t$$

This relation applies in the interval $0 < t < t_1$. At $t = t_1$ strain of σ_o/E is recovered, and $(\sigma_o/\eta)t_1$ is the *permanent* or *residual strain*. These results are indicated in Fig. 4-20(c). The solution exemplifies an elementary creep problem.

EXAMPLE 4-6

If a Maxwell solid is initially strained an amount ε_o causing an initial stress of σ_o, and the strain ε_o is maintained, how does the stress vary with time?

SOLUTION

Here the strain rate $\dot{\varepsilon} = 0$ since no change in strain is permitted. This fact simplifies Eq. 4-26. To determine the constant of integration, one notes that at $t = 0$ the stress is σ_o. Therefore, the governing differential equation is

$$\frac{d\sigma}{dt} + \frac{E}{\eta}\sigma = 0$$

Solving this equation with a constant of integration A,

$$\sigma = Ae^{-(E/\eta)t}, \quad \text{and, since} \quad \sigma(0) = \sigma_o, \quad \sigma = \sigma_o e^{-(E/\eta)t}$$

This result is plotted in Fig. 4-20(d). It is interesting to note how with time the stress gradually decreases, tending asymptotically toward zero. This situation is characteristic of an initially stressed bolt at high temperature which clamps rigid flanges of a machine or of a tendon in a prestressed concrete beam. As the material creeps, stress relaxation takes place. For this reason a Maxwell material sometimes is referred to as a *relaxing material*. This problem is of great practical importance in many applications.

The above procedures can be generalized to many more materials. A combination in series of the Maxwell and the Voigt-Kelvin models establishes the basic model, the Standard Solid, for studying linearly viscoelastic materials. Other combinations of springs and dashpots with different constants have been used effectively for representing high polymers, fibers, concrete, etc. Extensions to three-dimensional problems have been also achieved.* The extension of the theory to nonlinear viscoelastic materials is being actively pursued.

From a phenomenological point of view, for real materials the relaxation and creep curves must be considered fundamental properties of a given material and must be determined experimentally. In a relaxation

---

* Bland, D. R., *The Theory of Linear Viscoelasticity* (Long Island City, N.Y.: Pergamon Press, Inc., 1960), p. 19.

experiment a constant strain ε_o is maintained and the corresponding stress $\sigma(t)$ is determined. By dividing $\sigma(t)$ by ε_o the *relaxation modulus* $E(t)$ is obtained. A qualitative curve for such an experiment is shown in Fig. 4-21(a). If data from several relaxation experiments done at different constant strains ε_o give the same relaxation modulus $E(t)$, the material is *linearly viscoelastic*.

In a creep experiment a constant stress σ_o is maintained and the corresponding strain $\varepsilon(t)$ is obtained. By dividing $\varepsilon(t)$ by σ_o, one finds the *creep compliance* $J_c(t)$. A typical function of $J_c(t)$ is shown in Fig. 4-21(b). Again, if the creep-compliance curves for several experiments performed at different stress levels coincide, the viscoelastic material is *linear*. Stated another way, for linear, viscoelastic materials at any constant stress σ_o or strain ε_o, one has

$$\varepsilon(t) = \sigma_o J_c(t) \quad \text{and} \quad \sigma(t) = \varepsilon_o E(t) \quad (4\text{-}27)$$

For application of the above equations, it is important to note the Boltzmann* superposition principle, which is valid for a number of materials. This principle asserts that the strain at a given time is the sum of the strains caused by the loads applied independently for their respective durations of time. For example, if, as shown in Fig. 4-21(c), a stress σ_o is applied at $t = 0$, the strain at any time $t > 0$ is $\sigma_o J_c(t)$. Then, if at time t_1 another stress σ_1 is added, for $t > t_1$ the additional strain is $\sigma_1 J_c(t - t_1)$. For the second load application the same creep-compliance function applies, but its origin is moved to t_1. In general

$$\varepsilon(t) = \sigma_o J_c(t) + \sigma_1 J_c(t - t_1)$$
$$+ \sigma_2 J_c(t - t_2) + \cdots \quad (4\text{-}27\text{a})$$

Boltzmann's principle also applies if a succession of strains is applied to a material. For such

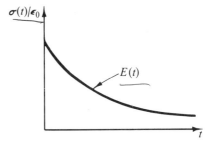

(a) Relaxation modulus

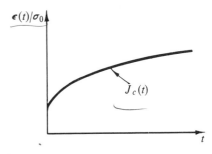

(b) Creep compliance function

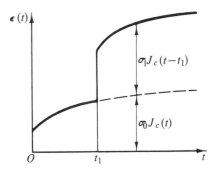

(c) Strain superposition according to Boltzmann's principle

Fig. 4-21. Typical behavior of viscoelastic materials.

---

* L. Boltzmann (1844–1906) was a distinguished physicist particularly known for his research on the kinetic theory of gases and in quantum mechanics and statistical mechanics. He taught at Graz and Vienna, Austria, and at Leipzig and Munich, Germany.

a case the relation analogous to Eq. 4-27a is*

$$\sigma(t) = \varepsilon_0 E(t) + \varepsilon_1 E(t - t_1) + \varepsilon_2 E(t - t_2) + \cdots \quad \text{(4-27b)}$$

Analogous relations to Eqs. 4-27a and b can be written for linear viscoelastic materials in pure shear.

The material constants for creep and relaxation are strongly affected by temperature. In this regard it is instructive to examine the

---

* If a continuous change in $\varepsilon(t)$ occurs, Eq. 4-27b can be written in the form of a Duhamel integral as

$$\sigma(t) = \int_{-\infty}^{t} E(t - t') \frac{d\varepsilon}{dt'} dt'$$

An analogous expression also applies to $\varepsilon(t)$:

$$\varepsilon(t) = \int_{-\infty}^{t} J_c(t - t') \frac{d\sigma}{dt'} dt'$$

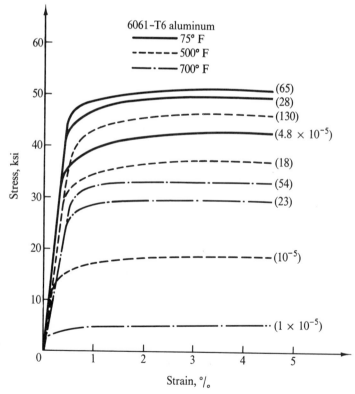

Fig. 4-22. Effect of strain rate and temperature on stress-strain curves for 6061-T6 aluminum.

experimentally determined stress-strain diagrams* for aluminum in Fig. 4-22. (The numbers in parentheses refer to the strain rates measured in inches per inch per second.) Here the pronounced effects of strain rates and temperature on the mechanical behavior of this material can be clearly seen. Conclusions for viscoelastic materials based on short-duration tests at one temperature can be grossly misleading.

Section 4-17
Deflection of axially loaded members

PART D
DEFORMATION OF AXIALLY LOADED MEMBERS

4-17. DEFLECTION OF AXIALLY LOADED MEMBERS

The method of determining deformations or deflections of axially loaded members is based on the previously discussed procedures and equations. To formulate this problem in general terms, consider the axially loaded bar in Fig. 4-23(a). In this bar the cross-sectional area varies along the length, and forces of various magnitudes are applied at several points. Now suppose that in this problem the change in length of the bar between two points A and B caused by the applied force is sought. The quantity wanted is the sum (or accumulation) of the deformations that take place in infinitesimal lengths of the rod. Therefore, if the amount of deformation that takes place in an arbitrary element of length dx is formulated, the sum or integral of this effect over the given length gives the quantity sought.

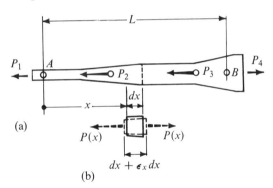

Fig. 4-23. An axially loaded bar.

An arbitrary element cut from the bar is shown in Fig. 4-23(b). From free-body considerations, this element is subjected to a pull $P(x)$, which, in general, is a variable quantity. The infinitesimal deformation du that takes place within this element upon application of the forces is equal to the strain ε_x multiplied by the length dx. The total deformation between any two given points on a bar is simply the sum of the element deformations. Therefore, the displacement u of any point on the bar is given by an integral of the infinitesimal displacements plus a constant of integration. This constant C_1 accounts for the given displacement at one of the

---

* K. G. Hoge, "Influence of Strain Rate on Mechanical Properties of 6061-T6 Aluminum Under Uniaxial and Biaxial States of Stress," *Experimental Mechanics*, **6**, no. 10 (April 1966), p. 204.

boundaries. The solution of the differential equation $du/dx = \varepsilon_x$ reads

$$u = \int_0^x du + C_1 = \int_0^x \varepsilon_x \, dx + C_1 \tag{4-28}$$

Note that the deflection of a rod is treated as a one-dimensional problem; the displacements of all points across a section are assumed to be the same. Therefore, a section initially perpendicular to the axis of a rod moves axially a distance u parallel to itself.*

The magnitude of the strain ε_x depends on the magnitude of the stress σ_x. The latter is found in general by dividing the variable force $P(x)$ by the corresponding area $A(x)$, i.e., $\sigma_x = P/A$. The relationship, either analytical or graphical, between ε_x and σ_x must be known to solve the problem. For linearly elastic materials, according to Hooke's law for uniaxial stress, $\varepsilon_x = \sigma_x/E$. Therefore, for this special case, applicable only within the linearly elastic range of the behavior of the material, one has

$$u = \int_0^x \frac{\sigma_x}{E} dx + C_1 = \int_0^x \frac{P(x)}{A(x)E} dx + C_1 \tag{4-29}$$

In problems where the area of a rod is variable, a proper function for it must be substituted into Eq. 4-29. In practice, it is sometimes sufficiently accurate in such problems to approximate the shape of a rod by a finite number of elements as shown in Fig. 4-24. The deflections for each of these elements are added to obtain the total deflection.

Instead of solving the first-order differential equation for u, as above, it is instructive to formulate this problem as a second-order equation. Such an equation for linearly elastic materials follows from two observations. First, since, in general,† $du/dx = \varepsilon = \sigma/E = P/(AE)$, one has

$$P = AE(du/dx) \tag{4-30}$$

Fig. 4-24

The second relation is based on the equilibrium requirements for an infinitesimal element of an axially loaded bar. For this purpose consider a typical element such as that in Fig. 4-25, where all forces are shown with a positive sense according to the previously adopted sign convention. Since $\Sigma F_x = 0$ or $dP + p_x \, dx = 0$,

$$\frac{dP}{dx} = -p_x \quad \left[\frac{\text{lb}}{\text{in.}}\right] \tag{4-31}$$

---

* If a rod is attached to a rigid support, Poisson's effect tends to change the transverse dimensions, and so this condition cannot be rigorously fulfilled. Fortunately, this introduces only a very small error in the response of axially loaded members, unless such members are very short in relation to their width. Problems of this type are considered in texts on the theory of elasticity.
† For simplicity the subscripts have been dropped.

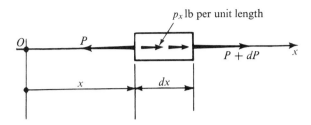

Fig. 4-25. Infinitesimal element of an axially loaded bar.

This equation states that the rate of change with x of the internal axial force P is equal to the negative of the applied force p_x. On this basis, assuming AE constant

$$\frac{d}{dx}\left(\frac{du}{dx}\right) = \frac{1}{AE}\frac{dP}{dx} \quad \text{or} \quad AE\frac{d^2u}{dx^2} = -p_x \quad (4\text{-}32)$$

This equation can be solved by using the procedures discussed in Chapter 2. Singularity functions can be used for concentrated forces. To solve a problem with Eq. 4-32, either u or P must be prescribed at each boundary. Equation 4-30 is used to give a condition dependent on P at a boundary. Positive answers indicate tensile forces and extensions, and vice versa. Statically indeterminate problems may be analyzed using this equation; a more complete discussion of this topic is postponed until Chapter 12.

The three following examples show applications of Eq. 4-29 or Eq. 4-32 or both. An additional example illustrates the solution of a deflection problem for an axially loaded rod of two different materials, including inelastic behavior. Application of Eq. 4-32 to situations requiring singularity functions is left to the reader.

EXAMPLE 4-7

Consider the rod AB of constant cross-sectional area A and of length L shown in Fig. 4-26(a). Determine the relative displacement of the end A with respect to B when a force P is applied; i.e., find the deflection of the free end caused by the application of a concentrated force P. The elastic modulus of the material is E.

SOLUTION

In this problem the rod may be treated as being weightless as only the effect of P on the deflection is investigated. Hence, no matter where a cut C-C is made through the rod, $P(x) = P$, Fig. 4-26(b). The infinitesimal elements, Fig. 4-26(c), are everywhere the same, subjected to a constant pull P. Likewise, $A(x)$ everywhere has

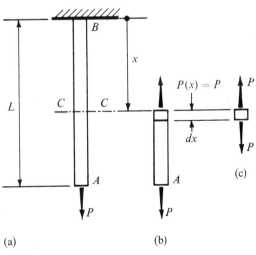

Fig. 4-26

a constant value A. Applying Eq. 4-29 and noting that at the origin of x the displacement u is zero, one has

$$u = \frac{P}{AE}\int_0^x dx + C_1 = \frac{Px}{AE} + C_1$$

Since $u(0) = 0$, $C_1 = 0$ and $u = Px/AE$. At the free end

$$u(L) \equiv \Delta = PL/AE \qquad (4\text{-}33)$$

This indicates that the deflection of the rod is directly proportional to the applied force and the length, and inversely proportional to A and E. This equation will be referred to in subsequent work. It and similar equations for more complicated cases are necessary in vibration analyses for determining the spring constant k, which represents the stiffness of a system and is defined as $k = P/\Delta$ (lb per inch). In this case $k = AE/L$. The constant k is also called the *stiffness influence coefficient*. The reciprocal of k defines the *flexibility (influence) coefficient* $f = k^{-1}$.

EXAMPLE 4-8

Determine the relative displacement of points A and D of the steel rod of variable cross-sectional area shown in Fig. 4-27(a) when it is subjected to the four concentrated forces P_1, P_2, P_3, and P_4. Let $E = 30 \times 10^6$ psi.

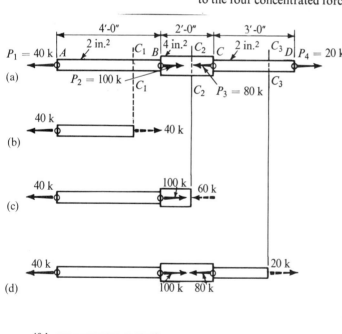

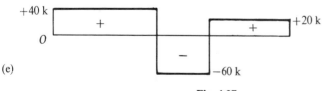

Fig. 4-27

SOLUTION

Section 4-17
Deflection of axially loaded members

In attacking such a problem, a check must first be made to ascertain that the body as a whole is in equilibrium, i.e., $\Sigma F_x = 0$. Here, by inspection it may be seen that such is the case. Next, the variation of P along the length of the bar must be studied. This may be done conveniently with the aid of sketches as shown in Fig. 4-27(b), (c), and (d), which show that no matter where a section C_1-C_1 is taken between points A and B, the force in the rod is $P = +40$ kips. Similarly, between B and C, $P = -60$ kips, and between C and D, $P = +20$ kips. An axial-force diagram for these quantities is in Fig. 4-27(e). The variation of A is in Fig. 4-27(a). Neither P nor A is a continuous function along the rod since both have jumps or *sudden* changes in their values. Hence, in integrating, unless singularity functions are used, the limits of integration must be "broken." Thus applying Eq. 4-29 and noting that for the origin at A the constant $C_1 = 0$, one has

$$u = \int_0^L \frac{P(x)\,dx}{A(x)E} = \int_A^B \frac{P_{AB}\,dx}{A_{AB}E} + \int_B^C \frac{P_{BC}\,dx}{A_{BC}E} + \int_C^D \frac{P_{CD}\,dx}{A_{CD}E}$$

In the last three integrals the respective P and A are constants between the limits shown. The subscripts of P and A denote the range of applicability of the function; thus P_{AB} applies in the interval AB, etc. These integrals revert to the solution of the previous example, i.e., Eq. 4-33. Applying it and substituting numerical values,

$$u = \sum \frac{PL}{AE} = +\frac{40{,}000(4)12}{2(30)10^6} - \frac{60{,}000(2)12}{4(30)10^6} + \frac{20{,}000(3)12}{2(30)10^6}$$

$$= +0.032 - 0.012 + 0.012 = +0.032 \text{ in.}$$

This operation adds, or superposes, the individual deformations of the three "separate" rods. Each of these rods is subjected to a constant force. The positive sign of the answer indicates that the rod elongates as a positive sign is associated with tensile forces. The equality of the absolute values of the deformations in lengths BC and CD is purely accidental. Note that in spite of the relatively large stresses present in the rod, the value of u is small. Finally, do not fail to observe that units of all quantities have been changed to be consistent. Forces originally given in kips have been changed into pounds, lengths into inches.

EXAMPLE 4-9

Find the deflection caused by its own weight of the free end A of the rod AB having a constant cross-sectional area A and weighing p_o lb per inch, Fig. 4-28(a).

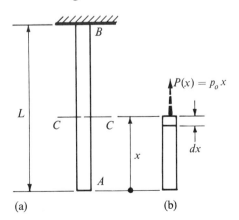

Fig. 4-28

SOLUTION

In this case $P(x)$ is variable. It is conveniently expressed as $p_o x$ if the origin is taken at A. Here again Eq. 4-29 can be applied:

$$u = \int_0^x \frac{P(x)\,dx}{A(x)E} + C_1 = \frac{1}{AE}\int_0^x p_o x\,dx + C_1 = \frac{p_o x^2}{2AE} + C_1$$

At the boundary B, where $x = L$, the displacement is zero, i.e., $u(L) = 0$. This condition must be used to evaluate the constant of integration: $C_1 = -p_o L^2/2AE$. Thus $u = -p_o(L^2 - x^2)/2AE$ and $u(0) = -p_o L^2/2AE$. The negative sign indicates that the displacement u is in the opposite direction to that of positive x. If W designates the total weight of the rod, the absolute maximum deflection is $WL/2AE$. Compare this expression with Eq. 4-33.

In this problem Eq. 4-32 instead of Eq. 4-29 could be applied. With the gravity load acting downward and with the positive x axis directed upward, the sign of the load in Eq. 4-32 must be negative, i.e., $AE\,d^2u/dx^2 = -(-p_o)$. As in the previous solution, one of the boundary conditions is $u(L) = 0$. The second one is $u'(0) = 0$, where $u' = du/dx$; this follows from the fact that at the free end $P = 0$. (See Eq. 4-30.)

If a concentrated force P, in addition to the bar's own weight, were acting on the bar AB at the end A, the total end deflection due to the two causes by superposition would be

$$|u| = \frac{PL}{AE} + \frac{WL}{2AE} = \frac{[P + (W/2)]L}{AE}$$

EXAMPLE 4-10

A 30-in.-long aluminum rod is enclosed within a steel-alloy tube, Figs. 4-29(a) and (b). The two materials are bonded together. If the stress-strain diagrams for the two materials can be idealized as shown, respectively, in Fig. 4-29(d), what end deflection will occur for $P_1 = 80$ kips and for $P_2 = 125$ kips? The cross-sectional areas of steel A_s and of aluminum A_a are the same and equal to 0.5 in.^{2}.

SOLUTION

By applying the method of sections, one can easily determine the axial force at an arbitrary section, Fig. 4-29(c). However, unlike the case in any problem considered so far, the manner in which the resistance to the force P is distributed between the two materials is not known. Thus, the problem is internally statically indeterminate. The requirements of equilibrium (statics) remain valid, but additional conditions are necessary to solve the problem. One of the auxiliary conditions comes from the requirements of the compatibility of deformations. However, since the requirements of statics involve forces and deformations involve displacements, a connecting condition based on the property of materials must be added.

Let subscripts a and s on P, ε, and σ identify these quantities as being for aluminum and steel, respectively. Then, noting that the applied force is supported by a force developed in steel and aluminum and that

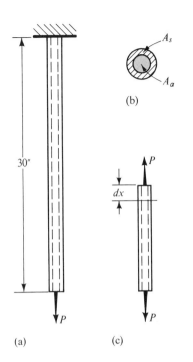
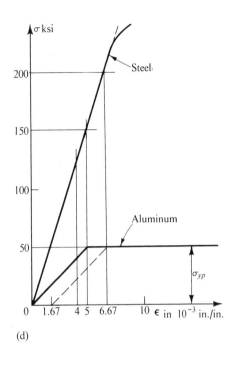

Fig. 4-29

at every section the displacement or the strain of the two materials is the same, and tentatively assuming elastic response of both materials, one has

Equilibrium: $\quad P_a + P_s = P_1 \quad$ or $\quad P_2$

Deformation: $\quad u_a = u_s \quad$ or $\quad \varepsilon_a = \varepsilon_s$

Material properties: $\quad \varepsilon_a = \sigma_a/E_a \quad$ and $\quad \varepsilon_s = \sigma_s/E_s$

By noting that $\sigma_a = P_a/A_a$ and $\sigma_s = P_s/A_s$, one can solve the three equations. From the diagram the elastic moduli are $E_s = 30 \times 10^6$ psi and $E_a = 10 \times 10^6$ psi. Thus

$$\varepsilon_a = \varepsilon_s = \frac{\sigma_a}{E_a} = \frac{\sigma_s}{E_s} = \frac{P_a}{A_a E_a} = \frac{P_s}{A_s E_s}$$

Hence $P_s = [A_s E_s/(A_a E_a)]P_a = 3P_a$, and $P_a + 3P_a = P_1 = 80$ k; therefore, $P_a = 20$ k, and $P_s = 60$ k.

Applying Eq. 4-33 to either material, the tip deflection for 80 kips will be

$$u = \frac{P_s L}{A_s E_s} = \frac{P_a L}{A_a E_a} = \frac{20(10^3)30}{0.5(10)10^6} = 0.120 \text{ in.}$$

129

This corresponds to a strain of $0.120/30 = 4 \times 10^{-3}$ in. per inch. In this range both materials respond elastically, which satisfies the material-property assumption made at the beginning of this solution. In fact, as may be seen from Fig. 4-29(d), since for the linearly elastic response the strain can reach 5×10^{-3} in. per inch for both materials, by direct proportion the applied force P can be as large as 100 kips.

At $P = 100$ kips the stress in aluminum reaches 50 ksi. According to the idealized stress-strain diagram no higher stress can be resisted by this material, although the strains may continue to increase. Therefore, beyond $P = 100$ kips, the aluminum rod can be counted upon to resist only $P_a = A_a \sigma_{yp} = 0.5 \times 50 = 25$ kips. The remainder of the applied load must be carried by the steel tube. For $P_2 = 125$ kips, 100 kips must be carried by the steel tube. Hence $\sigma_s = 100/0.5 = 200$ ksi. At this stress level $\varepsilon_s = 200/(30 \times 10^3) = 6.67 \times 10^{-3}$ in. per inch. Therefore, the tip deflection

$$u = \varepsilon_s L = 6.67 \times 10^{-3} \times 30 = 0.200 \text{ in.}$$

Note that it is not possible to determine u from the strain in aluminum since no unique strain corresponds to the stress of 50 ksi, which is all that the aluminum rod can carry. However, in this case the elastic steel tube contains the plastic flow. Thus, the strains in both materials are the same, i.e., $\varepsilon_s = \varepsilon_a = 6.67 \times 10^{-3}$ in. per inch, see Fig. 4-29(d).

If the applied load $P_2 = 125$ kips were removed, both materials in the rod would rebound elastically. Thus if one imagines the bond between the two materials broken, the steel tube would return to its initial shape. But a permanent set (stretch) of $(6.67 - 5) \times 10^{-3} = 1.67 \times 10^{-3}$ in. per inch would occur in the aluminum rod. This incompatibility of strain cannot develop if the two materials are bonded together. Instead, residual stresses develop, which maintain the same axial deformations in both materials. In this case, the aluminum rod remains slightly compressed, and the steel tube is slightly stretched. The solution of such statically indeterminate problems is considered in greater detail in Chapter 12. The small effect due to Poisson's ratio is neglected in the above discussion.

4-18. STRESS CONCENTRATIONS

From the preceding articles of this chapter it is seen that stresses are accompanied by deformations. If such deformations take place at the same uniform rate in adjoining elements, no additional stresses, other than those given for example by Eq. 3-5, occur. However, if the uniformity of the cross-sectional area of a member is interrupted or if the force is actually applied over a very small area, a perturbation in stresses takes place because the adjoining elements must be physically continuous in a deformed state. They must stretch or contract equal amounts at the adjoining sides of all particles. These deformations result from linear and shearing deformations involving the properties of materials E, G, and v and the applied forces. Methods of obtaining this disturbed-stress distribution are beyond the scope of this text. Such problems are treated in the mathematical theory of elasticity. Even by those advanced methods only the simpler cases can be solved; the mathematical difficulties become too great for many practically significant problems.* For the group of problems that are not tractable mathematically, special experimental techniques (mainly photoelasticity, briefly discussed in Chapter 10) have been developed to determine the actual stress distribution.

Section 4-18
Stress concentrations

Here it is significant to examine qualitatively the results of more advanced investigations. For example, in Fig. 4-30(a) a short block is shown loaded by a concentrated force P. This problem could be solved

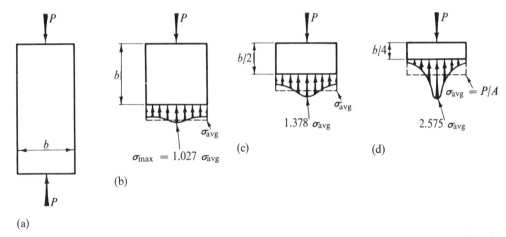

Fig. 4-30. Stress distribution near a concentrated force.

by using Eq. 3-5, i.e., $\sigma = P/A$. But is this answer really correct? Reasoning in a qualitative way, it is apparent that the strains must be maximum in the vicinity of the applied force, hence the corresponding stresses must also be maximum. That indeed is the answer given by the theory of elasticity.† The end results for normal stress distribution at various sections are shown in the adjoining stress-distribution diagrams, Figs. 4-30(b), (c), and (d). For present purposes, physical intuition is sufficient to justify these results. Note particularly the high peak of the normal stress at a section close to the applied force.‡ Also note how rapidly this peak smoothes out to a nearly uniform stress distribution at a

---

* Approximate numerical procedures formulated on the basis of finite elements or finite difference equations are now widely used for the solution of complex problems. Digital computers are indispensable in such work.

† S. Timoshenko and J. N. Goodier, *Theory of Elasticity* (2nd ed.) (New York: McGraw-Hill Book Company, 1951), p. 52. Figure 4-30 is adopted from this source.

‡ In a purely elastic material the stress is infinite right under a "concentrated" force.

*Chapter 4
Strain, constitutive
laws, and axial
deformation*

section below the top equal to the width of the bar. This illustrates the famed *St. Venant principle* of rapid dissipation of localized stresses. This principle asserts that the effect of forces or stresses applied over a small area may be treated as a statically equivalent system which, at a distance approximately equal to the width or thickness of a body, causes stress distribution which follows a simple law. Hence Eq. 3-5 is nearly true at a distance from the point of application of a concentrated force equal to the width of the member. Note also that at every level where the stress is investigated accurately, the average stress is still correctly given by Eq. 3-5. This follows since the equations of statics must always be satisfied. No matter how irregular the nature of the stress distribution at a given section through a member, an integral (or sum) of $\sigma\,dA$ over the whole area must be equal to the applied force.

Because of the great difficulty encountered in solving for the above-mentioned peak or local stresses, a convenient scheme has been developed in practice. This scheme consists simply of calculating the stress by the elementary equations (such as Eq. 3-5) and then multiplying the stress so computed by a number called the *stress-concentration factor*. In this text this number will be designated by K. The values of the stress-concentration factor depend only on the geometrical proportions of the member. These factors are available in technical literature in various tables and graphs.*

---

* R. J. Roark, *Formulas for Stress and Strain* (New York: McGraw-Hill Book Company, 1954).

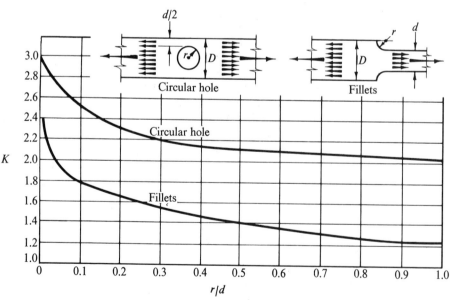

Fig. 4-31. Stress-concentration factors for flat bars in tension. (After M. M. Frocht, "Factors of Stress Concentration Photoelastically Determined," *Transactions of the American Society of Mechanical Engineers*, **57** (1935), A-67.)

Using this scheme, Eq. 3-5 may be rewritten

$$\sigma_{max} = K \frac{P}{A} \qquad (4\text{-}34)$$

From Fig. 4-30(d), at a depth below the top equal to one-quarter of the width of the member, $K = 2.575$. Hence $\sigma_{max} = 2.575\, \sigma_{av}$.

Two other particularly significant stress-concentration factors for flat axially loaded members are shown in Fig. 4-31. The corresponding factors which may be read from this graph represent a ratio of the peak of the actual stress in the net or small section of the member as shown in Fig. 4-32 to the average stress in the net section given by Eq. 3-5.

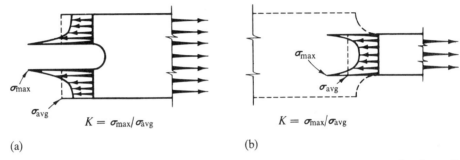

Fig. 4-32. Meaning of the stress-concentration factor K.

A considerable stress concentration also occurs at the root of threads. This depends to a large degree upon the sharpness of the cut. For an ordinary thread the stress-concentration factor is in the neighborhood of 2½. The application of Eq. 4-34 present no difficulties, providing proper graphs or tables of K are available.

EXAMPLE 4-12

Find the maximum stress due to stress concentration in the member AB in the forked end A in Example 3-3.

SOLUTION

Geometrical proportions:

$$\frac{\text{radius of the hole}}{\text{net width}} = \frac{3/16}{1/2} = 0.375$$

From Fig. 4-31:* $K \approx 2.2$ for $r/d = 0.375$

Average stress from Example 3-3: $\sigma_{av} = P/A_{net} = 11.2$ ksi

Maximum stress, Eq. 4-33: $\sigma_{max} = K\sigma_{av} = 2.2(11.2) = 24.6$ ksi

---

* Strictly speaking the stress concentration depends on the condition of the hole–whether it is empty or filled with a bolt or pin.

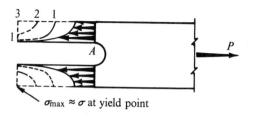

Fig. 4-33. Behavior of a flat bar of mild steel when stressed beyond the yield point.

This answer indicates that actually a large local increase in stress occurs at this hole, a fact that may be highly significant.

In considering stress-concentration factors in design, it must be remembered that their theoretical or photoelastic determination is based on the use of Hooke's law. If members are gradually stressed beyond the proportional limit of the material, these factors lose their significance. For example, consider a flat bar of mild steel, of the proportions shown in Fig. 4-33, that is subjected to a gradually increasing force P. The stress distribution will be geometrically similar to that shown in Fig. 4-32 until σ_{max} reaches the yield point of the material. However, with a further increase in the applied force, σ_{max} remains the same since a great deal of deformation can take place as the material yields. Therefore the stress at A remains virtually "frozen" at the same value. Nevertheless, for equilibrium, stresses acting over the net area must be high enough to resist P. As a result of this, the stress distribution begins to look something like that shown by line 1-1 in Fig. 4-33, then like 1-2, and finally like 1-3. Hence, for ductile materials prior to rupture, the local stress concentration is practically wiped out, and prior to necking a nearly uniform distribution of stress across the net section occurs.

The above argument is not quite as true for materials less ductile than mild steel. Nevertheless, the tendency is in that direction unless the material is unusually brittle, like glass or some alloy steels. The argument presented applies to situations where the force is gradually applied or is static in character. It is not applicable for fluctuating loads such as found in some machine elements. There the working stress level that is actually reached locally determines the fatigue behavior of the member. The maximum permissible stress is set from an *S-N* diagram (Art. 3-8). Failure of most machine parts can be traced to progressive cracking originating at points of high local stress. In machine design, then, stress concentrations are of paramount importance, although some machine designers feel that the theoretical concentrations are somewhat high. Apparently some tendency is present to smooth out the stress peaks even in members subjected to dynamic loads.

From the above discussion and accompanying charts it should be apparent why an experienced machine designer tries to "streamline" the junctures and transitions of elements that make up a structure.

PROBLEMS FOR SOLUTION

4-1. In two-dimensional problems the three strain components are ε_x, ε_y, and γ_{xy}. However, as may be seen from Eqs. 4-3 and 4-5, these three quantities are functions of only two displacement components u and v. Therefore the strains cannot be independent of one another and a relationship must exist between them. Show that such a relation is

$$\frac{\partial^2 \varepsilon_x}{\partial y^2} + \frac{\partial^2 \varepsilon_y}{\partial x^2} = \frac{\partial^2 \gamma_{xy}}{\partial x \partial y}$$

This is called the *condition of compatibility*; it assures that the displacements are single-valued.

4-2. The condition of compatibility for a two-dimensional case expressed in the above problem is in terms of strains. Using Hooke's law for the plane stress (Eq. 4-11 with $\sigma_z = 0$), show that the same condition in terms of the components of stress is

$$\nabla^2(\sigma_x + \sigma_y) = 0$$

where

$$\nabla^2 = \frac{\partial^2}{\partial x^2} + \frac{\partial^2}{\partial y^2}$$

To establish this relation, make use of the equations of equilibrium, Eq. 3-3, and assume that the body forces $X = Y = 0$.

4-3. Sometimes it is desirable to write the generalized Hooke's law given by Eq. 4-11 in an inverse form as

$$\sigma_x = \lambda e + 2\mu\varepsilon_x \qquad \sigma_y = \lambda e + 2\mu\varepsilon_y$$

$$\sigma_z = \lambda e + 2\mu\varepsilon_z$$

where λ and μ are the Lamé constants and $e = \varepsilon_x + \varepsilon_y + \varepsilon_z$.

By solving the first three parts of Eq. 4-11 simultaneously and using Eq. 4-12, show that in terms of the engineering constants

$$\lambda = \frac{\nu E}{(1+\nu)(1-2\nu)} \qquad \text{and} \qquad \mu = G$$

4-4. Consider an axisymmetrically strained body such as a circular tube under internal pressure. Assuming that in such a body only radial displacements can occur, show that in polar coordinates the radial and the tangential strains are, respectively,

$$\varepsilon_r = du/dr \qquad \text{and} \qquad \varepsilon_\theta = u/r$$

where u is the radial displacement.

4-5. A piece of 2-in.-by-10-in.-by-½-in. steel plate is subjected to uniformly distributed stresses along its edges (see figure). (a) If $P_x = 20$ kips and $P_y = 40$ kips, what change in thickness occurs due to the application of these forces? (b) To cause the same change in thickness as in (a) by P_x alone, what must be its magnitude? Let $E = 30 \times 10^6$ psi and $\nu = 0.25$.

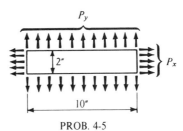

PROB. 4-5

4-6. An 8-ft-by-12-ft, ¼-in.-thick steel plate panel is subjected to a uniformly distributed loading p_x in the x direction, and p_y in the y direction (see figure). If the total change in length from the unstressed condition in the x direction is $+0.0768$ in., and in the y direction $+0.0864$ in., what are p_x and p_y in kips per foot? Let $E = 30 \times 10^6$ psi, and $G = 12 \times 10^6$ psi. *Ans.* 91.2 k per foot, 76.8 k per foot.

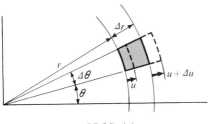

PROB. 4-4

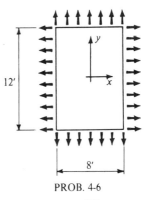

PROB. 4-6

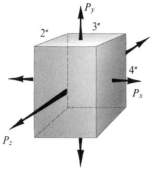

PROB. 4-7

4-7. A rectangular aluminum alloy block has the dimensions shown in the figure. The resultants of uniformly distributed stresses are $P_x = 40$ kips, $P_y = 48$ kips, and $P_z = 36$ kips. Determine the magnitude of a single system of tensile forces acting only in the x direction which would cause the same deformation in the x direction as the initial forces. Let $E = 10 \times 10^6$ psi, and $\nu = 0.25$.

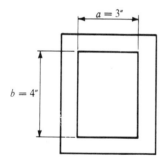

PROB. 4-8

4-8. A 3 in.-by-4-in., 0.040-in.-thick, rectangular brass diaphragm is stretched between a rigid frame made of Invar as shown in the figure. If a drop in temperature of 200°F occurs, determine the resulting normal stresses in the diaphragm. Assume that for the brass $E = 16 \times 10^6$ psi, $G = 6 \times 10^6$ psi, $\alpha = 11 \times 10^{-6}$ in. per inch per degree Fahrenheit, whereas for Invar the coefficient of thermal expansion is zero over the range of the temperature considered.

4-9. Rework the above problem if the Invar frame is replaced by a rigid steel frame.

Assume that for steel $E = 29 \times 10^6$ psi, and $\alpha = 6 \times 10^{-6}$ in. per inch per degree Fahrenheit.

4-10. Verify Eq. 4-21 using Eqs. 4-11 and 4-20.

4-11. Show that the strain energy density in terms of strains is

$$U_0 = \frac{\lambda e^2}{2} + \mu(\varepsilon_x^2 + \varepsilon_y^2 + \varepsilon_z^2)$$
$$+ \frac{\mu}{2}(\gamma_{xy}^2 + \gamma_{yz}^2 + \gamma_{zx}^2)$$

where the Lamé constants λ and μ are as defined in Prob. 4-3.

4-12. A 2-in. square alloy-steel bar 30 in. long is a part of a machine and must resist an axial energy load of 900 in-lb. What must the proportional limit of the steel be to safely resist the energy load elastically with a factor of safety of 4? What is the modulus of resilience for such a steel? Let $E = 30 \times 10^6$ psi.

4-13. A 40-in.-long steel rod of 2-in. diameter is subjected to an axial energy load of 36 in-lb that causes a tensile stress in the rod. (a) Determine the maximum tensile stress. $E = 30 \times 10^6$ psi. (b) If the same rod is machined down to a 1-in. diameter in the middle half of the bar, i.e., for a distance of 20 in., will the maximum stress increase or decrease and by how much?

4-14. A weight W falls freely along a rod until it strikes the nut at the bottom as illustrated. (a) Show that the dynamic or impact force P_{dyn} delivered to the rod is

$$P_{\mathrm{dyn}} = W(1 + \sqrt{1 + 2h/\Delta_{st}})$$

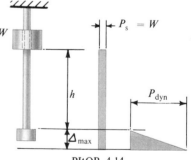

PROB. 4-14

where h is the distance of the free fall of the weight, and Δ_{st} is the static deflection of the rod caused by application of the weight W. In deriving this relation assume the following:

1. Materials behave elastically, and no dissipation of energy takes place at the point of impact or at the supports owing to local inelastic deformation of materials.

2. The inertia of a system resisting an impact may be neglected.

3. The deflection of a system is directly proportional to the magnitude of the applied force whether a force is dynamically or statically applied. (*Hint:* Equate the energy lost by the weight W to the strain energy in the rod acting as a spring; see figure.)

(b) Apply the equation found in (a) to the case where $W = 6$ lb, $h = 20$ in., and a $\tfrac{1}{2}$-in.-diameter steel rod is 30 in. long. Calculate the maximum stress in the rod caused by the falling weight. Let $E = 30 \times 10^6$ psi. *Ans.* 35.6 ksi.

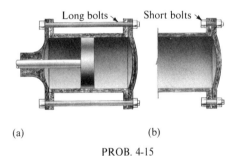

PROB. 4-15

4-15. Why are long bolts rather than short ones used in pneumatic cylinders and jack hammers (see figure)? Is it desirable to use bolts with upset* ends? Develop some analytical justification for your answers.

4-16. Suppose that a series of tension tests is performed on several identical specimens of Maxwell material. In each experiment the strain rate is held constant. (a) Sketch a family of stress-strain curves which would result from this series of tests. (b) In each case, what is the

---

* The diameter of the rod at the ends is enlarged by forging in order to maintain the nominal rod diameter at the root of threads.

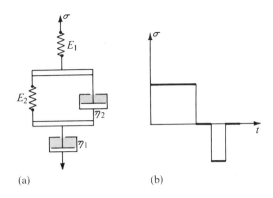

PROB. 4-17

elastic modulus E at $t = 0$? (c) Discuss the implication of the results.

4-17. A Standard Solid for a viscoelastic material is obtained by placing in series a Maxwell and a Kelvin unit as shown in figure (a). This model is capable of representing most of the essential features of viscoelastic behavior. Sketch the strain-time relationship which results from the stress-time input shown in figure (b).

4-18. In one of the California oil fields a very long steel drill pipe got stuck in hard clay (see figure). It was necessary to determine at what depth this occurred. The engineer on the job ordered the pipe subjected to a large upward tensile force. As a result of this operation the pipe came up elastically 2 ft. At the same time the pipe elongated 0.0014 in. in an 8-in. gage length. Approximately where was the pipe stuck? Assume that the cross-sectional area of

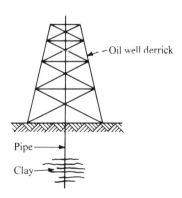

PROB. 4-18

the pipe was constant and that the media surrounding the pipe hindered elastic deformation of the pipe in a static test very little. Ans. 11,400 ft.

4-19. A stainless steel rod 30 ft long used in a control mechanism must transmit a tensile force of 980 lb without stretching more than 0.2 in. or exceeding an allowable stress of 20,000 psi. What must the diameter of the rod be? Give the answer to the nearest eighth of an inch. $E = 28 \times 10^6$ psi. Ans. $\frac{3}{8}$ in.

4-20. A solid cylinder of 2-in. diameter and 36-in. length is subjected to a tensile force of 30 kips. One part of this cylinder, L_1 long, is of steel; the other part, fastened to steel, is aluminum and is L_2 long. (a) Determine the lengths L_1 and L_2 so that the two materials elongate an equal amount. (b) What is the total elongation of the cylinder? $E_s = 30 \times 10^6$ psi; $E_a = 10 \times 10^6$ psi.

4-21. A round steel bar having a cross section of 0.5 in.² is attached at the top and is subjected to three axial forces, as shown in the figure. Find the deflection of the free end caused by these forces. Plot the axial force and the axial deflection diagrams. Let $E = 30 \times 10^6$ psi. Ans. $u_{\min} = 0$, $u_{\max} = 0.040$ in.

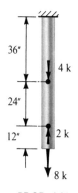

PROB. 4-21

4-22. Using Eq. 4-32 and singularity functions determine the general expression for the deflection u in the above problem, and calculate the maximum displacement u.

4-23. A bar of steel and a bar of aluminum have the dimensions shown in the figure.

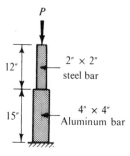

PROB. 4-23

Calculate the magnitude of the force P that will cause the total length of the two bars to decrease 0.010 in. Assume that the normal stress distribution over all cross sections of both bars is uniform and that the bars are prevented from buckling sidewise. Plot the axial deflection diagram. Let $E_s = 30 \times 10^6$ psi, and $E_a = 10 \times 10^6$ psi. Ans. 51.6 k.

4-24. A uniform timber pile which has been driven to a depth L in clay carries an applied load of F at the top. This load is resisted entirely by friction f along the pile, which varies in the parabolic manner shown in the figure. (a) Determine the total shortening of

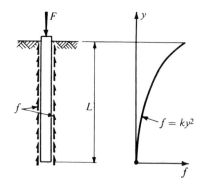

PROB. 4-24

the pile in terms of F, L, A and E. (b) If $P = 96$ kips, $L = 40$ ft, $A = 100$ in.², and $E = 1.5 \times 10^6$ psi, how much does such a pile shorten? (*Hint:* From the equilibrium requirement, first determine the constant k.) Ans. (a) $FL/(4AE)$.

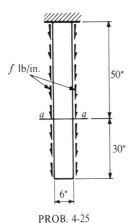

PROB. 4-25

4-25. A $\tfrac{1}{2}$-in.-thick steel plate 6 in. wide and 80 in. long is subjected to a set of uniformly distributed frictional forces along its two edges as shown in the figure. If, due to these forces, the total change in the transverse 6-in. dimension at level a-a is 600×10^{-6} in., determine the total elongation of the bar in the longitudinal direction. Let $E = 30 \times 10^6$ psi and $\nu = 0.25$. *Ans.* 0.0427 in.

4-26. Two bars are to be cut from a 1-in.-thick metal plate so that both bars have a constant thickness of 1 in. Bar A is to have a constant width of 2 in. throughout its entire length. Bar B is to be 3 in. wide at the top and 1 in. wide at the bottom. Each bar is to be subjected to the same load P. Determine the ratio L_A/L_B so

that both bars will stretch the same amount. Neglect the weight of the bar. *Ans.* 1.10.

4-27. The dimensions of a frustrum of a right circular cone supported at the large end on a rigid base are shown in the figure. Determine the deflection of the top due to the weight of the body. The unit weight of material is γ; the elastic modulus is E. (*Hint:* Consider the origin of the coordinate axes at the vertex of the extended cone.) *Ans.* $160\,\gamma/E$.

PROB. 4-27

4-28. If the cone of the previous problem is turned upside down, i.e., stood on its small end, what will the deflection of the top due to its own weight be?

4-29. Find the total elongation Δ of a slender elastic bar of constant cross-sectional area A such as shown in the figure if it is rotated in a horizontal plane with an angular velocity of ω radians per second. The unit weight of the material is γ. Neglect the small amount of extra material by the pin. (*Hint:* First find the stress at a section a distance r from the pin by integrating the effect of the inertial forces between r and L, see Example 3-7; then apply Eq. 4-29. Alternatively, use Eq. 4-32 with p_x as the body force.) *Ans.* $2\gamma\omega^2 L^3/3gE$.

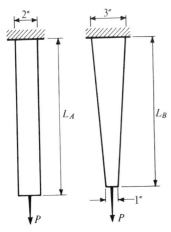

PROB. 4-26

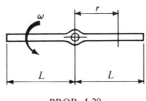

PROB. 4-29

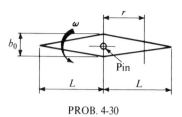

PROB. 4-30

4-30. A bar of constant thickness t is rhomb-shaped in plan as shown in the figure. Determine the total elongation of this bar caused by rotating it in a horizontal plane with an angular velocity ω around the pin. Other data are the same as in the previous problem.

4-31. Two wires are connected to a rigid bar as shown in the figure. The wire on the left is of steel having $A = 0.10$ in.², and $E = 30 \times 10^6$ psi. The aluminum-alloy wire on the right has $A = 0.20$ in.², and $E = 10 \times 10^6$ psi. If a weight $W = 2,000$ lb is applied as shown, how much will it deflect due to the stretch in the wires?

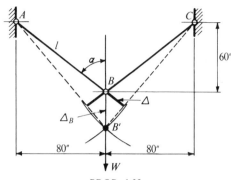

PROB. 4-32

4-33. A jib crane has the dimensions shown in the figure. Determine the minimum diameter d of the rod AB so that the vertical deflection of point B would not exceed 0.20 in. and the stress in it would be less than 20 ksi when the force $F = 10$ kips is applied. Assume that the axial deformation of BC is negligible and that the deflection is due entirely to the stretch of the rod AB. (See hint for the previous problem.)

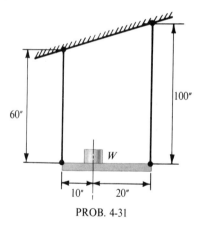

PROB. 4-31

4-32. Two identical wires ($A = 0.10$ in.², $E = 10 \times 10^6$ psi) are arranged as shown in the figure. Determine the deflection of joint B caused by the application of the load $W = 3$ kips. (*Hint:* (a) Calculate elongation Δ of each wire. (b) With the centers at A and C and using $(l + \Delta)$ as radii, locate point B' which is the deflected position of B. (c) Since the deformations are small, assume that $\Delta \approx \Delta_B \cos \alpha$. The deflections are greatly exaggerated on the diagram.)

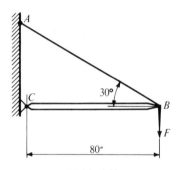

PROB. 4-33

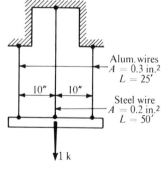

PROB. 4-34

4-34. If a load of 1 kip is applied to a rigid bar suspended by three wires as shown in the figure, what force will be resisted by each wire? The outside wires are aluminum ($E = 10^7$ psi). The inside wire is steel ($E = 30 \times 10^6$ psi). Initially there is no slack in the wire. (*Hint:* This is a statically indeterminate problem. A supplementary equation based on deformation must be formulated as in Example 4-10.) *Ans.* Each wire carries $\frac{1}{3}$ kip.

4-35. An elastic bar of constant cross section is built in at both ends as shown in the figure. Using Eq. 4-32 and singularity functions, determine the reactions and the axial force distribution in the bar. Plot the axial force and the axial deformation diagrams. Let $(a + b) = L$. (*Note:* This is a statically indeterminate problem.)

PROB. 4-35

4-36. What will be the deflection of the free end of the rod in Example 4-9 (Fig. 4-28) if, instead of Hooke's law, the stress-strain relationship is $\sigma = K\varepsilon^{1/n}$, where n is a number dependent on the properties of the material?

4-37. The bar shown in the figure has the stress-strain relationship

$$\sigma_x(x, y) = E_0[2 - (y/h)^2]\varepsilon_x(x, y)$$

Using equilibrium and the assumptions of geometry of deformation, (a) calculate and plot the stress distribution in the bar in terms of the load P and the cross-sectional dimensions, (b) calculate the elongation of the bar.

4-38. A rod of two different cross-sectional areas is made of soft copper and is subjected to a tensile load as shown in the figure. (a) Determine the elongation of the rod caused by the application of a force $P = 5$ kips. Assume that the axial stress-strain relationship is

$$\varepsilon = \sigma/16{,}000 + (\sigma/165)^3$$

where σ is in ksi. (b) Rework (a) assuming that the copper is a linearly elastic material with E equal to the tangent modulus at the

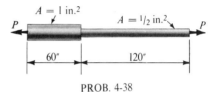

PROB. 4-38

origin of the σ-ε relationship in (a). (c) If the material initially behaves as in (a), what will be the residual elongation upon the removal of the force P? *Ans.* 0.028 in.

4-39. A flat bar 1 in. thick is originally 4 in. wide and 60 in. long. Then the middle half of the length of the bar is machined down from both sides to make the central portion of the bar 2 in. wide. If at this step the fillets are made so that $r/d = \frac{1}{2}$ (see Fig. 4-31), what axial force may be applied to the bar without exceeding the yield stress of 50 ksi, and what will be the total elongation of the bar? Let $E = 29 \times 10^6$ psi.

4-40. A long slot is cut out from a 1-in.-by-6-in. steel bar 10 ft long as shown in the figure. (a)

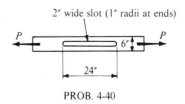

PROB. 4-40

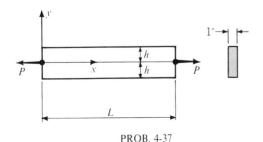

PROB. 4-37

Find the maximum stress if an axial force $P = 50$ kips is applied to the bar. Assume that the upper curve in Fig. 4-31 is applicable. (b) For the same case determine the total elongation of the rod. Neglect local effects of stress concentrations and assume that the reduced cross-sectional area extends for 2 ft. (c) Estimate the elongation of the same rod if $P = 160$ kips. Assume that steel yields 0.020 in. per inch at a stress of 40 ksi. (d) On removal of the load in (c), what is the residual deflection? Let $E = 30 \times 10^6$ psi. *Ans.* (a) 28.7 ksi; (b) 0.0367 in.; (c) 0.56 in.; (d) 0.448 in.

4-41. The bar shown in the figure is cut from a 1-in.-thick piece of steel. At the changes in section, approximate stress concentration

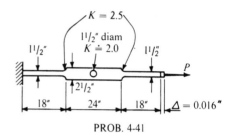

PROB. 4-41

factors are as indicated. A force P is applied producing a total change of length in the bar of 0.016 in. Determine the maximum stress in the bar caused by this force. Neglect the effect of the hole and the stress concentrations on the axial deformation. Let $E = 30 \times 10^6$ psi. *Ans.* $\sigma_{max} = 28,400$ psi.

Torsion 5

5-1. INTRODUCTION

In the preceding chapters, beside the general concepts of the mechanics of solid deformable bodies, the behavior of axially loaded rods was investigated in detail. By the application of the method of sections and by the assumption of equal strains in longitudinal fibers, a formula for stress in an axially loaded rod was developed. Then an expression was established for obtaining the axial deformation of members. In this chapter, similar relations for externally statically determinate members subjected only to torque about their longitudinal axes will be established. The investigation will be confined to the effect of a single type of action, i.e., of a torque causing twist or torsion in a member. Members subjected simultaneously to torque and bending, frequently occurring in practice, will be treated in Chapter 10. Statically indeterminate cases are discussed in Chapter 12.

A major part of this chapter is devoted to the treatment of members having circular and tubular cross-sectional areas. Solid noncircular sections are only briefly discussed. In practice, members that transmit torque, such as shafts of motors, torque tubes of power equipment, etc., are predominantly

Chapter 5
Torsion

circular or tubular in cross section. Thus, many of the important applications fall within the scope of the formulas developed.

5-2. APPLICATION OF METHOD OF SECTIONS

In analyzing members subjected to torque, the basic approach outlined in Art. 1-3 is followed. First, the system as a whole is examined for equilibrium, and then the method of sections is applied by passing a cutting plane perpendicular to the axis of the member. Everything to either side of a cut is then removed, and the internal or resisting torque necessary to maintain equilibrium of the isolated part is determined. For finding this internal torque in statically determinate members, only one equation of statics, $\Sigma M_x = 0$, where the x axis is directed along the member, is required. By applying this equation to an isolated part of a shaft, one finds the internal resisting torque developed within the member necessary to balance the externally applied torques. The external and internal torques are numerically equal but act in opposite directions.

In this chapter, shafts will be assumed "weightless" or supported at frequent enough intervals to make the effect of bending negligible. Axial forces that may also act simultaneously on the member are excluded for the present.

EXAMPLE 5-1

Find the internal torque at section a-a for the shaft shown in Fig. 5-1(a) and acted upon by the three torques indicated.

SOLUTION

The 300 in-lb torque at C is balanced by the two torques of 200 in-lb and 100 in-lb at A and B, respectively. Therefore, the body as a whole is in equilibrium. Next, by passing a cutting plane a-a perpendicular to the axis of the rod anywhere between A and B, a free body of a part of the

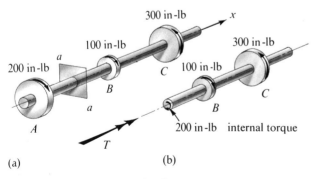

Fig. 5-1

shaft, shown in Fig. 5-1(b), is obtained. Whereupon, from $\Sigma M_x = 0$ or

externally applied torque = internal torque

Section 5-3
Basic assumptions

the conclusion is reached that the internal or resisting torque developed in the shaft between *A* and *B* is 200 in-lb. Using the right-hand screw rule, one may represent this torque acting on the section by the double-headed vector as shown. Similar considerations lead to the conclusion that the internal torque resisted by the shaft between *B* and *C* is 300 in-lb.

It may be seen intuitively that for a member of constant cross-sectional area the maximum internal torque causes the maximum stress and imposes the most severe condition on the material. Hence, in investigating a torsion member, several sections may have to be examined to determine the largest internal torque. A section where the largest internal torque is developed is the *critical section*. In Example 5-1 the critical section is anywhere between points *B* and *C*. If the torsion member varies in size, it is more difficult to decide where the material is critically stressed. Several sections may have to be investigated and stresses computed to determine the critical section. These situations are analogous to the case of an axially loaded rod, and means must be developed to determine stresses as a function of the internal torque and the size of the member. In the next several articles the necessary formulas are derived.

Members subjected to torque are very widely used as rotating shafts for transmitting power. For future reference, a formula will be established for the conversion of horsepower into torque acting through the shaft. By definition, 1 hp does work of 550 ft-lb per second, or 550(12)60 in-lb per minute. Likewise, it will be recalled from dynamics that power is equal to the torque multiplied by the angle, measured in radians, through which the torque rotates per unit of time. For a shaft rotating at *N* rpm, the angle is $2\pi N$ radians per minute. Hence, if a shaft were transmitting a constant torque *T* measured in inch-pounds, it would do $2\pi NT$ in-lb of work per minute. Equating this to the horsepower supplied

hp (550)12(60) [in-lb per minute] = $2\pi NT$[in-lb per minute]

$$T = 63{,}000 \text{ hp}/N \quad \text{in-lb} \tag{5-1}$$

where *N* is the number of revolutions of the shaft transmitting the horsepower (hp) per minute. This equation converts the horsepower delivered to the shaft into a constant torque acting through it as the power is applied.

5-3. BASIC ASSUMPTIONS

To establish a relation between the internal torque and the stresses it sets up in members with circular cross sections and round tubes, it is necessary to make several assumptions, the validity of which will be justified further on. These, in addition to homogeneity of material, are as follows:

1. A plane section of material perpendicular to the axis of a circular member remains plane after the torques are applied, i.e., no warpage or distortion of parallel planes normal to the axis of a member takes place.*

2. In a circular member subjected to torque, shearing strains γ vary linearly from the central axis. This assumption is illustrated in Fig. 5-2; it means that an imaginary plane such as AO_1O_3C moves to $A'O_1O_3C$ when the torque is applied. Alternatively, if an imaginary radius O_3C is considered fixed in direction, similar radii initially at O_2B and O_1A rotate to the respective new positions O_2B' and O_1A'. These radii remain straight.

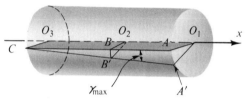

Fig. 5-2. Variation of strain in a circular member subjected to torque.

It must be emphasized that these assumptions hold only for solid or hollow circular members. For this class of members these assumptions work so well that they apply beyond the limit of the elastic behavior of a material. These assumptions will be used again in Art. 5-8, where stress distribution beyond the proportional limit is discussed. However, if attention is confined to the linearly elastic case, Hooke's law applies.

3. Thus it follows that shearing stress is proportional to shearing strain.

In the interior of a member it is difficult to justify the first two assumptions directly. However, after deriving stress and deformation formulas based on them, unquestionable agreement is found between measured and computed quantities. Moreover, their validity may be rigorously demonstrated by the methods of the theory of elasticity, which is based on the requirements of strain compatibility and the generalized Hooke's law.

5-4. THE TORSION FORMULA

In the elastic case, since stress is proportional to strain and the latter varies linearly from the center, stresses vary linearly from the central axis of a circular member. The stresses induced by the assumed defor-

---

* Actually it is also implied that parallel planes perpendicular to the axis remain a constant distance apart. This is not true if deformations are large. However, since the usual deformations are very small, stresses not considered here are negligible. For details see S. Timoshenko, *Strength of Materials*, (3rd. ed.) Part II, Advanced Theory and Problems, Chapter VI (Princeton, N.J.: D. Van Nostrand Co., Inc., 1956).

mations are shearing stresses and lie in the plane parallel to the section taken normal to the axis of a rod. The variation of shearing stress is illustrated in Fig. 5-3. Unlike the case in an axially loaded rod, this stress is not of uniform intensity. The maximum shearing stress occurs at points most remote from the center O and is designated τ_{max}. These points, such as point C in Fig. 5-3, lie at the periphery of a section at a distance c from the center. And, by virtue of a linear stress variation, at any arbitrary point a distance ρ from O, the shearing stress is $(\rho/c)\tau_{max}$.

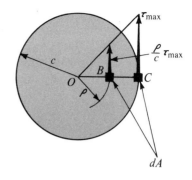

Fig. 5-3. Variation of stress in a circular member in the elastic range.

Once the stress distribution at a section is established, the resistance to torque in terms of stress can be expressed. The resistance to the torque so developed must be equivalent to the internal torque. Hence an equality may be formulated thus:

$$\int_A \underbrace{\underbrace{\frac{\rho}{c}\tau_{max}}_{\text{(stress)}} \underbrace{dA}_{\text{(area)}}}_{\text{(force)}} \underbrace{\rho}_{\text{(arm)}} = T$$

(torque)

where the integral sums up all torques developed on the cut by the infinitesimal forces acting at a distance ρ from the axis, O in Fig. 5-3, over the whole area A of the cross section, and where T is the resisting torque.

At any given section τ_{max} and c are constant, hence the above relation may be written

$$\frac{\tau_{max}}{c} \int_A \rho^2 \, dA = T \tag{5-2}$$

However, $\int \rho^2 \, dA$, the polar moment of inertia of a cross-sectional area, is a constant for a particular cross-sectional area. In this text it will be designated by J. For a circular section, $dA = 2\pi\rho \, d\rho$, where $2\pi\rho$ is the circumference of an annulus with a radius ρ of width $d\rho$. Hence

$$J = \int_A \rho^2 \, dA = \int_0^c 2\pi\rho^3 \, d\rho = \frac{\pi c^4}{2} = \frac{\pi d^4}{32} \tag{5-3}$$

where d is the diameter of a solid circular shaft. If c or d is measured in inches, J has the units of inches⁴.

By using the symbol J for the polar moment of inertia of a circular area, Eq. 5-2 may be written more compactly:

$$\tau_{max} = Tc/J \tag{5-4}$$

Chapter 5
Torsion

This is the famed *torsion formula** for circular shafts, which expresses the maximum shearing stress in terms of the resisting torque and the dimensions of a member. In applying this formula the internal torque T is usually expressed in inch-pounds, c in inches, and J in inches⁴. Such usage makes the units of the torsion shearing stress

$$\frac{[\text{in-lb}][\text{in.}]}{[\text{in.}^4]} = [\text{lb per in.}^2] \text{ or psi}$$

A more general relation than Eq. 5-4 for a shearing stress τ at any point a distance ρ from the center of a section is

$$\tau = \frac{\rho}{c}\tau_{max} = \frac{T\rho}{J} \qquad (5\text{-}5)$$

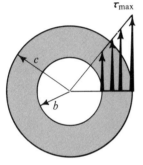

Equations 5-4 and 5-5 are applicable with equal rigor to circular tubes since the same assumptions as used in the above derivation apply. It is necessary, however, to modify J. For a tube, as may be seen from Fig. 5-4, the limits of integration for Eq. 5-3 extend from b to c. Hence for a circular tube

$$J = \int_A \rho^2\, dA = \int_b^c 2\pi\rho^3\, d\rho = \frac{\pi c^4}{2} - \frac{\pi b^4}{2} \qquad (5\text{-}6)$$

Fig. 5-4. Variation of stress in a hollow circular member in the elastic range.

or stated otherwise: J for a circular tube equals J for a solid shaft, using the outer diameter, minus J for a shaft having the inner diameter.

For thin tubes, if b is nearly equal to c, and $c - b = t$, the thickness of the tube, J reduces to a simple approximate expression:

$$J \approx 2\pi c^3 t \qquad (5\text{-}6a)$$

which is sufficiently accurate in many applications.

The basic concepts used in deriving the torsion formula for circular members may be recapitulated as follows:

1. *Equilibrium requirements* are used to determine the internal or resisting torque.
2. *Deformation* is assumed so that shearing strain varies linearly from the axis of the shaft.
3. *Material properties* in the form of Hooke's law are used to relate the assumed strain variation to stress.

The same concepts were used to solve Example 4-10 and were implied in the stress and deformation formulas for all axial-load problems.

---

* It was derived by C. A. Coulomb, a French engineer, in about 1775 in connection with his work on electric instruments. His name has been immortalized by its use for a practical unit of quantity in electricity.

They will be relied on again in treating the inelastic behavior of circular shafts subjected to torque. For the latter purpose only item 3 above must be modified.

*Section 5-5
Remarks on the torsion formula*

5-5. REMARKS ON THE TORSION FORMULA

So far the shearing stresses given by Eqs. 5-4 and 5-5 have been considered as acting only in the plane of a cut perpendicular to the axis of the shaft. There indeed they are acting to form a couple resisting the externally applied torques. However, to understand the problem further, an infinitesimal cylindrical element, shown in Fig. 5-5(b), is isolated from the member of Fig. 5-5(a). The shearing stresses acting in the planes perpendicular to the axis of the rod are known from Eq. 5-5. Their directions coincide with the direction of the internal resisting torque. (This should be clearly visualized by the reader.) On an adjoining parallel plane of a disc-like element these stresses act in opposite directions. However, they cannot exist alone, as was shown in Art. 3-3. Numerically equal shearing stresses must act on the axial planes (such as the planes *aef* and *bcg* in Fig. 5-5(b)) to fulfil the requirements of static equilibrium for an element.*

Shearing stresses acting in the axial planes follow the same variation in intensity as do the shearing stresses in the planes perpendicular to the axis of the rod. This variation of shearing stresses on the mutually perpendicular planes is shown in Fig. 5-5(c), where a portion of the shaft has been removed for the purposes of illustration.

---

* Note that maximum shearing stresses as shown diagrammatically in Fig. 5-5(a) actually act on planes perpendicular to the axis of the rod and on planes passing through the axis of the rod. The representation shown is purely schematic. The free surface of a shaft is free of all stresses.

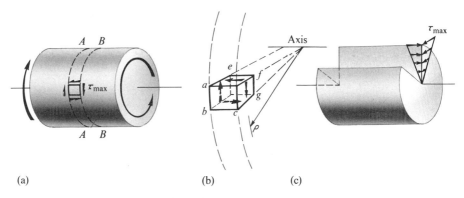

Fig. 5-5. Shearing stresses exist on mutually perpendicular planes in a shaft subjected to torque.

**Chapter 5
Torsion**

At this point it is instructive to recall the complete representation of the stress tensor in Eq. 3-1. In this problem the results of Eqs. 5-4 or 5-5 can be stated as*

$$\begin{pmatrix} 0 & \tau & 0 \\ \tau & 0 & 0 \\ 0 & 0 & 0 \end{pmatrix} \qquad (5\text{-}7)$$

Since only two elements of the stress tensor exist, no ambiguity is caused by having no subscripts on τ in Eq. 5-7. In general elasticity-problem subscripts such as x and θ are attached to τ to designate the axial and the tangential directions. Just as for a Cartesian element, for a cylindrical element $\tau_{x\theta} = \tau_{\theta x}$.

In an isotropic material it makes little difference in which direction the shearing stresses act. However, not all materials used in engineering applications are isotropic. For example, wood exhibits drastically different properties of strength in different directions. The shearing strength of wood on planes parallel to the grain is much less than that on planes perpendicular to the grain. Hence, although equal intensities of shearing stress exist on mutually perpendicular planes, wooden shafts of inadequate size fail longitudinally along axial planes. Wooden shafts are occasionally used in the process industries.

EXAMPLE 5-2

Find the maximum torsional shearing stress in the shaft AC shown in Fig. 5-1(a). Assume the shaft from A to C to be of $\frac{1}{2}$-in. diameter.

SOLUTION

From Example 5-1 the maximum internal torque resisted by this shaft is known to be 300 in-lb. Hence, $T = 300$ in-lb, and $c = d/2 = 0.25$ in.

From Eq. 5-3: $\quad J = \dfrac{\pi d^4}{32} = \dfrac{\pi (0.5)^4}{32} = 0.00614$ in.^{4}

From Eq. 5-4:

$$\tau_{\max} = \frac{Tc}{J} = \frac{(300)(0.25)}{0.00614} = 12{,}200 \text{ psi}$$

This maximum shearing stress at 0.25 in. from the axis of the rod acts in the plane of a cut perpendicular to the axis of the rod and along the longitudinal planes passing through the axis of the rod (Fig. 5-5(c)).

EXAMPLE 5-3

Consider a long tube of 1-in. outside diameter d_o and of 0.9-in. inside diameter d_i twisted about its longitudinal axis with a torque T of 400

---

* Some readers may find it interesting to look over Example 10-2 at this time.

in-lb. Determine the shearing stresses at the outside and the inside of the tube, Fig. 5-6.

Section 5-5
Remarks on the torsion formula

SOLUTION

From Eq. 5-6: $J = \dfrac{\pi(c^4 - b^4)}{2} = \dfrac{\pi(d_o^4 - d_i^4)}{32}$

$= \dfrac{\pi(1^4 - 0.9^4)}{32} = 0.0337 \text{ in.}^4$

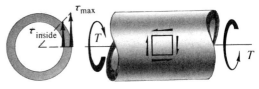

Fig. 5-6

From Eq. 5-4:
$$\tau_{max} = \dfrac{Tc}{J} = \dfrac{(400)(\frac{1}{2})}{0.0337} = 5{,}930 \text{ psi}$$

From Eq. 5-5:
$$\tau_{inside} = \dfrac{T\rho}{J} = \dfrac{400(0.9/2)}{0.0337} = 5{,}330 \text{ psi}$$

Since no material works at a low stress, it is important to note that a tube requires less material than a solid shaft to transmit a given torque at the same stress. By making the wall thickness small and the diameter large, nearly uniform shearing stress τ is obtained in the wall. This fact makes thin tubes suitable for experiments where a uniform "field" of pure shearing stress is wanted (Art. 4-15). To avoid local crimping or buckling, the wall thickness, however, cannot be excessively thin.

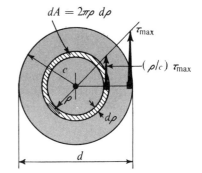

Fig. 5-7

EXAMPLE 5-4

Find the energy absorbed by an elastic circular rod subjected to a constant torque in terms of the maximum shearing stress and the volume of the material, Fig. 5-7.

SOLUTION

The shearing stress acting on an element at a distance ρ from the center of the cross section is $\tau_{max}\rho/c$. Then, using Eq. 4-18 with $\tau_{xy} = \tau$, $\gamma_{xy} = \tau/G$, and integrating over the volume V of the rod L in. long, one obtains

$$U = \int_V \dfrac{\tau^2}{2G} dV = \int_V \dfrac{\tau_{max}^2 \rho^2}{2Gc^2} 2\pi \rho \, d\rho \, L$$

$$= \dfrac{\tau_{max}^2}{2G} \dfrac{2\pi L}{c^2} \int_0^c \rho^3 \, d\rho = \dfrac{\tau_{max}^2}{2G} \dfrac{2\pi L}{c^2} \dfrac{c^4}{4}$$

$$= \dfrac{\tau_{max}^2}{2G} (\tfrac{1}{2} \text{ vol})$$

If instead of the solid shaft a thin-walled tube were used, then

$$U = \frac{\tau_{max}^2}{2G} \text{ (vol)}$$

For the same level of maximum stress, the uniformly stressed material absorbs energy more efficiently.

5-6. DESIGN OF CIRCULAR MEMBERS IN TORSION

In designing members for strength, allowable shearing stresses must be selected. These depend on the information available from experiments and on the intended application. Accurate information on the capacity of materials to resist shearing stresses comes from tests on thin-walled tubes. Solid shafting is employed in routine tests. Moreover, as torsion members are so often used in power equipment, many fatigue experiments are done. Characteristically, the shearing stress that a material can withstand is lower than the normal stress. The ASME (American Society of Mechanical Engineers) code of recommended practice for transmission shafting gives an allowable value in shearing stress of 8,000 psi for ordinary steel.* In practical designs, suddenly applied and shock loads warrant special considerations.

Once the torque to be transmitted by the shaft is known and the maximum shearing stress is selected, the proportions of the member become fixed. Thus from Eq. 5-4

$$\frac{J}{c} = \frac{T}{\tau_{max}} \tag{5-8}$$

where J/c is the *parameter* on which the elastic strength of a shaft depends. For an axially loaded rod such a parameter is the cross-sectional area of a member. For a solid shaft, $J/c = \pi c^3/2$, where c is the outside radius. By using this expression and Eq. 5-8 the required radius of a shaft may be determined. For a hollow shaft, a number of tubes can provide the same numerical value of J/c, so the problem has an infinite number of possible solutions.

EXAMPLE 5-5

Select a solid shaft for a 10-hp motor operating at 1,800 rpm. The maximum shearing stress is limited to 8,000 psi.

SOLUTION

From Eq. 5-1:

$$T = \frac{63{,}000 \text{ hp}}{N} = \frac{63{,}000(10)}{1{,}800} = 350 \text{ in-lb}$$

---

* Extensive recommendations for other materials may be found in machine-design books. For example, see V. M. Faires, *Design of Machine Elements* (4th ed.) (New York: The Macmillan Company, 1965), p. 580.

From Eq. 5-8:

$$\frac{J}{c} = \frac{T}{\tau_{max}} = \frac{350}{8,000} = 0.0438 \text{ in.}^3$$

$$\frac{J}{c} = \frac{\pi c^3}{2} \quad \text{or} \quad c^3 = \frac{2J}{\pi c} = \frac{2(0.0438)}{\pi} = 0.0279 \text{ in.}^3$$

Hence $\quad c = 0.303$ in. $\quad$ or $\quad d = 2c = 0.606$ in.

For practical purposes a ⅝-in. shaft would probably be selected.

EXAMPLE 5-6

Select solid shafts to transmit 200 hp each without exceeding a shearing stress of 10,000 psi. One of these shafts operates at 20 rpm and the other at 20,000 rpm.

SOLUTION

Subscript 1 applies to the low-speed shaft; 2 to the high-speed shaft.

From Eq. 5-1:

$$T_1 = \frac{(hp)(63,000)}{N_1} = \frac{200(63,000)}{20} = 630,000 \text{ in-lb}$$

Similarly $\quad T_2 = 630$ in-lb

From Eq. 5-8:

$$\frac{J_1}{c} = \frac{T_1}{\tau_{max}} = \frac{630,000}{10,000} = 63 \text{ in.}^3$$

$$\frac{J_1}{c} = \frac{\pi d_1^3}{16} \quad \text{or} \quad d_1^3 = \frac{16}{\pi}(63) = 322 \text{ in.}^3$$

and $\quad d_1 = 6.85$ in.

Similarly $\quad d_2 = 0.685$ in.

This example illustrates the reason for the modern tendency to use high-speed machines in mechanical equipment. The difference in size of the two shafts is striking. Further saving in the weight of the material may be effected by making use of hollow tubes.

5-7. ANGLE OF TWIST OF CIRCULAR MEMBERS

So far in this chapter, methods of determining stresses in solid and hollow circular shafts subjected to torque have been discussed. Now attention will be directed to the method of finding the angle of twist for

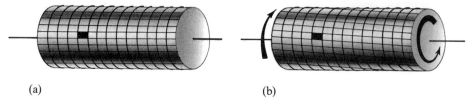

(a)   **(b)**

Fig. 5-8. Circular shaft (a) before (b) after torque is applied.

shafts subjected to torsional loading. The interest in this problem is at least three-fold. First, it is important to predict the twist of a shaft per se since at times it is not sufficient to design it only to be strong enough: it also must not deform excessively. Then, magnitudes of angular rotations of shafts are needed in the torsional vibration analysis of machinery, although this subject is not treated here. Finally, the angular twist of members is needed in dealing with statically indeterminate torsional problems, which are discussed in Chapter 12.

According to Assumption 1 stated in Art. 5-3, planes perpendicular to the axis of a circular rod do not warp. The elements of a shaft undergo deformation of the type shown in Fig. 5-8(b). The shaded element is shown in its undistorted form in Fig. 5-8(a). From such a shaft a typical element of length dx is shown isolated in Fig. 5-9.

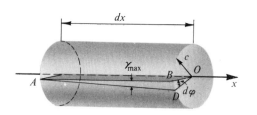

Fig. 5-9. An element of a circular shaft subjected to torque.

In the element shown a line or "fiber" such as AB initially is parallel to the axis of the shaft. After the torque is applied, it assumes a new position AD. At the same time, by virtue of Assumption 2, Art. 5-3, radius OB remains straight and rotates through a small angle $d\varphi$ to a new position OD.

Denoting the small angle DAB by γ_{max}, from geometry one has

$$\text{arc } BD = \gamma_{max}\, dx \quad \text{or} \quad \text{arc } BD = d\varphi\, c$$

where both angles are small and are measured in radians. Hence

$$\gamma_{max}\, dx = d\varphi\, c \tag{5-9}$$

γ_{max} applies only in the zone of an infinitesimal "tube" of uniform maximum shearing stress τ_{max}. Limiting attention to linearly elastic response makes Hooke's law applicable. Therefore, according to Eq. 4-11, the angle γ_{max} is proportional to τ_{max}, i.e., $\gamma_{max} = \tau_{max}/G$. Moreover, by Eq. 5-4, $\tau_{max} = Tc/J$. Hence $\gamma_{max} = Tc/(JG)$.* Substituting the latter

---

* The foregoing argument can be carried out in terms of any γ, which progressively becomes smaller as the axis of the rod is approached. The only difference in derivation consists in taking an arc corresponding to BD an arbitrary distance ρ from the center and using $T\rho/J$ instead of Tc/J for τ.

expression into Eq. 5-9 and canceling c,

$$\frac{d\varphi}{dx} = \frac{T}{JG} \quad \text{or} \quad d\varphi = \frac{T\,dx}{JG} \quad (5\text{-}10)$$

Section 5-7
Angle of twist of
circular members

This is the relative angle of twist of two adjoining sections an infinitesimal distance dx apart. To find the total angle of twist φ between any two sections A and B on a shaft, the rotations of all elements must be summed. Hence, the general expression for the angle of twist at any section for a shaft of a linearly elastic material is

$$\varphi = \int_0^x \frac{T(x)}{J(x)G}\,dx + C_1 \quad (5\text{-}11)$$

where the constant C_1 is the angle of twist at the origin. The internal torque T and the polar moment of inertia J may vary along the length of a shaft. The direction of the angle of twist φ coincides with the direction of the applied torque T.

Equation 5-11 is valid for both solid and hollow circular shafts, which follows from the assumptions used in the derivation. The angle φ is measured in radians. Note the great similarity of this relation to Eq. 4-29 for the deformation of axially loaded rods. Here $T(x)$ replaces $P(x)$, $J(x)$ replaces $A(x)$, and G is used in place of E.

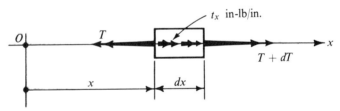

Fig. 5-10. Infinitesimal element of a shaft subjected to torque.

For constant JG Eq. 5-10 can be recast into a second-order differential equation. Preliminary to this step, consider an element, shown in Fig. 5-10, subjected to the end torques T and $T + dT$ and to an applied distributed torque t_x having the units of inch-pounds per inch. Since the right-hand screw rule for couples was used, all these quantities are shown in the figure as having a positive sense (for sign convention see Fig. 2-2). For equilibrium of this infinitesimal element

$$t_x\,dx + dT = 0 \quad \text{or} \quad dT/dx = -t_x$$

On differentiating Eq. 5-10 with respect to x, one has the required result:

$$JG\frac{d^2\varphi}{dx^2} = \frac{dT}{dx} = -t_x \quad (5\text{-}12)$$

155

The boundary conditions for this equation consist of specifying either φ or T at each boundary. From Eq. 5-10 it should be clear that $T = JG\, d\varphi/dx$. In this case, as in Eq. 4-32, singularity functions can be employed for concentrated moments. Equation 5-12 can be used in the solution of indeterminate problems. A more complete discussion of such problems is given in Chapter 12.

The following two examples illustrate applications of Eq. 5-11.

EXAMPLE 5-7

Find the relative rotation of section B-B with respect to section A-A of the solid shaft shown in Fig. 5-11 when a constant torque T is being transmitted through it. The polar moment of inertia of the cross-sectional area J is constant.

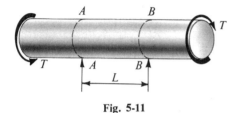

Fig. 5-11

SOLUTION

In this case $T(x) = T$ and $J(x) = J$; hence from Eq. 5-11

$$\Delta \varphi = \int_0^L \frac{T\,dx}{JG} = \frac{T}{JG}\int_0^L dx = \frac{TL}{JG} \quad (5\text{-}13)$$

Equation 5-13 is an important relation. It can be used in the design of shafts for stiffness, i.e., for limiting the amount of twist that may take place in their length. For such an application T, L, and G are known quantities, and the solution of Eq. 5-13 yields J. This fixes the size of the required shaft (see Eqs. 5-3 and 5-6). Note that for stiffness requirements, J, rather than J/c of the strength requirement, is the significant parameter. This equation is used in torsional vibration analyses. The term JG is referred to as the *torsional stiffness* of the shaft.

Another application of Eq. 5-13 is found in the laboratory. There a shaft may be subjected to a known torque T, J may be computed from the dimensions of the specimen, and the relative angular rotation φ between two planes a distance L apart may be measured. Then, by using Eq. 5-13, the shearing modulus of elasticity in the elastic range can be computed, i.e., $G = TL/J\varphi$.

In using Eq. 5-13 note particularly that the angle φ must be expressed in radians. Also observe the similarity of Eq. 5-13 to Eq. 4-33, $\Delta = PL/AE$, formerly derived for axially loaded rods.

EXAMPLE 5-8

Consider the stepped shaft shown in Fig. 5-12 attached to a wall at E, and determine the rotation of the end A when the two torques at B and at D are applied. Assume the shearing modulus G to be 12×10^6 psi, a typical value for steels.

SOLUTION

From Eq. 5-3:

$$J_{AB} = J_{BC} = \frac{\pi d^4}{32} = \frac{\pi 1^4}{32} = 0.0982 \text{ in.}^4$$

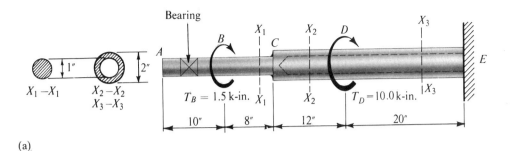

(a)

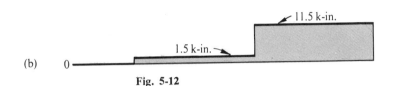

(b)

Fig. 5-12

From Eq. 5-6:

$$J_{CD} = J_{DE} = \frac{\pi}{32}(d_o^4 - d_i^4) = \frac{\pi}{32}(2^4 - 1^4) = 1.47 \text{ in.}^4$$

where subscripts indicate the range of applicability of a given value. Then by passing arbitrary sections X_1-X_1, X_2-X_2, and X_3-X_3, and each time considering a portion of the shaft to the left of such sections, the internal torques for the various intervals are found to be

$$T_{AB} = 0, \quad T_{BD} = T_{BC} = T_{CD} = 1.5 \text{ kip-in.}, \quad T_{DE} = 11.5 \text{ kip-in.}$$

The torque diagram corresponding to these quantities is in Fig. 5-12(b).

To find the rotation of the end A, Eq. 5-11 is applied with the limits of integration broken at points where T or J changes its value abruptly. Integrating from right to left since the right end is built in, one obtains $C_1 = 0$.

$$\varphi = \int_E^A \frac{T(x)\,dx}{J(x)G} = \int_E^D \frac{T_{DE}\,dx}{J_{DE}G} + \int_D^C \frac{T_{CD}\,dx}{J_{CD}G}$$

$$+ \int_C^B \frac{T_{BC}\,dx}{J_{BC}G} + \int_B^A \frac{T_{AB}\,dx}{J_{AB}G}$$

In the last group of integrals T's and J's are constant between the limits considered, so each integral reverts to a known solution, Eq. 5-13. Hence

$$\varphi = \frac{T_{DE}L_{DE}}{J_{DE}G} + \frac{T_{CD}L_{CD}}{J_{CD}G} + \frac{T_{BC}L_{BC}}{J_{BC}G} + \frac{T_{AB}L_{AB}}{J_{AB}G}$$

$$= \frac{11{,}500(20)}{1.47(12)10^6} + \frac{1{,}500(12)}{1.47(12)10^6} + \frac{1{,}500(8)}{0.0982(12)10^6} + 0$$

$$= 0.0130 + 0.0010 + 0.0102$$

$$= 0.0242 \text{ radian} \quad \text{or} \quad (360/2\pi)(0.0242) = 1.39°$$

The part *AB* of the shaft contributes nothing to the value of the angle φ as no internal torque acts through it. It rotates as much as the section at *B*. Little is contribtued to φ by the shaft from *C* to *D* because a small internal torque and a large *J* are associated with this segment. No doubt there is a disturbance in the strains at the step, but this local effect plays a small role in the overall rotation.

The angle computed would hold equally true for a relative rotation of sections for an analogous problem of a rotating shaft.

5-8. SHEARING STRESSES AND DEFORMATIONS IN CIRCULAR SHAFTS IN THE INELASTIC RANGE

The torsion formula for circular sections derived above is based on Hooke's law. Therefore, it applies only up to the point where the proportional limit of a material in shear is reached in the outer annulus of a shaft. Now the solution will be extended to include inelastic behavior of a material. As before the equilibrium requirements at a section must be met. The deformation assumption of linear strain variation from the axis remains applicable. Only the difference in material properties affects the solution.

A section through a shaft is shown in Fig. 5-13(a). The linear strain variation is shown schematically on the same figure. Some possible mechanical properties of materials in shear, obtained, for example, in experiments with thin tubes in torsion, are as shown in Figs. 5-13(b), (c), and (d). The corresponding shearing-stress distribution is shown to the right in each case. The stresses are determined from the strain. For example, if at an interior annulus the strain is *a*, Fig. 5-13(a), the corresponding stress is found from the stress-strain diagram. This procedure is applicable to solid shafts as well as to integral shafts made of concentric tubes of different materials, providing the corresponding stress-strain diagrams are used. The derivation for a linearly elastic material is simply a special case of this approach.

After the stress distribution is known, the torque *T* carried by these stresses is found as before, i.e.,

$$T = \int_A [\tau(dA)]\rho \qquad (5\text{-}14)$$

Either analytical or graphical procedures can be used for evaluating this integral.

Although the shearing-stress distribution after the elastic limit is exceeded is nonlinear and the elastic torsion formula Eq. 5-4 does not

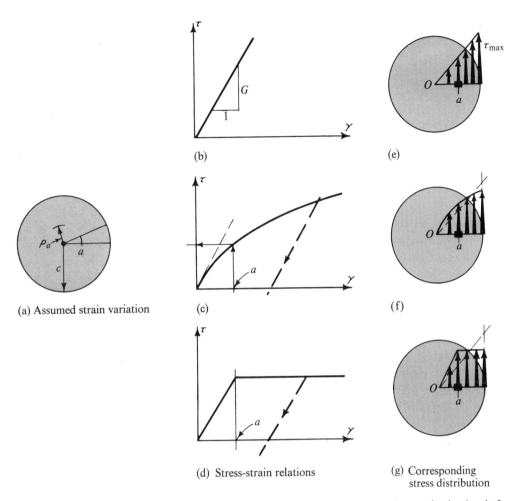

(a) Assumed strain variation

(b)

(c)

(d) Stress-strain relations

(e)

(f)

(g) Corresponding stress distribution

Fig. 5-13. Stresses in circular shafts.

apply, it is sometimes used at the outer edge of a shaft to calculate a fictitious stress for the ultimate torque. The computed stress is called the *modulus of rupture;* see the dashed lines on Figs. 5-13(f) and (g). It serves as a rough index of the ultimate strength of a material in torsion. For a thin-walled tube the stress distribution is very nearly the same regardless of the mechanical properties of the material, Fig. 5-14. For this reason experiments with thin-walled tubes are widely used in establishing the shearing stress-strain (τ-γ) diagrams.

If a shaft is strained into the plastic range and the applied torque is then removed, every

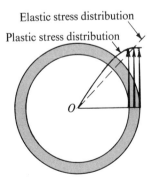

Fig. 5-14. For thin-walled tubes the difference between elastic and plastic stresses is small.

"imaginary" annulus rebounds elastically. Because of the differences in the strain paths, which cause permanent set in the material, residual stresses develop. This process will be illustrated in one of the examples which follow.

For determining the rate of twist of a circular shaft or tube, Eq. 5-9 can be used in the following form:

$$\frac{d\varphi}{dx} = \frac{\gamma_{\max}}{c} = \frac{\gamma_a}{\rho_a} \tag{5-15}$$

Here either the maximum shearing strain at c or the strain at ρ_a determined from the stress-strain diagram must be used.

EXAMPLE 5-9

A solid steel shaft of 1-in. diameter is so severely twisted that only a $\frac{1}{3}$-in.-diameter elastic core remains on the inside, Fig. 5-15(a). If the material properties can be idealized as shown in Fig. 5-15(b), what residual stresses and residual rotation will remain upon release of the applied torque?

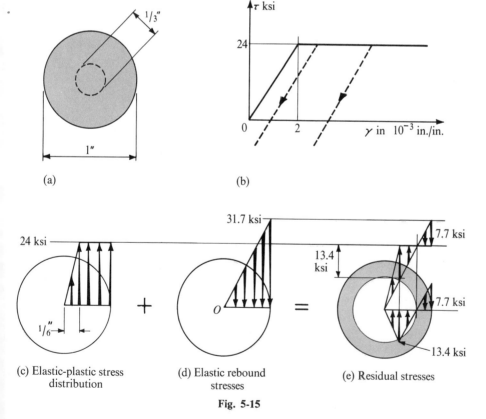

(a)

(b)

(c) Elastic-plastic stress distribution

(d) Elastic rebound stresses

(e) Residual stresses

Fig. 5-15

SOLUTION

To begin, the magnitude of the initially applied torque and the corresponding angle of twist must be determined. The stress distribution corresponding to the given condition is shown in Fig. 5-15(c). The stresses vary linearly from 0 to 24 ksi when $0 \leq \rho \leq \frac{1}{6}$ in.; the stress is a constant 24 ksi for $\rho > \frac{1}{6}$ in. Equation 5-14 can be used to determine the applied torque T. The release of the torque T causes elastic stresses, and Eq. 5-4 applies, Fig. 5-15(d). The difference between the two stress distributions, corresponding to no external torque, gives the residual stresses.

Section 5-8 Shearing stresses and deformations in circular shafts in the inelastic range

$$T = \int_A \tau \rho \, dA = \int_0^c 2\pi\tau\rho^2 \, d\rho = \int_0^{1/6} \left[\frac{\rho}{1/6} 24\right] 2\pi\rho^2 \, d\rho$$

$$+ \int_{1/6}^{1/2} (24) \, 2\pi\rho^2 \, d\rho = 0.17 + 6.05 = 6.22 \text{ kip-in.}$$

(Note the smallness of the contribution of the first integral.)

$$\tau_{max} = \frac{Tc}{J} = \frac{6.22 \times 0.5}{\pi/32} = 31.7 \text{ ksi}$$

At $\rho = 0.5$ in., $\tau_{residual} = 31.7 - 24.0 = 7.7$ ksi. Two diagrams of the residual stresses are shown in Fig. 5-15(e). For clarity the initial results are replotted from the horizontal line. In the entire shaded portion of the diagram, the residual torque is clockwise; an exactly equal residual torque acts in the opposite direction in the inner portion of the shaft.

The initial rotation is best determined by calculating the twist of the elastic core. At $\rho = \frac{1}{6}$ in., $\gamma = 2 \times 10^{-3}$. The elastic rebound of the shaft is given by Eq. 5-13. The difference between the inelastic and the elastic twists gives the residual rotation per inch of shaft. If the initial torque is re-applied in the same direction, the shaft responds elastically.

Inelastic: $\dfrac{d\varphi}{dx} = \dfrac{\gamma_a}{\rho_a} = \dfrac{2 \times 10^{-3}}{1/6} = 12 \times 10^{-3}$ per inch

Elastic: $\dfrac{d\varphi}{dx} = \dfrac{T}{JG} = \dfrac{6.22}{(\pi/32)12 \times 10^3} = 5.28 \times 10^{-3}$ per inch

Residual: $\dfrac{d\varphi}{dx} = (12 - 5.28)10^{-3} = 6.72 \times 10^{-3}$ radian per inch

EXAMPLE 5-10

Determine the torque carried by a solid circular shaft of mild steel when shearing stresses above the proportional limit are reached essentially everywhere. For mild steel, the shearing stress-strain diagram can be idealized to that shown in Fig. 5-16(a). The shearing yield-point stress τ_{yp} is to be taken as being the same as the proportional limit in shear τ_{pl}.

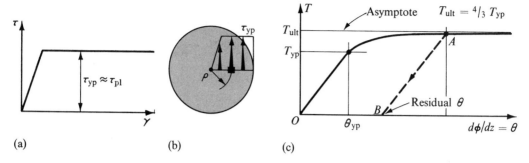

Fig. 5-16

SOLUTION

If a large torque is imposed on a member, large strains take place everywhere except near the center. Corresponding to the large strains for the idealized material considered, the yield-point shearing stress will be reached everywhere except near the center. However, the resistance to the applied torque offered by the material located near the center of the shaft is negligible as the corresponding ρ's are small, Fig. 5-16(b). (See the contribution to the torque T by the elastic action in Example 5-9.) Hence, with a sufficient degree of accuracy it can be assumed that a constant shearing stress τ_{yp} is acting everywhere on the section considered. The torque corresponding to this condition may be considered the ultimate or limit torque. (Figure 5-16(c) gives a firmer basis for this statement.) Thus

$$T_{ult} = \int_A \tau_{yp}\, dA\, \rho = \int_0^c 2\pi\rho^2 \tau_{yp}\, d\rho$$
$$= \frac{2\pi c^3}{3}\tau_{yp} = \frac{4}{3}\frac{\tau_{yp}}{c}\frac{\pi c^4}{2} = \frac{4}{3}\frac{\tau_{yp} J}{c} \quad (5\text{-}16)$$

Note that according to Eq. 5-4 the maximum elastic torque capacity of a solid shaft is $T_{yp} = \tau_{yp} J/c$. Therefore since T_{ult} is $4/3$ times this value, only $33\frac{1}{3}$ per cent of the torque capacity remains after τ_{yp} is reached at the extreme fibers of a shaft. A plot of torque T vs. θ, the angle of twist per unit distance, as full plasticity develops is in Fig. 5-16(c). Point A corresponds to the results found in the preceding example; line AB is the elastic rebound; and point B is the residual θ for the same problem.

It should be noted that in machine members, because of the fatigue properties of materials, the ultimate static capacity of the shafts as evaluated above is often of minor importance.

5-9. STRESS CONCENTRATIONS

Equations 5-4 and 5-8 apply to solid and tubular shafts only while the material behaves elastically. Moreover, the cross-sectional areas along the shaft should remain reasonably constant. If a gradual variation in the diameter takes place, the above equations give satisfactory solutions. On the other hand, for stepped shafts where the diameters of the adjoining portions change abruptly, large perturbations of shearing stresses take place. In such a case, at the juncture of the two parts near the center of a shaft, shearing stresses remain about the same as previously discussed. On the other hand, at extreme points from the center, high local shearing stresses occur. Methods of determining these local concentrations of stress are beyond the scope of this text. However, by forming a ratio of the true maximum shearing stress to the maximum stress given by Eq. 5-4, a torsional stress-concentration factor may be obtained. An analogous method was used for obtaining the stress-concentration factors in axially loaded members (Art. 4-18). The stress-concentration factors depend only on the geometry of the member. Stress-concentration factors for various proportions of stepped round shafts are in Fig. 5-17.*

Section 5-9
Stress concentrations

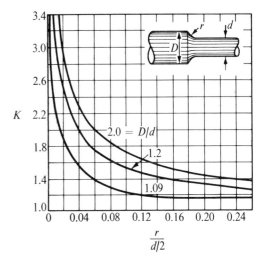

To obtain the actual stress at a geometrical discontinuity of a stepped shaft, a curve for a particular D/d is selected in Fig. 5-17. Then, the stress-concentration factor K corresponding to the given $r/(d/2)$ ratio is read from the curve. Finally, from the definition of K, the actual maximum shearing stress is obtained from the modified Eq. 5-4, i.e.,

$$\tau_{max} = K(Tc/J) \qquad (5\text{-}17)$$

where the shearing stress Tc/J is determined for the smaller shaft.

A study of stress-concentration factors in Fig. 5-17 emphasizes the need for a generous fillet radius r at all sections where a transition in the shaft diameter is made.

Fig. 5-17. Torsional stress-concentration factors in circular shafts of two diameters.

Considerable stress raisers also occur in shafts at oil holes and keyways. The latter are usually necessary for attaching pulleys and gears to a shaft. A shaft prepared for a key, Fig. 5-18, is no longer a circular member. A stress-concentration effect is particularly pronounced at the ends of keyways. Numerically, K for rectangular keyways is very high. For design details the reader is referred to books on machine design.

Because of some inelastic or nonlinear response in real materials, for reasons analogous to those pointed out in Art. 4-18, the theoretical stress concentrations based on the behavior of linearly elastic material tend to be high.

---

* This figure is adapted from a paper by L. S. Jacobsen, "Torsional-Stress Concentrations in Shafts of Circular Cross-section and Variable Diameter," *Transactions of the American Society of Mechanical Engineers*, **47** (1926), 632.

Fig. 5-18. Circular shaft with a keyway.

5-10. TWIST OF VISCOELASTIC CIRCULAR BARS

In some applications torsional members exhibit time-dependent behavior. For example, a bar subjected to a constant torque may continue to twist with elapsing time. This is a creep phenomenon. As an illustration of this behavior, consider a circular bar of a linear, viscoelastic material subjected at a time $t = 0$ to a torque $T(x)$ which remains constant thereafter.* The angle of twist φ of this bar is a function of the position x on the bar and of the time t, hence symbolically $\varphi(x, t)$. The angle of twist of this bar per unit length is $\theta(x, t) = \partial \varphi(x, t)/\partial x$, which is a more convenient quantity to consider in the discussion of this problem. On integrating θ along the bar, the total angle of twist φ can always be obtained.

For any circular bar of viscoelastic material subjected to a torque, the basic kinematic assumption that strains vary linearly from the axis of the bar remains valid. Therefore, recalling that ρ is the radial distance from the center of the shaft, one now has†

$$\gamma(x, t, \rho) = \theta(x, t)\rho \tag{5-18}$$

which states that for a given x and t the shearing strain γ is simply a linear function of ρ as in the linearly elastic case.

Next one turns attention to a constitutive relation for a *linear*, viscoelastic material and, by complete analogy with Eq. 4-27, writes the time-dependent shearing strain as

$$\gamma(t) = \tau_o \bar{J}_c(t) \tag{5-19}$$

where τ_o is a constant shearing stress, and $\bar{J}_c(t)$ is the creep compliance in shear. This expression can be recast as

$$\tau_o = \gamma(t)/\bar{J}_c(t) \tag{5-19a}$$

In a viscoelastic problem this equation plays the same role as does the ordinary stress-strain relation in the elastic problem. Therefore, if the stress τ_o for an element is a function of its position given by x and ρ, Eq. 5-19a can be generalized to

$$\tau_o(x, \rho) = \frac{\gamma(x, t, \rho)}{\bar{J}_c(t)} = \frac{\theta(x, t)}{\bar{J}_c(t)} \rho \tag{5-20}$$

where the last result is obtained by using the expression for γ given by Eq. 5-18.

---

* Analytically this can be expressed as $T(x, t) = T(x)H(t)$, where $H(t)$ is the Heaviside operator, which is zero for $t < 0$ and is unity for $t > 0$. As in this discussion a single input of the torque $T(x)$ is considered, this more complete notation is not used.

† Using the above notation, Eq. 5-9 yields $\gamma_{max} = \theta c$. For an arbitrary radius ρ, rather than c, this expression becomes $\gamma = \theta \rho$, or, in functional notation, $\gamma(x, \rho) = \theta(x) \rho$. In Eq. 5-18 the additional dependence of both γ and θ on the time t is included.

In Eq. 5-20 note especially that the shearing stress τ_o is independent of time and varies linearly with ρ precisely as it does in the linearly elastic case. Hence it can be concluded that the stress distribution in circular bars is the same for the linearly elastic as for the linearly viscoelastic materials and does not change with time.

Section 5-10 Twist of viscoelastic circular bars

Equation 5-20 for the shearing stress can be substituted into the equilibrium relation $T = \int \tau \rho \, dA$ in the same manner as has been used in deriving the elastic torsion formula, Eq. 5-4. In this way one obtains

$$T(x) = \int_A \tau_o(x, \rho)\rho \, dA = \int_A \frac{\theta(x, t)}{\bar{J}_c(t)} \rho^2 \, dA$$

$$= \frac{\theta(x, t)}{\bar{J}_c(t)} \int_A \rho^2 \, dA = \frac{\theta(x, t)}{\bar{J}_c(t)} J(x) \qquad (5\text{-}21)$$

where, as in the elastic solution, $J(x)$ is the polar moment of inertia of the cross-sectional area. It is a function of x as c may vary with x.

Recasting Eq. 5-21, one obtains

$$\theta(x, t) = T(x)\bar{J}_c(t)/J(x) \qquad (5\text{-}22)$$

which gives the angle of twist per unit length of the bar due to the applied torque as a function of time. At a time $t = 0$, since $\bar{J}_c(0) = G^{-1}$,

$$\theta(x, 0) = T(x)/(JG) \qquad (5\text{-}22a)$$

which completely agrees with Eq. 5-10 and gives the angle of twist per unit length for a linearly elastic bar. For this reason, if $\theta(x, 0) \equiv \theta_{el}(x)$, Eq. 5-22 can be rewritten as

$$\theta(x, t) = \theta_{el}(x)G\bar{J}_c(t) \qquad (5\text{-}23)$$

which shows particularly clearly the relationship between the elastic angle of twist θ_{el} and the angle θ at any time t for a bar of a material having the creep compliance $\bar{J}_c$, which must be determined experimentally. This solution is applicable to statically determinate as well as to statically indeterminate problems providing the material is *linearly* viscoelastic and the boundary conditions do not change with time.

It is interesting that by essentially the same reasoning it can be shown that if a constant angle of twist φ_0 is imposed at the free end of a circular shaft of a linear, viscoelastic material, at time $t = 0$,

$$T(t) = G(t)J\varphi_0/L \qquad (5\text{-}24)$$

where $G(t)$ is the relaxation modulus in shear. According to this relation, as in the case of axially loaded bars, the internal torque T decays with time.

It must be emphasized that the above two solutions are limited to step-function inputs of torque or angle of twist, respectively, at $t = 0$, which then remain constant with time. If a sequence of load or deformation inputs, respectively, is applied, it is necessary to use the Boltzmann superposition principle (Art. 4-16).

As pointed out in Art. 4-16, in practical applications great care in selecting the material constants in viscoelastic problems is essential. These are strongly dependent on temperature. Moreover, for some materials the idealization that the viscoelastic response is linear may not be sufficiently accurate.

5-11. SOLID NONCIRCULAR MEMBERS

The analytical treatment of solid noncircular members in torsion is beyond the scope of this book. Mathematically the problem is complicated.* The first two assumptions stated in Art. 5-3 do not apply for noncircular members. Sections perpendicular to the axis of a member warp when a torque is applied. The nature of the distortions that take place in a rectangular section may be seen from Fig. 5-19.† For a rectan-

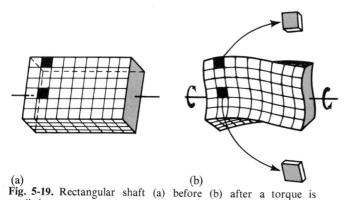

Fig. 5-19. Rectangular shaft (a) before (b) after a torque is applied.

gular member, oddly enough, the corner elements do not distort at all. Shearing stresses at the corners are zero, and they are maximum at the midpoints of the long sides. Figure 5-20 shows the shearing-stress distribution along three radial lines emanating from the center. Note particularly the difference in this stress distribution compared with that of a circular section. For the latter, the stress is a maximum at the most remote point, but for the former the stress is zero at the most remote point. This situation can be explained by considering a corner element as

---

* This problem remained unsolved until a famous French elastician, B. de St. Venant, developed a solution for such problems in 1853. The general torsion problem is sometimes referred to as the St. Venant problem.

† An experiment with a rubber eraser on which a rectangular grating is ruled demonstrates this type of distortion.

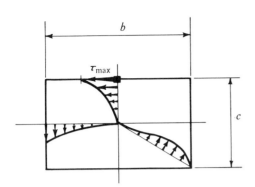
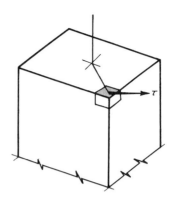

Fig. 5-20. Shearing-stress distribution in a rectangular shaft subjected to a torque.

Fig. 5-21. The shearing stress shown cannot exist.

shown in Fig. 5-21. If a shearing stress τ existed at the corner, it could be resolved into two components parallel to the edges of the bar. However, as shears always occur in pairs acting on mutually perpendicular planes, these components would have to be met by shears lying in the planes of the outside surfaces. The latter situation is impossible as outside surfaces are free of all stresses. Hence τ must be zero. Similar considerations can be applied to other points on the boundary. All shearing stresses in the plane of a cut near the boundaries act parallel to them.

Analytical solutions for torsion of rectangular, elastic members have been obtained.* The methods used are beyond the scope of this book. The final results of such analysis, however, are of interest. For the maximum shearing stress (see Fig. 5-20) and the angle of twist, these results can be put into the following form:

$$\tau_{max} = \frac{T}{\alpha bc^2} \quad \text{and} \quad \varphi = \frac{TL}{\beta bc^3 G} \tag{5-25}$$

T as before is the applied torque; b is the long side and c is the short side of the rectangular section. The values of parameters α and β depend upon the ratio b/c. A few of these values are recorded in the table below. For thin sections, when b is much greater than c, the values of α and β approach $\frac{1}{3}$.

TABLE OF COEFFICIENTS FOR RECTANGULAR SHAFTS

b/c	1.00	1.50	2.00	3.00	6.00	10.0	∞
α	0.208	0.231	0.246	0.267	0.299	0.312	0.333
β	0.141	0.196	0.229	0.263	0.299	0.312	0.333

---

* S. Timoshenko and J. N. Goodier, *Theory of Elasticity* (2nd. ed.) (New York: McGraw-Hill Book Company, 1951), p. 277.

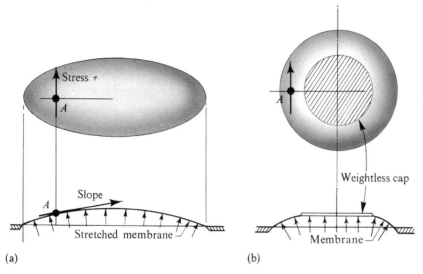

Fig. 5-22. Membrane analogy: (a) simply connected region, (b) multiply connected (tubular) region.

Formulas as above are available for many other types of cross-sectional areas in more advanced books. For cases which cannot be conveniently solved mathematically, a remarkable method has been devised.* It happens that the solution of the partial differential equation which must be solved in the elastic torsion problem is mathematically the same as the equation for a thin membrane, such as a soap film, lightly stretched over a hole. This hole must be geometrically similar to the cross section of the shaft being studied. Light air pressure must be kept on one side of the membrane. Then the following points can be shown.

1. The shearing stress at any point is proportional to the slope of the stretched membrane at the same point, Fig. 5-22.

---

*This analogy was introduced by a German engineering scientist, L. Prandtl, in 1903.

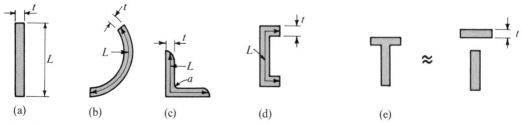

Fig. 5-23. Members of equal cross-sectional areas of the same thickness carrying the same torque.

2. The direction of a particular shearing stress at a point is at right angles to the slope of the membrane at the same point, Fig. 5-22.

3. Twice the volume enclosed by the membrane is proportional to the torque carried by the section.

Section 5-12 Thin-walled hollow members

The foregoing analogy is called the *membrane analogy*. In addition to its value in experimental applications, it is a very useful mental tool for visualizing stresses and torque capacities of members. For example, all the sections shown in Fig. 5-23 can carry approximately the same torque at the same maximum shearing stress (same maximum slope of the membrane) since the volume enclosed by the membranes would be nearly the same in all cases. (For all these shapes $b = L$, and $c = t$ in Eq. 5-25.) However, use of a little imagination will convince the reader that the contour lines of a soap film will "pile up" at a for the angular section. Hence high local stresses will occur at that point.

Another analogy, the *sand-heap analogy*, has been developed for plastic torsion.* Dry sand is poured onto a raised flat surface having the shape of the cross section of the member. The surface of the sand heap so formed assumes a constant slope. For example, a cone is formed on a circular disc, or a pyramid on a square base. The constant maximum slope of the sand corresponds to the limiting surface of the membrane in the previous analogy. The volume of the sand heap, hence its weight, is proportional to the fully plastic torque carried by a section. The other items in connection with the sand surface have the same interpretation as those in the membrane analogy.

5-12. THIN-WALLED HOLLOW MEMBERS

Unlike solid noncircular members, thin-walled tubes of any shape can be rather simply analyzed for the magnitude of the shearing stresses and the angle of twist caused by a torque applied to the tube. Thus, consider a tube of an arbitrary shape with varying wall thickness, such as shown in Fig. 5-24(a), subjected to a torque T. Isolate an element from this tube, as shown to an enlarged scale in Fig. 5-24(b). This element must be in equilibrium under the action of the forces F_1, F_2, F_3, and F_4. These forces are equal to the shearing stresses acting on the cut planes multiplied by the respective areas.

From $\Sigma F_x = 0$, $F_1 = F_3$; but $F_1 = \tau_2 t_2\, dx$, and $F_3 = \tau_1 t_1\, dx$, where τ_2 and τ_1 are shearing stresses acting on the respective areas $t_2\, dx$ and $t_1\, dx$. Hence, $\tau_2 t_2\, dx = \tau_1 t_1\, dx$, or $\tau_1 t_1 = \tau_2 t_2$. However, since the longitudinal cutting planes were taken an arbitrary distance apart, it follows from the above relations that the product of the shearing stress and the wall

---

* A. Nadai, *Theory of Flow and Fracture of Solids*, vol. 1 (2nd. ed.) (New York: McGraw-Hill Book Company, 1950).

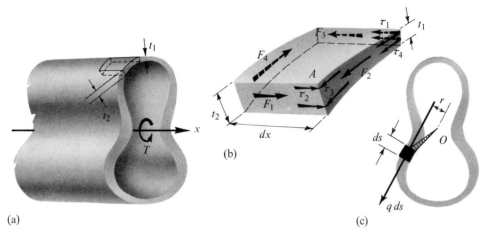

Fig. 5-24. Thin-walled member of variable thickness.

thickness is the same, i.e., constant, on any such planes. This constant will be denoted by q, and if the shearing stress is measured in pounds per square inch and the thickness of the tube in inches, q is measured in pounds per inch (lb per inch).

In Art. 3-3, Eq. 3-2, it was established that shearing stresses on mutually perpendicular planes are equal at a corner of an element. Hence at a corner such as A in Fig. 5-24(b), $\tau_2 = \tau_3$; similarly, $\tau_1 = \tau_4$. Therefore $\tau_4 t_1 = \tau_3 t_2$, or in general q is constant in the plane of a cut perpendicular to the axis of a member. On this basis an analogy may be formulated. The inner and outer boundaries of the wall can be thought of as being the boundaries of a channel. Then one can imagine a constant quantity of water steadily circulating in this channel. In this arrangement the quantity of water flowing through a plane across the channel is constant. Because of this analogy the quantity q has been termed the *shear flow*.

Next consider the cross section of the tube as shown in Fig. 5-24(c). The force per inch of the perimeter of this tube, by virtue of the previous argument, is constant and is the shear flow q. This shear flow multiplied by the length ds of the perimeter gives a force $q\,ds$ per differential length. The product of this infinitesimal force $q\,ds$ and r around some convenient point such as O, Fig. 5-24(c), gives the contribution of an element to the resistance of the applied torque T. Adding or integrating this,

$$T = \oint rq\,ds$$

where the integration process is carried around the tube along the center line of the perimeter.

Instead of carrying out the actual integration, a simple interpretation of the above integral is available. From Fig. 5-24(c) it can be seen that $r\,ds$ is twice the value of the shaded area of an infinitesimal triangle whose area is one-half the base times the altitude. Hence, inasmuch as q is a constant, the complete integral is twice the whole area A bounded by the center line of the perimeter of the tube. Therefore

$$T = 2Aq \quad \text{or} \quad q = T/(2A) \tag{5-26}$$

This equation* applies only to thin-walled tubes. The area A is approximately an average of the two areas enclosed by the inside and the outside surfaces of a tube, or, as noted above, it is an area enclosed by the center line of the wall's contour. Equation 5-26 is not applicable at all if the tube is slit.

Since for any tube the shear flow q given by Eq. 5-26 is constant, from the definition of shear flow, the shearing stress at any point of a tube where the wall thickness is t is

$$\tau = q/t \tag{5-27}$$

In the elastic range Eqs. 5-26 and 5-27 are applicable to any shape of tube. For inelastic behavior Eq. 5-27 applies only if the thickness t is constant. The analysis of tubes of more than one cell is beyond the scope of this book.

For linearly elastic material the angle of twist of a hollow tube may be found by applying the principle of conservation of energy. Since the tube itself is a linear spring, the external work is $T\theta/2$, where T is the applied torque and θ is an angle of twist in a unit distance, i.e. $\theta = d\varphi/dx$. The internal strain energy is given by Eq. 4-22, in which only one term for shearing strain energy is applicable in this case. On this basis, for a tube one unit long, one has

$$\tfrac{1}{2} T\theta = \tfrac{1}{2} T \frac{d\varphi}{dx} = \int_V \frac{\tau^2}{2G} dV$$

where for a unit distance the volume $dV = t\,ds$ and $\tau = T/2At$. Therefore

$$\theta = \frac{d\varphi}{dx} = \frac{T}{4A^2 G} \oint \frac{ds}{t} \tag{5-28}$$

EXAMPLE 5-11

Rework Example 5-3 using Eqs. 5-26 and 5-27. The tube has outside and inside radii of 0.5 in. and 0.45 in., respectively, and the applied torque is 400 in-lb.

---

* Equation 5-26 is sometimes called Bredt's formula in honor of the German engineer who developed it.

SOLUTION

The mean radius of the tube is 0.475 and the wall thickness is 0.05 in. Hence

$$\tau = \frac{q}{t} = \frac{T}{2At} = \frac{400}{2\pi(0.475)^2(0.05)} = 5{,}640 \text{ psi}$$

Note that by using Eqs. 5-26 and 5-27 only one shearing stress is obtained and that it is just about the average of the two stresses computed in Example 5-3. The thinner the walls, the more accurate the answer, or vice versa.

It is interesting to note that a rectangular tube, shown in Fig. 5-25, with a wall thickness of 0.05 in., for the same torque will have nearly the same shearing stress as the above circular tube. This is so because its enclosed area of 0.71 in.^{2} is about the same as $\pi(0.475)^2$, the A of the circular tube. However, some local stress concentrations will be present at the corners of a square tube.

By applying Eq. 5-28, one can compute the angle of twist per unit distance for either the circular or the square tube. For the first case a slightly more accurate answer may be found using Eq. 5-11. This is left for the reader to verify.

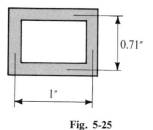

Fig. 5-25

PROBLEMS FOR SOLUTION

5-1. A solid circular shaft of 2-in. diameter is to be replaced by a hollow circular tube. If the outside diameter of the tube is limited to 3 in., what must be the thickness of the tube for the same linearly elastic material working at the same maximum stress? Determine the ratio of weights for the two shafts.

5-3. A motor, through a set of gears, drives a line shaft as shown in the figure, at 630 rpm. Thirty hp are delivered to a machine on the right; 90 hp on the left. Select a solid round shaft of the same size throughout. The allowable shearing stress is 5,750 psi. *Ans.* 2-in. diameter.

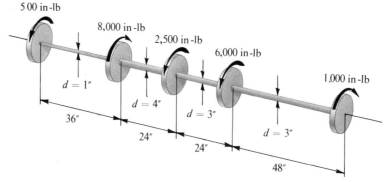

PROB. 5-2

5-2. The solid cylindrical shaft of variable size shown in the figure is acted upon by the torques indicated. What is the maximum torsional stress in the shaft and between what two pulleys does it occur?

5-4. (a) Determine the maximum shearing stress in the shaft subjected to the torques shown in the figure. (b) Find the angle of twist in degrees between the two ends. Let $G = 12 \times 10^6$ psi. *Ans.* (a) 900 psi, (b) 0.11°.

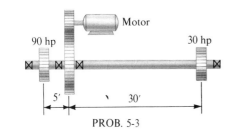

PROB. 5-3

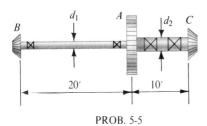

PROB. 5-4

5-5. A 100-hp motor is driving a line shaft through gear A at 26.3 rpm. Bevel gears at B and C drive rubber-cement mixers. If the power requirement of the mixer driven by gear B is 25 hp and that of C is 75 hp, what are the required shaft diameters? The allowable shearing stress in the shaft is 6,000 psi. A sufficient number of bearings is provided to avoid bending. If G is 12×10^6 psi, what is the angle of twist under load in the left section of the shaft? State answer in degrees. *Ans.* $d_1 = 3.71$ in., $d_2 = 5.34$ in., and $\varphi = 3.72°$.

PROB. 5-5

5-6. What must the length of a 0.2-in.-diameter aluminum wire be so that it could be twisted through one complete revolution without exceeding a shearing stress of 6,000 psi? $G = 3.84 \times 10^6$ psi.

5-7. A hollow steel rod 6 in. long is used as a torsional spring. The ratio of inside to outside diameters is $1/2$. The required stiffness for this spring is $1/12$ of a degree per 1 inch-pound of torque. Determine the outside diameter of this rod. $G = 12 \times 10^6$ psi. *Ans.* 0.25 in.

5-8. A hollow circular shaft of linearly elastic material of length L has an outside diameter $d_0 = 4$ in., and inside diameter $d_i = 3$ in. Determine the minimum diameter d for a solid shaft of the same material to replace the hollow shaft so that in the new shaft neither the maximum stress nor the angle of twist exceed the same quantities in the original design.

5-9. Two gears are attached to two 2-in.-diameter steel shafts as shown in the figure. The gear at B has an 8-in. pitch diameter; the gear at C a 16-in. pitch diameter. Through what angle will the end A turn if at A a torque of 5,000 in-lb is applied and the end D of the second shaft is prevented from rotating? $G = 12 \times 10^6$ psi. *Ans.* 11°.

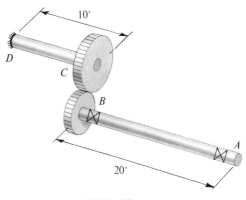

PROB. 5-9

5-10. A stepped steel shaft ($E = 29 \times 10^6$ psi, $G = 12 \times 10^6$ psi) as shown in the figure is subjected to a torque $T_1 = 7,500 \pi$ lb-in. and $T_2 = 2,500 \pi$ lb-in. For this shaft $a = 15$ ft, $b = 5$ ft, and the diameter d_2 from B to C is 2 in. Find the minimum permissible diameter d_1 for the shaft from A to B if the allowable

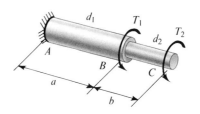

PROB. 5-10, 5-11, 5-22

shearing stress is 6,000 psi and the total twist between A and C is limited to 3°. *Ans.* 3.64 in.

5-11. A stepped shaft of linearly elastic material as shown in the figure has the following dimensions: $a = 30$ in., $b = 20$ in., $d_1 = 2$ in., and $d_2 = 1\frac{1}{2}$ in. If a torque T_1 is applied at B, what must the torque T_2 be to have no rotation at C? Plot the torque $T(x)$ and the angle-of-twist $\varphi(x)$ diagrams. (*Note:* The solution of this problem constitutes a solution of a statically indeterminate problem for a shaft built in at both ends.)

5-12. A solid, tapered steel shaft is rigidly fastened to a fixed support at one end and is subjected to a torque T at the other end (see figure). Find the angular rotation of the free end if $d_1 = 6$ in.; $d_2 = 2$ in.; $L = 20$ in.; and $T = 27,000$ in-lb. Assume that the usual assumptions of strain in prismatic circular shafts subjected to torque apply and let $G = 12 \times 10^6$ psi. *Ans.* 0.263°.

PROB. 5-12

5-13. A 6-in.-diameter shaft of a linearly elastic material has in it a conical bore which is 24 in. long as shown in the figure. The shaft is rigidly attached to a fixed support at one end and is subjected to a torque T at the free end. Determine the maximum angular rotation of the shaft.

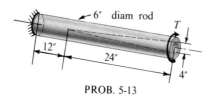

PROB. 5-13

5-14. A hollow, tapered torque-tube of a linearly elastic material has the dimensions shown in the figure. Determine the relative rotations of the ends due to a unit torque. (Such calculations are frequently required in

PROB. 5-14

vibration analysis.) Tube wall thickness throughout is $1/(2\pi)$ in. *Ans.* $30/G$.

5-15. The loading on a control torque-tube for an aileron of an airplane may be idealized by a uniformly varying torque $t_x = kx$ in-lb per inch, where k is a constant (see figure). Determine the angle of twist of the free end. Assume JG to be constant. Solve the problem using Eq. 5-11 or 5-12 as directed.

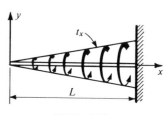

PROB. 5-15

5-16. Assume that during a drilling operation a shaft of constant torsional rigidity JG is loaded by a concentrated torque $T_1 = -1,000$ in-lb and a distributed torque $t_x = 100$ in-lb per inch as shown in the figure. Find the angular rotation of the free end. Use singularity functions and Eq. 5-12. Plot the torque $T(x)$ and the angle-of-twist $\varphi(x)$ diagrams.

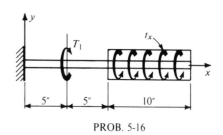

PROB. 5-16

5-17. Using singularity functions and Eq. 5-12 determine the reactions at the built-in ends caused by the application of the torque T_1, see figure. Plot the torque $T(x)$ and the angle-of-twist $\varphi(x)$ diagrams. (*Note:* This is a

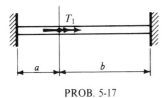

PROB. 5-17

statically indeterminate problem and only kinematic boundary conditions are used in its solution.)

5-18. A steel coupling is forged integrally with the shaft as shown in the figure. After machining, eight holes on a 9-in.-diameter bolt circle are to be provided for $1\frac{1}{4}$-in. bolts. (a) If this coupling will operate at 200 rpm, what horsepower can be transmitted through the bolts? Assume that the capacity of the bolts depends on the allowable shearing stress of 6,000 psi. (b) Using the same allowable shearing stress as for the bolts, what should the size of the main shaft be for balanced design? (*Note:* Concentrate the areas of the bolts at their centers.) *Ans.* $T = 265$ k-in.

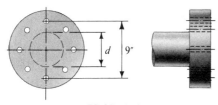

PROB. 5-18

5-19. A circular shaft is made by shrinking an aluminum tube onto a brass rod to form a section of two materials, which then act together as a unit (see figure). (a) If, due to the application of a torque T, a shearing stress of 10 ksi occurs in the outer fibers of the shaft, what is the magnitude of the torque T? (b) If the shaft is 36 in. long, what will the angle of twist be due to the torque T? For aluminum let $E = 10 \times 10^6$ psi, $G = 4 \times 10^6$ psi; for brass $E = 16 \times 10^6$ psi and $G = 6 \times 10^6$ psi. *Ans.* (a) 2,090 k-in., (b) 0.018 radian.

5-20. Rework Example 5-9 (Fig. 5-15) assuming that the solid steel shaft is 2 in. in diameter with a 1-in.-diameter inner elastic core.

5-21. Find the required fillet radius for the juncture of a 6-in.-diameter shaft with a 4-in.-diameter segment if the shaft transmits 110 hp at 100 rpm and the maximum shearing stress is limited to 8,000 psi.

5-22. A stepped shaft such as shown in the figure is subjected to the action of two torques T_1 and T_2. It is known that in magnitude T_1 is four times larger than T_2 and acts in the opposite direction to that shown in the figure. The distance $a = 200$ in., $b = 100$ in., $d_1 = 2$ in., and $d_2 = 1$ in. If under simultaneous application of T_1 and T_2 the free end of the shaft rotates through 0.0625 radian, what is the maximum shearing stress at the step in the shaft? Assume stress concentration factor $K = 1.4$, $E = 29 \times 10^6$ psi, and $G = 12 \times 10^6$ psi. *Ans.* 8.4 ksi. (See figure for Prob. 5-10.)

5-23. Assume that a circular shaft of a linearly viscoelastic material of length L is rigidly held at one end and at a time t_1 is subjected to a torque T_1 at the free end. At a time $t_2 > t_1$ the applied torque is increased to a total of $1\frac{1}{2}\ T_1$. At a time $t_3 > t_2 > t_1$ the applied torque is entirely removed. (a) Assuming that the material is Maxwellian, sketch a diagram showing the angle of twist of the free end as a function of time. (b) For the same conditions of input sketch another diagram for a material having the properties of a Standard Solid (see Prob. 4-17).

5-24. A rectangular bar of a linearly elastic material having the cross-sectional dimensions of a by $2a$ (see figure) is to be replaced by a solid circular bar of the same material. Determine the minimum diameter d of a bar so

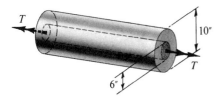

PROB. 5-19

PROBS. 5-24, 5-28

that for an applied torque neither the maximum shearing stress nor the angle of twist would exceed the corresponding quantities in the original design.

5-25. Compare the torsional strength and stiffness of thin-walled tubes of circular cross section of linearly elastic material with and without a longitudinal slot (see figure). *Ans.* $3R/t$, $t^2/(3R^2)$.

PROB. 5-25

5-26. The cross section of an 18-in., 45.8-lb channel may be idealized as consisting of two $5/8$-in.-by-$3\frac{3}{4}$-in. flange plates and a $\frac{1}{2}$-in. by $17\frac{3}{8}$-in. web plate. (See figure and Table 5 in the Appendix.) Determine the ratio of the applied torque T carried by each part of the section, assuming elastic behavior and neglecting stress concentrations. (*Hint:* Each flange carries a torque T_1; the web carries T_2. Hence, according to the membrane analogy, $2T_1 + T_2 = T$. Moreover, all parts of the cross section twist through the same angle φ given by Eq. 5-25. By equating the angle of twist of the web to that of a flange, a ratio of T_1 to T_2 can be found. Simultaneous solution of the two equations leads to the required result.)

PROB. 5-26

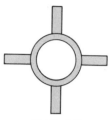

PROB. 5-27

5-27. An agitator shaft acting as a torsional member is made by welding to a circular pipe four rectangular bars as shown in the figure. The pipe is of 4-in. outside diameter and is $\frac{1}{2}$ in. thick; each of the rectangular bars is $\frac{3}{4}$ in. by 2 in. If the maximum elastic shearing stress, neglecting the stress concentrations, is limited to 8 ksi, what torque T can be applied to this member? (*Note:* See hint for the above problem. Here, $4T_{bar} + T_{tube} = T$. Also equate angles of twist for the tube to that of a paddle.)

5-28. Using the sand-heap analogy, determine the ultimate torsional moment of resistance for a rectangular section of a by $2a$ (see figure). (*Hint:* First, using the analogy, verify Eq. 5-16 for a solid circular shaft, where the height of the heap is $c\tau_{yp}$. Twice the volume included by the heap yields the required results.) *Ans.* $5a^3\tau_{yp}/6$.

5-29. Solve Prob. 5-28 for a section bounded by an equilateral triangle with the length of each side equal to a.

5-30 through 5-32. For members having the cross sections shown in the figures, find the maximum shearing stresses and angles of twist per unit length due to an applied torque of 1,000 in-lb in each case. Neglect stress concentrations. Where applicable, comment on the advantage gained by the increase in the wall thickness over part of the cross section.

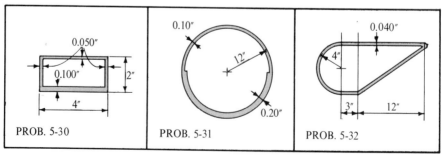

PROB. 5-30   PROB. 5-31   PROB. 5-32

Bending stresses in beams 6

6-1. INTRODUCTION

The system of forces that may exist at a section of a beam was discussed in Chapter 2. It was found to consist of an axial force, a shearing force, and a bending moment. The effect on a member of one of these forces, the axial force, was discussed in Chapters 3 and 4. In this chapter another element of the force system that may be present at a section of a beam, the internal bending moment, will be considered. Since in some cases a segment of a beam may be in equilibrium under the action of a moment alone, a condition called *pure bending* or *flexure*, this in itself represents a complete problem. This chapter relates the internal bending moment to the stresses it causes in a beam. Both the linearly elastic and the inelastic behavior of beams are considered. If, in addition to the internal bending moment, an axial force and a shear also act simultaneously, complex stresses arise. These will be treated in Chapters 8, 9, and 10. Discussion of the deflection of beams due to bending based on the fundamental deformation assumption introduced here is postponed until Chapter 11. Derivation of the formula for the elastic strain energy due to bending of beams is delayed until Chapter 13, where it is used in deflection calculations.

Most of this chapter is devoted to methods

*Chapter 6
Bending stresses
in beams* of determining the elastic and the inelastic, or plastic, stress distribution caused by bending moments in straight beams made of homogeneous materials. Discussions of beams made from two or more different materials, curved beams, and stress concentrations are also included.

6-2. SOME IMPORTANT LIMITATIONS OF THE THEORY

Just as in the case of axially loaded rods and in the torsion problem, all forces applied to a beam will be assumed to be delivered without shock or impact. Moreover, all the beams will be assumed to be stable under the applied forces. A similar point was brought out in Chapter 3, where it was indicated that a rod acting in compression cannot be too slender, or its behavior will not be governed by the usual compressive strength criterion. In such cases the stability of the member becomes important. As an example, consider the possibility of using a sheet of paper on edge as a beam. Such a beam has a substantial depth, but even if it is used to carry a force over a small span, it will buckle sidewise and collapse. The same phenomenon may take place in more substantial members which may likewise collapse under an applied force. Such unstable beams do not come within the scope of this chapter. All the beams considered here will be assumed to be sufficiently stable laterally by virtue of their proportions or to be thoroughly braced in the transverse direction. A better understanding of this important phenomenon will result after the study of the chapter on columns. It is indeed fortunate that the majority of beams used in structural framing and as machine elements are such that the flexural theory developed here is applicable; the theory governing the stability of members is more complex.

6-3. THE BASIC KINEMATIC ASSUMPTION

In the technical theory of bending, to establish the relation among the applied bending moment, the cross-sectional properties of a member, and the internal stresses and deformations, the approach applied earlier is again employed. This requires first that a plausible deformation assumption reduce the internally statically indeterminate problem to a determinate one; second, that the deformations causing strains be related to stresses through the appropriate stress-strain relations; and, finally, that the equilibrium requirements of external and internal forces be met. The key kinematic assumption for the deformation of a beam as used in the technical theory is discussed in this article. It is important to note that a generalization of this assumption forms the basis for the theories of plates and shells. The apparent mathematical complexity of the latter subjects is due to their two- and three-dimensional aspects, which require the use of the partial differential relations. On the other hand, the static beam problem resolves itself into a dependence on only one independent variable. This will become especially apparent during the study of Chapter 11 on the deflection of beams, where an ordinary differential equation suffices.

Section 6-3
The basic kinematic assumption

For the present consider a horizontal prismatic beam having a cross section with a vertical axis of symmetry, Fig. 6-1(a). Let the line through the centroid of all cross sections be referred to as the axis of the beam. Next, imagine a number of planes passing through the beam perpendicular to its axis, and a number of horizontal planes. In a side view these planes form a rectangular grid, Fig. 6-1(a). When such a beam is subjected to positive bending moments M at its ends as in Fig. 6-1(b), the beam bends, the planes perpendicular to the beam axis tilt slightly, and

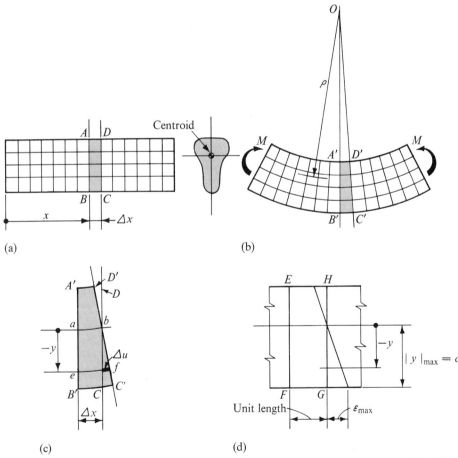

Fig. 6-1. The basic deformation assumption in bending.

the horizontal planes curve. Lines such as *AB* and *DC* remain straight.*
This observation forms the basis for the fundamental hypothesis† of the flexure theory. It may be stated thus: *Plane sections through a beam taken normal to its axis remain plane after the beam is subjected to bending.*

As demonstrated in texts on the theory of elasticity, this assumption is completely true for elastic, rectangular members in pure bending. If shears also exist, a small error is introduced.‡ Practically speaking, however, this assumption is generally applicable with a high degree of accuracy whether the material behaves elastically or plastically providing the depth in the beam is small in relation to its span. In this chapter the stress analysis of all beams is based on this assumption. As this assumption also establishes the radius of curvature ρ of a beam, Fig. 6-1(b), it will be utilized in Chapter 11 for calculations of beam deflections due to bending.

For the purposes of stress analysis it is necessary to restate the basic kinematic assumption in a slightly different manner. Thus, for example, consider the initially undeformed element *ABCD*, Fig. 6-1(a), and compare it with the deformed element $A'B'C'D'$, Fig. 6-1(b). Since in this member all similar elements between the initially vertical planes experience the same deformation, for convenience an element with a vertical plane $A'B'$ is selected for discussion. An enlarged view of this element is shown in Fig. 6-1(c). From this diagram it is seen that the fibers or "filaments" of the beam along a surface such as *ab* do not change in length. Moreover, as the element selected was an arbitrary one, fibers free of stress and strain exist continuously over the whole length and width of the beam. These fibers lie in a surface called the *neutral surface* of the beam. Its intersection with a right section through the beam is termed the *neutral axis* of the beam. Either term implies a location of zero stress or strain in the member subjected to bending.

The precise location of the neutral surface in a beam will be determined in subsequent articles. Locating it requires consideration of the stress-strain relations of the material and the equilibrium requirements of the whole member. Here a study of the nature of the strains in fibers parallel to the neutral surface is made.

Consider a generic (typical) fiber *ef* parallel to the neutral surface and located at a distance $-y$ from it. During bending the fiber elongates

---

* This can be demonstrated by using a rubber model with a ruled grating drawn on it. Alternatively, thin vertical rods passing through the rubber block can be used. In the immediate vicinity of the applied moments the deformation is more complex. However, in accord with the St. Venant's principle (Art. 4-18), this is only a local phenomenon which rapidly dissipates.

† This hypothesis with an inaccuracy was first introduced by Jacob Bernoulli (1645–1705), a Swiss mathematician. In the correct form it dates back to the writings of the French engineering educator M. Navier (1785–1836).

‡ See the discussion in Chapter 7 in connection with Fig. 7-12.

an amount Δu. If this elongation is divided by the initial length Δx of the fiber, according to Eq. 4-3, the linear strain ε_x in that fiber is obtained, i.e.,

*Section 6-4
The elastic flexure formula*

$$\varepsilon_x = \lim_{\Delta x \to 0} \frac{\Delta u}{\Delta x} = \frac{du}{dx} \qquad (6\text{-}1)$$

Since, however, the elongations of different fibers of initially the same length vary linearly from the neutral axis, Fig. 6-1(c), the fundamental assumption may be restated thus: *In a beam subjected to bending, strains in its fibers vary linearly, or directly, as their respective distances from the neutral surface.* Analytically this condition can be expressed as $\varepsilon_x = by$, where b is a constant. This situation is analogous to the one found earlier in the torsion problem where the shearing strains very linearly from the axis of a circular shaft. In a beam, strains vary linearly from the neutral surface. This variation is represented diagrammatically in exaggerated form in Fig. 6-1(d). These linear strains are associated with stresses which act normal to the section of a beam.

In the next article straight elastic beams with an axis of symmetry are considered. The applied bending moments are assumed to lie in a plane containing this axis of symmetry and the beam axis. For simplicity in making sketches, the axis of symmetry will be taken vertically. Several cross-sectional areas of beams satisfying these conditions are shown in Fig. 6-2. A generalization of the problem to include arbitrary cross-sectional areas is made in Art. 6-5.

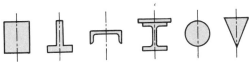

Fig. 6-2. Beam cross sections with a vertical axis of symmetry.

6-4. THE ELASTIC FLEXURE FORMULA

In the preceding article it was shown that in a bent beam strains vary linearly from the neutral axis. Such a variation of linear strain ε_x is schematically shown in Fig. 6-3(a) for a beam bent around the horizontal axis. By limiting the discussion to linearly elastic material with σ_x as the only nonzero stress, according to Hooke's law, $\sigma_x = E\varepsilon_x$. Therefore for an elastic beam the normal stresses σ_x resulting from bending also must vary linearly as their respective distances from the neutral axis. This is shown diagrammatically in Fig. 6-3(b); it can be expressed analytically as

$$\sigma_x = By \qquad (6\text{-}2)$$

where B is a constant.* The variable y can assume both positive and negative values.

The constant B in Eq. 6-2 can be related to the applied bending

---

* From the previous article it can be seen that $B = bE$.

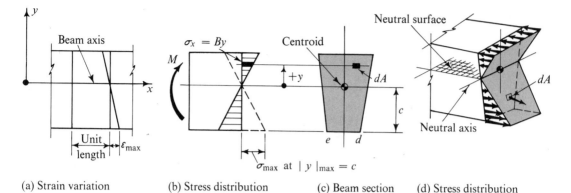

Fig. 6-3. Elastic beam segment in pure flexure.

moment and the cross-sectional properties of a beam. This relation constitutes the elastic flexure formula; it can be derived from equilibrium considerations of a beam segment.

Two conditions of equilibrium for a beam segment such as shown in Fig. 6-3(b) lead to the required results. One of these conditions is that the sum of all forces in the x direction must equal zero, i.e.,

$$\Sigma F_x = 0 \rightarrow +, \qquad \int_A \sigma_x \, dA = 0$$

where the subscript A indicates that the summation of the infinitesimal forces must be carried out over the entire cross-sectional area A of the beam. The above equation with the aid of Eq. 6-2 can be rewritten as

$$\int_A By \, dA = B \int_A y \, dA = 0$$

Since, however, in a stressed beam the constant B cannot be zero, it follows that

$$\int_A y \, dA = 0$$

But by definition

$$\int_A y \, dA = \bar{y} A$$

where $\bar{y}$ is the distance from a base line (the neutral axis in the case considered) to the centroid of the area A, so $\bar{y} A = 0$. Then since A is not zero, $\bar{y}$ must be. Therefore, the distance from the neutral axis to the centroid of the area must be zero, and so the neutral axis must pass through the centroid of the cross-sectional area of the beam. Hence the neutral axis may be easily determined for any beam by simply finding the centroid of its cross-sectional area.

The second condition of equilibrium useful in this problem is that the sum of all moments around the z axis must vanish. For the beam segment of Fig. 6-3(b) this yields

*Section 6-4
The elastic flexure formula*

$$\sum M_z = 0 \circlearrowleft +, \qquad M + \int_A (\sigma_x \, dA) y = 0$$

where y is the moment arm from the neutral axis to a generic infinitesimal force ($\sigma_x \, dA$) acting on an element of area dA. Initially all quantities under the integral are assumed to be positive. On substituting Eq. 6-2 into the above equation, after some minor simplifications, one obtains

$$M = -B \int_A y^2 \, dA$$

The integral depends only on the geometrical properties of the cross-sectional area. In mechanics this integral is called the *moment of inertia* of the cross-sectional area about the centroidal axis when y is measured from such an axis. It is a definite constant for any particular area, and it will be designated by I. With this notation $M = -BI$, and the constant $B = -M/I$. By substituting this value of the constant B into Eq. 6-2, the elastic *flexure formula** for beams is obtained:

$$\sigma_x = -\frac{My}{I} \qquad (6\text{-}3)$$

This equations shows that for a positive bending moment M and positive y's the normal stresses σ_x are compressive; for negative y's, the stresses are tensile. An illustration of this type of stress distribution is shown in Fig. 6-3(d). In many applications the direction of the normal stresses is known from the context of the problem, and in such cases the subscript on σ is superfluous.

As at a given section through a beam both M and I are constant, the normal stress σ_x reaches its largest value when the absolute value of y is a maximum. It is customary to designate this value of $|y|_{\max}$ by c. It is also a common practice to dispense with the sign of Eq. 6-3 as the sense of the normal stress can be found by inspection. At a given section the normal stresses must build up a couple statically equivalent to the resisting bending moment, the sense of which is known. On this basis

$$\sigma_{\max} = \frac{Mc}{I} \qquad (6\text{-}4)$$

---

* It took nearly two centuries to develop this seemingly simple expression. The first attempts to solve the flexure problem were made by Galileo in the seventeenth century. In the form in which it is used today, the problem was solved in the early part of the nineteenth century. Generally, Navier is credited for this accomplishment. However, some maintain that credit should go to Coulomb, who also derived the torsion formula.

Chapter 6
Bending stresses in beams

Equations 6-3 and 6-4 are of unusually great importance in the mechanics of solids. In these formulas, M is the internal or resisting bending moment, which is equal numerically to the external moment at the section where the stresses are sought. In the English system of units it is best to express the bending moment in inch-pound units for use in these formulas. The distance y from the neutral axis of the beam to the point on a section where the normal stress σ_x is wanted is measured perpendicular to the neutral axis and should be expressed in inches. When it reaches its maximum value (measured either up or down) it corresponds to c, and as y approaches this maximum value, the normal stress σ_x approaches σ_{max}. In this equation I is the moment of inertia of the whole cross-sectional area of the beam about its neutral axis. To avoid confusion with the polar moment of inertia, I is sometimes referred to as the rectangular moment of inertia. It has the dimensions of [in.^{4}]. Its evaluation for various areas will be discussed in Art. 6-6. The use of units as indicated above makes the units of stress σ, [in-lb][in.]/[in.^{4}] = [lb per in.^{2}], pounds per square inch or psi.

It should be emphasized that σ_x as given by Eq. 6-3 is the only stress which results from pure bending of a beam. Therefore, in the matrix representation of the stress tensor, one has

$$\begin{pmatrix} \sigma_x & 0 & 0 \\ 0 & 0 & 0 \\ 0 & 0 & 0 \end{pmatrix}$$

As will be pointed out in Chapters 9 and 10, this stress may be transformed or resolved into stresses acting along different sets of coordinate axes.

In concluding this discussion it is interesting to note that due to Poisson's ratio, the compressed zone of a beam expands laterally;* the tensile zone contracts. The strains in the y and z directions are $\varepsilon_y = \varepsilon_z = -\nu\varepsilon_x$, where, further, $\varepsilon_x = \sigma_x/E$, and σ_x is given by Eq. 6-3. This is in complete agreement with the rigorous solution. Poisson's effect, as may

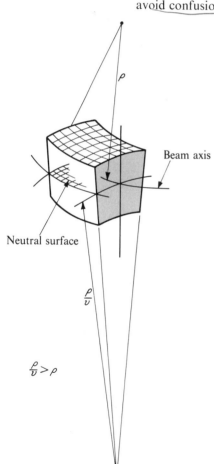

Fig. 6-4. Segment of bent, rectangular beam.

---
* An experiment with an ordinary rubber eraser is recommended!

be shown by the methods of elasticity, deforms the neutral axis into a curve of large radius; and the neutral surface becomes curved in two opposite directions, Fig. 6-4. In the previous treatment the neutral surface was assumed to be curved in one direction only. These interesting details are not of much significance in most practical problems.

Section 6-5 Pure bending of beams with unsymmetrical section

Again the reader should note that in deriving the basic flexure formula, the same concepts were applied as were encountered previously. These may be summarized as follows:

1. *Deformation* was assumed giving the linear variation of strain from the neutral axis.
2. *Properties of materials* were used to relate strain and stress.
3. *Equilibrium* conditions were used to locate the neutral axis and to determine the internal stresses.

6-5. PURE BENDING OF BEAMS WITH UNSYMMETRICAL SECTION

Pure bending of elastic beams having an axis of symmetry was discussed in the preceding article. The applied moments were assumed to act in the plane of symmetry. These limitations, although expedient in developing the flexural theory, are too severe and may be greatly relaxed. The same formulas can be used for any beam in pure bending, providing the bending moments are applied in a plane parallel to either principal axis of the cross-sectional area. The previous derivation can be repeated identically. Stresses vary linearly from the neutral axis passing through the centroid. As before, the stress on any elementary area dA, Fig. 6-5, is $\sigma_x = By$. Hence $By\,dA$ is an infinitesimal force acting on an element. The sum of the moments of these internal forces around the z axis develops the internal moment. However, as symmetry is lacking, these internal forces may build up a moment around the y axis. This must be reconciled.

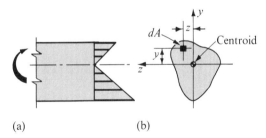

Fig. 6-5. A beam with an unsymmetrical cross-sectional area.

The moment arms of forces acting on infinitesimal areas around the y axis are equal to z. Thus a possible moment M_{yy} around the y axis is

$$M_{yy} = \int_A By\,dA\,z = B\int_A yz\,dA$$

The last integral represents the product of inertia of the cross-sectional area. It is equal to zero if the axes selected are the principal axes of the area. Since these axes are the ones used here, $M_{yy} = 0$, and thus the formulas derived earlier apply to a beam with any shape of cross section.

*Chapter 6
Bending stresses
in beams*

If a bending moment is applied without being parallel to either principal axis, the procedures discussed in Chapter 8 must be followed.

6-6. COMPUTATION OF THE MOMENT OF INERTIA

In applying the flexure formula, the moment of inertia I of the cross-sectional area about the neutral axis must be determined. Its value is defined by the integral of $y^2\,dA$ over the entire cross-sectional area of a member, and it must be emphasized that for the flexure formula the moment of inertia must be computed around the neutral axis of the cross-sectional area. This axis, according to Art. 6-4, passes through the centroid of the cross-sectional area. For symmetrical sections the neutral axis is perpendicular to the axis of symmetry. Such an axis is one of the *principal axes** of the cross-sectional area. Most readers should already be familiar with the method of determining the moment of inertia I. However, the necessary procedure is reviewed below.

The first step in evaluating I for an area is to find the centroid of the area. An integration of $y^2\,dA$ is then performed with respect to the horizontal axis passing through the area's centroid. Actual integration over areas is necessary for only a few elementary shapes, such as rectangles, triangles, etc. After this is done, most cross-sectional areas used in practice may be broken down into a combination of these simple shapes. Values of moments of inertia for some simple shapes may be found in any standard civil or mechanical engineering handbook (also see Table 2 in the Appendix). To find I for an area composed of several simple shapes, the *parallel-axis theorem* (sometimes called the *transfer formula*) is necessary; the development of it follows.

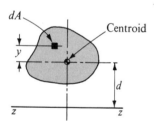

Fig. 6-6. Shaded area used in deriving the parallel-axis theorem.

The area shown in Fig. 6-6 has a moment of inertia I_o around the horizontal axis passing through its own centroid, i.e., $I_o = \int y^2\,dA$, where y is measured from the centroidal axis. The moment of inertia I_{zz} of the same area around another horizontal axis z-z by definition is

$$I_{zz} = \int_A (d + y)^2\,dA$$

where as before y is measured from the axis through the centroid. Squaring the quantities in the parentheses and placing the constants outside the integral signs

$$I_{zz} = d^2 \int_A dA + 2d \int_A y\,dA + \int_A y^2\,dA = Ad^2 + 2d \int_A y\,dA + I_o$$

---

* By definition the principal axes are those about which the rectangular moment of inertia is a maximum or a minimum. Such axes are always mutually perpendicular. The product of inertia, defined by $\int yz\,dA$ vanishes for the principal axes (see Fig. 6-5). An axis of symmetry of a cross-section is always a principal axis. For further details see any book on engineering mechanics.

However, since the axis from which y is measured passes through the centroid of the area, $\int y\, dA$ or $\bar{y}A$ is zero. Hence

$$I_{zz} = I_o + Ad^2 \qquad (6\text{-}5)$$

Section 6-6
Computation of the moment of inertia

This is the parallel-axis theorem. It can be stated as follows: The moment of inertia of an area around any axis is equal to the moment of inertia of the same area around a parallel axis passing through the area's centroid, plus the product of the same area and the square of the distance between the two axes.

The following examples illustrate the method of computing I directly by integration for two simple areas. Then an application of the parallel-axis theorem to a composite area is given. Values of I for commercially fabricated steel beams, angles, and pipes are given in Tables 3 to 8 of the Appendix.

EXAMPLE 6-1

Find the moment of inertia around the horizontal axis passing through the centroid for the rectangular area shown in Fig. 6-7.

SOLUTION

The centroid of this section lies at the intersection of the two axes of symmetry. Here it is convenient to take dA as $b\, dy$. Hence

$$I_{zz} = I_o = \int_A y^2\, dA = \int_{-h/2}^{+h/2} y^2 b\, dy = b\left.\frac{y^3}{3}\right|_{-h/2}^{+h/2} = \frac{bh^3}{12} \qquad (6\text{-}6)$$

Similarly $\qquad I_{yy} = b^3h/12$

These expressions are used frequently, as rectangular beams are commonly employed in practice.

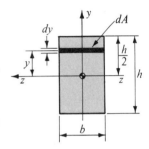

Fig. 6-7

EXAMPLE 6-2

Find the moment of inertia about a diameter for a circular area of radius c, Fig. 6-8.

SOLUTION

Since there is some chance of confusing I with J for a circular section, it is well to refer to I as the *rectangular* moment of inertia of the area in this case.

To find I for a circle, first note that $\rho^2 = z^2 + y^2$, as may be seen from the figure. Then using the definition of J, noting the symmetry around both axes, and using Eq. 5-3

$$J = \int_A \rho^2\, dA = \int_A (y^2 + z^2)\, dA = \int_A y^2\, dA + \int_A z^2\, dA$$

$$= I_{zz} + I_{yy} = 2I_{zz} \qquad (6\text{-}7)$$

$$I_{zz} = I_{yy} = J/2 = \pi c^4/4$$

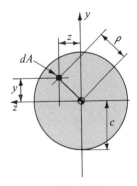

Fig. 6-8

In mechanical applications circular shafts often act as beams, hence Eq. 6-7 will be found useful. For a tubular shaft, the moment of inertia of the hollow interior must be subtracted from the above expression.

EXAMPLE 6-3

Determine the moment of inertia I around the horizontal axis for the area shown in Fig. 6-9, for use in the flexure formula.

SOLUTION

As the moment of inertia wanted is for use in the flexure formula, it must be obtained around the axis through the centroid of the area. Hence the centroid of the area must be found first. This is most easily done by treating the entire outer section and deducting from it the hollow interior. For convenience, the work is carried out in tabular form. Then the parallel-axis theorem is used to obtain I.

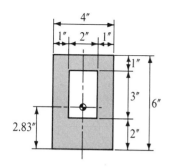

Fig. 6-9

Area	A [in.²]	y [in.] (from bottom)	Ay
Entire area	4(6) = 24	3	72
Hollow interior	−2(3) = −6	3.5	−21
	$\sum A$ = 18 in.²		$\sum Ay$ = 51 in.³

$$\bar{y} = \frac{\sum Ay}{\sum A} = \frac{51}{18} = 2.83 \text{ in. from bottom}$$

For entire area:

$$I_o = \frac{bh^3}{12} = \frac{4(6)^3}{12} = 72.00 \text{ in.}^4$$

$$Ad^2 = 24(3.00 - 2.83)^2 = \underline{0.69 \text{ in.}^4}$$

$$I_{zz} = 72.69 \text{ in.}^4$$

For hollow interior:

$$I_o = \frac{bh^3}{12} = \frac{2(3)^3}{12} = 4.50 \text{ in.}^4$$

$$Ad^2 = 6(3.50 - 2.83)^2 = \underline{2.69 \text{ in.}^4}$$

$$I_{zz} = 7.19 \text{ in.}^4$$

For composite section: $I_{zz} = 72.69 - 7.19 = 65.50$ in.⁴

Note particularly that in applying the parallel-axis theorem, each element of the composite area contributes two terms to the total I. One term is the moment of inertia of an area around its own centroidal axis, the other term is due to the transfer of its axis to the centroid of the whole area. Methodical work is the prime requisite in solving such problems correctly.

6-7. REMARKS ON THE FLEXURE FORMULA

The bending stress at any point of a beam section is given by Eq. 6-3, $\sigma_x = -My/I$. The largest stress at the same section follows from this relation by taking $|y|$ at a maximum, which leads to Eq. 6-4 $\sigma_{max} = Mc/I$.

In most practical problems the maximum stress given by Eq. 6-4 is the quantity sought; thus it is desirable to make the process of determining $\sigma_{\max}$ as simple as possible. This can be accomplished by noting that both I and c are constants for a given section of a beam. Hence I/c is a constant. Moreover, since this ratio is only a function of the cross-sectional dimensions of a beam, it can be uniquely determined for any cross-sectional area. This ratio is called the *elastic section modulus* of a section and will be designated by S. With this notation Eq. 6-4 becomes

*Section 6-7
Remarks on the
flexure formula*

$$\sigma_{\max} = \frac{Mc}{I} = \frac{M}{I/c} = \frac{M}{S} \qquad (6\text{-}8)$$

or stated otherwise

$$\text{maximum bending stress} = \frac{\text{bending moment}}{\text{elastic section modulus}}$$

If the moment of inertia I is measured in inches^4 and c in inches, S is measured in inches^3. Likewise, if M is measured in inch-pounds, the units of stress, as before, become pounds per square inch. It bears repeating that the distance c as used here is measured from the neutral axis to the most remote fiber of the beam. This makes $I/c = S$ a minimum, and consequently M/S gives the maximum stress. The efficient sections for resisting bending have as large an S as possible for a given amount of material. This is accomplished by locating as much of the material as possible far from the neutral axis.

The use of the term *elastic section modulus* in Eq. 6-8 corresponds somewhat to the use of the area term A in Eq. 3-5 ($\sigma = P/A$). However, only the maximum flexural stress on a section is obtained from Eq. 6-8, but the stress computed from Eq. 3-5 holds true across the whole section of a member.

Equation 6-8 is widely used in practice because of its simplicity. To facilitate its use, section moduli for many manufactured cross sections are tabulated in handbooks. Values for a few steel sections are given in Tables 3 to 8 in the Appendix. Equation 6-8 is particularly convenient for the design of beams. Once the maximum bending moment for a beam is determined and an allowable stress is decided upon, Eq. 6-8 may be solved for the required section modulus. This information is sufficient to select a beam. However, a detailed consideration of beam design will be delayed until Chapter 10. This is necessary inasmuch as a shearing force, which in turn causes stresses, usually also acts at a beam section. The interaction of the various kinds of stresses must be considered first to gain complete insight into the problem.

The application of the flexure formulas to particular problems should cause little difficulty if the meaning of the various terms occurring in them has been thoroughly understood. The following two examples illustrate investigations of bending stresses at specific sections.

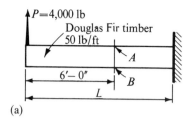 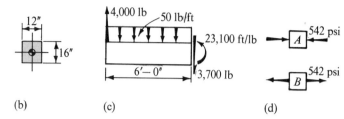

Fig. 6-10

EXAMPLE 6-4

A 12-in.-by-16-in. (full-sized), wooden cantilever beam weighing 50 lb per foot carries an upward concentrated force of 4,000 lb at the end, as shown in Fig. 6-10(a). Determine the maximum bending stresses at a section 6 ft from the free end.

SOLUTION

A free-body diagram for a 6-ft segment of the beam is shown in Fig. 6-10(c). To keep this segment in equilibrium requires a shear of $4,000 - 50(6) = 3,700$ lb and a bending moment of $4,000(6) - 50(6)3 = 23,100$ ft-lb at the cut section. Both these quantities are shown with their proper sense in Fig. 6-10(c). By inspecting the cross-sectional area, the distance from the neutral axis to the extreme fibers is seen to be 8 in., hence $c = 8$ in. This is applicable to both the tension and the compression fibers.

From Eq. 6-6: $\quad I_{zz} = \dfrac{bh^3}{12} = \dfrac{12(16)^3}{12} = 4,095$ in.^{4}

From Eq. 6-4: $\quad \sigma_{max} = \dfrac{Mc}{I} = \dfrac{23,100(12)8}{4,095} = \pm 542$ psi

From the sense of the bending moment shown in Fig. 6-10(c) the top fibers of the beam are seen to be in compression, and the bottom ones in tension. In the answer given, the positive sign applies to the tensile stress, the negative sign applies to the compressive stress. Both of these stresses decrease at a linear rate toward the neutral axis where the bending stress is zero. The normal stresses acting on infinitesimal elements at A and B are shown in Fig. 6-10(d). It is important to learn to make such a representation of an element as it will be frequently used in Chapters 8, 9, and 10.

ALTERNATE SOLUTION

If only the maximum stress is desired, the equation involving the section modulus may be used. The section modulus for a rectangular section in algebraic form is

$$S = \dfrac{I}{c} = \dfrac{bh^3}{12} \dfrac{2}{h} = \dfrac{bh^2}{6} \qquad (6\text{-}9)$$

In this problem, $S = 12(16)^2/6 = 512$ in.^{3}, and by Eq. 6-8

$$\sigma_{max} = \frac{M}{S} = \frac{23{,}100(12)}{512} = 542 \text{ psi}$$

In either solution, do not fail to notice that the bending moment substituted into the equations has the units of inch-pounds.

Section 6-7
Remarks on the
flexure formula

EXAMPLE 6-5

Find the maximum tensile and compressive stresses acting normal to the section *A-A* of the machine bracket shown in Fig. 6-11(a) caused by the applied force of 8 kips.

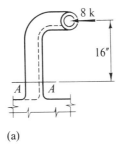
(a)

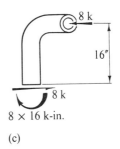
(c)

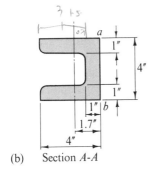
(b)   Section *A-A*

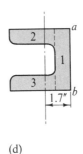
(d)

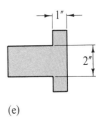
(e)

Fig. 6-11

SOLUTION

The shear and bending moment of proper magnitude and sense to maintain the segment of the member in equilibrium are shown in Fig. 6-11(c). Next the neutral axis of the beam must be located. This is done by locating the centroid of the area shown in Fig. 6-11(b) (also see Fig. 6-11(d)). Then the moment of inertia about the neutral axis is computed. In both these calculations the legs of the cross section are assumed rectangular, neglecting fillets. Then, keeping in mind the sense of the resisting bending moment and applying Eq. 6-4, one obtains the desired values.

Area Number	A [in.²]	y [in.] (from ab)	Ay
1	4.0	0.5	2.0
2	3.0	2.5	7.5
3	3.0	2.5	7.5
	$\sum A = 10.0$ in.²		$\sum Ay = 17.0$ in.³

$$\bar{y} = \frac{\sum Ay}{\sum A} = \frac{17.0}{10.0} = 1.70 \text{ in.} \quad \text{from the line } ab$$

$$I = \sum (I_o + Ad^2) = \frac{4(1)^3}{12} + 4(1.2)^2 + \frac{(2)1(3)^3}{12} + 2(3)(0.8)^2$$
$$= 14.43 \text{ in.}^4$$

$$\sigma_{\max} = \frac{Mc}{I} = \frac{(8)16(2.3)}{14.43} = 20.4 \text{ ksi} \quad \text{(compression)}$$

$$\sigma_{\max} = \frac{Mc}{I} = \frac{(8)16(1.7)}{14.43} = 15.1 \text{ ksi} \quad \text{(tension)}$$

These stresses vary linearly toward the neutral axis and vanish there. If for the same bracket the direction of the force P were reversed, the sense of the above stresses would also reverse. The results obtained would be the same if the cross-sectional area of the bracket were made T-shaped as shown in Fig. 6-11(e). The properties of this section about the significant axis are the same as those of the channel. Both these sections have an axis of symmetry.

The above example shows that members resisting flexure may be proportioned so as to have a different maximum stress in tension than in compression. This is significant for materials having different strengths in tension and compression. For example, cast iron is strong in compression and weak in tension. Thus, the proportions of a cast-iron member may be so set as to have a low maximum tensile stress. The potential capacity of the material may thus be better utilized. This matter will be further considered in the chapter on the design of beams.

6-8. INELASTIC BENDING OF BEAMS

The elastic flexure formula derived earlier is valid only while stress is proportional to strain. A more general theory will now be discussed for material that does not obey Hooke's law.

The basic kinematic assumption of the flexure theory, as stated in Art. 6-3, asserts that plane sections through a beam taken normal to its axis remain plane after the beam is subjected to bending. This assumption remains applicable if the material behaves inelastically. With no further assumptions, it means that strains in the fibers of a beam subjected to bending vary directly as their respective distances from the neutral axis. With this as a basis, together with equilibrium requirements and non-linear stress-strain relations, the generalized theory of flexure is constructed.

Section 6-8
Inelastic bending
of beams

Consider a segment of a prismatic beam subjected to bending moments. The cross-sectional area of this beam has a vertical axis of symmetry, Fig. 6-12(c). The linear variation of the strains from the neutral axis is diagrammatically represented for such a beam in Fig. 6-12(b). At the neutral axis, which as yet is undetermined, the strain is zero. Strains at points away from the neutral axis correspond to the horizontal distances from the line ab to the line cd in Fig. 6-12(b). For example, the strain of a fiber at a distance $-y_2$ from the neutral axis is ε_2. Such distances define the axial strain of every fiber in a beam.

To make the argument general, the material will be assumed to have a different stress-strain curve in tension and compression. A possible curve for such a material is shown in Fig. 6-12(a). Such curves may be obtained from axial-load experiments.

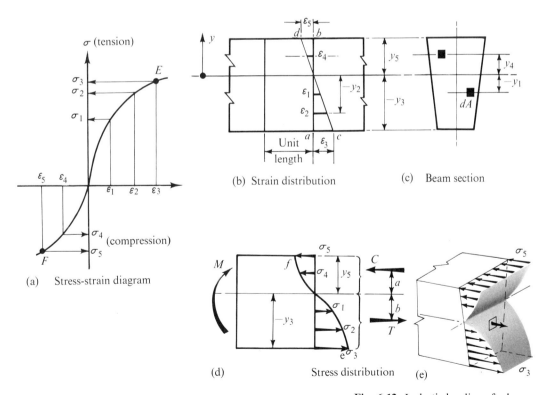

Fig. 6-12. Inelastic bending of a beam.

*Chapter 6
Bending stresses
in beams*

If the Poisson effect is neglected, the longitudinal fibers of a beam in bending behave independently. Each one of these may be thought of as an infinitesimal axially loaded rod, stressed to a level dependent upon its strain. Since the variation of the strain in a beam is set by the assumption, the stress pattern may be formulated from the stress-strain curve, Fig. 6-12(a). For example, corresponding to the tensile strain ε_1, at a distance $-y_1$ from the neutral axis, a tensile stress σ_1 acts in the beam. Similarly, ε_4 is associated with $-\sigma_4$, a compressive stress. The same thing applies to any other fiber of the beam. This determines the stress distribution shown in Figs. 6-12(d) and (e); it resembles the shape of the stress-strain curve (compare EF with ef by turning it clockwise through 90°).

Since the beam acts in pure bending, the same equations of statics will be used here as were used in establishing the elastic flexure formula. The two applicable relations as before are

$$\sum F_x = 0 \quad \text{or} \quad \int_A \sigma_x \, dA = 0 \tag{6-10}$$

$$\sum M_z = 0 \quad \text{or} \quad -\int_A \sigma_x y \, dA = M \tag{6-11}$$

where σ_x is a normal stress acting on an infinitesimal element dA of the cross-sectional area A of the beam, and y is the distance from the neutral axis to an element dA. It should be noted that for a positive M the integral in Eq. 6-11 becomes positive since for positive y's the stress σ_x is negative, whereas for negative y's the stress σ_x is positive.

The solution of the most general problem in inelastic bending, i.e., the satisfaction of the equilibrium Eqs. 6-10 and 6-11, requires a trial-and-error procedure. Initially the location of the neutral axis is unknown. A possible method consists of assuming a strain distribution, thus locating a trial neutral axis and giving the stress distribution shown in Fig. 6-12(d). Such trials must be continued until the sum of the forces C on the compression side of the beam is equal to the sum of the forces T on the tension side of the beam. When such a condition is fulfilled, the neutral axis of the beam is located. Note particularly that in inelastic flexure the neutral axis of a beam may not coincide with the centroidal axis of the cross-sectional area. It does so only if the cross-sectional area has two axes of symmetry and the stress-strain diagram is identical in tension and compression.

After the neutral axis is located and the magnitudes of C and T are known, their line of action may be determined. This is possible since the stress distribution on the cross-sectional area is known. Finally, the resisting moment is $T(a + b)$ or $C(a + b)$. The foregoing process is equivalent to the integration indicated by Eq. 6-11. However, the resisting moment so computed, based upon assumed strains, may not be equal to the applied moment. Hence, the process must be repeated by initially

*Section 6-8
Inelastic bending
of beams*

assuming greater or smaller strain at the extreme fibers until the resisting moment becomes equal to the applied moment.

The foregoing method of solving a general problem is tedious, and accelerated procedures for arriving at a solution have been developed.* However, the above discussion should be sufficient to give a picture of the behavior of a beam in flexure beyond the elastic limit. As a simple example, consider a beam of rectangular cross section subjected to bending. Let the stress-strain diagram of the beam material be alike in tension and compression as shown in Fig. 6-13(a). Then, as progressively increasing bending moments are applied to the beam, the strains will increase as exemplified by ε_1, ε_2, and ε_3 in Fig. 6-13(b). Corresponding to these strains and their linear variation from the neutral axis, the stress distribution will look as shown in Fig. 6-13(c). The neutral axis coincides with the centroidal axis in this case, as the section has two axes of symmetry and the stress-strain diagram is alike in tension and compression.

If σ_3 corresponds to the ultimate strength of the material in axial tension, the ultimate bending moment which the beam is capable of resisting may be predicted. It is associated with the stress distribution

---

* A. Nadai, *Theory of Flow and Fracture of Solids*, vol. I (New York: McGraw-Hill Book Company, 1950), p. 356.

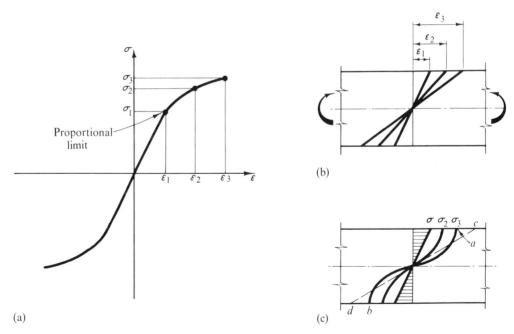

Fig. 6-13. Rectangular beam in bending exceeding the proportional limit of the material.

given by the curved line *ab* shown in Fig. 6-13(c). An equivalent resistance to the bending moment based on the assumption of linear stress distribution from the neutral axis is shown by the line *cd* in the same figure. Since both of these stress distributions supposedly resist the same moment and in the latter case lower stresses act near the neutral axis, higher stresses must act near the outer fibers. The stress in the extreme fibers, computed on the basis of the elastic flexure formula for the experimentally determined ultimate bending moment, is called the *rupture modulus* of the material in bending. It is higher than the true stress. For materials whose stress-strain diagrams approach a straight line all the way up to the ultimate strength, the discrepancy between the true maximum stress and the rupture modulus is small. On the other hand, the discrepancy is large for materials with a pronounced curvature in the stress-strain curve.

As another important example of inelastic bending, consider a rectangular beam of elastic-plastic material, Fig. 6-14. In such an idealization of material behavior a sharp separation of the member into distinct elastic and plastic zones is possible. For example, if the strain in the extreme fibers is double that at the beginning of yielding, only the middle half of the beam remains elastic, Fig. 6-14(a). In this case the outer quarters of the beam yield. The magnitude of the moment M_1 corresponding to this condition can be readily computed (see Example 6-7). At higher strains the elastic zone or core diminishes. Stress distribution corresponding to this situation is shown in Figs. 6-14(b) and (c).

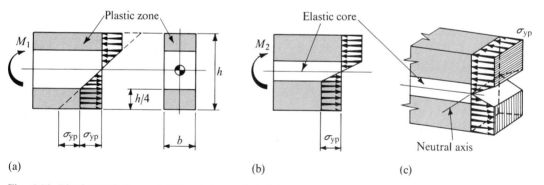

Fig. 6-14. Elastic-plastic beam at different stages of straining.

EXAMPLE 6-6

Determine the plastic or the ultimate capacity in flexure of a mild steel beam of rectangular cross section. Consider the material to be ideally elastic-plastic.

SOLUTION

The idealized stress-strain diagram is in Fig. 6-15(a). It is assumed that the material has the same properties in tension and compression. The strains that can take place during yielding are much greater than the

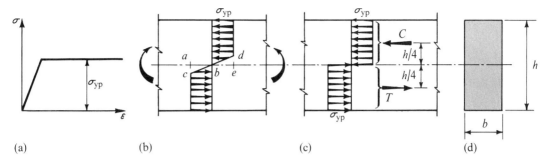

Fig. 6-15

maximum elastic strain (15 to 20 times the latter quantity). Therefore, since unacceptably large deformations of the beam occur along with very large strains, the plastic moment may be taken as the ultimate moment.

The stress distribution shown in Fig. 6-15(b) applies after a large amount of deformation takes place. In computing the resisting moment the stresses corresponding to the triangular areas *abc* and *bde* may be neglected without unduly impairing the accuracy. They contribute little resistance to the applied bending moment because of their short moment arms. Hence the idealization of the stress distribution to that shown in Fig. 6-15(c) is permissible and has a simple physical meaning. The whole upper half of the beam is subjected to a uniform compressive stress σ_{yp}, while the lower half is all under a uniform tension σ_{yp}. That the beam is divided evenly into a tension and a compression zone follows from symmetry. Numerically

$$C = T = \sigma_{yp}(bh/2), \quad \text{i.e., (stress)} \times \text{(area)}$$

Each one of these forces acts at a distance $h/4$ from the neutral axis. Hence the plastic or ultimate resisting moment of the beam is

$$M_p \equiv M_{ult} = C\left(\frac{h}{4} + \frac{h}{4}\right) = \sigma_{yp}\frac{bh^2}{4}$$

where b is the breadth of the beam and h is its height.

The same solution may be obtained by directly applying Eqs. 6-10 and 6-11. Noting the sign of stresses, one can conclude that Eq. 6-10 is satisfied by taking the neutral axis through the middle of the beam. By taking $dA = b\,dy$ and noting the symmetry around the neutral axis, one changes Eq. 6-11 to

$$M_p \equiv M_{ult} = -2\int_0^{h/2}(-\sigma_{yp})yb\,dy = \sigma_{yp}bh^2/4 \quad \textbf{(6-12)}$$

The resisting bending moment of a beam of rectangular section when the outer fibers just reach σ_{yp}, as given by the elastic flexure formula, is

$M_{yp} = \sigma_{yp}I/c = \sigma_{yp}(bh^2/6)$ therefore $M_p/M_{yp} = 1.50$

The ratio M_p/M_{yp} depends only on the cross-sectional properties of a member and is called the *shape factor*. The shape factor above for the rectangular beam shows that M_{yp} may be exceeded by 50 per cent before the ultimate capacity of a rectangular beam is reached.

For static loads such as occur in buildings, ultimate capacities can be detmerined using plastic moments. The procedures based on such concepts are referred to as *the plastic method of analysis* or *design*. For such work *plastic section modulus Z* is defined as follows:

$$M_p = \sigma_{yp}Z \quad (6\text{-}13)$$

For the rectangular beam analyzed above $Z = bh^2/4$.

The *Steel Construction Manual** provides a table of plastic section moduli for many common steel shapes. An abridged list of plastic section moduli for steel sections is given in Table 9 of the Appendix. For a given M_p and σ_{yp} the solution of Eq. 6-13 for Z is very simple.

The method of limit or plastic analysis is unacceptable in machine design in situations where fatigue properties of the material are important.

EXAMPLE 6-7

Find the residual stresses in a rectangular beam upon removal of the ultimate bending moment.

SOLUTION

The stress distribution associated with an ultimate moment is shown in Fig. 6-16(a). The magnitude of this moment has been determined in the preceding example and is $M_p = \sigma_{yp}bh^2/4$. Upon release of this plastic moment M_p every fiber in the beam can rebound elastically. Neglecting Bauschinger's effect (see Fig. 4-16), one sees that the elastic range during

---

* American Institute of Steel Construction, *AISC Steel Construction Manual* (New York: AISC, Inc., 1963), pp. 2–6 through 2–9.

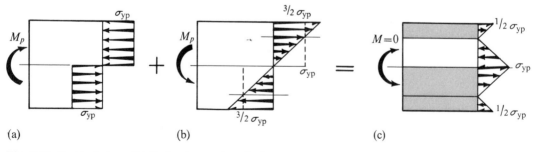

Fig. 6-16. Residual stress distribution in a rectangular bar.

the unloading is double that which could take place initially. Therefore, since $M_{yp} = \sigma_{yp}bh^2/6$ and the moment being released is $\sigma_{yp}(bh^2/4)$ or $1.5M_{yp}$, the maximum stress calculated on the basis of elastic action is $\tfrac{3}{2}\sigma_{yp}$ as shown in Fig. 6-16(b). Superimposing the initial stresses at M_p with the elastic rebound stresses due to the release of M_p, one finds the residual stresses, Fig. 6-16(c). Note that both tensile and compressive longitudinal microresidual stresses remain in the beam. The tensile zones are shaded in the figure. If such a beam were machined by gradually reducing its depth, the release of the residual stresses would cause undesirable deformations of the bar.

Section 6-8
Inelastic bending
of beams

EXAMPLE 6-8

Determine the moment resisting capacity of an elastic-plastic rectangular beam.

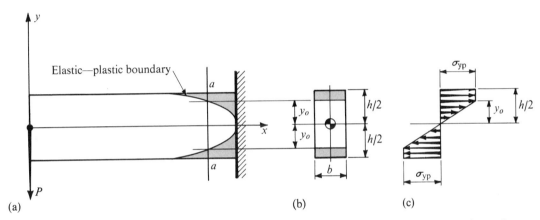

Fig. 6-17. Elastic-plastic cantilever.

SOLUTION

To make the problem more definite consider a cantilever loaded as in Fig. 6-17(a). If the beam is made of ideal elastic-plastic material and the applied force P is large enough to cause yielding, plastic zones will be formed (shown shaded in the figure). At an arbitrary section a-a the corresponding stress distribution will be as shown in Fig. 6-17(c). The elastic zone extends over the depth of $2y_o$. Noting that within the elastic zone the stresses vary linearly and that everywhere in the plastic zone the axial stress is σ_{yp}, one finds that the resisting moment M is

$$M = -2\int_0^{y_o}\left(-\frac{y}{y_o}\sigma_{yp}\right)(b\,dy)y - 2\int_{y_o}^{h/2}(-\sigma_{yp})(b\,dy)y$$

$$= \sigma_{yp}\frac{bh^2}{4} - \sigma_{yp}\frac{by_o^2}{3} = M_p - \sigma_{yp}\frac{by_o^2}{3} \qquad (6\text{-}14)$$

where the last simplification is done in accordance with Eq. 6-12. It is interesting to note that, in this general equation, if $y_o = 0$, the moment capacity becomes equal to the plastic or ultimate moment. On the other hand, if $y_o = h/2$, the moment reverts back to the limiting elastic case where $M = \sigma_{yp} b h^2/6$. When the applied bending moment along the span is known, the elastic-plastic boundary can be determined by solving Eq. 6-14 for y_o. As long as an elastic zone or core remains, the plastic deformations cannot progress without a limit. This is a case of contained plastic flow. This was encountered earlier in Examples 4-10 and 5-9.

6-9. STRESS CONCENTRATIONS

The flexure theory developed in the preceding articles applies only to beams of constant cross-sectional area. Such beams may be referred to as *prismatic* beams. If the cross-sectional area of the beam varies gradually, no significant deviation from the stress pattern discussed earlier takes place. However, if notches, grooves, bolt holes, or an abrupt change in the cross-sectional area of the beam occurs, high local stresses arise. This situation is analogous to the ones discussed earlier for axial and torsion members. Again it is very difficult to obtain analytical expressions for the actual stress. Much of the information regarding the actual stress distribution comes from accurate photoelastic experiments.

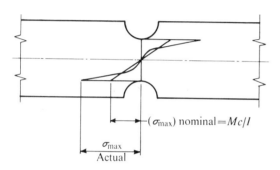

$K = (\sigma_{max})$ actual$/(\sigma_{max})$ nominal

Fig. 6-18. Stress-concentration factor in pure bending.

Fortunately, as in the other cases discussed, only the geometric proportions of the member affect the local stress pattern. Moreover, since interest is in the maximum stress, the idea of the stress-concentration factor may be used to advantage. The ratio K of the actual maximum stress to the nominal maximum stress in the minimum section as given by Eq. 6-4 is defined as the stress-concentration factor in bending. This concept is illustrated in Fig. 6-18. Hence, in general,

$$(\sigma_{max})_{actual} = K(Mc/I) \qquad (6\text{-}15)$$

Figures 6-19 and 6-20 are plots of stress-concentration factors for two representative cases.* The factor K, depending on the proportions of the member, may be obtained from these diagrams. A study of these graphs indicates the desirability of generous fillets and the elimination of sharp notches to reduce local stress concentrations. These remedies are highly desirable in machine design. In structural work, particularly where ductile materials are used and the applied forces are not fluctuating, stress concentrations are ignored.

---

* These figures are reproduced from M. M. Frocht, "Factors of Stress Concentration Photoelastically Determined," *Trans. ASME*, **57** (1935), A-67.

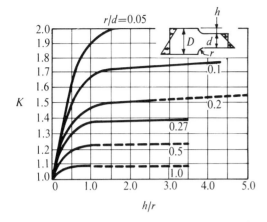
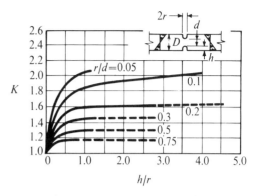

Fig. 6-19. Stress-concentration factors in pure bending for flat bars with various fillets.

Fig. 6-20. Stress-concentration factors in bending for grooved flat bars.

If the cross-sectional area of a beam is irregular itself, stress concentrations also occur. This becomes particularly significant if the cross-sectional area has re-entrant angles. For example, high localized stresses occur at the point where the flange and the web* of an *I* beam meet. To minimize these, commercially rolled shapes have a generous fillet at all such points.

In addition to stress concentrations caused by changes in the cross-sectional area of a beam, another effect is significant. Forces often are applied over a limited area of a beam. Moreover, the reactions act only locally on a beam at the points of support. In the previous treatment, all such forces were idealized as concentrated forces. In practice the average bearing pressure between the member delivering such a force and the beam are computed at the point of contact of such forces with the beam. This bearing pressure or stress acts normal to the neutral surface of a beam and is at right angles to the bending stresses discussed in this chapter. A more detailed study of the effect of such forces shows that they cause a disturbance of all stresses on a local scale, and the bearing pressure as normally computed is a crude approximation. The stresses at right angles to the flexural stresses behave more nearly as shown in Fig. 4-30. An investigation of the disturbance caused in the bending-stress distribution by the bearing stresses is beyond the scope of this book.†

The reader must remember that the stress-concentration factors

---

* The *web* is a thin vertical part of a beam. Thin horizontal parts of a beam are called *flanges*.
† By virtue of the St. Venant principle (Art. 4-18), at distances away from the concentrated forces comparable with the cross-sectional dimensions of a member, the formulas developed in this text are accurate, but the usual formulas are not applicable for short, stubby beams such as gear teeth.

apply only while the material behaves elastically. Inelastic behavior of material tends to reduce these factors.

6-10. BEAMS OF TWO MATERIALS

So far, the beams analyzed were assumed to be of one homogeneous material. Important uses of beams made of several different materials occur in practice. Beams of two materials are especially common. Wooden beams are often reinforced by metal straps, and concrete beams are reinforced with steel rods. The fundamental theory underlying the elastic analysis of such beams will be discussed in this article. Extension of the analysis into the inelastic range follows the procedures discussed in Art. 6-8. A solution for a beam behaving inelastically is given in one of the examples that follow.

Consider a symmetrical beam of two materials with a cross section as shown in Fig. 6-21(a). The outer material 1 has an elastic modulus E_1, and the modulus of the inner material 2 is E_2. If such a beam is subjected to bending, the basic deformation assumption used in the flexure theory remains valid. Plane sections at right angles to the axis of a beam remain plane. Therefore the strains must vary linearly from the neutral axis, as shown in Fig. 6-21(b). For the elastic case, stress is proportional to strain, and the stress distribution, assuming $E_1 > E_2$, is as shown in Fig. 6-21(c). Note that at the surfaces of contact of the two materials a break in the intensity of stress is indicated. Although the strain in both materials at such surfaces is equal, a greater stress develops in the stiffer material. The stiffness of a material is measured by the elastic modulus E. The foregoing information is sufficient to solve any beam problem of two (or more) materials by using a trial-and-error solution similar to the one discussed in Art. 6-8. However, a considerable simplification of the procedure is possible. In formally applying $\Sigma F_x = 0$ to locate the neutral axis and $\Sigma M_z = 0$ to obtain the resisting moment, only the correct magnitudes and locations of the resisting forces (not stresses) are significant. The new technique consists of constructing a section of one material on which the resisting forces are the same as on the original section. Such a section is termed an *equivalent* or *transformed cross-sectional area*. After a beam of several materials is reduced to an equivalent beam of one material, the usual elastic flexure formula applies.

The transformation of a section is accomplished by changing dimensions of a cross section parallel to the neutral axis in the ratio of elastic moduli of the materials. For example, if the equivalent section is wanted in material 1, the dimensions corresponding to material 1 do not change. The horizontal dimensions of material 2 are changed by a ratio n, where $n = E_2/E_1$, Fig. 6-21(d). On the other hand, if the transformed section is to be of material 2, the horizontal dimension of the other material is changed by a ratio $n_1 = E_1/E_2$, Fig. 6-21(e). The ratio n_1 is the reciprocal of n.

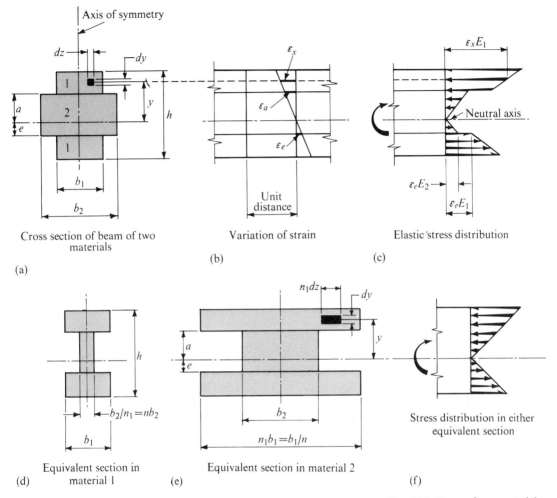

Fig. 6-21. Beam of two materials.

The legitimacy of transforming sections is seen by comparing the forces acting on the original and on the equivalent sections. The force from a known strain ε_x acting on an elementary area $dz\, dy$ in Fig. 6-21(a) is $\varepsilon_x E_1\, dz\, dy$. The same element of area in Fig. 6-21(e) is $n_1\, dz\, dy$. The force acting on it is $\varepsilon_x E_2 n_1\, dx\, dz$. However, from the definition of n_1, $E_1 = n_1 E_2$. So the forces acting on both elements are the same, and both, by virtue of their location, contribute equally to the resisting moment.

In a beam with a transformed area, strains and stresses vary linearly from its neutral axis. The stresses calculated in the usual manner are correct for the material of which the transformed section is made. For the other material the computed stress must be multiplied by the ratio n or n_1 of the transformed to the actual area. For example, the force acting on $n_1\, dz\, dy$ in Fig. 6-21(e) actually acts on $dz\, dy$ of the real material.

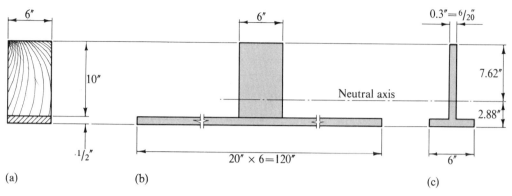

Fig. 6-22

EXAMPLE 6-9

Consider a composite beam of the cross-sectional dimensions shown in Fig. 6-22(a). The upper 6-in.-by-10-in. (full-sized) part is wood, $E_w = 1.5 \times 10^6$ psi; the bottom $\frac{1}{2}$-in.-by-6-in. strap is steel, $E_s = 30 \times 10^6$ psi. If this beam is subjected to a bending moment of 20,000 ft-lb around a horizontal axis, what are the maximum stresses in the steel and wood?

SOLUTION

The ratio of the elastic moduli $E_s/E_w = 20$. Hence, using a transformed section of wood, the width of the bottom strip is $6(20) = 120$ in. The transformed area is shown in Fig. 6-22(b). Its centroid and moment of inertia around the centroidal axis are

$$y = \frac{6(10)5 + (0.5)120(10.25)}{6(10) + (0.5)120} = 7.62 \text{ in.} \quad \text{from the top}$$

$$I_{zz} = \frac{6(10)^3}{12} + (6)10(2.62)^2 + \frac{120(0.5)^3}{12} + (0.5)120(2.63)^2 = 1{,}328 \text{ in.}^4$$

The maximum stress in the wood is

$$(\sigma_w)_{max} = \frac{Mc}{I} = \frac{(20{,}000)12(7.62)}{1{,}328} = 1{,}380 \text{ psi}$$

The maximum stress in the steel is

$$(\sigma_s)_{max} = n\sigma_w = 20\frac{(20{,}000)12(2.88)}{1{,}328} = 10{,}400 \text{ psi}$$

ALTERNATE SOLUTION

A transformed area in terms of steel may be used instead. Then the equivalent width of wood is $b/n = 6/20$, or 0.3 in. This transformed area is shown in Fig. 6-22(c).

$$y = \frac{(0.3)10(5.5) + 6(0.5)(0.25)}{(0.3)10 + 6(0.5)} = 2.88 \text{ in.} \quad \text{from the bottom}$$

*Section 6-10
Beams of two materials*

$$I_{zz} = \frac{(0.3)10^3}{12} + (0.3)10(2.62)^2 + \frac{6(0.5)^3}{12} + (0.5)6(2.63)^2 = 66.5 \text{ in.}^4$$

$$(\sigma_s)_{\max} = \frac{(20,000)12(2.88)}{66.5} = 10,400 \text{ psi}$$

$$(\sigma_w)_{\max} = \frac{\sigma_s}{n} = \left(\frac{1}{20}\right)\frac{(20,000)12(7.62)}{66.5} = 1,380 \text{ psi}$$

Note that if the transformed section is an equivalent wooden section, the stresses in the actual wooden piece are obtained directly. Conversely, if the equivalent section is steel, stresses in steel are obtained directly. The stress in a material stiffer than the material of the transformed section is increased since to cause the same unit strain a higher stress is required.

EXAMPLE 6-10

Determine the maximum stress in the concrete and the steel for a reinforced-concrete beam with the section shown in Fig. 6-23(a) if it is subjected to a positive bending moment of 50,000 ft-lb. The reinforcement consists of two #9 steel bars. (These bars are $1\frac{1}{8}$ in. in diameter and have a cross-sectional area of 1 in.²) Assume the ratio of E for steel to that of concrete to be 15, i.e., $n = 15$.

SOLUTION

Plane sections are assumed to remain plane in a reinforced-concrete beam. Strains vary linearly from the neutral axis as shown in Fig. 6-23(b) by the line ab. A transformed section in terms of concrete is used to solve this problem. However, concrete is so weak in tension that there is no assurance that minute cracks will not occur in the tension zone of the beam. For this reason no credit is given to concrete for resisting

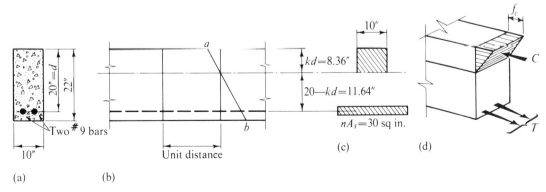

Fig. 6-23

tension. On the basis of this assumption, concrete in the tension zone of a beam only holds the reinforcing steel in place.* Hence in this analysis it virtually does not exist at all, and the transformed section assumes the form shown in Fig. 6-23(c). The cross section of concrete has its own shape above the neutral axis; below it no concrete is shown. Steel, of course, can resist tension, so it *is* shown as the transformed concrete area. For computation purposes, the steel is located by a single dimension from the neutral axis to its centroid. There is a negligible difference between this distance and the true distances to the various steel fibers. The crude placement of rods on the job concurs with this practice.

So far the idea of the neutral axis has been used, but its location is unknown. However, it is known that this axis coincides with the axis through the centroid of the transformed section. It is further known that the first (or statical) moment of the area on one side of a centroidal axis is equal to the first moment of the area on the other side. Thus, let kd be the distance from the top of the beam to the centroidal axis as shown in Fig. 6-23(c), where k is an unknown ratio† and d is the distance from the top of the beam to the center of the steel. An algebraic restatement of the foregoing locates the neutral axis, about which I is computed and stresses are determined as in the preceding example.

$$\underbrace{10(kd)}_{\text{concrete area}} \underbrace{(kd/2)}_{\text{arm}} = \underbrace{30}_{\text{transformed steel area}} \underbrace{(20 - kd)}_{\text{arm}}$$

$$5(kd)^2 = 600 - 30(kd)$$

$$(kd)^2 + 6(kd) - 120 = 0$$

Hence $\quad kd = 8.36$ in. $\quad$ and $\quad 20 - kd = 11.64$ in.

$$I = \frac{10(8.36)^3}{12} + 10(8.36)\left(\frac{8.36}{2}\right)^2 + 0 + 30(11.64)^2 = 6{,}020 \text{ in.}^4$$

$$(\sigma_c)_{\max} = \frac{Mc}{I} = \frac{(50{,}000)12(8.36)}{6{,}020} = 833 \text{ psi}$$

$$\sigma_s = n\frac{Mc}{I} = \frac{15(50{,}000)12(11.64)}{6{,}020} = 17{,}400 \text{ psi}$$

ALTERNATE SOLUTION

After kd is determined, instead of computing I, a procedure evident from Fig. 6-23(d) may be used. The resultant force developed by the stresses acting in a "hydrostatic" manner on the compression side of the beam

---

* Actually it is used to resist shear and provide fireproofing for the steel.
† This conforms with the usual notation used in books on reinforced concrete. In this text h is generally used to represent the height or depth of the beam.

must be located $kd/3$ below the top of the beam. Moreover, if b is the width of the beam, this resultant force $C = \frac{1}{2}(\sigma_c)_{max}b(kd)$ (average stress times area). The resultant tensile force T acts at the center of the steel and is equal to $A_s\sigma_s$, where A_s is the cross-sectional area of the steel. Then if jd is the distance between T and C, and since $T = C$, the applied moment M is resisted by a couple equal to Tjd or Cjd.

Section 6-10 Beams of two materials

$$jd = d - kd/3 = 20 - (8.36/3) = 17.21 \text{ in.}$$

$$M = Cjd = \tfrac{1}{2} b(kd)(\sigma_c)_{max}(jd)$$

$$(\sigma_c)_{max} = \frac{2M}{b(kd)(jd)} = \frac{2(50{,}000)12}{10(8.36)(17.21)} = 833 \text{ psi}$$

$$M = T(jd) = A_s\sigma_s jd$$

$$\sigma_s = \frac{M}{A_s(jd)} = \frac{(50{,}000)12}{2(17.21)} = 17{,}400 \text{ psi}$$

Both methods naturally give the same answer. The second method is more convenient in practical applications. Since steel and concrete have different allowable stresses, the beam is said to have balanced reinforcement when it is designed so that the respective stresses are at their allowable level simultaneously. Note that the beam shown would become virtually worthless if the bending moments were applied in the opposite direction.

EXAMPLE 6-11

Determine the ultimate moment carrying capacity for the reinforced-concrete beam of the preceding example. Assume that the steel reinforcement yields at 40,000 psi and that the ultimate strength of concrete $f'_c = 2{,}500$ psi.

SOLUTION

When the reinforcing steel begins to yield, large deformations commence. This is taken to be the ultimate capacity of steel: hence $T_{ult} = A_s\sigma_{yp}$.

At the ultimate moment, experimental evidence indicates that the compressive stresses can be approximated by the rectangular stress block shown in Fig. 6-24. It is customary to assume the average stress in this compressive stress block to be $0.85f'_c$. On this basis, keeping in mind that $T_{ult} = C_{ult}$, one has

$$T_{ult} = \sigma_{yp}A_s = 40{,}000 \times 2 = 80{,}000 \text{ lb} = C_{ult}$$

$$k'd = \frac{C_{ult}}{0.85f'_c b} = \frac{80{,}000}{0.85 \times 2{,}500 \times 10} = 3.77 \text{ in.}$$

$$M_{ult} = T_{ult}\left(d - \frac{k'd}{2}\right) = 80{,}000\left(20 - \frac{3.77}{2}\right)\frac{1}{12} = 121{,}000 \text{ ft-lb}$$

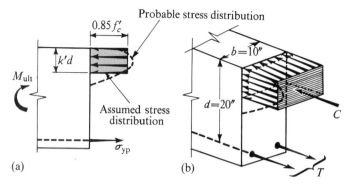

Fig. 6-24

6-11. CURVED BEAMS

The flexure theory for curved bars is developed in this article. Attention is confined to beams having an axis of symmetry of the cross section, with this axis lying in one plane along the length of the beam. Only the elastic case is treated,* with the usual proviso that the elastic modulus is the same in tension and compression.

Consider a curved member such as shown in Figs. 6-25(a) and (b). The outer fibers are at a distance of r_o from the center of curvature O. The inner fibers are at a distance of r_i. The distance from O to the centroidal

---

* For plastic analysis of curved bars see for example H. D. Conway, "Elastic-Plastic Bending of Curved Bars of Constant and Variable Thickness," *Journal of Applied Mechanics*, **27**, no. 4 (Dec. 1960), 733–34.

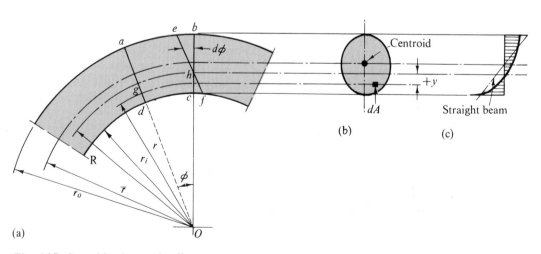

Fig. 6-25. Curved bar in pure bending.

axis is $\bar{r}$. The solution* of this problem is again based on the familiar assumption: Sections perpendicular to the axis of the beam remain plane after a bending moment M is applied. This is diagrammatically represented by the line ef in relation to an element of the beam $abcd$. The element is defined by the central angle ϕ.

Although the basic deformation assumption is the same as for straight beams, and, from Hooke's law, the normal stress $\sigma = E\varepsilon$, a difficulty is encountered. The initial length of a beam fiber such as gh depends upon the distance r from the center of curvature. Thus, although the total deformation of beam fibers (described by the small angle $d\phi$) follows a linear law, strains do not. The elongation of a generic fiber gh is $(R - r)\,d\phi$, where R is the distance from O to the neutral surface (not yet known), and its initial length is $r\phi$. The strain ε of any arbitrary fiber is $(R - r)(d\phi)/r\phi$, and the normal stress σ on an element dA of the cross-sectional area is

$$\sigma = E\varepsilon = E\frac{(R - r)\,d\phi}{r\phi} \tag{6-16}$$

For future use note also that

$$\frac{\sigma r}{R - r} = \frac{E\,d\phi}{\phi} \tag{6-17}$$

Equation 6-16 gives the normal stress acting on an element of area of the cross section of a curved beam. The location of the neutral axis follows from the condition that the summation of the forces acting perpendicular to the section must be equal to zero, i.e.

$$\sum F_n = 0, \qquad \int_A \sigma\,dA = \int_A \frac{E(R - r)\,d\phi}{r\phi}\,dA = 0$$

However, since E, R, ϕ, and $d\phi$ are constant at any one section of a stressed beam, they may be taken outside the integral sign and a solution for R obtained. Thus:

$$\frac{E\,d\phi}{\phi}\int_A \frac{R - r}{r}\,dA = \frac{E\,d\phi}{\phi}\left[R\int_A \frac{dA}{r} - \int_A dA\right] = 0$$

$$R = \frac{A}{\int_A dA/r} \tag{6-18}$$

where A is the cross-sectional area of the beam and R locates the neutral

---

* This approximate solution was developed by E. Winkler in 1858. The exact solution of the same problem by the methods of the mathematical theory of elasticity is due to M. Golovin, who solved it in 1881.

axis. Note that the neutral axis so found does not coincide with the centroidal axis. This differs from the situation found to be true for straight elastic beams.

Now that the location of the neutral axis is known, the equation for the stress distribution is obtained by equating the external moment to the internal resisting moment built up by the stresses given by Eq. 6-16. The summation of moments is made around the z axis, which is normal to the plane of the figure shown in Fig. 6-25(a).

$$\sum M_z = 0, \quad M = \int_A \underbrace{\sigma\, dA}_{\text{(force)}} \underbrace{(R - r)}_{\text{(arm)}} = \int_A \frac{E(R - r)^2\, d\phi}{r\phi}\, dA$$

Again remembering that E, R, ϕ, and $d\phi$ are constant at a section, by using Eq. 6-17, and performing the algebraic steps indicated, the following is obtained:

$$M = \frac{E\, d\phi}{\phi} \int_A \frac{(R - r)^2}{r}\, dA = \frac{\sigma r}{R - r} \int_A \frac{(R - r)^2}{r}\, dA$$

$$= \frac{\sigma r}{R - r} \int_A \frac{R^2 - Rr - Rr + r^2}{r}\, dA$$

$$= \frac{\sigma r}{R - r} \left[R^2 \int_A \frac{dA}{r} - R \int_A dA - R \int_A dA + \int_A r\, dA \right]$$

Here, since R is a constant, the first two integrals vanish as may be seen from the bracketed expression appearing just before Eq. 6-18. The third integral is A, and the last integral, by definition, is $\bar{r}A$. Hence

$$M = \frac{\sigma r}{R - r} (\bar{r}A - RA)$$

whence the normal stress acting on a curved beam at a distance r from the center of curvature is

$$\sigma = \frac{M(R - r)}{rA(\bar{r} - R)} \qquad (6\text{-}19)$$

If positive y is measured toward the center of curvature from the neutral axis, and $\bar{r} - R = e$, Eq. 6-19 may be written in a form which more closely resembles the flexure formula for straight beams

$$\sigma = \frac{My}{Ae(R - y)} \qquad (6\text{-}20)$$

These equations indicate that the stress distribution in a curved bar follows a hyperbolic pattern. The maximum stress is always on the inside (concave) side of the beam. A comparison of this result with the one that follows from the formula for straight bars is shown in Fig. 6-25(c). Note particularly that in the curved bar the neutral axis is pulled toward the center of the curvature of the beam. This results from the higher stresses developed below the neutral axis. The theory developed applies of course only to elastic stress distribution and only to beams in pure bending. For a consideration of situations where an axial force is also present at a section see Art. 8-2.

EXAMPLE 6-12

Compare stresses in a 2-in.-by-2-in. rectangular bar subjected to end couples of 13,333 in-lb in the three special cases: (a) straight beam, (b) beam curved to a radius of 10 in. along the centroidal axis, i.e., $\bar{r} = 10$ in., Fig. 6-26(a), and (c) beam curved to $\bar{r} = 3$ in.

SOLUTION

Case (a) follows directly by applying Eqs. 6-9 and 6-8 in that order.

$$S = \frac{bh^2}{6} = \frac{2(2)^2}{6} = 1.33 \text{ in.}^3$$

$$\sigma_{max} = \frac{M}{S} = \frac{13,333}{1.33} = \pm 10,000 \text{ psi or } 10 \text{ ksi}$$

This result is shown in Fig. 6-26(c). $\bar{r} = \infty$ since a straight bar has an infinite radius of curvature.

To solve parts (b) and (c) the neutral axis must be located first. This is found in general terms by integrating Eq. 6-18. For the rectangular section, the elementary area is taken as $b\, dr$, Fig. 6-26(b). The integration is carried out between the limits r_i and r_o, the inner and outer radii, respectively.

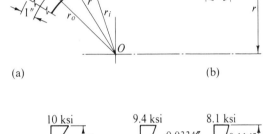

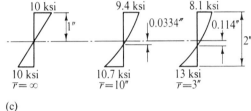

Fig. 6-26

$$R = \frac{A}{\int_A dA/r} = \frac{bh}{\int_{r_i}^{r_o} b\, dr/r} = \frac{h}{\int_{r_i}^{r_o} dr/r} \quad (6\text{-}21)$$

$$= \frac{h}{|\ln r|_{r_i}^{r_o}} = \frac{h}{\ln\left(\dfrac{r_o}{r_i}\right)} = \frac{h}{2.3026 \log\left(\dfrac{r_o}{r_i}\right)}$$

where h is the depth of the section, ln is the natural logarithm, and log is a logarithm with a base of 10 (common logarithm).

For Case (b), $h = 2$ in., $\bar{r} = 10$ in., $r_i = 9$ in., and $r_o = 11$ in. The solution is obtained by evaluating Eqs. 6-21 and 6-19. Subscript i refers to the normal stress σ of the inside fibers; o of the outside fibers.

$$R = \frac{2}{2.3026 \log (11/9)} = \frac{2}{2.3026 (\log 11 - \log 9)} = 9.9666 \text{ in.}$$

$$e = \bar{r} - R = 10 - 9.9666 = 0.0334 \text{ in.}$$

$$\sigma_i = \frac{M(R - r_i)}{r_i A(\bar{r} - R)} = 13,333 \frac{9.9666 - 9}{9(4)(0.0334)} = 10,700 \text{ psi}$$

$$\sigma_o = \frac{M(R - r_o)}{r_o A(\bar{r} - R)} = 13,333 \frac{9.9666 - 11}{11(4)(0.0334)} = -9,400 \text{ psi}$$

The negative sign of σ_o indicates a compressive stress. These quantities and the corresponding stress distribution are shown in Fig. 6-26(c); $\bar{r} = 10$ in.

Case (c) is computed in the same way. Here $h = 2$ in., $\bar{r} = 3$ in., $r_i = 2$ in., and $r_o = 4$ in. Results of the computation are shown in Fig. 6-26(c).

$$R = \frac{2}{\ln (4/2)} = \frac{2}{\ln 2} = \frac{2}{0.6931} = 2.886 \text{ in.}$$

$$e = 3 - 2.886 = 0.114 \text{ in.}$$

$$\sigma_i = 13,333 \frac{0.886}{2(0.114)4} = 13,000 \text{ psi}$$

$$\sigma_o = 13,333 \frac{-1.114}{4(0.114)4} = -8,140 \text{ psi}$$

Several important conclusions, generally true, may be reached from the above example. First, *the usual flexure formula is reasonably good for beams of considerable curvature.* Only 7 per cent error in the maximum stress occurs in Case (b) for $\bar{r}/h = 5$, an error tolerable for most applications. For greater ratios of $\bar{r}/h$ this error diminishes. As the curvature of the beam increases, the stress on the concave side rapidly increases over the one given by the usual flexure formula. When $\bar{r}/h = 1.5$ a 30 per cent error occurs. Second, the evaluation of the integral for R over the cross-sectional area may become very complex. Finally, calculations of R must be very accurate since differences between R and numerically comparable quantities are used in the stress formula.

The last two difficulties prompted the development of other methods of solution. One such method consists of expanding certain terms of the solution into a series,[*] another of building up a solution on the basis of a special transformed section. Yet another device consists of working "in reverse." Curved beams of various cross sections, curvatures, and applied moments are analyzed for stress; then these quantities are divided

---

[*] S. Timoshenko, *Strength of Materials* (3rd ed.), Part I (Princeton, N.J.: D. Van Nostrand Co., Inc., 1955), p. 369 and p. 373.

by a flexural stress that would exist for the same beam if it were straight. These ratios are then tabulated.* Hence, conversely, if stress in a curved beam is wanted, it is given as

$$\sigma = K(Mc/I) \tag{6-22}$$

where the coefficient K is obtained from a table or a graph and Mc/I is computed as in the usual flexure formula.

An expression for the distance from the center of curvature to the neutral axis of a curved beam of circular cross-sectional area is given below for future reference:

$$R = (\bar{r} + \sqrt{\bar{r}^2 - c^2})/2 \tag{6-23}$$

where $\bar{r}$ is the distance from the center of curvature to the centroid and c is the radius of the circular cross-sectional area.

PROBLEMS FOR SOLUTION

6-1. According to the National Lumber Manufacturers Association the dressed size of a 6-in.-by-10-in. timber is $5\frac{1}{2}$ in. by $9\frac{1}{2}$ in. (see Table 10 in the Appendix). Verify the moment of intertia I given in the table for such a member, and, if the allowable stress for a certain grade of wood is 1,200 psi, determine its capacity in bending.

6-2. Properties of steel, wide-flange beams are given in Table 4 of the Appendix. Consider the lightest member listed there, which is 8 WF 17, and verify the moments of inertia around the x axis and the y axis for it. If the allowable stress is 24 ksi, find the flexural capacity of this member around each of the two axes.

6-3. If the cross-sectional area of a bar has the form of an ellipse, show that $I_{xx} = \pi a b^3/4$. The equation for an ellipse is $x^2/a^2 + y^2/b^2 = 1$.

6-4. For a linearly elastic material, at the same maximum stress for a square member in the two different positions shown in the figure, determine the ratio of the bending moments. Bending takes place around the horizontal axis.

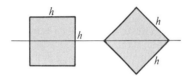

PROB. 6-4

6-5. A corrugated steel plate 0.20 in. thick has the cross section shown in the figure. (a) Calculate the section modulus per foot of width of this plate. (b) If such a plate is to be used to support a uniformly distributed load on a simple span, what load per square foot of projected area can be carried? Assume that the maximum bending stress controls the design and that $\sigma_{\text{allow}} = 18{,}000$ psi. The span $L = 6$ ft. (*Hint:* See Example 2-6, Fig. 2-22.)

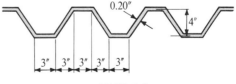

PROB. 6-5

---

* R. J. Roark, *Formulas for Stress and Strain* (4th. ed.) (New York: McGraw-Hill Book Company, 1965), Table VII, p. 165.

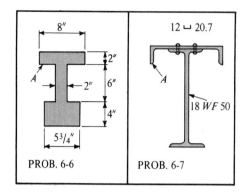

PROB. 6-6   PROB. 6-7

6-6 and 6-7. (a) For the cross-sectional areas with the dimensions shown in the figures, determine the moment of inertia for each section with respect to the horizontal centroidal axis. (b) If the maximum elastic stress due to bending around the horizontal axis is $+20{,}000$ psi, determine the stress at points A.

6-8. In a small dam a typical vertical beam is subjected to the hydrostatic loading shown in the figure. Determine the stress at point D of section a-a due to the bending moment. *Ans.* 820 psi.

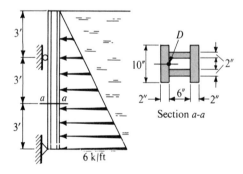

PROB. 6-8

6-9. Consider the data given in Prob. 2-35 and select a standard, steel I beam based on the maximum bending moment (see Table 3 in the Appendix). Assume $\sigma_{\text{allow}} = 24$ ksi and neglect the weight of the beam. (*Note:* In actual design of such members stresses due to shear, which are discussed in the next chapter, are also investigated. Usually, however, as will be shown in Chapter 10, the flexural requirements govern the selection of a member.)

6-10. For the data of Prob. 2-47 select a WF section (see Table 4 of the Appendix), other conditions as in the preceding problem. (*Note:* Consider $|M|_{\max}$ for selecting the section.)

6-11. For the data of Prob. 2-46 select a round bar based on the maximum bending moment. Use an allowable bending stress of 10,000 psi. (See note in Prob. 6-9 above.)

6-12. A T beam shown in the figure is made of a material the behavior of which may be idealized as having a tensile proportional limit of 3,000 psi and a compressive proportional limit of 6,000 psi. With a factor of safety of $1\frac{1}{2}$ on the initiation of yielding, find the magnitude of the largest force F which may be applied to this beam in a downward direction as well as in an upward direction. Base answers only on the consideration of the maximum bending stresses caused by F.

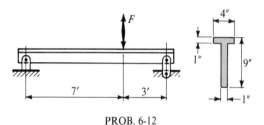

PROB. 6-12

6-13. A T beam is to be made of a linearly elastic material which can develop a maximum compressive stress of 15,000 psi, and a maximum tensile stress of 5,000 psi. This beam, with the cross section shown in the figure, is to carry a positive pure bending moment M. It is desired to achieve a balanced design so that the largest possible bending stresses are reached simultaneously. (a) What should be the dimension of the flange width w of the beam? (b) What is the moment capacity M of this beam? *Ans.* (a) 9 in.; (b) 60 k-in.

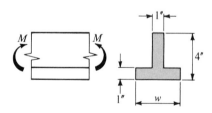

PROB. 6-13

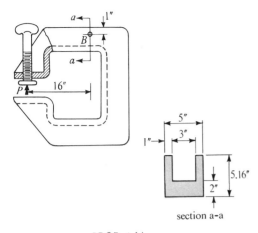

PROB. 6-14

6-14. As the screw of a large, steel C clamp, such as shown in the figure, is tightened down upon an object, the strain in the horizontal direction due to bending only is being measured by a strain gage at point B. If a strain of 900×10^{-6} in. per inch is noted, what is the load on the screw corresponding to the value of the observed strain? Let $E = 30 \times 10^6$ psi. *Ans.* 26.9 k.

6-15. A jib crane has the dimensions shown in the figure. At point A on the outside of the vertical member an electric strain gage is located and indicates a positive strain of 600×10^{-6} when a vertical force P of unknown magnitude is applied. If the vertical member of the crane is a 2-in., standard, steel pipe, what is the magnitude of the applied force P? Let $E = 29 \times 10^6$ psi. (See Table 8 in the Appendix.) *Ans.* 407 lb.

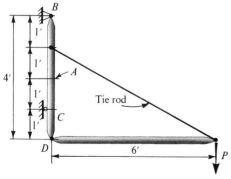

PROB. 6-15

6-16. A 32-lb, structural T beam (ST 8 WF) used as a cantilever is loaded as shown in the figure. Compute the magnitude of the load P which caused the following longitudinal strains in the beam: At gage no. 1, an elongation of 527×10^{-6} in. per inch.; and at gage no. 2, a shortening of 73×10^{-6} in. per inch. For this T beam, $I = 48.3$ in.^{4} around the neutral axis, and $E = 30 \times 10^6$ psi. *Ans.* 2.9 k.

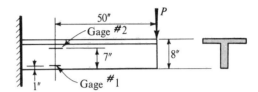

PROB. 6-16

6-17. Consider a linearly elastic beam subjected to a bending moment M around one of its principal axes for which the moment of inertia of the cross-sectional area around that axis is I. Show that for such a beam the normal force F acting on any part of the cross-sectional area A_1 is

$$F = MQ/I$$

where
$$Q = \int_{A_1} y \, dA = \bar{y} A_1$$

and $\bar{y}$ is the distance from the neutral axis of the cross section to the centroid of the area A_1 as shown in the figure.

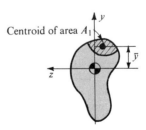

PROB. 6-17

6-18. A beam having a solid rectangular cross section with the dimensions shown in the figure is subjected to a positive bending moment of 12,000 ft-lb acting around the horizontal axis. (a) Find the compressive

215

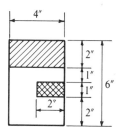

PROB. 6-18

force acting on the shaded area of the cross section developed by the bending stresses. (b) Find the tensile force acting on the cross-hatched area of the cross-section.

6-19. Two 2-in.-by-6-in., full-sized wooden planks are glued together to form a T section as shown in the figure. If a positive bending moment of 2,270 ft-lb is applied to such a beam acting around a horizontal axis, (a) find the stresses at the extreme fibers ($I = 136$ in.^{4}), (b) calculate the total compressive force developed by the normal stresses above the

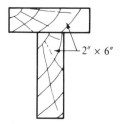

PROB. 6-19

neutral axis because of the bending of the beam, (c) find the total force due to the tensile bending stresses at a section and compare it with the result found in (b). *Ans.* (b) 5,000 lb.

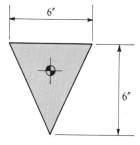

PROB. 6-20

6-20. A beam has the cross section of an isosceles triangle as shown in the figure and is subjected to a negative bending moment of 36 in-kips around a horizontal axis. Determine the location and magnitude of the resultant tensile and compressive forces acting on the section. (See Table 2 in the Appendix.) *Ans.* 10.67 k.

6-21. By integration, determine the force developed by the bending stresses and its position acting on the shaded area of the cross section of the beam shown in the figure if the beam is subjected to a negative bending moment of 30,000 in-lb acting around the horizontal axis. *Ans.* 12.7 k.

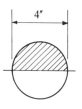

PROB. 6-21

6-22. Find the largest bending moment which an 8-in.-by-6-in.-by-1-in. angle may carry without exceeding a stress of 20 ksi. (*Hint:* the minimum radius of gyration for the angle is given in Table 7 of the Appendix. By definition, $I_{\min} = Ar_{\min}^2$, where A is the cross-sectional area. Moreover $I_{\min} + I_{\max} = I_{xx} + I_{yy}$, hence $I_{\max}$ may be obtained.) *Ans.* 358 k-in.

6-23. Show that the maximum bending stress for a beam of rectangular cross section is $\sigma_{\max} = (Mc/I)[(2n+1)/(3n)]$ if, instead of Hooke's law, the stress-strain relationship is $\sigma^n = K\varepsilon$, where K is a constant and n is a number dependent on the properties of the material.

6-24. Assume that a beam with a symmetrical cross section is made of a homogeneous, isotropic material having the stress-strain law $|\sigma| = E\varepsilon^2$. If such a beam is subjected to a pure bending moment M in its plane of symmetry, show that $\sigma = My^2/I_3$, where I_3 is the third moment of area about the neutral axis. Also show how one can locate the position of the neutral axis. Consider that ordinary assumptions of the technical beam theory apply.

6-25. A beam with a rectangular cross section is subjected to pure bending. The stress-strain diagram for the material of the beam is to be idealized as elastic–perfectly plastic. If the maximum flexural strain in the beam is two times the yield-point strain for the material, what is the ratio of the applied bending moment to the bending moment at yield? Solve the problem directly without the use of Eq. 6-14. *Ans.* 1.38.

6-26. (a) If a rectangular beam of an elastic-plastic material is subjected to a moment $1\tfrac{3}{8}$ times greater than M_{yp}, how much of the beam remains elastic? (b) What will be the residual stress pattern upon the release of the moment in (a)?

6-27. For a beam of an elastic-plastic material having a circular cross section determine the shape factor $k = M_p/M_{yp}$. *Ans.* 1.70.

6-28. For a beam of an elastic-plastic material having a square cross section and being bent around one of the diagonals (see Prob. 6-4), determine the shape factor $k = M_p/M_{yp}$. *Ans.* 2.

6-29. Verify the plastic modulus for the 8 WF 17 section given in Table 9 of the Appendix. Also find the shape factor.

6-30 and 6-31. Find the ratios M_{ult}/M_{yp} for mild steel beams resisting bending around the

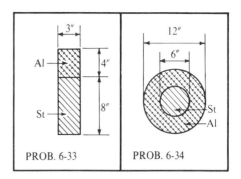

PROB. 6-33   PROB. 6-34

6-33 and 6-34. Composite beams having the cross-sectional dimensions shown in the figures are subjected to positive bending moments of 60-kip-ft each. Materials are fastened together so that the beams act as a unit. Determine the maximum bending stress in each material. $E_s = 30 \times 10^6$ psi; $E_a = 10 \times 10^6$ psi. (*Hint:* For Prob. 6-34 see Prob. 6-3.) *Ans. Prob.* 6-34. $\sigma_s = \pm 5.65$ ksi.

6-35. A beam having a rectangular cross section, b by h, is bent around the horizontal axis. The beam material exhibits linearly elastic properties in tension and ideally plastic properties in compression, see figure. If the maximum stress level in tension and compression is $|\sigma_o|$, i.e., numerically the same, what moment M is carried by the beam in terms of σ_o, b, and h? *Ans.* $(\tfrac{11}{54})bh^2\sigma_o$.

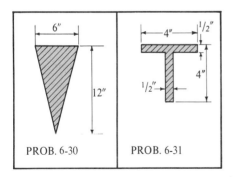

PROB. 6-30   PROB. 6-31

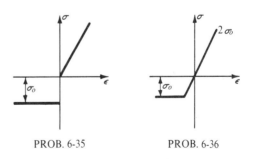

PROB. 6-35   PROB. 6-36

horizontal axes and having the cross-sectional dimensions shown in the figures. Assume the idealized stress-strain diagram used in Example 6-6, Fig. 6-15. *Ans. Prob.* 6-30. 2.34; *Prob.* 6-31. 1.80.

6-32. Rework Prob. 6-31 using the cross-sectional dimensions given in Prob. 6-19. *Ans.* 1.76.

6-36. Assume that a material has a stress-strain diagram as shown in the figure with $2\sigma_o$ being the ultimate strength in tension. Determine the ultimate moment M for a rectangular beam, b by $h = 2c$, bent around the horizontal axis. *Ans.* $1.11bc^2\sigma_o$.

6-37. A 6-in.-by-12-in., rectangular section is subjected to a positive bending moment of

217

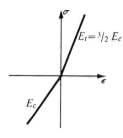

PROB. 6-37

180,000 ft-lb around the "strong" axis. The material of the beam is nonisotropic and is such that the modulus of elasticity in tension is $1\frac{1}{2}$ times as great as in compression, see figure. If the stresses do not exceed the proportional limit, find the maximum tensile and compressive stresses in the beam. *Ans.* 16.7 ksi, -13.6 ksi.

6-38 and 6-39. Determine the allowable bending moment around horizontal neutral axes for the composite beams of wood and steel having the cross-sectional dimensions shown in the figures. Materials are fastened together so that they act as a unit. $E_s = 30 \times 10^6$ psi; $E_w = 1.2 \times 10^6$ psi. The allowable bending stresses are $\sigma_s = 20,000$ psi and $\sigma_w = 1,200$ psi. *Ans. Prob. 6-38.* 450 k-in.

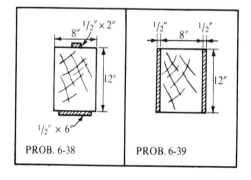

PROB. 6-38   PROB. 6-39

6-40. A 5-in.-thick concrete slab is longitudinally reinforced with steel bars as shown in the figure. (a) Determine the allowable bending moment per 1-ft width of this slab. Assume $n = 12$ and that the allowable stresses for steel and concrete are 18,000 psi and 900 psi, respectively. (b) Find the ultimate moment capacity per foot of width of the slab if for steel $\sigma_{yp} = 40$ ksi and for concrete $f'_c = 3,000$ psi. (*Note:* #3 bars are $\frac{3}{8}$ in. in diameter having $A = 0.11$ in.² each.)

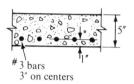

#3 bars
3" on centers

PROB. 6-40

6-41. (a) A beam has a cross section as shown in the figure, and is subjected to a positive bending moment which causes a tensile stress in the steel of 18,000 psi. If $n = 10$, what is the value of the bending moment? (b) If $\sigma_{yp} = 50$ ksi, and $f'_c = 4,000$ psi, what is the ultimate moment capacity of the section? *Ans.* (a) 131.6 k-ft.

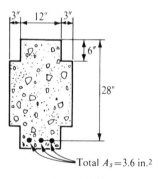

PROB. 6-41

6-42. Rework Example 6-12 by changing h to 4 in.

6-43. Derive Eq. 6-23.

6-44. What is the largest bending moment which may be applied to a curved bar, such as shown in Fig. 6-25(a), with $\bar{r} = 3$ in., if it has a circular cross-sectional area of 2-in. diameter and the allowable stress is 12 ksi?

Shearing stresses in beams 7

7-1. INTRODUCTION

It was shown in Chapter 2 that in a planar problem three elements of a force system may be necessary at a section of a beam to maintain the segment in equilibrium. These are an axial force, a shearing force, and a bending moment. The stress caused by an axial force was investigated in Chapter 3. In Chapter 6 the nature of the stresses caused by a bending moment in a beam was discussed. The stresses in a beam caused by the shearing force will be investigated in this chapter.

In all the previous derivations of the stress distribution in a member, the same reasoning was employed. First, a strain distribution was assumed; second, the relationship between stress and strain was brought in; and, finally, equilibrium conditions were used to establish the desired relations. However, the development of the expression linking the shearing force and the cross-sectional area of a beam to the stress follows a different path. The same procedure cannot be employed as no simple assumption for the strain distribution due to the shearing force can be made. Instead, an indirect approach is used. The stress distribution caused by flexure, as determined in the preceding chapter,

*Chapter 7
Shearing stresses
in beams*

is assumed. It, together with equilibrium requirements, resolves the problem of the shearing stresses. The consideration of beam deflection due to shear is postponed until Chapter 13.

In this chapter the investigation of stresses will be limited to those in straight beams. The analysis of shearing stresses in curved beams is beyond the scope of this text. In the early part of this chapter only beams with a symmetrical cross section will be considered and the applied forces will be assumed to act in the plane containing an axis of symmetry and the axis of the beam. An illustration of shearing stress distribution in an elastic-plastic beam is also given. In addition to shearing stresses, the related problem of interconnection requirements for fastening together several longitudinal elements of built-up beams will also be considered.

7-2. SOME PRELIMINARIES

Because the approach required in the solution for shearing stresses in beams is different from that which was encountered before, to begin with a general description of the method to be employed will be given.

First it is necessary to recall that in beams a relationship exists between a shear V at a section and a change in bending moment M. Thus according to Eq. 2-5

$$dM = -V\,dx \quad \text{or} \quad dM/dx = -V \tag{7-1}$$

So, if a shear is acting at a section of a beam, there will be a different bending moment at an adjoining section. When shear is present, the difference between the bending moments on the adjoining sections is equal to $-V\,dx$. If no shear acts at the adjoining sections of a beam, no

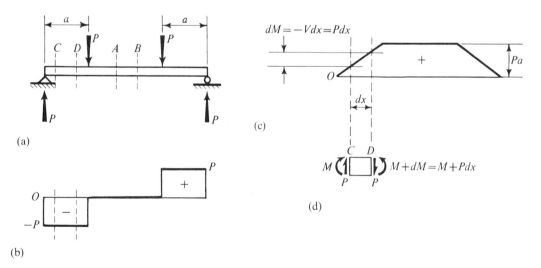

Fig. 7-1. Relation between shear and bending-moment diagrams for the loading shown.

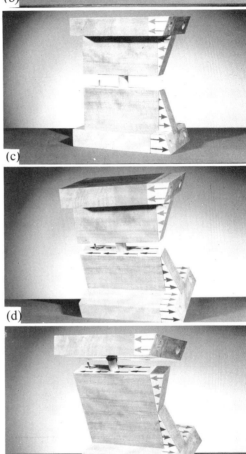

change in the bending moment occurs. Conversely, the rate of change of the bending moment along the beam is numerically equal to the shear. Thus, although shear is treated in this chapter as an independent action on a beam, it is inseparably linked with a change in the bending moment along the length of the beam.

To emphasize the meaning of Eq. 7-1 further, the shear and bending diagrams of Example 2-9 are reproduced in Fig. 7-1. At any two sections such as A and B taken through the beam anywhere between the applied forces P, the bending moment is the same. No shear acts at these sections. On the other hand, at any two sections such as C and D near the support, a change in the bending moment does take place. Shearing forces act at these sections as in Fig. 7-1(d). In the subsequent discussion, the possibility of equal, as well as of different, bending moments on two adjoining sections through a beam must be appreciated.

Next, before the detailed analyses, a study of a sequence of photographs of a model (Fig. 7-2) may prove helpful. The model represents a segment of an I beam. In Fig. 7-2(a), in addition to the beam itself, blocks simulating stress distribution caused by bending moments may be seen. The

Fig. 7-2. Shear flow model of an I beam. (a) Beam segment with bending stresses shown by blocks. (b) Shearing force transmitted through dowel. (c) For determining the force on a dowel only change in moment is needed. (d) The shearing force divided by the area of the cut yields shearing stress. (e) Horizontal cut below the flange for determining the shearing stress. (f) Vertical cut through the flange for determining the shearing stress.

Chapter 7
Shearing stresses
in beams

moment on the right is assumed to be larger than the one on the left. This system of forces is in equilibrium providing vertical shears V (not seen in this view) also act on the beam segment. By separating the model along the neutral surface, one obtains two separate parts of the beam segment as in Fig. 7-2(b). Either one of these parts alone again must be in equilibrium.

If in an actual beam the upper and the lower segments of Fig. 7-2(b) are connected by a dowel or a bolt, the axial forces on either the upper or the lower part caused by the bending moment stresses must be maintained in equilibrium by a force in the dowel. The force which must be resisted can be evaluated by summing the forces in the axial direction caused by bending stresses. In performing such a calculation either the upper or the lower part of the beam segment can be used. The horizontal force transmitted by the dowel is the force needed to balance the net force caused by the bending stresses acting on the two adjoining sections. Alternatively, by subtracting the same bending stress on both ends of the segment, the same results can be obtained. This is shown schematically in Fig. 7-2(c), where assuming a zero bending moment on the left, only the normal stresses due to the increment in moment within the segment need be shown acting on the right.

If initially the *I* beam considered is one piece requiring no bolts or dowels, an imaginary plane can be used to separate the beam segment into two parts as shown in Fig. 7-2(d). As before, the net force which must be developed across the cut area to maintain equilibrium can be determined. Dividing this force by the area of the imaginary horizontal cut gives average shearing stresses acting in this plane. In the analysis it is again expedient to work with the change in bending moment rather than with the total moments on the end sections.

After the shearing stresses on one of the planes (the horizontal one in Fig. 7-2(d)) are found, shearing stresses on mutually perpendicular planes of an infinitesimal element also become known since $\tau_{xy} = \tau_{yx}$. This approach establishes the shearing stresses in the plane of the beam section taken normal to its axis.

The process discussed above is quite general; two additional illustrations of separating the segment of the beam are in Figs. 7-2(e) and (f). In the first case, the imaginary horizontal plane separates the beam just below the flange. Either the upper or the lower part of this beam can be used in calculating the shearing stresses in the cut. In Fig. 7-2(f), the imaginary vertical plane cuts off a part of the flange. This permits calculation of shearing stresses lying in a vertical plane in the figure.

Before finally proceeding with the development of equations for determining the shearing stresses in connecting bolts and in beams, an intuitively evident example is worthy of note. Consider a wooden plank placed on top of another as in Figs. 7-3(a) and (b). If these planks act as a beam and are not interconnected, sliding at the surfaces of their contact will take place. The tendency for this sliding may be visualized by considering the two loaded planks shown in Fig. 7-3(b). The interconnection of these planks with nails or glue is necessary to make them act as an integral beam. In the next article an equation will be derived for determining the required interconnection between the component parts of a beam to make them act as a unit. In the following article this equation will be modified to yield shearing stresses in initially solid beams.

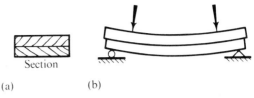

Fig. 7-3. Separate planks not fastened together slide on each other when loaded.

7-3. SHEAR FLOW

Consider a beam made from several continuous planks whose cross section is shown in Fig. 7-4(a). For simplicity the beam has a rectangular cross section, but such a limitation is not necessary. To make this beam act as an integral member, it is assumed that the planks are fastened at intervals by vertical bolts. An element of this beam isolated by

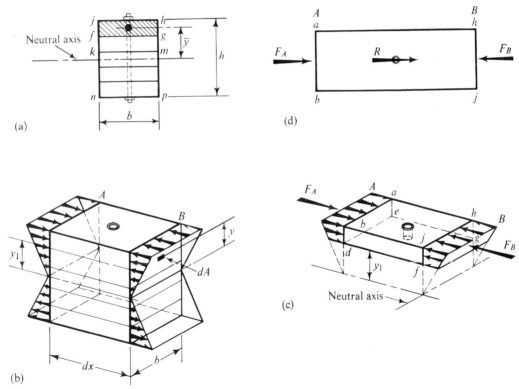

Fig. 7-4. Elements for deriving an expression for the shear flow in a beam.

two parallel sections, both of which are perpendicular to the axis of the beam, is shown in Fig. 7-4(b).

If the element shown in Fig. 7-4(b) is subjected to a bending moment $+M_A$ at end A and to $+M_B$ at end B, bending stresses which act normal to the sections are developed. These bending stresses vary linearly from their respective neutral axes, and at any point at a distance y from the neutral axis are $-M_B y/I$ on the B end, and $-M_A y/I$ on the A end.

From the beam element, Fig. 7-4(b), isolate the top plank as in Fig. 7-4(c). The fibers of this plank nearest the neutral axis are located by the distance y_1. Then, since stress times area is equal to a force, the forces acting perpendicular to the ends A and B of this plank may be determined. At the end B the force acting on an infinitesimal area dA at a distance y from the neutral axis is $(-M_B y/I)\, dA$. The total force acting on the area $fghj$ (A_{fghj}) is the integral of these elementary forces over this area. Denoting the total force acting normal to the area $fghj$ by F_B and remembering that, at a section, M_B and I are constants, one obtains the following relation:

$$F_B = \int_{\substack{\text{area}\\fghj}} -\frac{M_B y}{I}\, dA = -\frac{M_B}{I}\int_{\substack{\text{area}\\fghj}} y\, dA = -\frac{M_B Q}{I} \qquad (7\text{-}2)$$

where

$$Q = \int_{\substack{\text{area}\\fghj}} y\, dA = A_{fghj}\bar{y} \qquad (7\text{-}3)$$

The integral defining Q is the first or the statical moment of area $fghj$ around the neutral axis. By definition $\bar{y}$ is the distance from the neutral axis to the centroid of A_{fghj}.* Illustrations of the manner of determining Q are in Fig. 7-5. Equation 7-2 provides a convenient means of calculating the longitudinal force acting normal to any selected part of the cross-sectional area.

---

* Area $fgpn$ and its $\bar{y}$ may be used for finding $|Q|$.

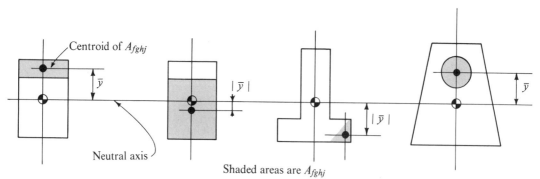

Fig. 7-5. Meaning of terms for finding $|Q|$.

By considering next end A of the element in Fig. 7-4, one can express the total force acting normal to the area $abde$ as

*Section 7-3
Shear flow*

$$F_A = -\frac{M_A}{I}\int_{\substack{\text{area}\\abde}} y\, dA = -\frac{M_A Q}{I} \qquad (7\text{-}4)$$

where the meaning of Q is the same as that in Eq. 7-2 since for prismatic beams an area such as $fghj$ is equal to the area $abde$. Hence if the moments at A and B were equal, it would follow that $F_A = F_B$, and the bolt shown in the figure would perform a nominal function of keeping the planks together and would not be needed to resist any known forces.

On the other hand, if M_A is not equal to M_B, which is always the case when shears are present at the adjoining sections, F_A is not equal to F_B. More push (or pull) develops on one end of a "plank" than on the other, as different normal stresses act on the section from the two sides. Thus if $M_A \neq M_B$, equilibrium of the horizontal forces in Fig. 7-4(c) may be attained only by developing a horizontal resisting force R in the bolt. If $M_B > M_A$, then $F_B > F_A$, and $F_A + R = F_B$, Fig. 7-4(d). The force $F_B - F_A = R$ tends to shear the bolt in the plane of the plank $edfg$.* If the shearing force acting across the bolt at the level km, Fig. 7-4(a), were to be investigated, the two upper planks should be considered as one unit.

If $M_A \neq M_B$ and the element of the beam is only dx long, the bending moments on the adjoining sections change by an infinitesimal amount. Thus if the bending moment at A is M_A, the bending moment at B is $M_B = M_A + dM$. Likewise, in the same distance dx the longitudinal forces F_A and F_B change by an infinitesimal force dF, i.e., $F_B - F_A = dF$. By substituting these relations into the expressions for F_B and F_A found above, with areas $fghj$ and $abde$ taken equal, one obtains an expression for the differential longitudinal push (or pull) dF:

$$dF = F_B - F_A = \left(-\frac{M_A + dM}{I}\right)Q - \left(-\frac{M_A}{I}\right)Q = -\frac{dM}{I}Q$$

In the final expression for dF the actual bending moments at the adjoining sections are eliminated. Only the difference in the bending moments dM at the adjoining sections remains in the equation.

Instead of working with a force dF which is developed in a distance dx, it is more significant to obtain a similar force per unit of beam length. This quantity is obtained by dividing dF by dx. Physically this quantity represents the difference between F_B and F_A for an element of the beam 1 in. long. The quantity dF/dx will be designated by q and will be referred

---

* The forces $(F_B - F_A)$ and R are not collinear, but the element shown in Fig. 7-4(c) is in equilibrium. To avoid ambiguity, shearing forces acting in the vertical cuts are omitted from the diagram.

to as the *shear flow*. Since force is usually measured in pounds, shear flow q has units of pounds per inch. Then, recalling that $dM/dx = -V$, one obtains the following expression for the shear flow in beams:

$$q = \frac{dF}{dx} = -\frac{dM}{dx}\frac{1}{I}\int_{area} y\, dA = \frac{V A_{fghj}\bar{y}}{I} = \frac{VQ}{I} \quad (7\text{-}5)$$

In this equation I stands for the moment of inertia of the entire cross-sectional area around the neutral axis, just as it does in the flexure formula from which it came. The total shearing force at the section investigated is represented by V, and the integral of $y\, dA$ for determining Q extends only over the cross-sectional area of the beam to one side of this area at which q is investigated.

In retrospect, note carefully that Eq. 7-5 was derived on the basis of the elastic flexure formula, but no term for a bending moment appears in the final expressions. This resulted from the fact that only the change in the bending moments at the adjoining sections had to be considered, and the latter quantity is linked with the shear V. The shear V was substituted for $-dM/dx$, and this masks the origin of the established relations. Equation 7-5 is very useful in determining the necessary interconnection between the elements making up a beam. This will be illustrated by examples.

EXAMPLE 7-1

Two long wooden planks form a T section of a beam as shown in Fig. 7-6(a). If this beam transmits a constant vertical shear of 690 lb, find the necessary spacing of the nails between the two planks to make the beam act as a unit. Assume that the allowable shearing force per nail is 150 lb.

SOLUTION

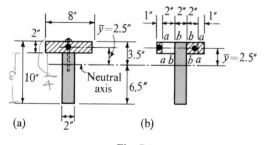

Fig. 7-6

In attacking such problems the analyst must ask: What part of a beam has a tendency to slide longitudinally from the remainder? Here it is the plane of contact of the two planks; Eq. 7-5 must be applied to determine the shear flow in this plane. To do this the neutral axis of the whole section and its moment of inertia around the neutral axis must be found. Then as V is known and Q is defined as the statical moment of the area of the upper plank around the neutral axis, q may be determined. The distance y_c from the top to the neutral axis is

$$y_c = \frac{2(8)1 + 2(8)6}{2(8) + 2(8)} = 3.5 \text{ in.}$$

$$I = \frac{8(2)^3}{12} + (2)8(2.5)^2 + \frac{2(8)^3}{12} + (2)8(2.5)^2 = 291 \text{ in.}^4$$

$$Q = A_{fghj}\bar{y} = (2)8(3.5 - 1) = 40 \text{ in.}^3$$

$$q = \frac{VQ}{I} = \frac{690(40)}{291} = 95 \text{ lb per in.}$$

Thus, a force of 95 lb must be transferred from one plank to the other in every linear inch along the length of the beam. However, from the data given, each nail is capable of resisting a force of 150 lb, hence one nail can take care of $150/95 = 1.59$ linear inches along the length of the beam. As shear remains constant at the consecutive sections of the beam, the nails should be spaced throughout at 1.59-in. intervals. In a practical problem a 1.5-in. spacing would probably be used.

SOLUTION FOR AN ALTERNATE ARRANGEMENT OF PLANKS

If, instead of using the two planks as above, a beam of the same cross section were made from five pieces, Fig. 7-6(b), a different nailing schedule would be required.

To begin, the shear flow between one of the 1-in.-by-2-in. pieces and the remainder of the beam is found, and although the contact surface a-a is vertical, the procedure is the same as before. The push or pull on an element is built up in the same manner as formerly:

$$Q = A_{fghj}\bar{y} = (1)2(2.5) = 5 \text{ in.}^3$$

$$q = \frac{VQ}{I} = \frac{690(5)}{291} = 11.8 \text{ lb per in.}$$

If the same nails as before are used to join the 1-in.-by-2-in. piece to the 2-in.-by-2-in. piece, they may be $150/11.8 = 12.7$ in. apart. This nailing applies to both sections a-a.

To determine the shear flow between the 2-in.-by-10-in. vertical piece and either one of the 2-in.-by-2-in. pieces, the whole 3-in.-by-2-in. area must be used to determine Q. It is the difference of pushes (or pulls) on this whole area that causes the unbalanced force which must be transferred at the surface b-b:

$$Q = A_{fghj}\bar{y} = (3)2(2.5) = 15 \text{ in.}^3$$

$$q = \frac{VQ}{I} = \frac{690(15)}{291} = 35.6 \text{ lb per in.}$$

Space nails at $150/35.6 = 4.2$ in. or, in practice, 4-in. intervals along the length of the beam in both sections b-b. These nails could be driven in first and then the 1-in.-by-2-in. pieces put on.

EXAMPLE 7-2

A simple beam on a 20-ft span carries a load of 200 lb per foot including its own weight. The beam cross section is to be made from several full-sized wooden pieces as in Fig. 7-7(a). Specify the spacing of the $\frac{1}{2}$-in.

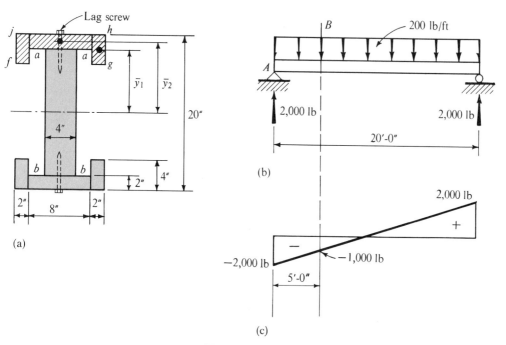

Fig. 7-7

lag screws shown which is necessary to fasten this beam together. Assume that one ½-in. lag screw, as determined by laboratory tests, is good for 500 lb when transmitting a lateral load parallel to the grain of the wood. For the entire section I is equal to 6,060 in.[4]

SOLUTION

To find the spacing of the lag screws, the shear flow at section a-a must be determined. The loading on the given beam is shown in Fig. 7-7(b); to show the variation of the shear along the beam, the shear diagram is constructed in Fig. 7-7(c). Then, to apply the shear flow formula, Q must be determined. This is done by considering the shaded area to one side of the cut a-a in Fig. 7-7(a). The statical moment of this area is most conveniently computed by multiplying the area of the two 2-in.-by-4-in. pieces by the distance from their centroid to the neutral axis of the beam and adding to this product a similar quantity for the 2-in.-by-8-in. piece. The largest shear flow occurs at the supports, as the largest vertical shears V of 2,000 lb act there:

$$Q = A_{fghj}\bar{y} = \sum A_i \bar{y}_i = 2A_1\bar{y}_1 + A_2\bar{y}_2$$

$$= 2(2)4(8) + 2(8)9 = 272 \text{ in.}^3$$

$$q = \frac{VQ}{I} = \frac{2,000(272)}{6,060} = 90 \text{ lb per in.}$$

At the supports the spacing of the lag screws must be 500/90 = 5.56 in. apart. This spacing of the lag screws applies only at a section where the shear V is equal to 2,000 lb. Similar calculations for a section where $V = 1,000$ lb gives $q = 45$ lb per inch, and the spacing of the lag screws becomes 500/45 = 11.12 in. Thus it is proper to specify the use of ½-in. lag screws at 5½ in. on centers for a distance of 5 ft nearest both the supports and at 11 in. for the middle half of the beam. A greater refinement in making the transition from one spacing of fastenings to another may be desirable in some problems. The same spacing of lag screws should be used at the section *b-b* as at the section *a-a*.

In a manner analogous to the above, the spacing of rivets or bolts in fabricated beams made from continuous angles and plates, Fig. 7-8, may be determined. Welding requirements are established similarly. The nominal shearing stress in a rivet is determined by dividing the total shearing force transmitted by the rivet (shear flow times spacing of the rivets) by the cross-sectional area of the rivet.

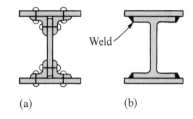

Fig. 7-8. Typical beam sections consisting of several components: (a) plate girder, (b) *I* beam reinforced with plates.

7-4. THE SHEARING STRESS FORMULA FOR BEAMS

The shearing stress formula for beams may be obtained from the shear flow formula. Analogously to the earlier procedure, an element of a beam may be isolated between two adjoining sections taken perpendicular to the axis of the beam. Then by passing another imaginary section through this element parallel to the axis of the beam, a new element is obtained, which corresponds to the element of one "plank" used in the earlier derivations. A side view of such an element is shown in Fig. 7-9(a), where the imaginary longitudinal cut is made at a distance y_1 from the neutral axis.* The cross-sectional area of the beam is shown in Fig. 7-9(c).

If shearing forces exist at the sections through the beam, a different bending moment acts at section A than at B. Hence more push or pull is developed on one side of the area *fghj* than on the other, and, as before, the difference in the longitudinal forces in a distance dx is

$$dF = -\frac{dM}{I} \int_{\substack{\text{area} \\ fghj}} y\, dA = -\frac{dM}{I} A_{fghj} \bar{y} = -\frac{dM}{I} Q$$

The force equilibrating dF is developed in the plane of the longi-

---

*Since $dM/dx = -V$, for a positive V the change in moment $dM = -V\, dx$. For this reason $M_A > M_B$ and the magnitudes of the normal stresses in Fig. 7-9(a) are shown accordingly.

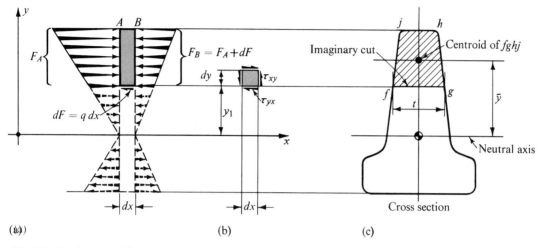

Fig. 7-9. Derivation of shearing stresses τ_{xy} and τ_{yx} in a beam.

tudinal cut taken parallel to the axis of the beam.* Therefore, assuming that the shearing stress is uniformly distributed across the cut of width t, the shearing stress in the longitudinal plane may be obtained by dividing dF by the area $t\,dx$. This yields the horizontal shearing stress τ_{yx}. However, since in an infinitesimal element numerically equal shearing stresses act on the mutually perpendicular planes, the shearing stress τ_{xy} in the plane of the vertical section right at the longitudinal cut, Fig. 7-9(b), also becomes known. On this basis

$$\tau_{xy} = \tau_{yx} = \frac{dF}{dx\,t} = -\frac{dM}{dx}\frac{A_{fghj}\bar{y}}{It}$$

This equation may be simplified further by again recognizing that $dM/dx = -V$. Therefore, alternatively,

$$\tau_{xy} = \tau_{yx} = \frac{V A_{fghj}\bar{y}}{It} = \frac{q}{t} = \frac{VQ}{It} \tag{7-6}$$

This is the formula for determining the shearing stresses in beams.† It gives them right at the longitudinal cut. As before, V is the total shearing force at a section, and I is the moment of inertia of the whole

---

* It is interesting to note an alternative reason for the appearance of $\bar{y}$ in the above equation. As $\bar{y}$ locates the fiber which is at an average distance from the neutral axis in the shaded area of a section, $(dM)\bar{y}/I$ gives the average normal stress acting on this area.

† This formula was derived by D. I. Jouravsky in 1855. Its development was prompted by observing horizontal cracks in wood ties on several of the railroad bridges between Moscow and St. Petersburg.

cross-sectional area about the neutral axis. Here Q is the statical moment of the *partial* area of the cross section to one side of the imaginary longitudinal cut around the neutral axis, and $\bar{y}$ is the distance from the neutral axis of the beam to the centroid of the partial area A_{fghj}. Finally, t is the width of the imaginary longitudinal cut, which is usually equal to the thickness or width of the member. In terms of the matrix representation of the stress tensor, Eq. 7-6 has the following meaning:

Section 7-4
The shearing stress formula for beams

$$\begin{pmatrix} 0 & \tau_{xy} \\ \tau_{yx} & 0 \end{pmatrix} \qquad (7\text{-}6a)$$

where τ_{xy} and τ_{yx} are the vertical and the horizontal shearing stresses, respectively; they are numerically equal. The vertical stresses act in the plane of the transverse section through a beam. They develop the vertical shear V, and thus the requirement of statics $\Sigma F_y = 0$ is satisfied. The validity of this statement for a special case will be illustrated in Example 7-3. In many applications of Eq. 7-6 the subscripts on τ are superfluous and can be omitted.

Care must be exercised in making the longitudinal cuts preparatory for use in Eq. 7-6. The proper sectioning of some cross-sectional areas of beams is shown in Figs. 7-10(a), (b), (d), and (e). The use of inclined cutting planes should be avoided unless the cut is made across a small thickness. For thin members Eq. 7-6 may be used to determine the shearing stresses with a cut such as *f-g* of Fig. 7-10(b). These shearing stresses act in a vertical plane and are directed perpendicularly to the plane of the paper. They act in entirely different planes than those obtained by making horizontal cuts, such as *f-g* in Figs. 7-10(a) and (d). (Also see Fig. 7-2(f).) Matching shearing stresses act horizontally, Fig. 7-10(c). Such stresses do not contribute directly to the resistance of the vertical shear V; they will be discussed further in Art. 7-6.

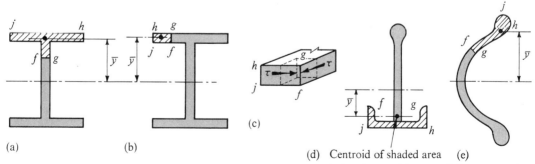

Fig. 7-10. Proper sectioning for partial areas of the cross section for computing the shearing stresses in beams.

The application of Eq. 7-6 to two particularly important types of cross-sectional areas of beams now will be illustrated.

EXAMPLE 7-3

Derive an expression for the shearing stress distribution in a beam of solid rectangular cross section transmitting a vertical shear V.

SOLUTION

The cross-sectional area of the beam is in Fig. 7-11(a). A longitudinal cut through the beam at a distance y_1 from the neutral axis isolates the partial area $fghj$ of the cross section. Here $t = b$ and the infinitesimal

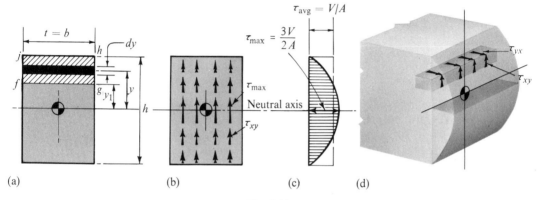

Fig. 7-11

area of the cross section may be conveniently expressed as $b\,dy$. By applying Eq. 7-6 the horizontal shearing stress is found at the level y_1 of the beam. At the same cut, numerically equal vertical shearing stresses act in the plane of the cross section.

$$\tau_{xy} = \tau_{yx} = \frac{VQ}{It} = \frac{V}{It}\int_{\substack{\text{area}\\fghj}} y\,dA = \frac{V}{Ib}\int_{y_1}^{h/2} by\,dy$$

$$= \frac{V}{I}\left.\frac{y^2}{2}\right|_{y_1}^{h/2} = \frac{V}{2I}\left[\left(\frac{h}{2}\right)^2 - y_1^2\right]$$

Thus, in a beam of rectangular cross section both the horizontal and the vertical shearing stresses vary parabolically. The maximum value of the shearing stress is obtained when y_1 is equal to zero. In the plane of the cross section, Fig. 7-11(b), this is diagrammatically represented by $\tau_{\max}$ at the neutral axis of the beam. At increasing distances from the neutral axis, the shearing stresses gradually diminish. At the upper and lower boundaries of the beam, the shearing stresses cease to exist as $y_1 = \pm h/2$. These values of the shearing stresses at the various levels of the beam may be represented by the parabola shown in Fig.

7-11(c). An isometric view of the beam with horizontal and vertical shearing stresses is shown in Fig. 7-11(d).

*Section 7-4
The shearing stress formula for beams*

To satisfy the condition of statics $\Sigma F_y = 0$, at a section of the beam the sum of all the vertical shearing stresses τ_{xy} times their respective areas dA must be equal to the vertical shear V. That this is the case may be shown by integrating $\tau_{xy} \, dA$ over the whole cross-sectional area A of the beam, using the general expression for τ_{xy} found above.

$$\int_A \tau_{xy} \, dA = \frac{V}{2I} \int_{-h/2}^{+h/2} \left[\left(\frac{h}{2}\right)^2 - y_1^2 \right] b \, dy_1 = \frac{Vb}{2I} \left[\left(\frac{h}{2}\right)^2 y_1 - \left(\frac{y_1^3}{3}\right) \right]_{-h/2}^{+h/2}$$

$$= \frac{Vb}{(2bh^3/12)} \left[\left(\frac{h}{2}\right)^2 h - \frac{2}{3}\left(\frac{h}{2}\right)^3 \right] = V$$

As the derivation of Eq. 7-6 was indirect, this proof showing that the shearing stresses integrated over the section equal the vertical shear is reassuring. Moreover, since an agreement in signs is found, this result indicates that *the direction of the shearing stresses at the section through a beam is the same as that of the shearing force V*. This fact may be used to determine the sense of the shearing stresses.

As noted above, the maximum shearing stress in a rectangular beam occurs at the neutral axis, and for this case the general expression for $\tau_{\max}$ may be simplified since $y_1 = 0$.

$$\tau_{\max} = \frac{Vh^2}{8I} = \frac{Vh^2}{8bh^3/12} = \frac{3}{2}\frac{V}{bh} = \frac{3}{2}\frac{V}{A}$$

where V is the total shear and A is the entire cross-sectional area. The same result may be obtained more directly if it is noted that to make $VQ/(It)$ a maximum, Q must attain its largest value since in this case V, I, and t are constants. From the property of the statical moments of areas around a centroidal axis, the maximum value of Q is obtained by considering one-half the cross-sectional area around the neutral axis of the beam. Hence, alternately,

$$\tau_{\max} = \frac{VQ}{It} = \frac{V(bh/2)(h/4)}{(bh^3/12)b} = \frac{3}{2}\frac{V}{A} \tag{7-7}$$

Since beams of rectangular cross-sectional area are used frequently in practice, Eq. 7-7 is very useful. It is widely used in the design of wooden beams as the shearing strength of wood on planes parallel to the grain is small. Thus, although equal shearing stresses exist on mutually perpendicular planes, wooden beams have a tendency to split longitudinally along the neutral axis. Note that the maximum shearing stress is $1\frac{1}{2}$ times as great as the average shearing stress V/A.

EXAMPLE 7-4

Rework the preceding example using differential equations of equilibrium. For convenience assume the beam to be 1 in. wide.

SOLUTION

From the point of view of elasticity, internal stresses and strains in beams are statically indeterminate. However, in the technical theory discussed here, the introduction of a kinematic hypothesis that plane sections remain plane after bending changes this situation. Here, in Eq. 6-3, it is asserted that in a beam $\sigma_x = -My/I$. Therefore, one part of Eq. 3-3—that giving the differential equation of equilibrium for a two-dimensional problem with a body force $X = 0$—suffices to solve for the unknown shearing stress. From the conditions of no shearing stress at the top and the bottom boundaries, $\tau_{yx} = 0$ at $y = \pm h/2$, the constant of integration is found.

From Eq. 3-3:
$$\frac{\partial \sigma_x}{\partial x} + \frac{\partial \tau_{xy}}{\partial y} = 0$$

But $\sigma_x = -\dfrac{My}{I}$, hence $\dfrac{\partial \sigma_x}{\partial x} = -\dfrac{\partial M}{\partial x}\dfrac{y}{I} = \dfrac{Vy}{I}$

and Eq. 3-3 becomes
$$\frac{Vy}{I} + \frac{d\tau_{xy}}{dy} = 0$$

Upon integrating
$$\tau_{xy} = -\frac{Vy^2}{2I} + C_1$$

Since $\tau_{xy}(\pm h/2) = 0$, one has $C_1 = +\dfrac{Vh^2}{8I}$

and
$$\tau_{xy} = \tau_{yx} = +\frac{V}{2I}\left[\left(\frac{h}{2}\right)^2 - y^2\right]$$

This agrees with the result found earlier since here $y = y_1$.

According to Hooke's law, shearing deformations must be associated with shearing stresses. Therefore the shearing stresses given by the above relation must cause shearing deformations. As shown in Fig. 7-12, maximum shearing distortions occur at $y = 0$, and no distortions take place at $y = \pm h/2$. This warps the initially plane section through the beam and contradicts the basic assumption of the technical bending theory. However, by the methods of elasticity it can be shown that these shearing distortions of the plane sections are negligibly small for slender members; the technical theory is completely adequate if the length of a member is at least two to three times greater than its total depth. This conclusion is of far-reaching importance since it means that the existence of a shear at a section does not invalidate the expressions for bending stresses derived earlier. As noted before, at the point of load application as well as at a rigidly built-in end, additional local disturbances of stresses occur.

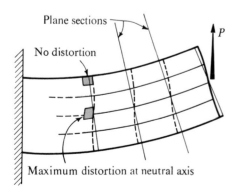

Fig. 7-12. Shearing distortions of a beam.

EXAMPLE 7-5

Using the technical theory, determine the shearing stress distribution due to shear V in the elastic-plastic zone of a rectangular beam.

SOLUTION

The situation in the problem occurs, for example, in a cantilever loaded as shown in Fig. 7-13(a). In the elastic-plastic zone, the external bending moment $M = -Px$, whereas, according to Eq. 6-14, the internal resisting moment $M = M_p - \sigma_{yp} b y_0^2/3$. Upon noting that y_o varies with x and differentiating the above equations, one notes the following equality:

$$\frac{dM}{dx} = -P = -\frac{2 b y_o \sigma_{yp}}{3} \frac{dy_o}{dx}$$

This relation will be needed later. First, however, proceeding as in the elastic case, consider the equilibrium of a beam element as shown in Fig. 7-13(b). Larger longitudinal forces act on the right side of this element than on the left. By separating it at the neutral axis and equating the force at the cut to the difference in the longitudinal force, one obtains

$$\tau_o \, dx \, b = \sigma_{yp} \, dy_o b/2$$

where b is the width of the beam. After substituting dy_o/dx from the relation found earlier and eliminating b, one finds the maximum horizontal shearing stress τ_o:

$$\tau_o = \frac{\sigma_{yp}}{2} \frac{dy_o}{dx} = \frac{3P}{4 b y_o} = \frac{3}{2} \frac{P}{A_o} \tag{7-8}$$

where A_o is the cross-sectional area of the elastic part of the cross section. The shearing stress distribution for the elastic-plastic case is shown in Fig. 7-13(c). This can be contrasted with that for the elastic case, shown in Fig. 7-13(d). Since equal and opposite normal stresses

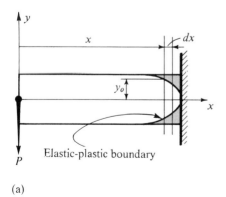

(a)

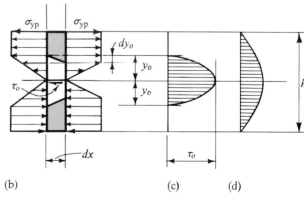

(b)   (c)   (d)

Fig. 7-13. Shearing stress distribution in a rectangular, elastic-plastic beam.

occur in the plastic zones, no unbalance in longitudinal forces occurs and no shearing stresses are developed.

This elementary solution has been refined by using a more carefully formulated criterion of yielding caused by the simultaneous action of normal and shearing stresses.* Some fundamental aspects of the interaction of such stresses will be considered in Chapter 9.

EXAMPLE 7-6

An I beam is loaded as in Fig. 7-14(a). If it has the cross section shown in Fig. 7-14(c), determine the shearing stresses at the levels indicated. Neglect the weight of the beam.

SOLUTION

A free-body diagram of a segment of the beam is in Fig. 7-14(b). It is seen from this diagram that the vertical shear at every section is 50 kips. Bending moments do not enter directly into the present problem. The shear flow at the various levels of the beam is computed in the table below using Eq. 7-5. Since $\tau = q/t$ (Eq. 7-6), the shearing stresses are obtained by dividing the shear flows by the respective widths of the beam.

$$I = \frac{6(12)^3}{12} - \frac{(5.5)(11)^3}{12} = 254 \text{ in.}^4$$

For use in Eq. 7-5 the ratio $V/I = -50,000/254 = -197 \text{ lb/in.}^4$

Level	A_{fghj}*		$\bar{y}$**	$Q = A_{fghj}\bar{y}$	$q = VQ/I$	t	τ, psi
1-1	0		6	0	0	6.0	0
2-2	(0.5)6	= 3.00	5.75	17.25	−3,400	6.0 0.5	−570 −6,800
3-3	(0.5)6 (0.5)(0.5)	= 3.00 = 0.25	5.75 5.25	$\left.\begin{array}{r}17.25\\1.31\end{array}\right\}18.56$	−3,650	0.5	−7,300
4-4	(0.5)6 (0.5)(5.5)	= 3.00 = 2.75	5.75 2.75	$\left.\begin{array}{r}17.25\\7.56\end{array}\right\}24.81$	−4,890	0.5	−9,780

* A_{fghj} is the partial area of the cross section above a given level in square inches.
** $\bar{y}$ is the distance from the neutral axis to the centroid of the partial area in inches.

The negative signs of τ show that, for the section considered, the stresses act downward on the right face of the elements. The sense of the shearing stresses acting on the section coincides with the sense of the shearing force V. For this reason a strict adherence to the sign convention is often unnecessary. It is always true that $\int_A \tau \, dA$ is equal to V and has the same sense.

---

* D. C. Drucker, "The Effect of Shear on the Plastic Bending of Beams," *Journal of Applied Mechanics*, **23** (1956), pp. 509–14.

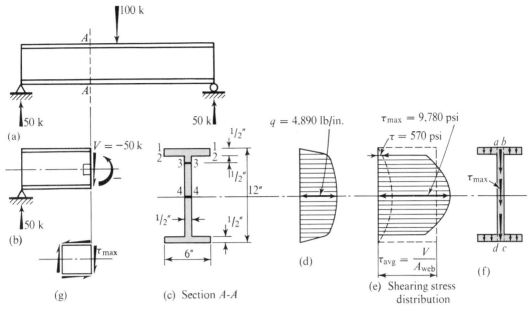

Fig. 7-14

Note that at the level 2-2 two widths are used to determine the shearing stress—one just above the line 2-2, and one just below. A width of 6 in. corresponds to the first case, and 0.5 in. to the second. This transition point will be discussed in the next article. The results obtained, which by virtue of symmetry are also applicable to the lower half of the section, are plotted in Fig. 7-14(d) and (e). By a method similar to the one used in the preceding example, it may be shown that the curves in Fig. 7-14(e) are parts of a second-degree parabola.

The variation of the shearing stress indicated by Fig. 7-14(e) may be interpreted as is shown in Fig. 7-14(f). The maximum shearing stress occurs at the neutral axis; the vertical shearing stresses throughout the web of the beam are nearly of the same magnitude. The shearing stresses occurring in the flanges are very small. For this reason the maximum shearing stress in an *I* beam is often approximated by dividing the total shear V by the cross-sectional area of the web (area *abcd* in Fig. 7-14(f)). Hence

$$(\tau_{max})_{approx} = V/A_{web} \tag{7-9}$$

In the example considered this gives

$$(\tau_{max})_{approx} = \frac{50{,}000}{(0.5)12} = 8{,}330 \text{ psi}$$

This stress differs by about 15 per cent from the one found by the accurate formula. For most cross sections a much closer approximation

*Chapter 7
Shearing stresses
in beams*

to the true maximum shearing stress may be obtained by dividing the shear by the web area between the flanges only. For the above example this procedure gives a stress of 9,091 psi, which is an error of only about 8 per cent. It should be clear from the above that division of V by the whole cross-sectional area of the beam to obtain the shearing stress is not permissible.

An element of the beam at the neutral axis is shown in Fig. 7-14(g). At levels 3-3 and 2-2, bending stresses, in addition to the shearing stresses, act on the vertical faces of the elements. No shearing stresses and only bending stresses act on the elements at level 1-1.

The maximum shearing stress was found to be at the neutral axis in both of the above examples. This is not always so. For example, if the sides of the cross-sectional area are not parallel, as for a triangular section, τ is a function of Q and t, and the maximum shearing stress occurs midway between the apex and the base, which does not coincide with the neutral axis.

7-5. LIMITATIONS OF THE SHEARING STRESS FORMULA

The shearing stress formula for elastic beams is based on the elastic flexure formula. Hence all of the limitations imposed on the flexure formula apply. The material is assumed to be elastic with the same elastic modulus in tension as in compression. The theory developed applies only to straight beams. Moreover, there are additional limitations which are not present in the flexure formula. Some of these will be discussed now.

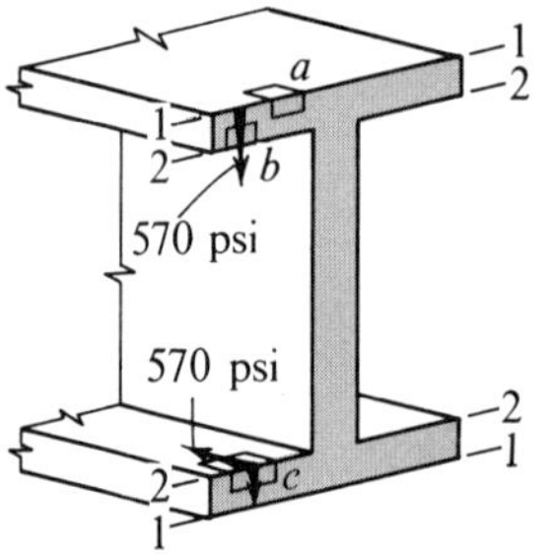

Consider a section through the I beam analyzed in Example 7-6. Some of the results of this analysis are reproduced in Fig. 7-15. The shearing stresses computed for level 1-1 apply to the infinitesimal element a. The vertical shearing stress is zero for this element. Likewise, no shearing stresses exist on the top plane of the beam. This is as it should be since the top surface of the beam is free of stress. Thus the conditions at this boundary are satisfied. A similar situation was found at the boundaries of a rectangular section.*

Fig. 7-15. Boundary conditions are not satisfied by the flange elements at the levels 2-2.

A different situation is discovered when the shearing stresses determined for the I beam at levels 2-2 are scrutinized. The shearing

---

* Rigorously some error may be shown to exist even in this case. Stresses given by Eq. 7-6 were established on the basis of equilibrium conditions of an element without reference to the conditions of compatibility of all components of strain (Poisson effect, etc.). For more details see A. E. H. Love, *Mathematical Theory of Elasticity* (4th Ed.), Chapter XV, p. 329. (New York: Dover Publications, Inc., 1944).

stresses were found to be 570 psi for elements such as b or c in the figure. This requires matching horizontal shearing stresses on the inner planes of the flanges. However, the latter planes must be free of the shearing stresses as they are free boundaries of the beam. This leads to a contradiction which cannot be resolved by the methods of the technical mechanics of solids. The more advanced techniques of the mathematical theory of elasticity must be used to obtain a correct solution.

Fortunately, the above defect of the shearing stress formula for beams is not too serious. The significant shearing stresses occur in the web, and, for all practical purposes, are correctly given by Eq. 7-6. No appreciable error is involved by using the relations derived in this chapter for thin-walled members, and the majority of beams used where shear is significant belong to this group.

In mechanical applications circular shafts frequently are used as beams. Hence beams having a solid circular cross section form an important class. These beams are not "thin-walled." An examination of the boundary conditions for a circular member, Fig. 7-16(a), leads to the conclusion that when shearing stresses are present they must act tangent to the boundary. As no matching shearing stress can exist on the free surface of the beam, no shearing stress component can act normal to the boundary. However, according to Eq. 7-6, vertical shearing stresses of equal intensity act at every level such as ac in Fig. 7-16(b). This is incompatible with the boundary conditions for elements such as a and c, and the solution indicated by Eq. 7-6 is inconsistent. The maximum shearing stresses, however, occurring at the neutral axis satisfy the boundary conditions and are very near their true value (within about 5 per cent).*

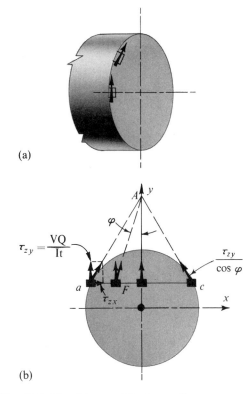

Fig. 7-16. Plausible modification of a solution based on the shearing stress formula to satisfy the boundary conditions.

7-6. FURTHER REMARKS ON THE DISTRIBUTION OF SHEARING STRESSES

In an I beam the existence of shearing stresses lying in a longitudinal cut such as c-c in Fig. 7-17(a) was indicated in Art. 7-2. These shearing stresses act perpendicular to the plane of the paper. Their magnitude may be found by applying Eq. 7-6, and their sense follows by considering the bending moments at the adjoining sections through the beam. In lieu of setting up a special sign convention, this proves to be the

---

* Love, *Mathematical Theory of Elasticity*, p. 348.

best approach. For example, if for the beam segment in Fig. 7-17(b) positive bending moments increase toward the reader, larger normal forces act on the nearer cross section. For the elements shown, $\tau t\, dx$ or $q\, dx$ must aid the smaller force acting on the partial area of the cross section. This determines the sense of the shearing stresses in the longitudinal cuts. Numerically equal shearing stresses act on the mutually perpendicular planes of an infinitesimal element, and the shearing stresses on such planes either meet or part with their directional arrowheads at a corner. In this manner the sense of the shearing stresses in the plane of the cross section becomes known.

The magnitude of the shearing stresses varies for the different vertical cuts. For example, if the cut c-c in Fig. 7-17(a) is at the edge of the beam, the shaded area of the beam's cross section is zero. However, if the thickness of the flange is constant, and the cut c-c is made progressively closer to the web, the shaded area increases from zero at a linear rate.

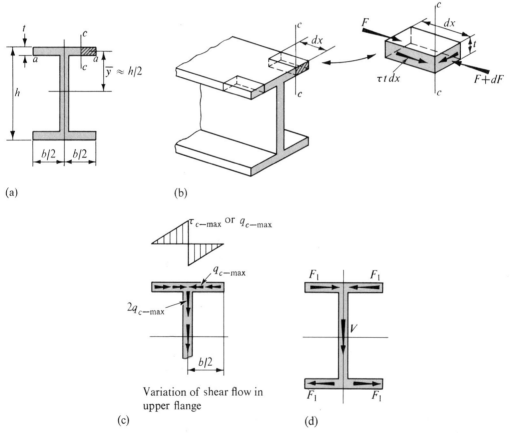

Fig. 7-17. Existence of shearing forces in the flange of an I beam which act perpendicularly to the axis of symmetry.

Moreover, as $\bar{y}$ remains constant for any such area, Q also increases linearly from zero toward the web. Therefore, since V and I are constant at any section through the beam, the shear flow $q_c = VQ/I$ follows the same variation. If the thickness of the flange remains the same, the shearing stress $\tau_c = VQ/It$ varies similarly. The same variation of q_c and τ_c exists on either side of the axis of symmetry of the cross section. However, these quantities in the plane of the cross section act in opposite directions on the two sides as may be determined by isolating another flange element to the left side of the web in Fig. 7-17(b). The variation of these shearing stresses or shear flows is represented in Fig. 7-17(c), where it is assumed that the web is very thin.

In common with all stresses, the shearing stresses in Fig. 7-17(c), when integrated over the area on which they act, are equivalent to a force. The magnitude of the horizontal force F_1 for one-half the flange, Fig. 7-17(d), is equal to the average shearing stress multiplied by one-half the flange area, i.e.,

$$F_1 = (\tau_{c\text{-max}}/2)(bt/2) \quad \text{or} \quad F_1 = (q_{c\text{-max}}/2)(b/2)$$

These horizontal forces act in the upper and lower flanges. Because of the symmetry of the cross section, these equal forces occur in pairs and oppose each other; thus they cause no apparent external effect.

To determine the shear flow at the juncture of the flange and the web (cut *a-a* in Fig. 7-17(a)), the whole area of the flange times $\bar{y}$ must be used in computing the value of Q. However, since in finding $q_{c\text{-max}}$ one-half the flange area times the same $\bar{y}$ has already been used, the sum of the two horizontal shear flows coming in from opposite sides gives the vertical shear flow at the cut *a-a*. Hence, figuratively speaking, the horizontal shear flows turn through 90° and merge to become the vertical shear flow. Then the shear flows at the various horizontal cuts through the web may be determined in the manner explained in the preceding articles. Moreover, as the resistance to the vertical shear V in thin-walled I beams is developed mainly in the web, it is so shown in Fig. 7-17(d). The sense of the shearing stresses and shear flows in the web coincides with the direction of the shear V. Note that the vertical shear flow "splits" upon reaching the lower flange. This is represented in Fig. 7-17(d) by the two forces F_1, which are the result of the horizontal shear flows in the flanges.

The shearing forces which act at a section of an I beam are shown in Fig. 7-17(d), and, for equilibrium, the applied vertical forces must act through the centroid of the cross-sectional area to be coincident with V. If the forces are so applied, no torsion of the member will occur. This is true for all sections having cross-sectional areas with an axis of symmetry. Thus, to avoid torsion of such members, the applied forces must act in the plane of symmetry of the cross section and the axis of the beam. A beam with an unsymmetrical section will be discussed next.

*Section 7-6
Further remarks on the distribution of shearing stresses*

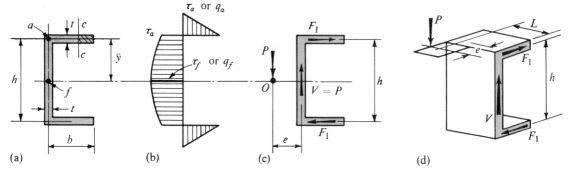

Fig. 7-18. Derivation for locating the shear center of a channel.

7-7. SHEAR CENTER

Consider a beam whose cross section is a channel, Fig. 7-18(a). The walls of this channel are assumed to be so thin that all computations may be based on the dimensions to the center line of the walls. Bending of this channel takes place around the horizontal axis, and although this cross section does not have a vertical axis of symmetry, it will be assumed that the bending stresses are given by the usual flexure formula. Assuming further that this channel resists a vertical shear, one knows that the bending moments will vary from one section through the beam to another.

By taking an arbitrary cut as c-c in Fig. 7-18(a), q and τ may be found in the usual manner. Along the horizontal legs of the channel, these quantities vary linearly from the free edge, just as they do for one side of the flange in an I beam. The variation of q and τ is parabolic along the web. The variation of these quantities is shown in Fig. 7-18(b), where they are plotted along the center line of the channel section.

The average shearing stress $\tau_a/2$ multiplied by the area of the flange gives a force $F_1 = (\tau_a/2)bt$, and the sum of the vertical shearing stresses over the area of the web is the shear

$$V = \int_{-h/2}^{+h/2} \tau \, t \, dy$$

These shearing forces acting in the plane of the cross section are shown in Fig. 7-18(c) and indicate that a force V and a couple $F_1 h$ are developed at the section through the channel. Physically there is a tendency for the channel to twist around some longitudinal axis. To prevent twisting and thus maintain the applicability of the initially assumed bending stress distribution, the external forces must be applied in such a manner as to balance the internal couple $F_1 h$. For example, consider the segment of a cantilever beam of negligible weight, shown in Fig. 7-18(d), to which a vertical force P is applied parallel to the web at a distance e from the center line of the web. To maintain this applied force in equilibrium, an equal and opposite shearing force V must be developed in the web. Likewise, to eliminate twisting of the channel, the couple Pe must equal the couple $F_1 h$. At the same section through the channel, the bending moment PL is resisted by the usual flexural stresses (not shown in the figure).

An expression for the distance e, locating the plane in which the force P must be applied to eliminate twist in the channel, may now be obtained. Thus, remembering that $F_1 h = Pe$ and $P = V$

$$e = \frac{F_1 h}{P} = \frac{(\tfrac{1}{2})\tau_a bth}{P} = \frac{bth}{2P}\frac{VQ}{It} = \frac{bth}{2P}\frac{Vbt(h/2)}{It} = \frac{b^2 h^2 t}{4I} \quad (7\text{-}10)$$

Note that the distance e is independent of the magnitude of the applied force P as well as of its location along the beam. The distance e is a property of a section and is measured outward from the center of the web to the applied force.

A similar investigation may be made to locate the plane in which the horizontal forces must be applied to eliminate twist in the channel. However, for the channel considered, by virtue of symmetry, it may be seen that this plane coincides with the neutral plane of the former case. The intersection of these two mutually perpendicular planes with the plane of the cross section locates a point which is called the *shear center*. It is designated by the letter O in Fig. 7-18(c). The shear center for any cross section lies on a longitudinal line parallel to the axis of the beam. Any transverse force applied through the shear center does not cause torsion of the beam. A detailed investigation of this problem shows that when a member of any cross-sectional area is twisted, the twist takes place around the shear center, which remains fixed. For this reason, the shear center is sometimes called the *center of twist*.

For cross-sectional areas having one axis of symmetry, the shear center is always located on the axis of symmetry. For those which have two axes of symmetry, the shear center coincides with the centroid of the cross-sectional area. This is the case for the I beam which was considered in the previous article.

The exact location of the shear center for unsymmetrical cross sections of thick material is difficult to obtain and is known only in a few cases. If the material is thin, as has been assumed in the preceding discussion, relatively simple procedures may always be devised to locate the shear center of the cross section. The usual method consists of determining the shearing forces, as F_1 and V above, at a section, and then finding the location of the external force necessary to keep these forces in equilibrium.

EXAMPLE 7-7

Find the approximate location of the shear center for a beam with the cross section of the channel shown in Fig. 7-19.

Section 7-7
Shear center

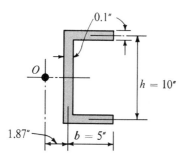

Fig. 7-19

SOLUTION

Instead of using Eq. 7-10 directly, some further simplifications may be made. The moment of inertia of a thin-walled channel around its neutral axis may be found with sufficient accuracy by neglecting the moment of inertia of the flanges around their own axes (only!). This expression for I may then be substituted into Eq. 7-10, and, after simplifications, a formula for e of channels is obtained.

$$I \approx I_{\text{web}} + (Ad^2)_{\text{flanges}} = \frac{th^3}{12} + 2bt\left(\frac{h}{2}\right)^2 = \frac{th^3}{12} + \frac{bth^2}{2} \quad (7\text{-}11)$$

$$e = \frac{b^2h^2t}{4I} = \frac{b^2h^2t}{4\left(\dfrac{bth^2}{2} + \dfrac{th^3}{12}\right)} = \frac{b}{2 + h/(3b)}$$

Equation 7-11 shows that when the width of flanges b is very large, e approaches its maximum value of $b/2$. When h is very large, e approaches its minimum value of zero. Otherwise, e assumes an intermediate value between these two limits. For the numerical data given in Fig. 7-19

$$e = \frac{5}{2 + 10/(3 \times 5)} = 1.87 \text{ in.}$$

Hence the shear center O is $1.87 - 0.05 = 1.82$ in. from the outside vertical face of the channel. The answer would not be improved if Eq. 7-10 were used in the calculations.

EXAMPLE 7-8

Find the approximate location of the shear center for the cross section of the I beam in Fig. 7-20. Note that the flanges are unequal.

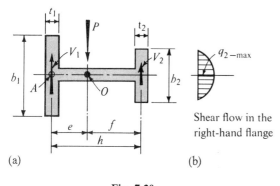

Fig. 7-20 (a) (b) Shear flow in the right-hand flange

SOLUTION

This cross section has a horizontal axis of symmetry, and the shear center is located on it; where remains to be answered. The applied force P causes significant bending and shearing stresses only in the flanges, and the contribution of the web to the resistance of the applied force P is negligible.

Let the shearing force resisted by the left flange of the beam be V_1, and that by the right flange, V_2. For equilibrium, $V_1 + V_2 = P$. Likewise, for no twist of the section, from $\Sigma M_A = 0$, $Pe = V_2 h$ (or $Pf = V_1 h$). Thus only V_2 remains to be determined to solve the problem. This may be done by noting that the right flange is actually an ordinary rectangular beam. The shearing stress (or shear flow) in such a beam is distributed parabolically, Fig. 7-20(b), and since the area of a parabola is two-thirds of the base times the maximum altitude, $V_2 = \frac{2}{3} b_2 (q_2)_{max}$. However, since the total shear $V = P$, by Eq. 7-5, $(q_2)_{max} = VQ/I = PQ/I$, where Q is the statical moment of the upper half of the right flange and I is the moment of inertia of the whole section. Hence

Section 7-7
Shear center

$$Pe = V_2 h = \tfrac{2}{3} b_2 (q_2)_{max} h = \frac{\tfrac{2}{3} h b_2 P Q}{I}$$

(7-12)

$$e = \frac{2hb_2}{3I} Q = \frac{2hb_2}{3I} \frac{b_2 t_2}{2} \frac{b_2}{4} = \frac{h}{I} \frac{t_2 b_2^3}{12} = \frac{hI_2}{I}$$

where I_2 is the moment of inertia of the right flange around the neutral axis. Similarly, it may be shown that $f = hI_1/I$, where I_1 applies to the left flange. If the web of the beam is thin, as originally assumed, $I \approx I_1 + I_2$, and $e + f = h$ as is to be expected.

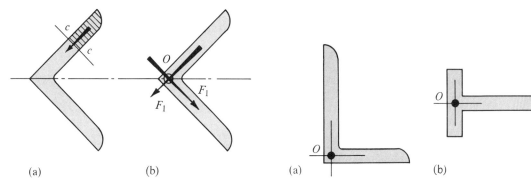

Fig. 7-21. Shear center for a symmetrical angle (equal legs) located by O.

Fig. 7-22. Shear center for the sections shown located by O.

A similar analysis leads to the conclusion that the shear center for a symmetrical angle is located at the intersection of the center lines of its legs, as in Figs. 7-21(a) and (b) since the shear flow at every section, as c-c, is directed along the center line of a leg. These shear flows yield two identical forces F_1 in the legs. The vertical components of these forces equal the vertical shear applied through O. An analogous situation is found for any angle or T section, as in Figs. 7-22(a) and (b). The location of the shear center for various members is particularly important in flight structures; for further details the reader is referred to books on this subject.*

---

* E. F. Bruhn, *Analysis and Design of Flight Vehicle Structures* (Cincinnati, Ohio: Tri-State Offset Co., 1965). Also, Paul Kuhn, *Stresses in Aircraft and Shell Structures* (New York: McGraw-Hill Book Company, 1956).

PROBLEMS FOR SOLUTION

7-1. A $\tfrac{1}{2}$-in.² brass bar is made from five 0.10-in.-by-0.5-in. strips and is riveted together with $\tfrac{1}{8}$-in.-diameter rivets $\tfrac{3}{4}$ in. on centers, see figure. If the allowable shearing stress in the rivets is 4,000 psi, what vertical shear may be applied to this bar acting as a cantilever? Would the performance of the bar improve if it were turned 90°?

PROB. 7-1

7-2. A 15-ft wooden beam overhangs 10 ft and carries a concentrated force $P = 950$ lb at the end, see figure. The beam is made up of full-sized, 2-in.-thick boards nailed together with nails which have a shear resistance of 96 lb each. The moment of inertia of the whole cross section is approximately 1,900 in.⁴ (a) What should the longitudinal spacing of the nails be connecting board A with boards B and C in the region of high shear? (b) For the same region, what should be the longitudinal spacing of the nails connecting board D with boards B and C? In calculations neglect the weight of the beam. *Ans.* (a) 1.6 in., (b) 8 in.

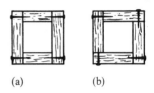

(a)   (b)

PROB. 7-3

nailing for transmitting shear. (b) If the shear to be transmitted by this member is 620 lb, what must the nail spacing be for the best design? The nailing is to be done with 16 d (16 penny) box nails that are good for 50 lb each in shear.

7-4. A simply supported beam has a cross section consisting of a 12⊔20.7 and an

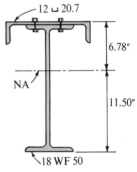

PROB. 7-4

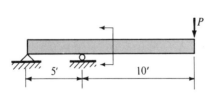

PROB. 7-2

7-3. A 10-in., square box beam is to be made from four wood pieces 2 in. thick. Two possible designs are considered as shown in the figure. Moreover, the design shown in (a) can be turned 90° in the application. (a) Select the design requiring the minimum amount of

18 WF 50 fastened together by $\tfrac{3}{4}$ in.-diameter bolts spaced longitudinally 6 in. apart in each row as shown in the figure. If this beam is loaded with a downward concentrated force of 112 kips in the middle of the span, what is the shearing stress in the bolts? Neglect the

weight of the beam. The moment of inertia I of the whole member around the neutral axis is 1,120 in.^{4} *Ans.* 12.4 ksi.

7-5. A *T*-flange girder is used to support a 200 kip load in the middle of a 24-ft simple span. The dimensions of the girder are given in the figure in a cross-sectional view. If the $\frac{7}{8}$-in.-diameter bolts are spaced 5 in. apart longitudinally, what shearing stress will be developed in the bolts by the applied loading? The moment of inertia of the girder around the neutral axis is approximately 11,000 in.^{4} *Ans.* 15.8 ksi.

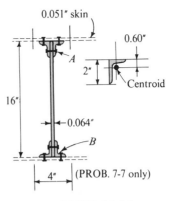

PROBS. 7-6, 7-7

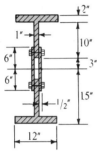

PROB. 7-5

7-6. An aircraft shear-resistant* plate girder is made up of four 2-in.-by-1$\frac{1}{2}$-in.-by-$\frac{1}{8}$-in. angles ($A = 0.42$ in.^{2} each for angle) and a 0.064-in. web as shown in the figure. Neglecting the skin, the moment of inertia of this section is 114.3 in.^{4} Rivets A are $\frac{3}{16}$ in. in diameter and are spaced longitudinally 1$\frac{1}{4}$ in. apart in each row. These rivets are good for 800 lb each in single shear. If 6,000-lb shear is to be transmitted by this section, what factor of safety, if any, is available for the rivets? In the calculations make no reduction in areas for the rivet holes.

7-7. Assume that in the above problem 4 in. of the skin may be included at the top and at the bottom in the moment of inertia of the section. Then, assuming 1$\frac{1}{4}$-in. spacing for rivets A, and 2-in. spacing for each row

---

* In aircraft design in some plate girders webs are permitted to wrinkle resulting in the so-called *semitension field beams*. To differentiate between the beams, if the web does not wrinkle, the term *shear resistant* is used.

of rivets B, calculate the shearing stresses in the rivets due to a shear $V = 6,000$ lb.

7-8. A beam is loaded so that the moment diagram for it varies as shown in the figure. (a) Find the maximum longitudinal shearing force in the $\frac{1}{2}$-in.-diameter bolts spaced 12 in. apart. (b) Find the maximum shearing stress in the glued joint.

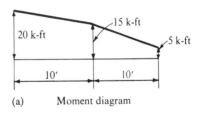

(a)   Moment diagram

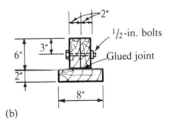

(b)

PROB. 7-8

7-9. A wooden *I* beam is made with a narrow lower flange because of space limitations, as shown in the figure. The lower flange is fastened to the web with nails spaced longitudinally 1.5 in. apart, while the vertical boards in the lower flange are glued in place.

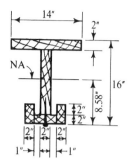

PROB. 7-9

Determine the stress in the glued joints and the force carried by each nail in the nailed joint if the beam is subjected to a vertical shear of 6 kips. The moment of inertia for the whole section around the neutral axis is 2,640 in.[4] *Ans.* 60 psi, 565 lb.

7-10. A cast iron beam has the cross-sectional dimensions shown in the figure. If the allowable stresses are 7 ksi in tension, 30 ksi in compression, and 8 ksi in shear, what is the maximum allowable shear and the maximum allowable bending moment for this beam? Consider only the vertical loading of the beam and confine calculations at the holes to section *a-a*. *Ans.* 51.1 k, 197 k-in.

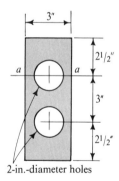

2-in.-diameter holes

PROB. 7-10

7-11. A steel welded box girder having the dimensions shown in the figure has to transmit a vertical shear force $V = 360$ k. Determine the shearing stresses at the sections *a*, *b*, and *c*. For this section $I = 45,000$ in.[4]

7-12. A beam is fabricated by slotting 4-in. standard steel pipes longitudinally and then securely welding them to a 23-in.-by-$\frac{3}{8}$-in. web plate as shown in the figure. I of the composite section around the neutral axis is 1,018 in.[4] If at a certain section this beam transmits a vertical shear of 40 kips, determine the shearing stress in the pipe and in the web plate at a level 10 in. above the neutral axis. *Ans.* 1,500 psi, 634 psi.

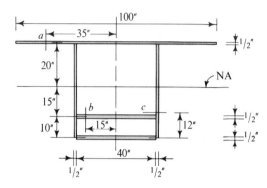

PROB. 7-11

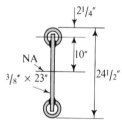

PROB. 7-12

7-13. Show that a formula, analogous to Eq. 7-7, for beams having a solid circular cross section of area A is $\tau_{max} = 4V/3A$.

7-14. Show that a formula, analogous to Eq. 7-7, for thin-walled circular tubes acting as beams having a net cross-sectional area A is $\tau_{max} = 2V/A$.

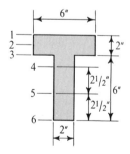

PROB. 7-15

7-15. A cast iron beam has a T section as shown in the figure ($I = 136$ in.^{4}). If this beam transmits a vertical shear of 54.4 kips, find the shearing stresses at the levels indicated. Report the results on a plot similar to the one shown in Fig. 7-14(e). *Ans.* $\tau_{\max} = 5{,}000$ psi.

7-16. A beam has a cross-sectional area in the form of an isosceles triangle for which the base b is equal to one-half its height h. (a) Using calculus and the conventional stress analysis formula, determine the location of the maximum shearing stress caused by a vertical shear V. Sketch the manner in which the shearing stress varies across the section. (b) If $b = 3$ in., $h = 6$ in., and $\tau_{\max}$ is limited to 100 psi, what is the maximum vertical shear V that this section may carry? *Ans.* (a) $h/2$; (b) 600 lb.

PROB. 7-16

7-17. A beam is made up of four 2-in.-by-4-in., full-sized Douglas Fir pieces which are glued to a 1-in.-by-18-in. Douglas Fir plywood web as shown in the figure. Determine the maximum allowable shear and the maximum allowable bending moment which this section can carry if the allowable bending stress is 1,500 psi; the allowable shearing stress in wood is 80 psi, and the allowable shearing stress in the glued joints is 40 psi.

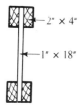

PROB. 7-17

7-18. A $1\frac{1}{2}$-in.-by-2-in. rectangular bar is attached to a channel section by $\frac{3}{8}$-in. machine screws at 6 in. on centers as shown in section in the figure. For the whole section,

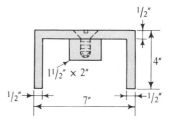

PROB. 7-18

around the horizontal neutral axis, $I = 11.20$ in.^{4} If the section transmits a vertical shear of 4 kips, (a) determine the shearing stress in the machine screws; (b) determine the shearing stress at the horizontal juncture of the web with the flanges; (c) find the maximum shearing stress. *Ans.* (b) 1,250 psi.

7-19. A 14 WF 87 beam supports a uniformly distributed load of 4 kips per foot, including its own weight, as shown in the figure. Using Eq. 7-6, determine the shearing stresses acting on the elements at A and B. Show the sense of the computed quantities on infinitesimal elements. If bending stresses also act on these elements, determine these, and indicate them acting on the elements.

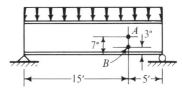

PROB. 7-19

7-20. A T beam is loaded by a force $P = 1{,}360$ lb as shown in the figure. From this beam isolate a segment 10-in.-by-5-in.-by-2-in., shown hatched in the figure. Then, on a free-body diagram of this segment, indicate the location, magnitude, and sense of all resultant

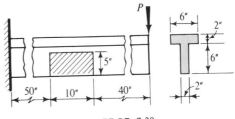

PROB. 7-20

forces acting on it caused by the bending and shearing stresses. Neglect the weight of the beam.

7-21. An *I* beam has the dimensions shown in the figure. In service this beam may be subjected to shearing forces either in the *y* or in the *z* directions, i.e., not simultaneously. On the assumption of linearly elastic behavior, compare the shear capacity of the beam in the two directions. *Ans.* $V_y = 0.782 V_z$.

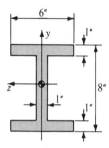

PROB. 7-21

7-22. A beam having the cross section with the dimensions shown in the figure transmits a vertical shear $V = 7$ kips applied through the shear center. Determine the shearing stresses at sections *A*, *B*, and *C*. *I* around the neutral axis is 35.7 in.⁴ The thickness of the material is $\frac{1}{2}$ in. throughout. *Ans.* 392 psi; 2,000 psi; 2,900 psi.

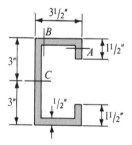

PROB. 7-22

7-23. A beam having a cross section with the dimensions shown in the figure is in a region where there is a constant, positive vertical shear of 24 kips. (a) Calculate the shear flow *q* acting at each of the five sections indicated in the figure. (b) Assuming a positive bending moment of 240 kip-in. at one section and a

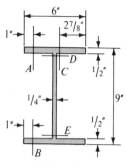

PROB. 7-23

larger moment at the adjoining section 1 in. away, draw isometric sketches of each segment of the beam isolated by the sections 1 in. apart and the five sections (*A*, *B*, *C*, *D*, and *E*) shown in the figure, and on the sketches indicate all forces acting on the segments. Neglect vertical shearing stresses in the flanges.

7-24. An 8 WF 31 beam is reinforced with two 10-in.-by-$\frac{1}{2}$-in. cover plates as shown in the

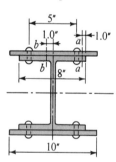

PROB. 7-24

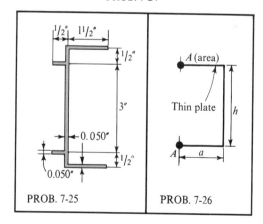

PROB. 7-25   PROB. 7-26

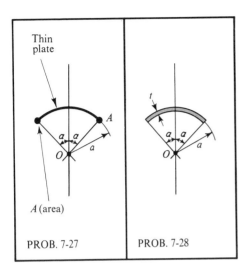

figure. The longitudinal spacing of the ½-in. rivets is 2 in., center to center. The beam transmits a vertical shear $V = 25$ k. Determine the shearing stresses in the flange and cover plate for the section *a-a* and *b-b*. (*Note:* For the determination of the shearing stresses include the entire cross section without reduction for the rivet holes.)

7-25 through 7-28. Determine the location of the shear center for the beams having the cross-sectional dimensions shown in the figures. In Probs. 7-26 and 7-27 assume that the cross-sectional area of the plate is negligible in comparison with the cross-sectional areas A of the flanges. *Ans. Prob.* 7-27. $e = (\alpha/\sin \alpha)a$ from 0; *Prob.* 7-28. $I = a^3 t(\alpha - \sin \alpha \cos \alpha)$ and $e = [(\sin \alpha - \alpha \cos \alpha)/(\alpha - \sin \alpha \cos \alpha)]2a$.

8 Compound stresses

8-1. INTRODUCTION

All classical stress analysis formulas of the technical theory of deformable bodies resulting from a single element of a force system acting at a section of a member have been established in the preceding chapters. For linearly elastic material these are summarized in the table on the next page, where for completeness the expressions for the elastic deformations are also included.

No such convenient table can be given for inelastically behaving members because of the larger variety and complexity of the constitutive relations. In inelastic problems individual cases must be analyzed using the basic kinematic assumptions together with the appropriate stress-strain relations and equilibrium equations.

In this chapter attention will be directed to problems where several elements of a force system occur at a section of a member simultaneously. The complete problem will be more fully discussed in Chapters 9 and 10. In this chapter a more limited objective is pursued. To begin, the normal stresses which arise from the simultaneous action of axial force and bending are considered. This is followed by a discussion of the normal stresses

BASIC STRESS AND DEFORMATION RELATIONS

Loading	Section	Elastic Stress	Elastic Deformation
Axial	Any	$\sigma = \dfrac{P}{A}$	$\dfrac{du}{dx} = \dfrac{P}{AE}$
Torsional	Circular	$\tau = \dfrac{T\rho}{J}$	$\dfrac{d\varphi}{dx} = \dfrac{T}{JG}$
	Rectangular	$\tau_{max} = \dfrac{T}{\alpha bc^2}$	$\dfrac{d\varphi}{dx} = \dfrac{T}{\beta bc^3 G}$
	Closed Thin-walled Tubular	$\tau = \dfrac{T}{2At}$	$\dfrac{d\varphi}{dx} = \dfrac{T}{4A^2 G}\oint \dfrac{ds}{t}$
Bending	Any	$\sigma = -\dfrac{My}{I}$	$\dfrac{d^2v}{dx^2} = \dfrac{M}{EI}$
	(Principal axes must be used)		(See Chapter 11)
	Symmetrical Curved bars	$\sigma = \dfrac{My}{Ae(R-y)}$	----------------
Shearing of beams	Any	$\tau = \dfrac{VQ}{It}$	(See Chapter 13)

Section 8-2 Superposition and its limitation

caused by unsymmetrical bending and an axial force. Then problems are discussed in which shearing stresses due to torque and direct shear occur simultaneously. Finally, at the end of the chapter, a special topic, a closely coiled helical spring, is considered.

8-2. SUPERPOSITION AND ITS LIMITATION

The basic stress analysis developed in this text thus far is completely predicated on small deformations of members. Situations such as occur in flexible rods, Fig. 8-1, are considered in Chapter 14. Superposition of several separately applied forces is not applicable if, for example, deflections significantly change the bending moments calculated on the basis of undeformed members. In Fig. 8-1(b), because of deflection v, an additional bending moment Pv is developed. In many problems, however, the deformation effect on stresses is small and can be neglected. This will be assumed in this chapter.

In members in which the overall deformations are small in the sense discussed above,

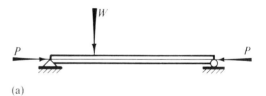

(a)

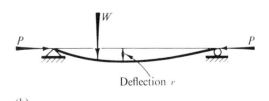

(b)

Fig. 8-1. Deflection in axially compressed beams induces an increase in bending moments.

*Chapter 8
Compound stresses*

superposition of the effects of separately applied forces is permissible. In considering this it is more basic to superpose strains than to superpose stresses as this enables one to treat both the elastic and the inelastic cases.

For a member simultaneously subjected to an axial force P and a bending moment M, strain superposition is shown schematically in Fig. 8-2. For clarity the strains are greatly exaggerated. Because of an axial force P a plane section perpendicular to the beam axis moves along it parallel to itself, Fig. 8-2(a). Because of a moment M applied around one of the principal axes a plane section rotates, Fig. 8-2(b). Superposition of strains due to P and M moves a plane section axially and rotates it as shown in Fig. 8-2(c). Note that if the axial force P causes a larger strain than that caused by M, the combined strains due to P and M will not change their sign within the member.

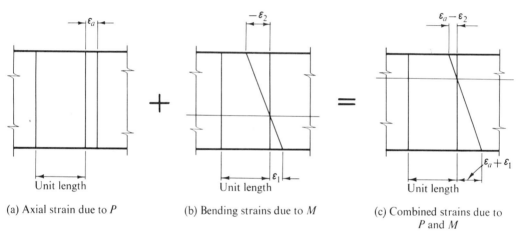

(a) Axial strain due to P (b) Bending strains due to M (c) Combined strains due to P and M

Fig. 8-2. Combined strains.

In addition to the moment which causes rotation of a plane section such as shown in Fig. 8-2, another moment acting around the other principal axis, the vertical axis in the diagram, can be applied. This second moment rotates the plane section around the vertical axis. The axial strain combined with strains caused by rotating the plane section around both principal axes is the most general case in an axially loaded bent member.

By supplementing the above basic kinematic assumptions with the stress-strain relations and conditions of equilibrium, one can solve either elastic or inelastic problems. Except for the case of symmetrical sections, however, only linearly elastic problems will be considered here. The more general cases of inelastic behavior, although susceptible to the same type of analysis, are very cumbersome. Discussion of combined shearing stresses also will be limited to linearly elastic cases.

In linearly elastic problems, a linear relationship exists between stress and strain. Therefore, unlike the case in inelastic problems, not only strains but also stresses can be superposed. This means that if on the same element and for the same coordinate system two sets of stresses are known, algebraic addition of the components of the stress tensor is possible, just as it is for vector components. This follows from the fact that components of the stress tensor located in identical positions in the matrix are associated with the same elements of area, and, except for sense, act in the same direction. For example, superposition of the primed and the double primed set of stresses for a two-dimensional problem would result in

Section 8-2 Superposition and its limitation

$$\begin{pmatrix} \sigma'_x & \tau'_{xy} \\ \tau'_{yx} & \sigma'_y \end{pmatrix} + \begin{pmatrix} \sigma''_x & \tau''_{xy} \\ \tau''_{yx} & \sigma''_y \end{pmatrix} = \begin{pmatrix} (\sigma'_x + \sigma''_x) & (\tau'_{xy} + \tau''_{xy}) \\ (\tau'_{yx} + \tau''_{yx}) & (\sigma'_y + \sigma''_y) \end{pmatrix} \quad (8\text{-}1)$$

Formulas such as those summarized in the preceding article are used to find the stress components in the first two matrices. Based on the above discussion it is important to note that <u>*superposition of stresses is applicable only in elastic problems where deformations are small.*</u>

Three examples illustrating solutions for stress distribution in symmetrical members subjected to axial loads and bending moments follow. The solution of an elastic-plastic problem is given as the third example of this group.

EXAMPLE 8-1

A 2-in.-by-3-in., 60-in.-long bar of negligible weight is loaded as shown in Fig. 8-3(a). Determine the maximum tensile and compressive stresses acting normal to the section through the beam. Assume elastic response of the material.

SOLUTION

To emphasize the method of superposition this problem is solved by dividing it into two parts. In Fig. 8-3(b) the bar is shown subjected only to the axial force, and in Fig. 8-3(c) the same bar is shown subjected only to the transverse force. For the axial force the normal stress throughout the length of the bar is

$$\sigma = \frac{P}{A} = \frac{6{,}000}{2(3)} = +1{,}000 \text{ psi} \quad \text{(tension)}$$

This result is indicated in Fig. 8-3(d). The normal stresses due to the transverse force depend on the magnitude of the bending moment, and the maximum bending moment occurs at the applied force. As the left reaction is 600 lb, $M_{max} = 600(15) = 9{,}000$ in-lb. From the flexure formula, the maximum stresses at the extreme fibers caused by this moment are

$$\sigma = \frac{Mc}{I} = \frac{6M}{bh^2} = \pm 3{,}000 \text{ psi}$$

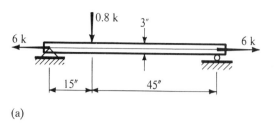

(a)

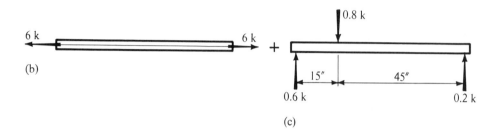

(b) (c)

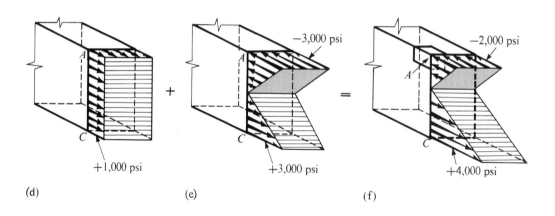

(d) (e) (f)

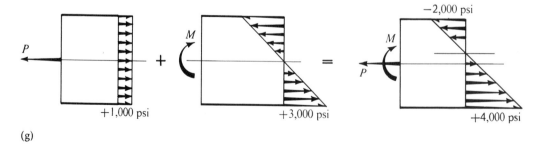

(g)

Fig. 8-3

These stresses act normal to the section of the beam and decrease linearly toward the neutral axis as in Fig. 8-3(e). Then, to obtain the compound stress for any particular element, bending stresses must be added algebraically to the direct tensile stress. Thus, as may be seen from Fig. 8-3(f), at point A the resultant normal stress is 2,000-psi compression, and at C it is 4,000-psi tension.

Side views of the stress vectors as commonly drawn are in Fig. 8-3(g). Although in this problem the given axial force is larger than the transverse force, bending causes higher stresses. However, the reader is cautioned not to regard slender members, particularly compression members, in the same light (see Fig. 8-1(b)).

Note that in the final result, the line of zero stress, which is located at the centroid of the section for flexure, moves upward. Also note that the local stresses, caused by the concentrated force, which act normal to the top surface of the beam, were not considered. Generally these stresses are treated independently as local bearing stresses.

A typical application of Eq. 8-1 to an element at A gives

$$\begin{pmatrix} +1{,}000 & 0 & 0 \\ 0 & 0 & 0 \\ 0 & 0 & 0 \end{pmatrix} + \begin{pmatrix} -3{,}000 & 0 & 0 \\ 0 & 0 & 0 \\ 0 & 0 & 0 \end{pmatrix} = \begin{pmatrix} -2{,}000 & 0 & 0 \\ 0 & 0 & 0 \\ 0 & 0 & 0 \end{pmatrix} \text{ psi}$$

The stress distribution shown in Figs. 8-3(f) and (g) would change if, for example, instead of the axial tensile forces of 6 kips applied at the ends, compressive forces of the same magnitude were acting on the member. The maximum tensile stress would be reduced to 2,000 psi from 4,000 psi, which would be desirable in a beam made of a material weak in tension and carrying a transverse load. This idea is utilized in prestressed construction. Tendons made of high-strength steel rods or cables passing through a beam with anchorages at the ends are used to precompress concrete beams. Such artifically applied forces inhibit the development of tensile stresses. Prestressing also has been used in racing-car frames.

Section 8-2
Superposition and its limitation

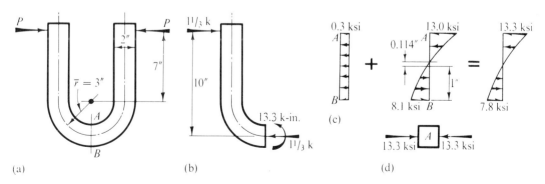

Fig. 8-4

Chapter 8
Compound stresses

EXAMPLE 8-2

A 2-in.-by-2-in. elastic bar bent into a U shape as in Fig. 8-4(a) is acted upon by two opposing forces P of $1\frac{1}{3}$ kips each. Determine the maximum normal stress occurring at the section A-B.

SOLUTION

The section to be investigated is in the curved region of the bar, but this makes no essential difference in the procedure. First, a segment of the bar is taken as a free body, as shown in Fig. 8-4(b). At section A-B the axial force, applied at the centroid of the section, and the bending moment necessary to maintain equilibrium are determined. Then, each element of the force system is considered separately. The stress caused by the axial forces is

$$\sigma = \frac{P}{A} = \frac{1.33}{2(2)} = 0.3 \text{ ksi} \quad \text{(compression)}$$

and is shown diagrammatically in the first sketch of Fig. 8-4(c). The normal stresses caused by the bending moment may be obtained by using Eq. 6-19. However, for this bar, bent to a 3-in. radius, the solution is already known from Example 6-12. The stress distribution corresponding to this case is shown in the second sketch of Fig. 8-4(c). By superposing the results of these two solutions, the compound stress distribution is obtained. This is shown in the third sketch of Fig. 8-4(c). The maximum stress occurs at A and is a compressive stress of 13.3 ksi. An isolated element for the point A is shown in Fig. 8-4(d). Shearing stresses are absent at section A-B as no shearing force is necessary to maintain equilibrium of the segment shown in Fig. 8-4(b). The relative insignificance of the stress caused by the axial force is striking.

Problems similar to the above commonly occur in machine design. Hooks, C clamps, frames of punch presses, etc. illustrate the variety of situations to which the foregoing methods of analysis must be applied.

EXAMPLE 8-3

Consider a rectangular elastic-plastic beam bent around the horizontal axis and simultaneously subjected to an axial tensile force. Determine the magnitudes of the axial forces and moments associated with the stress distributions shown in Figs. 8-5(a), (b), and (e).

SOLUTION

The stress distribution shown in Fig. 8-5(a) corresponds to the limiting elastic case, where the maximum stress is at the point of impending yielding. For this case the stress-superposition approach can be used. Hence

$$\sigma_{\max} = \sigma_{yp} = \frac{P_1}{A} + \frac{M_1 c}{I} \tag{8-2}$$

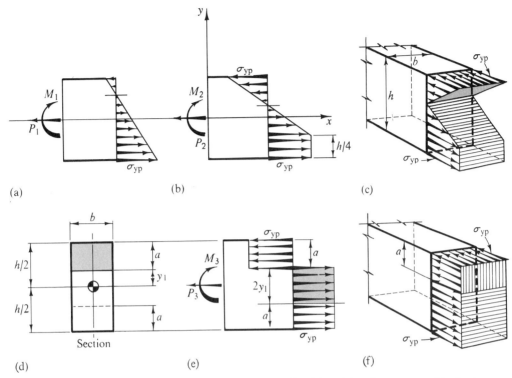

Fig. 8-5. Combined axial and bending stresses: (a) elastic stress distribution; (b) and (c) elastic-plastic stress distribution; (e) and (f) fully plastic stress distribution.

The force P at yield can be defined as $P_{yp} = A\sigma_{yp}$; from Eq. 6-8 the moment at yield is $M_{yp} = (I/c)\sigma_{yp}$. Dividing Eq. 8-2 by σ_{yp} and making use of the relations for P_{yp} and M_{yp}, one obtains

$$\frac{P_1}{P_{yp}} + \frac{M_1}{M_{yp}} = 1 \tag{8-3}$$

This establishes a relationship between P_1 and M_1 so that the maximum stress just equals σ_{yp}. A plot of this equation corresponding to the case of impending yield is in Fig. 8-6. Plots of such relations are called *interaction curves*.

The stress distribution shown in Figs. 8-5(b) and (c) occurs after yielding has taken place in the lower quarter of the beam. With this stress distribution given, one can determine directly the magnitudes of P and M from the conditions of equilibrium. If on the other hand P and M were given, since superposition does not apply, a cumbersome process would be necessary to determine the stress distribution.

For the stresses given in Figs. 8-5(b) and (c), one simply applies Eqs. 6-10 and 6-11 developed for inelastic bending of beams, except that in Eq. 6-10 the sum of the normal stresses must equal the axial force P. Noting that in the elastic zone the stress can be expressed algebraically as $\sigma = (\sigma_{yp}/3) - [8\sigma_{yp}y/(3h)]$ and that in the plastic zone $\sigma = \sigma_{yp}$, one has

$$P_2 = \int_A \sigma\, dA = \int_{-h/4}^{+h/2} \frac{\sigma_{yp}}{3}\left(1 - \frac{8y}{h}\right)b\, dy + \int_{-h/2}^{-h/4} \sigma_{yp} b\, dy = \sigma_{yp}\frac{bh}{4}$$

$$M_2 = -\int_A \sigma y\, dA = -\int_{-h/4}^{+h/2} \frac{\sigma_{yp}}{3}\left(1 - \frac{8y}{h}\right) yb\, dy - \int_{-h/2}^{-h/4} \sigma_{yp} yb\, dy$$

$$= \frac{3}{16}\sigma_{yp}bh^2$$

Note that the axial force found above exactly equals the force acting on the plastic area of the section. The moment M_2 is greater than $M_{yp} = \sigma_{yp}bh^2/6$ and less than $M_{ult} = M_p = \sigma_{yp}bh^2/4$ (see Eq. 6-12).

The axial force and moment corresponding to the fully plastic case shown in Figs. 8-5(e) and (f) are simple to determine. As may be seen from Fig. 8-5(e) the axial force is developed by σ_{yp} acting on the area $2y_1 b$. Because of symmetry, these stresses make no contribution to the moment. Forces acting on the top and the bottom areas $ab = [(h/2) - y_1]b$, Fig. 8-5(d), form a couple with a moment arm of $h - a = (h/2) + y_1$. Therefore

$$P_3 = 2y_1 b\sigma_{yp} \quad \text{or} \quad y_1 = P_3/(2b\sigma_{yp})$$

and

$$M_3 = ab\sigma_{yp}(h - a) = \sigma_{yp}b\left(\frac{h^2}{4} - y_1^2\right) = M_p - \sigma_{yp}by_1^2$$

$$= \frac{3M_{yp}}{2} - \frac{P_3^2}{4b\sigma_{yp}}$$

Then dividing by $M_p = 3M_{yp}/2 = \sigma_{yp}bh^2/4$ and simplifying, one obtains

$$\frac{2M_3}{3M_{yp}} + \left(\frac{P_3}{P_{yp}}\right)^2 = 1 \qquad (8\text{-}4)$$

This is a general equation for the interaction curve for P and M necessary to achieve the fully plastic condition in a rectangular member (see Fig. 8-6). Unlike the equation for the elastic case, the relation is nonlinear.

8-3. SKEW BENDING

In Chapter 6, on the flexure of beams, it was emphasized that the derived flexure formula is applicable only if the bending moment acts around one or the other of the principal axes of the cross section. Since the plane of the applied moment M may be inclined with respect to the principal axes,

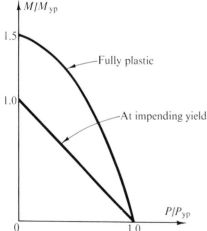

Fig. 8-6. Interaction curves for P and M for a rectangular member

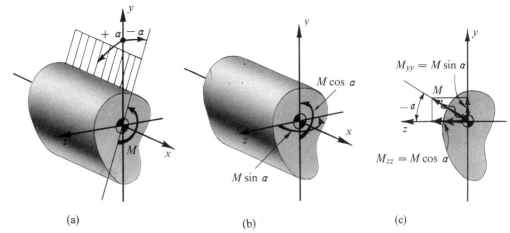

Fig. 8-7. (a) Bending moment in a plane which does not coincide with either principal axis; (b) and (c) bending-moment components in the planes of principal axes.

it is necessary to consider a more general case. Such a case is shown in Fig. 8-7(a) and is called *skew bending*.* The bending plane of M is located by an angle α which is positive when measured from the y axis toward the z axis in a counterclockwise direction.

To solve the stated problem, the applied moment M is resolved into two components acting in the planes of the principal axes. For the negative α shown in Fig. 8-7(a), the bending moment components acting around both the z and the y axes are positive (see Fig. 2-2). The one around the z axis is $M \cos \alpha$, and the one around the y axis is $M \sin \alpha$. Figures 8-7(b) and (c) show alternative representations of these positive moment components.

The elastic flexure formula previously derived can be applied to each one of the moment components acting around a principal axis, and the combined stress follows by superposition. An example of superposition is in Fig. 8-8, where for simplicity a rectangular section is shown. Analogous results hold true in general and one has†

---

* In many books such bending is called *unsymmetrical*. However, as the problem considered is more general than something lacking symmetry, the word *skew* is used in this text. This corresponds to the use of the words *schiefe* in German and *kosoi* in Russian, which mean inclined or skew.

† It is possible to derive the flexure formula for arbitrarily directed y and z axes. Such a formula, equivalent to Eq. 8-5, is

$$\sigma_x = -\frac{M_{zz}I_{yy} - M_{yy}I_{yz}}{I_{yy}I_{zz} - I_{yz}^2} y + \frac{M_{yy}I_{zz} - M_{zz}I_{yz}}{I_{yy}I_{zz} - I_{yz}^2} z \qquad \textbf{(8-5a)}$$

where I_{yy} and I_{zz} are moments of inertia, and I_{yz} is the product of inertia. For principal axes, $I_{yz} = 0$, and the above equation reverts to Eq. 8-5. For further details see, for example, D. J. Peery, *Aircraft Structures* (New York: McGraw-Hill Book Company, 1950).

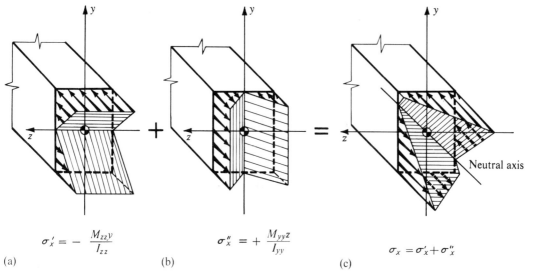

(a) $\sigma'_x = -\dfrac{M_{zz}y}{I_{zz}}$ (b) $\sigma''_x = +\dfrac{M_{yy}z}{I_{yy}}$ (c) $\sigma_x = \sigma'_x + \sigma''_x$

Fig. 8-8. Superposition of elastic bending stresses.

$$\sigma_x = -\frac{M_{zz}y}{I_{zz}} + \frac{M_{yy}z}{I_{yy}} \tag{8-5}$$

where the subscripts yy and zz on M and I refer to the respective principal axes of the cross-sectional area around which bending takes place. Note that the first term on the right side giving the stresses caused by bending around the z axis is negative just as Eq. 6-3 from which it comes. On the other hand, the second term, although analogous to Eq. 6-3, is taken positive to obtain the correspondence in sign between the normal stresses and the sense of the positive moment acting around the y axis. On this basis, in applying Eq. 8-5, if positive signs are associated with all quantities in conformity with the coordinate axes, positive results indicate tensile stresses; negative, compressive stresses. In most problems by thinking in terms of the physical action on the member, one can directly assign the sign of each term in Eq. 8-5, although the availability of the sign convention is desirable.

If, in general, the applied moment M acts in a plane making a positive angle α with the y axis, the bending-moment components are $M_{yy} = -M \sin \alpha$ and $M_{zz} = M \cos \alpha$, and Eq. 8-5 can be stated as

$$\sigma_x = -M\left(\frac{y}{I_{zz}}\cos \alpha + \frac{z}{I_{yy}}\sin \alpha\right) \tag{8-6}$$

From this relation an equation locating the neutral axis can be found by setting $\sigma_x = 0$. This yields

$$y = -z(I_{zz}/I_{yy}) \tan \alpha \tag{8-7}$$

A study of this equation using the procedures of analytic geometry shows that for skew bending, unless $I_{zz} = I_{yy}$, the neutral axis is not perpendicular to the plane of the applied moment. The neutral axis is, however, a straight line, and the "plane section" rotates around it. As in symmetrical bending, the largest stress occurs at the most remote point from the neutral axis. Note, however, that in skew bending the neutral axis does not coincide with either one of the principal axes and it is not located at right angles to the bending plane.

The analysis of inelastic beams for skew bending is very cumbersome and is beyond the scope of this text.*

Section 8-3
Skew bending

EXAMPLE 8-4

A 4-in.-by-6-in. (actual size) wooden beam shown in Fig. 8-9(a) is used to support a uniformly distributed load of 1,000 lb (total) on a simple span of 10 ft. The applied load acts in a plane making an angle of 30°

---

* M. S. Aghbabian and E. P. Popov, "Unsymmetrical Bending of Rectangular Beams Beyond the Elastic Limit," *Proceedings, First U.S. National Congress of Applied Mechanics*, 1951, pp. 579–84 (published by ASME).

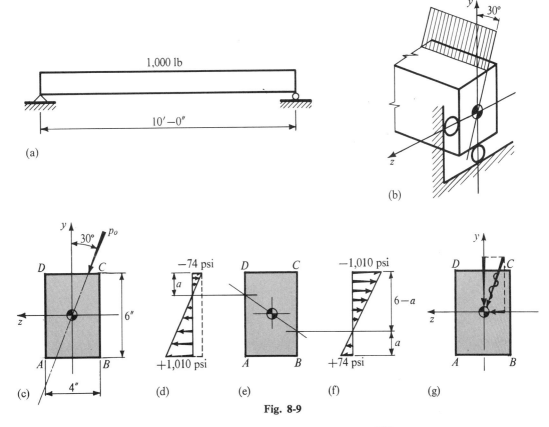

Fig. 8-9

with the vertical, as shown in Fig. 8-9(b) and again in Fig. 8-9(c). Calculate the maximum bending stress at midspan, and, for the same section, locate the neutral axis. Neglect the weight of the beam.

SOLUTION

The maximum bending in the plane of the applied load occurs at the midspan, and according to Example 2-6 it is equal to $p_o L^2/8$ or $WL/8$, where W is the total load on the span L. Hence

$$M = WL/8 = 1,000(10)/8 = 1,250 \text{ ft-lb}$$

Here $\alpha = -30°$, and the moment components acting around their respective axes are

$$M_{zz} = M \cos \alpha = 1,250(\sqrt{3}/2)12 = 13,000 \text{ in-lb}$$

$$M_{yy} = -M \sin \alpha = -1,250(-0.5)12 = 7,500 \text{ in-lb}$$

By considering the nature of the flexural stress distribution about both principal axes of the cross section, one may conclude that the maximum tensile stress occurs at A. The value of this stress follows by applying Eq. 8-5 with $y = c_1 = -3$ in., and $z = c_2 = +2$ in. Stresses at the other corners of the cross section are similarly determined.

$$\sigma_A = -\frac{M_{zz}c_1}{I_{zz}} + \frac{M_{yy}c_2}{I_{yy}} = \frac{13,000(3)}{4(6)^3/12} + \frac{7,500(2)}{6(4)^3/12}$$

$$= +542 + 468 = +1,010 \text{ psi} \quad \text{(tension)}$$

$$\sigma_B = +542 - 468 = +74 \text{ psi} \quad \text{(tension)}$$

$$\sigma_C = -542 - 468 = -1,010 \text{ psi} \quad \text{(compression)}$$

$$\sigma_D = -542 + 468 = -74 \text{ psi} \quad \text{(compression)}$$

To locate the neutral axis the stress distribution diagrams along the sides in Fig. 8-9(d) or (f) can be used. From similar triangles, $a/(6 - a) = 74/1,010$, or $a = 0.41$ in. This locates the neutral axis in Fig. 8-9(e). Alternatively, Eq. 8-7 with $\alpha = -30°$ can be used.

When skew bending of a beam is caused by applied transverse forces, as in the above example, an equivalent procedure is usually more convenient. The applied forces are first resolved into components which act parallel to the principal axes of the cross-sectional area. Then the bending moments caused by these components around the respective axes are computed for use in the flexure formula. For the above example, such components of the applied load are shown in Fig. 8-9(g). To avoid torsional stresses the applied transverse forces must act through the shear center. For bilaterally symmetrical sections, e.g., a rectangle, a circle, an I beam, etc., the shear center coincides with the centroid of the cross

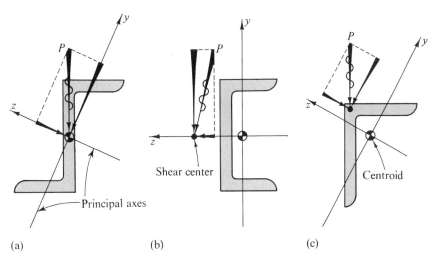

Fig. 8-10. Forces applied through shear center cause no torsion.

section. For other cross sections, such as channels, angles, Z sections, etc., the shear center lies elsewhere (see Art. 7-7). In such problems the transverse force must be applied at the shear center to avoid torsional stresses. This approach is illustrated in Fig. 8-10. Otherwise, in addition to the bending stresses, the torsional stresses must be investigated. In such cases the applied torque equals the applied force multiplied by its moment arm measured from the shear center.

8-4. ECCENTRICALLY LOADED MEMBERS

Occasionally situations arise where a force P acting parallel to the axis of the member is applied eccentrically with respect to the centroidal axis of the member, Fig. 8-11(a). By applying two equal and opposite forces P at the centroid, as shown in Fig. 8-11(b), the problem is changed to that of an axially applied force P and skew bending in the plane of the applied force P and the axis of the member. This skew bending moment can be further resolved into the components $M_{yy} = Pz_o$ acting around the y axis, and $M_{zz} = -Py_o$ acting around the z axis, Figs. 8-11(d) and (e). Then, the compound normal stress at any point (y, z) of the cross section, for an eccentrically loaded member, can be found by simply adding an axial stress term to Eq. 8-5. Hence

$$\sigma_x = \frac{P}{A} - \frac{M_{zz}y}{I_{zz}} + \frac{M_{yy}z}{I_{yy}} \qquad (8\text{-}8)$$

where P is taken positive for tensile forces. The remainder of the sign convention is the same as that for Eq. 8-5. Providing the y and the z axes

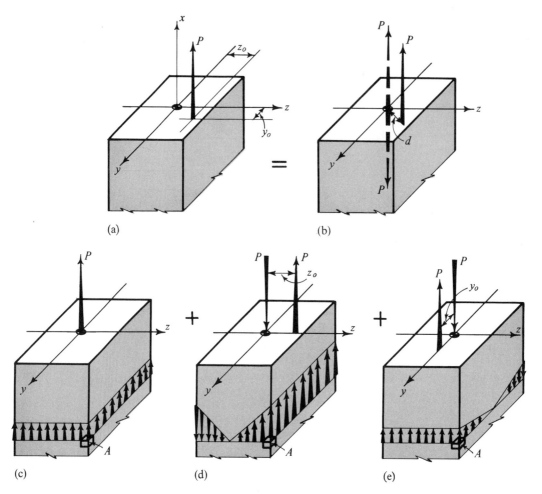

Fig. 8-11. Resolution of a problem into three problems, each one of which may be solved by the methods previously discussed.

are the principal axes, Eq. 8-8 is applicable to prismatic members of any cross-sectional shape.

For a given loading condition Eq. 8-8 can be rewritten as

$$\sigma_x = A + By + Cz \tag{8-9}$$

where A, B, and C are constants. This is seen to be an equation of a plane; it clearly shows the nature of stress distribution. For the linearly elastic case under discussion, dividing through Eq. 8-9 by the elastic modulus E recovers the basic kinematic assumption of the technical theory, i.e.,

$$\varepsilon_x = a + by + cz \tag{8-10}$$

where a, b, and c are constants.

In some eccentrically loaded members it is possible to locate the line of zero stress within the cross-sectional area of a member by determining a line where $\sigma_x = 0$. This line is analogous to the neutral axis occurring in pure bending. Unlike the former case, however, with $P \neq 0$ this line does not pass through the centroid of a section. For large axial loads and small moments, it lies outside the cross section. Its significance lies in the fact that the normal stresses vary linearly from it.

This method is applicable for compression members providing their length is small in relation to their transverse dimensions. Slender bars in compression require special treatment (Chapter 14). Also, near the point of application of the force, the analysis developed here is incorrect. There the stress distribution is greatly disturbed and is similar to a local stress concentration (see Art. 4-18 and especially Fig. 4-30).

Section 8-4
Eccentrically loaded members

EXAMPLE 8-5

Find the stress distribution at the section $ABCD$ for the block shown in Fig. 8-12(a) if $P = 14.4$ kips. At the same section, locate the line of zero stress. Neglect the weight of the block.

SOLUTION

The forces acting on the section $ABCD$, Fig. 8-5(c), are $P = -14.4$ kips, $M_{yy} = -14.4(6) = -86.4$ kip-in., and $M_{zz} = -14.4(3 + 3) = -86.4$

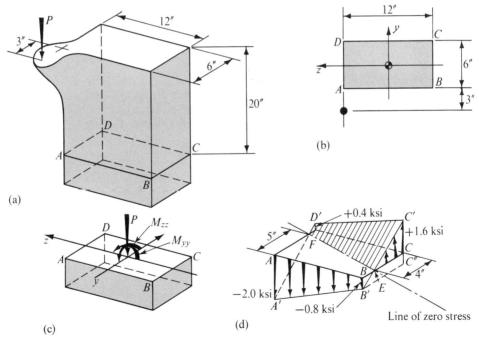

Fig. 8-12

kip-in. The cross section of the block $A = 6(12) = 72$ in.^{2}, and the respective section moduli are $S_{zz} = 12(6)^2/6 = 72$ in.^{3} and $S_{yy} = 6(12)^2/6 = 144$ in.^{3} Hence, using a relation equivalent to Eq. 8-8 gives the compound normal stresses for the corner elements:

$$\sigma = \frac{P}{A} \mp \frac{M_{zz}}{S_{zz}} \pm \frac{M_{yy}}{S_{yy}} = -\frac{14.4}{72} \pm \frac{86.4}{72} \mp \frac{86.4}{144} = -0.2 \pm 1.2 \mp 0.6$$

Here the units of stress are kips per square inch. The sense of the forces shown in Fig. 8-12(c) determines the signs of stresses. Therefore, if the subscript of the stress signifies its location, the corner normal stresses are:

$$\sigma_A = -0.2 - 1.2 - 0.6 = -2.0 \text{ ksi}$$

$$\sigma_B = -0.2 - 1.2 + 0.6 = -0.8 \text{ ksi}$$

$$\sigma_C = -0.2 + 1.2 + 0.6 = +1.6 \text{ ksi}$$

$$\sigma_D = -0.2 + 1.2 - 0.6 = +0.4 \text{ ksi}$$

These stresses are shown in Fig. 8-12(d). The ends of these four stress vectors at A', B', C', and D' lie in the plane $A'B'C'D'$. The vertical distance between the planes $ABCD$ and $A'B'C'D'$ defines the compound stress at any point on the cross section. The intersection of the plane $A'B'C'D'$ with the plane $ABCD$ locates the line of zero stress FE.

By drawing a line $B'C''$ parallel to BC, similar triangles $C'B'C''$ and $C'EC$ are obtained: thus the distance $CE = [1.6/(1.6 + 0.8)]6 = 4$ in. Similarly, the distance AF is found to be 5 in. Points E and F locate the line of zero stress.

EXAMPLE 8-6

Find the zone over which the vertical downward force P_o may be applied to the rectangular weightless block shown in Fig. 8-13(a) without causing any tensile stresses at the section A-B.

SOLUTION

The force $P = -P_o$ is placed at an arbitrary point in the first quadrant of the y-z coordinate system shown. Then the same reasoning used in the preceding example shows that with this position of the force the greatest tendency for a tensile stress exists at A. With $P = -P_o$, $M_{zz} = +P_o y$ and $M_{yy} = -P_o z$, setting the stress at A equal to zero fulfills the limiting condition of the problem. Using Eq. 8-8 allows the stress at A to be expressed as:

$$\sigma_A = 0 = \frac{(-P_o)}{A} - \frac{(P_o y)(-b/2)}{I_{zz}} + \frac{(-P_o z)(-h/2)}{I_{yy}}$$

or

$$-\frac{P_o}{A} + \frac{P_o y}{b^2 h/6} + \frac{P_o z}{bh^2/6} = 0$$

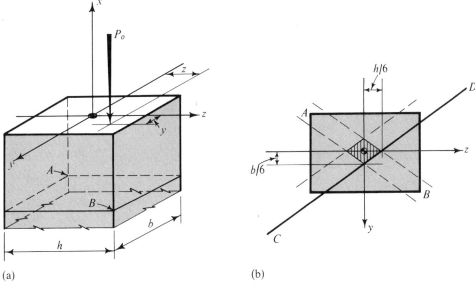

Fig. 8-13

Simplifying $[z/(h/6)] + [y/(b/6)] = 1$, which is an equation of a straight line. It shows that when $z = 0$, $y = b/6$; and when $y = 0$, $z = h/6$. Hence this line may be represented by the line CD in Fig. 8-13(b). A vertical force may be applied to the block anywhere on this line and the stress at A will be zero. Similar lines may be established for the other three corners of the section; these are shown in Fig. 8-13(b). If the force P is applied on any one of these lines or on any line parallel to such a line toward the centroid of the section, there will be no tensile stress at the corresponding corner. Hence the force P may be applied anywhere within the shaded area in Fig. 8-13(b) without causing tensile stress at any of the four corners or anywhere else. This zone of the cross-sectional area is called the *kern* of a section.

If for a rectangular block the location of the force P is limited to one of the lines of symmetry, the maximum eccentricity $e = h/6$ to give zero stress along one of the edges, Figs. 8-14(a) and (b). This leads to a practical rule, much used in the past by designers of masonry structures: If the resultant of vertical forces acts within the middle third of a rectangular section, there is no tension in the material at that section. If, further, the applied load P acts outside the middle third and the contact surfaces cannot transmit tensile forces, one has the case shown in Figs. 8-14(c) and (d). Here, assuming elastic action, the normal stress at B may be expressed as

$$\sigma_B = -\frac{P}{xb} + P\left(\frac{x}{2} - a\right)\frac{6}{bx^2} = 0$$

where $(x/2) - a$ is the eccentricity of the applied force with respect to

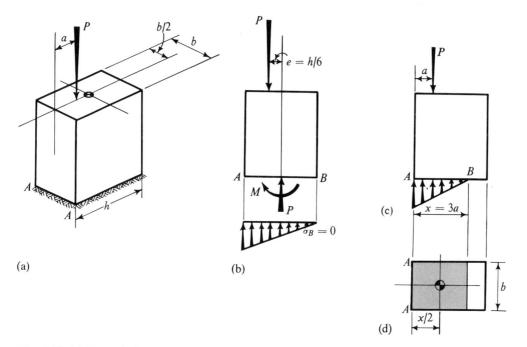

Fig. 8-14. (a) Eccentrically loaded block; (b) location of P to cause zero stress at B; (c) elastic stress distribution between two surfaces which are unable to transmit tensile forces.

the centroidal axis of the shaded contact area, and $bx^2/6$ is its section modulus. Solving for x, one finds that $x = 3a$; the pressure distribution will be "triangular" as in Fig. 8-14(c) (why?). As a decreases, the intensity of pressure on the line A-A increases; when a is zero, the block becomes unstable. Such problems are important in the design of foundations.

8-5. SUPERPOSITION OF SHEARING STRESSES

In the preceding part of the chapter superposition of the normal stresses σ_x was the principal concern. In problems where both the elastic torsional and direct shearing stresses can be determined, the compound shearing stress also may be found by superposition. This corresponds to superposition of the off-diagonal stresses in Eq. 8-1. Here attention will be directed to instances where the shearing stresses being superposed not only act on the same element of area but also have the same line of action.* Only elastic stresses fall within the scope of this treatment.

---

* Noncolinear shearing stresses acting on the same element of area can be added vectorially.

EXAMPLE 8-7

Section 8-5
Superposition of shearing stresses

Find the maximum shearing stress due to the applied forces in the plane A-B of the $\frac{1}{2}$ in.-diameter, high-strength shaft in Fig. 8-15(a).

SOLUTION

The free body of a segment of the shaft is shown in Fig. 8-15(b). The system of forces at the cut necessary to keep this segment in equilibrium consists of a torque $T = 200$ in-lb, a shear $|V| = 60$ lb, and a bending moment $M = 240$ in-lb.

Because of the torque T, the shearing stresses in the cut A-B vary linearly from the axis of the shaft and reach the maximum value given by Eq. 5-4, $\tau_{max} = Tc/J$. These maximum shearing stresses, agreeing in sense with the resisting torque T, are shown at points A, B, D, and E in Fig. 8-15(c).

The "direct" shearing stresses caused by the shearing force V may be obtained by using Eq. 7-6, $\tau = VQ/(It)$. For the elements A and

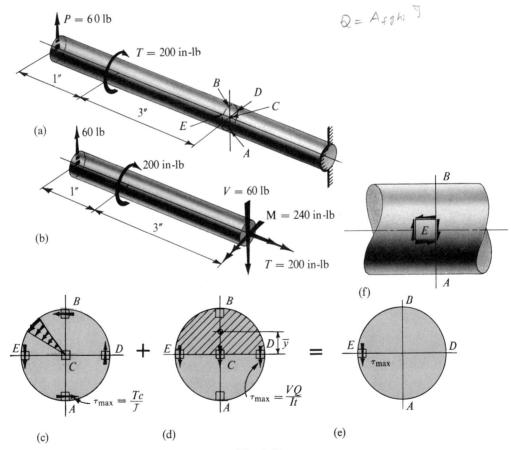

Fig. 8-15

B, Fig. 8-15(d), $Q = 0$, hence $\tau = 0$. The shearing stress reaches its maximum value at the level ED. To determine this, consider Q equal to the shaded area in Fig. 8-15(d) multiplied by the distance from its centroid to the neutral axis. The latter quantity is $\bar{y} = 4c/(3\pi)$, where c is the radius of the cross-sectional area. Hence $Q = (\pi c^2/2)[4c/(3\pi)] = 2c^3/3$. Moreover, since $t = 2c$, and $I = J/2 = \pi c^4/4$, the maximum direct shearing stress is

$$\tau_{max} = \frac{VQ}{It} = \frac{V}{2c}\frac{2c^3}{3}\frac{4}{\pi c^4} = \frac{4V}{3\pi c^2} = \frac{4V}{3A}$$

where A is the entire cross-sectional area of the rod. In Fig. 8-15(d) this shearing stress is shown acting downward on the elementary areas at E, C, and D. This direction agrees with the direction of the shear V.

To find the maximum compound shearing stress in the plane A-B, the stresses shown in Figs. 8-15(c) and (d) are superposed. Inspection shows that the maximum shearing stress is at E since in the two diagrams the shearing stresses at E have the same direction and sense. There are no direct shearing stresses at A and B, and at C there is no torsional shearing stress. The two shearing stresses have an opposite sense at D. The five points A, B, C, D, and E thus considered for the compound shearing stress are all that may be adequately treated by the methods developed in this text. However, this procedure selects the elements where the maximum shearing stresses occur.

$$J = \frac{\pi d^4}{32} = \frac{\pi (0.5)^4}{32} = 0.00614 \text{ in.}^4 \quad \text{and} \quad I = \frac{J}{2} = 0.00307 \text{ in.}^4$$

$$A = \pi d^2/4 = 0.196 \text{ in.}^2$$

$$(\tau_{max})_{torsion} = \frac{Tc}{J} = \frac{200(0.25)}{0.00614} = 8{,}150 \text{ psi}$$

$$(\tau_{max})_{direct} = \frac{VQ}{It} = \frac{4V}{3A} = \frac{4(60)}{3(0.196)} = 408 \text{ psi}$$

$$\tau_E = 8{,}150 + 408 = 8{,}560 \text{ psi}$$

A planar representation of the shearing stress at E with the matching stresses on the longitudinal planes is shown in Fig. 8-15(f). No normal stress acts on this element as it is located on the neutral axis.

8-6. STRESSES IN CLOSELY COILED HELICAL SPRINGS

Helical springs, such as the one shown in Fig. 8-16(a), are often used as elements of machines. With certain limitations, these springs may be analyzed for elastic stresses by a method similar to the one used in the preceding example. The discussion will be limited* to springs manufactured from rods or wires of circular cross section. Moreover, any one coil of such a spring will be assumed to lie in a plane which is nearly perpendicular to the axis of the spring. This requires that the adjoining coils be close together. With this limitation, a section taken perpendicular to the axis of the spring's rod becomes nearly vertical.† Hence to maintain equilibrium of a segment of the spring, only a shearing force $V = F$ and a torque $T = F\bar{r}$ are required at all sections through the rod, Fig. 8-16(b).‡ Note that $\bar{r}$ is the distance from the axis of the spring to the centroid of the rod's cross-sectional area.

The maximum shearing stress at an arbitrary section through the rod could be obtained as in the preceding example by superposing the torsional and the direct shearing stresses. This maximum shearing stress occurs at the inside of the coil at point E, Fig. 8-16(b). However, in the analysis of springs it has become customary to assume that the shearing stress caused by the direct shearing force is uniformly distributed over the cross-sectional area of the rod. Hence, the nominal direct shearing stress for any point on the cross section is $\tau = F/A$. Superposition of this stress and the torsional shearing stress at E gives the maximum compound shearing stress. Thus since $T = F\bar{r}$, $d = 2c$, and $J = \pi d^4/32$

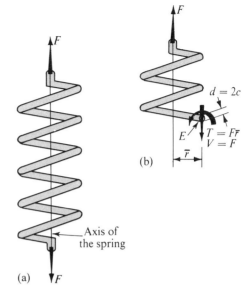

Fig. 8-16. Closely coiled helical spring.

$$\tau_{max} = \frac{F}{A} + \frac{Tc}{J} = \frac{Tc}{J}\left(\frac{FJ}{ATc} + 1\right) = \frac{16F\bar{r}}{\pi d^3}\left(\frac{d}{4\bar{r}} + 1\right) \quad (8\text{-}11)$$

---

* For a complete discussion on springs see A. M. Wahl, *Mechanical Springs* (Cleveland, Ohio: Penton Publishing Co., 1944).

† This eliminates the necessity of considering an axial force and a bending moment at the section taken through the spring.

‡ In previous work it has been reiterated that if a shear is present at a section, a change in the bending moment must take place along the member. Here a shear acts at every section of the rod, yet no bending moment nor a change in it occurs. This is so only because the rod is curved. An element of the rod viewed from the top is shown in the figure. At both ends the torques are equal to $F\bar{r}$ and act in the directions shown. The component of these vectors toward the axis of the spring O, resolved at the point of intersection of the vectors, $2F\bar{r}\,d\phi/2 = F\bar{r}\,d\phi$, opposes the couple developed by the vertical shears $V = F$, which are $\bar{r}\,d\phi$ apart.

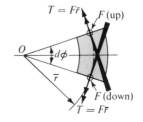

It is seen from this equation that as the diameter of the rod d becomes small in relation to the coil radius $\bar{r}$, the effect of the direct shearing stress becomes small. On the other hand, if the reverse is true, the first term in the parenthesis becomes important. In the latter case the results indicated by Eq. 8-11 are considerably in error, and Eq. 8-11 should not be used as it is based on the torsion formula for straight rods. As d becomes numerically comparable to $\bar{r}$, the length of the inside fibers of the coil differs greatly from the length of the outside fibers, and the strain assumption used in deriving the classical torsion formula is not applicable.

The spring problem has been solved exactly* by the methods of the mathematical theory of elasticity, and, although these results are complicated, for any one spring they may be made to depend on a single parameter $m = 2\bar{r}/d$, which is called the *spring index*. Thus Eq. 8-11 may be rewritten as

$$\tau_{\max} = K \frac{16F\bar{r}}{\pi d^3} \qquad (8\text{-}12)$$

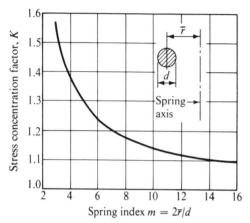

Fig. 8-17. Stress-concentration factor for helical round-wire compression or tension springs.

where K may be interpreted as a stress-concentration factor for closely coiled helical springs made from circular rods. A plot of K vs. the spring index is in Fig. 8-17.† For heavy springs the spring index is small, and the stress-concentration factor K becomes very important. For all cases the factor K accounts for the correct amount of direct shearing stress. Very high stresses are commonly allowed in springs because high-strength materials are used in their fabrication. For good-quality spring steel, working shearing stresses range anywhere from 30,000 psi to 100,000 psi.

8-7. DEFLECTION OF CLOSELY COILED HELICAL SPRINGS

As the subject of closely coiled helical springs was introduced above, for completeness, their deflection will be discussed in this article. Attention will be confined to closely coiled helical springs with a large spring index, i.e., the diameter of the wire will be assumed small in comparison with the radius of the coil. This permits the treatment of an element of a spring

---

* O. Goehner, "Die Berechnung Zylindrischer Schraubenfedern," *Zeitschrift des Vereins deutscher Ingenieure*, **76**, no. 1 (March 1932), p. 269.

† An analytical expression which gives the value of K within 1 or 2 per cent of the true value is frequently used. This expression in terms of the spring index m is $K_1 = [(4m - 1)/(4m - 4)] + (0.615/m)$. It was derived by A. M. Wahl on the basis of some simplifying assumptions and is known as the *Wahl correction factor* for curvature in helical springs.

between two closely adjoining sections through the wire as a straight circular bar in torsion. The effect of direct shear on the deflection of the spring will be ignored. This is usually permissible as the latter effect is small.

Consider a helical spring such as shown in Fig. 8-18. A typical element AB of this spring is subjected throughout its length to a torque $T = F\bar{r}$. This torque causes a relative rotation between the two adjoining planes A and B; with sufficient accuracy the amount of this rotation may be obtained by using Eq. 5-10, $d\varphi = T\,dx/(JG)$, for straight circular bars. Here the applied torque $T = F\bar{r}$, dx is the length of the element, G is the shearing modulus of elasticity, and J is the polar moment of inertia of the cross-sectional area of the wire.

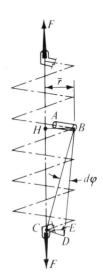

Fig. 8-18. Diagram used in deriving the expression for the deflection of a helical spring.

If the plane of the wire A is imagined fixed, the rotation of the plane B is given by the foregoing expression. The contribution of this element to the movement of the force F at C is equal to the distance BC multiplied by the angle $d\varphi$, i.e., $CD = BC\,d\varphi$. However, since the element AB is small, the distance CD is also small, and this distance may be considered perpendicular (although it is an arc) to the line BC. Moreover, only the vertical component of this deflection is significant as in a spring consisting of many coils, for any element on one side of the spring there is a corresponding equivalent element on the other. The diametrically opposite elements of the spring balance out the horizontal component of the deflection and permit only the vertical deflection of the force F. Therefore, by finding the vertical increment ED of the deflection of the force F due to an element of the spring AB and summing such increments for all elements of the spring, the deflection of the whole spring is obtained.

From similar triangles CDE and CBH

$$\frac{ED}{CD} = \frac{HB}{BC} \quad \text{or} \quad ED = \frac{CD}{BC} HB$$

However, $CD = BC\,d\varphi$, $HB = \bar{r}$, and ED may be denoted by $d\Delta$ as it represents an infinitesimal vertical deflection of the spring due to rotation of an element AB. Thus $d\Delta = \bar{r}\,d\varphi$ and

$$\Delta = \int d\Delta = \int \bar{r}\,d\varphi = \int_0^L \bar{r}\,\frac{T\,dx}{JG} = \frac{TL\bar{r}}{JG}$$

However, $T = F\bar{r}$, and for a closely coiled spring the length L of the wire may be taken with sufficient accuracy as $2\pi\bar{r}N$, where N is the number of

Chapter 8
Compound stresses

live or active coils of the spring. Thus the deflection Δ of the spring is

$$\Delta = 2\pi F \bar{r}^3 N/JG \qquad (8\text{-}13)$$

or if the value of J for the wire is substituted

$$\Delta = \frac{64 F \bar{r}^3 N}{G d^4} \qquad (8\text{-}14)$$

Equations 8-13 and 8-14 give the deflection of a closely coiled helical spring along its axis when such a spring is subjected to either a tensile or a compressive force F. In these formulas the effect of the direct shearing stress on the deflection is neglected, i.e., they give only the effect of torsional deformations.

The behavior of a spring may be conveniently defined by a force required to deflect the spring 1 in. This quantity is known as the *spring constant*. From Eq. 8-14 the spring constant k for a helical spring made from a wire with a circular cross section is

$$k = \frac{F}{\Delta} = \frac{G d^4}{64 \bar{r}^3 N} \quad \left[\frac{\text{lb}}{\text{in.}}\right] \qquad (8\text{-}15)$$

PROBLEMS FOR SOLUTION

8-1. An offset link of a machine is made of a 1-in. round bar and is shaped as shown in the figure. If the distance $e = 1$ in. and the allowable stress is 10,000 psi, what force P can be applied?

PROB. 8-1

8-2. An offset link is similar to the one shown in the preceding problem but is larger and has a cross section in the form of a T, see figure. At the ends of the link the tensile forces P are applied 4 in. above the bottom of the flanges, and the offset $e = 2\frac{1}{2}$ in. from the line of action of the forces. Find the maximum stress if $P = 40$ kips and the material behaves elastically.

8-3. A cast iron frame for a punch press has the proportions shown in the figure. What force P may be applied to this frame controlled by the stresses in the sections such as a–a, if the allowable stresses are 4,000 psi in tension and 12,000 psi in compression? *Ans.* 9.12 k.

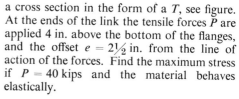

PROB. 8-3

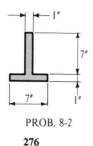

PROB. 8-2

8-4. An inclined beam having a cross section of 6 in. in width by 12 in. in depth supports a vertical force as shown in the figure for Prob. 2-6. Determine the maximum stress acting normal to the section a–a.

8-5. A 12 I 35.0 steel beam (see Table 3 in the Appendix) supports a force $P = 40$ kips as shown in the figure. Neglecting the weight of the beam, find the largest normal stress acting at section a–a. Assume linearly elastic behavior of the material.

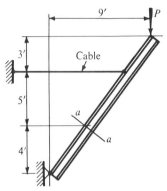

PROB. 8-5

8-6. Compute the maximum compressive stress acting normal to the section a–a for the structure shown in the figure. The post AB has a 12-in.-by-12-in. cross section. Neglect the weight of the structure. *Ans.* -174 psi.

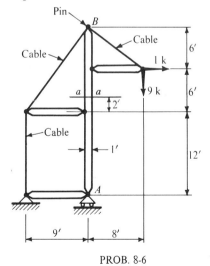

PROB. 8-6

PROB. 8-7

8-7. A bracket has the dimensions shown in the figure. Find the maximum stress acting normal to section a–a caused by the application of $P = 150$ lb. The inclined bar is $\frac{1}{2}$ in. by 2 in., and is laterally braced to prevent buckling. Assume that the bar behaves elastically.

8-8. Calculate the maximum compressive stress acting on section a–a caused by the applied load for the structure shown in the figure. The cross section at section a–a is that of a solid circular bar of 2-in. diameter. *Ans.* $-1,110$ psi.

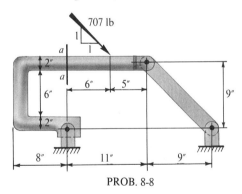

PROB. 8-8

8-9. A factory stairway having the centerline dimensions shown in the figure is made from two 9-in., 13.4-lb steel channels on edge separated by treads framing into them. The loading on each channel, including its own weight, is estimated to be 200 lb per foot of horizontal projection. Assuming that the lower end of the stairway is pinned and that the wall provides only horizontal support at the top, find the largest normal stress in the channels 5 ft above the floor level. *Ans.* $-2,620$ psi.

277

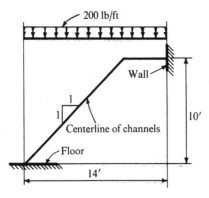

PROB. 8-9

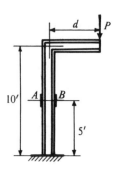

PROB. 8-11

8-10. A jib crane is made from an 8-in., 18.4-lb, steel *I* beam and a high-strength steel rod as shown in the figure. (a) Find the location of the movable load *P* that would cause the largest bending moment in the beam. Neglect the weight of the beam. (b) Using the load location found in (a), how large may the load *P* be? Assume that the effect of shear in the beam is not significant, and let the allowable normal stress in the beam be 18,000 psi. Comment on the accuracy of the criterion established in (a).

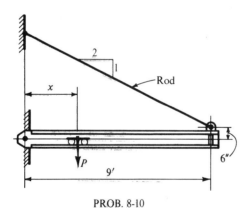

PROB. 8-10

8-11. A steel frame fabricated from 8 WF 17 steel sections supports a load *P* at a distance *d* from the center of the vertical column as shown in the figure. On the outside of the column at a distance 5 ft from the ground the following strains were measured: at A, $\varepsilon = 200 \times 10^{-6}$ in. per inch; and at B, $\varepsilon = -600 \times 10^{-6}$ in. per inch. What are the magnitudes of the load *P* and the distance *d*? Let $E = 30 \times 10^6$ psi.

8-12. A rectangular beam, such as shown in Fig. 8-5, of linearly elastic-plastic material is subjected to a positive bending moment M_2 and an axial force P_2. (a) If the strain at the top surface just reaches ε_{yp} and the strain at the bottom is $3\varepsilon_{yp}$, what force P_2 and moment M_2 act on the beam? (b) If the beam is initially subjected to the forces in (a), what will be the residual stress pattern upon their removal? Let $b = \frac{1}{2}$ in., $h = 1$ in., $\sigma_{yp} = 40$ ksi, and $E = 29 \times 10^6$ psi.

8-13. An *I* beam of linearly elastic-plastic material has the dimensions shown in the figure. When this beam is subjected simultaneously to an axial force and a bending moment, the strain at the juncture of the web with the top flange just reaches ε_{yp} in compression. The strain is zero 4 in. below the top of the beam, see figure. (a) Determine the axial force and the moment corresponding to the above conditions of strain. Assume $\sigma_{yp} = 50$ ksi, and $E = 29 \times 10^6$ psi. (b) Find and sketch the residual stress distribution which results after the release of the forces in (a).

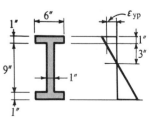

PROB. 8-13

8-14. Consider the beam of the previous example experiencing the large strain shown in the figure for this problem. (a) Determine the axial force and the bending moment corresponding to the given strain. Neglect the correction due to the small elastic zone, and consider all material yielding either in tension or in compression at the stress of 50 ksi. (b) Find and sketch the residual stress distribution which results after the release of the forces in (a).

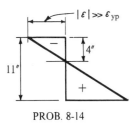

PROB. 8-14

8-15. A T beam of linearly elastic-plastic material has the dimensions shown in the figure. (a) If the strain is $-\varepsilon_{yp}$ at the top of the flange and is zero at the juncture of the web with the flange, what axial force P and bending moment M act on the beam? Assume $\sigma_{yp} = 36$ ksi. (b) Determine the residual stress pattern which would develop upon removal of the forces in (a).

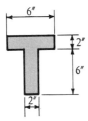

PROB. 8-15

8-16. A steel hook, having the proportions in the figure, is subjected to a downward load of 19 kips. The radius of the centroidal curved axis is 6 in. Determine the maximum stress in this hook.

8-17. If a hook similar to the one shown in Prob. 8-16 has a circular cross section of 1-in. radius and a 3-in. radius of the curved centroidal axis, what force P may be applied to the hook without exceeding a stress of 12,000 psi?

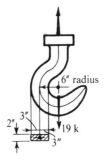

PROB. 8-16

8-18. A steel bar of 2-in. diameter is bent into a nearly complete circular ring of 12-in. outside diameter as shown in the figure. (a) Calculate the maximum stress in this ring caused by applying two 2,000-lb forces at the open end. (b) Find the ratio of the maximum stress found in (a) to the largest compressive stress acting normal to the same section. *Ans.* (a) $+30.8$ ksi.

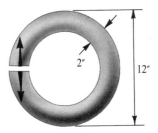

PROB. 8-18

8-19. A tilted, simple supported beam with a depth to width ratio of 2 to 1 is to span 12 ft and is to carry a uniformly distributed load of 1,000 lb per lineal foot, including its own weight, applied as shown in the figure. (a)

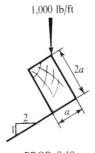

PROB. 8-19

279

Determine the required dimensions of the beam so that the maximum stress due to bending does not exceed 1,500 psi. (b) Locate the neutral axis of the beam and show its position on the sketch.

8-20. A full-sized, 2-in.-by-4-in. cantilever projects 4 ft from a wall in a tilted position as shown in the figure. At the free end a vertical force of 100 lb is applied which acts through the centroid of the section. Determine the maximum flexural stress, caused by the applied force, in the beam at the built-in end and locate the neutral axis. Neglect the weight of the beam. *Ans.* ±1,423 psi, 3.34 in.

PROB. 8-20

8-21. A 6-in.-by-6-in.-by-½-in. steel angle with one of its legs placed in a horizontal position and its other leg directed downward is used as a cantilever 70.7 in. long, see figure. If an upward force of 1,000 lb is applied at the end of this cantilever through the shear center, what are the maximum tensile and compressive stresses at the built-in end? Neglect the weight of the angle. (See hint for Prob. 6-22.) *Ans.* 14.7 ksi, −18.4 ksi.

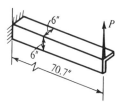

PROB. 8-21

8-22. If the block shown in Fig. 8-12 is made from steel weighing 0.283 lb per cubic inch, find the magnitude of the force P necessary to cause zero stress at D. Neglect the weight of the small bracket supporting the load. For the same condition, locate the line of zero stress at the section $ABCD$. *Ans.* 203 lb.

8-23. A cast iron block is loaded as shown in the figure. Neglecting the weight of the block, determine the stresses acting normal to a section taken 18 in. below the top and locate the line of zero stress.

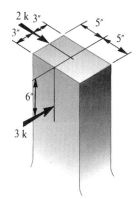

PROB. 8-23

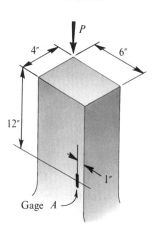

PROB. 8-24

8-24. An aluminum-alloy block is loaded as shown in the figure. The application of this load produces a tensile strain of 500×10^{-6} in. per inch at A as measured by means of an electrical strain gage. Compute the magnitude of the applied force P. Let $E = 10 \times 10^6$ psi. *Ans.* 30 k.

8-25. A short block has cross-sectional dimensions in plan view as shown in the figure. Determine the range along the line A–A over

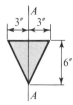

PROB. 8-25

which a downward vertical force could be applied to the top of the block without causing any tension at the base. Neglect the weight of the block. *Ans.* Between 3 in. and $4\frac{1}{2}$ in. from the apex.

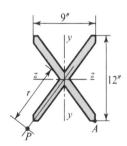

PROB. 8-26

8-26. A short compression member has the proportions shown in the figure; its $A = 72.9$ in.^{2}, $I_{zz} = 1{,}199$ in.^{4}, and $I_{yy} = 633$ in.^{4} Determine the distance r along the diagonal where a longitudinal force P should be applied so that point A lies on the line of zero stress. Neglect the weight of the member. *Ans.* 53 in.

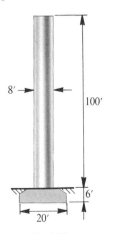

PROB. 8-28

8-27. Determine the kern for a member having a solid, circular cross section. *Ans.* $c/4$.

8-28. An 8-ft-diameter steel stack, partially lined with brick on the inside, together with a 20-ft-by-20-ft concrete foundation pad weighs 160,000 lb. This stack projects 100 ft above the ground level, as shown in the figure, and is anchored to the foundation. If the horizontal wind pressure is assumed to be 20 lb per square foot of the projected area of the stack and the wind blows in the direction parallel to one of the sides of the square foundation, what is the maximum foundation pressure? *Ans.* 1.21 k per square foot.

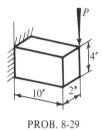

PROB. 8-29

8-29. A rectangular cantilever 10 in. long is loaded with $P = 10$ kips at the free end as shown in the figure. Determine the maximum shearing stress at the built-in end due to the direct shear and the torque. Show the result on a sketch analogous to Fig. 8-15(e).

8-30. A helical compression spring is made from $\frac{1}{8}$-in.-diameter phosphor-bronze wire and has an outside diameter of $1\frac{1}{4}$ in. If the allowable shearing stress is 30,000 psi, what force may be applied to this spring? Correct the answer for stress concentrations.

8-31. A helical spring is made of $\frac{1}{2}$-in.-diameter steel wire by winding it on a 5-in.-diameter mandrel. If there are 10 active coils, what is the spring constant? $G = 12 \times 10^6$ psi. What force must be applied to the spring to elongate it $1\frac{3}{4}$ in?

8-32. A helical valve spring is made of $\frac{1}{4}$-in.-diameter steel wire and has an outside diameter of 2 in. In operation the compressive force applied to this spring varies from 20 lb minimum to 70 lb maximum. If there are eight active coils, what is the valve lift (or travel) and what is the maximum shearing stress in the spring when in operation? $G = 11.6 \times 10^6$ psi. *Ans.* 0.38 in.

8-33. A heavy helical steel spring is made from a 1-in.-diameter rod and has an outside diameter of 9 in. As originally manufactured,

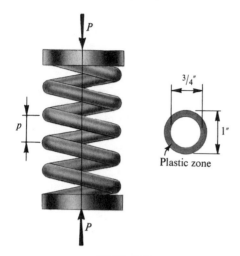

PROB. 8-33

it has the pitch $p = 3\frac{1}{2}$ in., see figure. If a force P, of such magnitude that the rod's outer annulus $\frac{1}{8}$ in. thick becomes plastic, is applied to this spring, estimate the reduction of the pitch of the spring on removal of the load. Assume linearly elastic-plastic material with $\tau_{yp} = 50$ ksi, with a $G = 12 \times 10^3$ ksi. Neglect the effects of stress concentration and of the direct shear on deflection. (*Hint:* See Examples 5-9 and 5-10.)

8-34. If the shearing stress in the bolts governs the allowable load P which may be applied to the connection shown in the figure, what is the

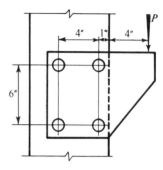

PROB. 8-34

allowable force P? The bolts are $\frac{3}{4}$ in. and the allowable shearing stress is 15 ksi. (*Hint:* Concentrate the bolt areas at their respective centers. Assume that a force applied at the centroid of the bolt areas is distributed equally among the bolts. The torque resistance is found as for a coupling, Prob. 5-18.) *Ans.* 10 kips.

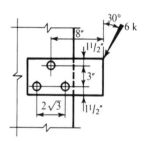

PROB. 8-35

8-35. Determine the maximum shearing stress in the rivets in a bracket loaded as shown in the figure. All rivets are 1 in. in diameter. (*Note:* See hint for the above problem.) *Ans.* 9,140 psi.

Transformation of stress and strain; yield and fracture criteria

9

9-1. INTRODUCTION

From the preceding chapters it should be apparent that in many instances both normal and shearing stresses act on an element of a body simultaneously. For example, a typical element A of a member subjected to axial and transverse forces, shown in Fig. 9-1(a), experiences a normal stress σ_x due to axial pull and bending, and also shearing stress τ_{xy} due to shear. Using the procedures developed thus far, one must take the planes which isolate this element either parallel or perpendicular to the axis of the member. On the other hand, it is possible to describe the state of stress at a point in terms of stresses acting on any inclined plane as shown in Fig. 9-1(c). Such stresses are equivalent descriptions of the state of stress at a point since these stresses, regardless of the planes on which they act, maintain the equilibrium of the element. The laws for transforming given stresses into equivalent stresses acting on any plane through a given point will be the principal topic in Part A of this chapter. The planes where either the normal or the shearing stresses reach their maximum intensity will receive special attention as the stresses associated with these planes have a particularly significant effect on materials.

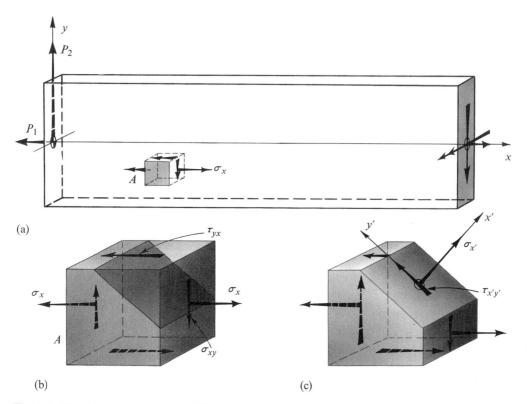

Fig. 9-1. State of stress at a point on different planes.

In Part B of this chapter, a topic parallel to the above for transforming strains associated with one set of axes to a different set of axes will be discussed. Additional information on stress-strain relations for linearly elastic material also will be given. This is supplementary to some of the items discussed in Chapter 4.

The conclusion of this chapter (Part C) is devoted to the discussion of mechanical properties of materials under biaxial and triaxial states of stress. Widely accepted yield criteria which are the basis of the laws of plasticity for ductile materials are emphasized. This part of the chapter supplements the discussion on constitutive relations for the uniaxial stresses of Chapter 4.

PART A
TRANSFORMATION OF STRESS

9-2. THE BASIC PROBLEM

It was pointed out in Chapter 3 that stress is a second-rank tensor. Since on the other hand vectors are only first-rank tensors, the laws of vector addition do not apply to stresses. However, it is possible to multiply stresses by the respective areas on which they act to obtain forces, which are vectors and consequently can be added or subtracted vectorially. It is in this manner that the problem of combining the normal with the shearing stresses is solved. This procedure will be first illustrated on a numerical example. Then the developed approach will be generalized to obtain algebraic relations for a stress transformation, which enable one to obtain stresses on any inclined plane from a given state of stress. The methods used in these derivations do not involve properties of a material. Therefore, providing the initial stresses are given, the derived relations are applicable whether the material behaves elastically or plastically.

*Section 9-2
The basic problem*

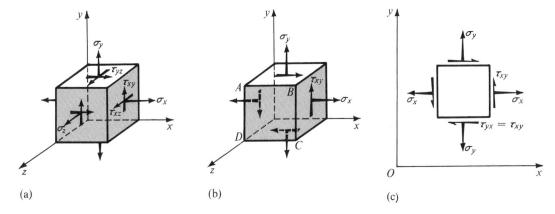

Fig. 9-2. Representation of stresses acting on an element.

In this text, in deriving the laws of transformation of stress at a point, complete generality will be avoided. Instead of treating a general three-dimensional state of stress,* such as shown in Fig. 9-2(a), elements with stresses as shown in Fig. 9-2(b) will be considered. In practical applications this type of stress is particularly significant since usually it is possible to select at an outer boundary of a member one face of an element, such as $ABCD$ in Fig. 9-2(b), which is free of significant surface stresses. On the other hand, the stresses acting on such elements right at the surface of a body are the highest ones stressed in a direction parallel to the surface. As before, for simplicity, the stresses acting on such elements will be shown as in Fig. 9-2(c).

EXAMPLE 9-1

Let the state of stress for an element be as shown in Fig. 9-3(a). An alternative representation of the state of stress at the same point may be given on an infinitesimal wedge with an angle of $\alpha = 22\frac{1}{2}°$ as in Fig.

---

* For a more general treatment of stress transformation the reader is referred to books on elasticity or plasticity. In the derivation developed here, in addition to σ_x, the normal stress σ_y is considered. The situation of having two normal stresses will be encountered in the next chapter in connection with thin shells.

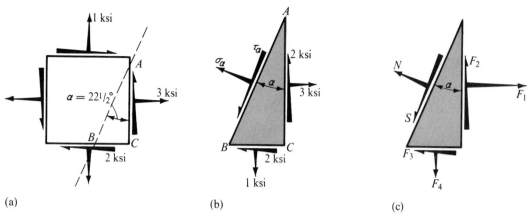

Fig. 9-3

9-3(b). Find the stresses which must act on the plane AB of the wedge to keep the element in equilibrium.

SOLUTION

The wedge ABC is a part of the element in Fig. 9-3(a); therefore the stresses on the faces AC and BC are known. The unknown normal and shearing stresses acting on the face AB are designated in the figure by σ_α and τ_α, respectively. Their sense is assumed arbitrarily.

To determine σ_α and τ_α, for convenience only, let the area of the face defined by the line AB be 1 in.² Then the area corresponding to the line AC is equal to (1) cos α = 0.924 in.²; and that to BC is equal to (1) sin α = 0.383 in.² (More rigorously, the area corresponding to the line AB should be taken as dA, but this quantity cancels out in the subsequent algebraic expressions.) Forces F_1, F_2, F_3, and F_4, Fig. 9-3(c), can be obtained by multiplying the stresses by their respective areas. The unknown equilibrant forces N and S act respectively normal and tangential to the plane AB. Then, applying the equations of static equilibrium to the forces acting on the wedge gives the forces N and S.

$$F_1 = 3(0.924) = 2.78 \text{ kips} \qquad F_2 = 2(0.924) = 1.85 \text{ kips}$$
$$F_3 = 2(0.383) = 0.766 \text{ kips} \qquad F_4 = 1(0.383) = 0.383 \text{ kips}$$

$\sum F_N = 0$, $\qquad N = F_1 \cos \alpha - F_2 \sin \alpha - F_3 \cos \alpha + F_4 \sin \alpha$

$\qquad\qquad\qquad = 2.78(0.924) - 1.85(0.383)$

$\qquad\qquad\qquad\qquad\qquad\qquad\qquad - 0.766(0.924) + 0.383(0.383)$

$\qquad\qquad\qquad = 1.29 \text{ kips}$

$\sum F_S = 0$, $\qquad S = F_1 \sin \alpha + F_2 \cos \alpha - F_3 \sin \alpha - F_4 \cos \alpha$

$\qquad\qquad\qquad = 2.78(0.383) + 1.85(0.924)$

$\qquad\qquad\qquad\qquad\qquad\qquad\qquad - 0.766(0.383) - 0.383(0.924)$

$\qquad\qquad\qquad = 2.12 \text{ kips}$

The forces N and S act on the plane defined by AB, which was initially assumed to be 1 in.² Their positive signs indicate that their assumed directions were chosen correctly. Dividing these forces by the area on which they act, the stresses acting on the plane AB are obtained. Thus $\sigma_\alpha = 1.29$ ksi and $\tau_\alpha = 2.12$ ksi and act in the direction shown in Fig. 9-3(b).

Section 9-3
Equations for the transformation of plane stress

The foregoing procedure accomplished a remarkable thing. It transformed the description of the state of stress from one set of planes to another. Either system of stresses pertaining to an infinitesimal element describes the state of stress at the same point of a body.

The procedure of isolating a wedge and using the equations of the equilibrium of forces to determine stresses on inclined planes is fundamental. Ordinary sign conventions of statics suffice to solve any problem. The reader is urged to return to this approach whenever questions arise regarding the more advanced procedures developed in the remainder of this chapter.

9-3. EQUATIONS FOR THE TRANSFORMATION OF PLANE STRESS

Two algebraic expressions, one for the normal stress and one for the shearing stress, can be developed to give these stresses in terms of the initially known stresses and of an angle of inclination of the plane being investigated. The dependence of the stresses on the inclination of the plane thus becomes clearly apparent. The derivatives of these algebraic expressions with respect to the angle of inclination, when set equal to zero, locate the planes on which either the normal or the shearing stress

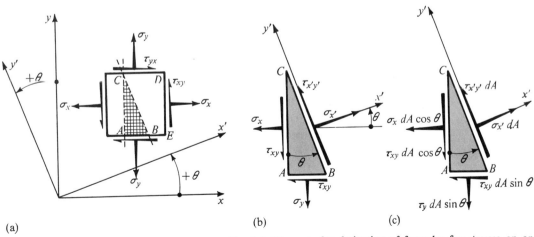

Fig. 9-4. Elements for derivation of formulas for stresses on an inclined plane.

reaches a maximum or minimum value. The stresses on these planes are of great importance in predicting the behavior of a given material.

The algebraic equations will be developed using an element, shown in Fig. 9-4(a), in a state of general plane stress. The normal tensile stresses are positive, and the compressive stresses are negative. Positive shearing stress is defined as *acting upward on the right face DE of the element*. The senses of the other shearing stresses follow from the equilibrium requirements. This sign convention for stresses is the one adopted in Chapter 3; it is in complete agreement with the positive direction of axial force and shear in bars as defined in Chapter 2. Here the transformation of stresses is sought from the xy system of coordinate axes to the $x'y'$ system. The angle θ which locates the x' axis is positive when measured from the x axis toward the y axis in a counterclockwise direction.

By passing a plane BC normal to the x' axis through the element, the wedge in Fig. 9-4(b) is isolated. The plane BC makes an angle θ with the vertical axis, and, if this plane has an area dA, the areas of the faces AC and AB are $dA \cos \theta$ and $dA \sin \theta$, respectively. Multiplying the stresses by their respective areas, a diagram with the forces acting on the wedge can be constructed, Fig. 9-4(c). Then, by applying the equations of static equilibrium to the forces acting on the wedge, stresses $\sigma_{x'}$ and $\tau_{x'y'}$ are obtained:

$$\sum F_{x'} = 0, \quad \sigma_{x'} dA = \sigma_x dA \cos \theta \cos \theta + \sigma_y dA \sin \theta \sin \theta$$
$$+ \tau_{xy} dA \cos \theta \sin \theta + \tau_{xy} dA \sin \theta \cos \theta$$

$$\sigma_{x'} = \sigma_x \cos^2 \theta + \sigma_y \sin^2 \theta + 2\tau_{xy} \sin \theta \cos \theta$$

$$= \sigma_x \frac{(1 + \cos 2\theta)}{2} + \sigma_y \frac{(1 - \cos 2\theta)}{2} + \tau_{xy} \sin 2\theta$$

$$\sigma_{x'} = \frac{\sigma_x + \sigma_y}{2} + \frac{\sigma_x - \sigma_y}{2} \cos 2\theta + \tau_{xy} \sin 2\theta \qquad (9\text{-}1)$$

Similarly, from $\sum F_{y'} = 0$,

$$\tau_{x'y'} = -\frac{\sigma_x - \sigma_y}{2} \sin 2\theta + \tau_{xy} \cos 2\theta \qquad (9\text{-}2)$$

Equations 9-1 and 9-2 are the general expressions for the normal and the shearing stress, respectively, on any plane located by the angle θ and caused by a known system of stresses. These equations are the equations for transformation of stress from one set of coordinate axes to another. Note particularly that σ_x, σ_y, and τ_{xy} are initially known stresses.

9-4. PRINCIPAL STRESSES

Often interest centers on the determination of the largest possible stress as given by Eqs. 9-1 and 9-2, and the planes on which such stresses occur will be found first. To find the plane for a maximum or a minimum normal stress, Eq. 9-1 is differentiated with respect to θ and the derivative set equal to zero, i.e.,

Section 9-4
Principal stresses

$$\frac{d\sigma_{x'}}{d\theta} = -\frac{\sigma_x - \sigma_y}{2} 2\sin 2\theta + 2\tau_{xy}\cos 2\theta = 0$$

Hence

$$\tan 2\theta_1 = \frac{\tau_{xy}}{(\sigma_x - \sigma_y)/2} \qquad (9\text{-}3)$$

plane for a maximum or a minimum normal stress

where the subscript of the angle θ is used to designate the angle which defines the plane of the maximum or minimum normal stress. Equation 9-3 has two roots since the value of the tangent of an angle in the diametrically opposite quadrants is the same, as may be seen from Fig. 9-5. These

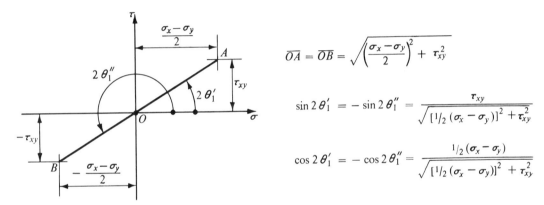

$$\overline{OA} = \overline{OB} = \sqrt{\left(\frac{\sigma_x - \sigma_y}{2}\right)^2 + \tau_{xy}^2}$$

$$\sin 2\theta_1' = -\sin 2\theta_1'' = \frac{\tau_{xy}}{\sqrt{[1/2\,(\sigma_x - \sigma_y)]^2 + \tau_{xy}^2}}$$

$$\cos 2\theta_1' = -\cos 2\theta_1'' = \frac{1/2\,(\sigma_x - \sigma_y)}{\sqrt{[1/2\,(\sigma_x - \sigma_y)]^2 + \tau_{xy}^2}}$$

Fig. 9-5. Angle functions for principal stresses.

roots are 180° apart, and, as Eq. 9-3 is for a double angle, the roots of θ_1 are 90° apart. One of these roots locates a plane on which the maximum normal stress acts; the other locates the corresponding plane for the minimum normal stress. To distinguish between these two roots, a prime and double prime notation is used.

Before evaluating the above stresses, carefully observe that if the location of planes on which no shearing stresses act is wanted, Eq. 9-2 must be set equal to zero. This yields the same relation as that in Eq. 9-3. Hence an important conclusion is reached: On planes on which maximum or minimum normal stresses occur, there are no shearing stresses. These planes are called the *principal planes* of stress, and the stresses acting on these planes—the maximum and minimum normal stresses—are called the *principal stresses.*

The magnitudes of the principal stresses may be obtained by substituting the values of the sine and cosine functions corresponding to the double angle given by Eq. 9-3 into Eq. 9-1. After this is done and the results are simplified, the expression for the maximum normal stress (denoted by σ_1) and the minimum normal stress (denoted by σ_2) becomes

$$(\sigma_{x'})^{\max}_{\min} = \sigma_{1 \text{ or } 2} = \frac{\sigma_x + \sigma_y}{2} \pm \sqrt{\left(\frac{\sigma_x - \sigma_y}{2}\right)^2 + \tau_{xy}^2} \qquad (9\text{-}4)$$

where the positive sign in front of the radical must be used to obtain σ_1, and the negative sign to obtain σ_2. The planes on which these stresses act can be determined by using Eq. 9-3. A particular root of Eq. 9-3 substituted into Eq. 9-1 will check the result found from Eq. 9-4 and at the same time will locate the plane on which this principal stress acts.

9-5. MAXIMUM SHEARING STRESSES

If σ_x, σ_y, and τ_{xy} are known for an element, the shearing stress on any plane defined by an angle θ is given by Eq. 9-2, and a study similar to the one made above for the normal stresses may be made for the shearing stress. Thus, similarly, to locate the planes on which the maximum or the minimum shearing stresses act, Eq. 9-2 must be differentiated with respect to θ and the derivative set equal to zero. When this is carried out and the results are simplified, the operations yield

$$\tan 2\theta_2 = -\frac{(\sigma_x - \sigma_y)/2}{\tau_{xy}} \qquad (9\text{-}5)$$

where the subscript 2 is attached to θ to designate the plane on which the shearing stress is a maximum or a minimum. Like Eq. 9-3, Eq. 9-5 has two roots, which again may be distinguished by attaching to θ_2 a prime or a double prime notation. The two planes defined by this equation are mutually perpendicular. Moreover, the value of $\tan 2\theta_2$ given by Eq. 9-5 is a negative reciprocal of the value of $\tan 2\theta_1$ in Eq. 9-3. Hence the roots for the double angles of Eq. 9-5 are 90° away from the corresponding roots of Eq. 9-3. This means that the angles which locate the planes of maximum or minimum shearing stress form angles of 45° with the planes of the principal stresses. A substitution into Eq. 9-2 of the sine and cosine functions corresponding to the double angle given by Eq. 9-5 and determined in a manner analogous to that in Fig. 9-5 gives the maximum and the minimum values of the shearing stresses. These, after simplifications, are

$$\tau^{\max}_{\min} = \pm\sqrt{\left(\frac{\sigma_x - \sigma_y}{2}\right)^2 + \tau_{xy}^2} \qquad (9\text{-}6)$$

Thus, the maximum shearing stress differs from the minimum shearing stress only in sign. Moreover, since the two roots given by Eq. 9-5 locate planes 90° apart, this result also means that the numerical values of the shearing stresses on the mutually perpendicular planes are the same. This concept was repeatedly used after being established in Art. 3-3. In this derivation the difference in sign of the two shearing stresses arises from the convention for locating the planes on which these stresses act. From the physical point of view these signs have no meaning and for this reason the largest shearing stress regardless of sign will be called the *maximum shearing stress*.

The definite sense of the shearing stress may always be determined by direct substitution of the particular root of θ_2 into Eq. 9-2. A positive shearing stress indicates that it acts in the direction assumed in Fig. 9-4(b), and vice versa. The determination of the maximum shearing stress is of utmost importance for materials which are weak in shearing strength. This will be discussed later in the chapter.

Unlike the principal stresses, for which no shearing stresses occur on the principal planes, the maximum shearing stresses act on planes which are usually not free of normal stresses. Substitution of θ_2 from Eq. 9-5 into Eq. 9-1 shows that the normal stresses which act on the planes of the maximum shearing stresses are

$$\sigma' = \frac{\sigma_x + \sigma_y}{2} \tag{9-7}$$

Therefore a normal stress acts simultaneously with the maximum shearing stress unless $\sigma_x + \sigma_y$ vanishes.

If σ_x and σ_y in Eq. 9-6 are the principal stresses, τ_{xy} is zero and Eq. 9-6 simplifies to

$$\tau_{max} = \frac{\sigma_1 - \sigma_2}{2} \tag{9-8}$$

EXAMPLE 9-2

For the state of stress in Example 9-1, reproduced in Fig. 9-6(a), (a) rework the previous problem for $\theta = -22\frac{1}{2}°$, using the general equations for the transformation of stress; (b) find the principal stresses and show their sense on a properly oriented element; and (c) find the maximum shearing stresses with the associated normal stresses and show the results on a properly oriented element.

SOLUTION

Case (a). By directly applying Eqs. 9-1 and 9-2 for $\theta = -22\frac{1}{2}°$, with $\sigma_x = +3$ ksi, $\sigma_y = +1$ ksi, and $\tau_{xy} = +2$ ksi, one has

$$\sigma_{x'} = \frac{3+1}{2} + \frac{3-1}{2} \cos(-45°) + 2 \sin(-45°)$$

$$= 2 + 1(0.707) - 2(0.707) = +1.29 \text{ ksi}$$

$$\tau_{x'y'} = -\frac{3-1}{2} \sin(-45°) + 2 \cos(-45°)$$

$$= +1(0.707) + 2(0.707) = +2.12 \text{ ksi}$$

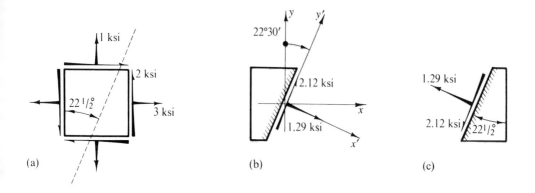
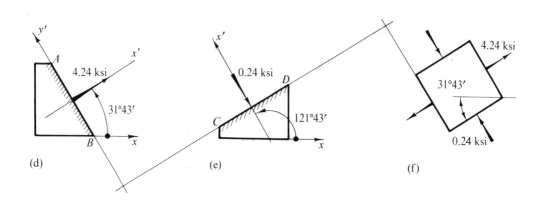
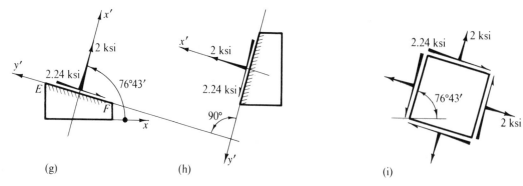

Fig. 9-6

The positive sign of $\sigma_{x'}$ indicates tension; whereas the positive sign of $\tau_{x'y'}$ indicates that the shearing stress acts in the $+y'$ direction, as shown in Fig. 9-4(b). These results are shown in Fig. 9-6(b) as well as in Fig. 9-6(c).

Section 9-5 Maximum shearing stresses

Case (b). The principal stresses are obtained by means of Eq. 9-4. The planes on which the principal stresses act are found by using Eq. 9-3.

$$\sigma_{1 \text{ or } 2} = \frac{3+1}{2} \pm \sqrt{\left(\frac{3-1}{2}\right)^2 + 2^2} = 2 \pm 2.24$$

$$\sigma_1 = +4.24 \text{ ksi} \quad \text{(tension)}, \quad \sigma_2 = -0.24 \text{ ksi} \quad \text{(compression)}$$

$$\tan 2\theta_1 = \frac{\tau_{xy}}{(\sigma_x - \sigma_y)/2} = \frac{2}{(3-1)/2} = 2$$

$$2\theta_1 = 63°26' \quad \text{or} \quad 63°26' + 180° = 243°26'$$

Hence $\quad \theta_1' = 31°43' \quad$ and $\quad \theta_1'' = 121°43'$

This locates the two principal planes AB and CD, Figs. 9-6(d) and (e), on which σ_1 and σ_2 act. On which one of these planes the principal stresses act is unknown. So, Eq. 9-1 is solved by using, for example, $\theta_1' = 31°43'$. The stress found by this calculation is the stress which acts on the plane AB. Then, since $2\theta_1' = 63°26'$,

$$\sigma_{x'} = \frac{3+1}{2} + \frac{3-1}{2} \cos 63°26' + 2 \sin 63°26' = +4.24 \text{ ksi} = \sigma_1$$

This result, besides giving a check on the previous calculations, shows that the maximum principal stress acts on the plane AB. The complete state of stress at the given point in terms of the principal stresses is shown in Fig. 9-6(f).

Case (c). The maximum shearing stress is found by using Eq. 9-6. The planes on which these stresses act are defined by Eq. 9-5. The sense of the shearing stresses is determined by substituting one of the roots of Eq. 9-5 into Eq. 9-2. Normal stresses associated with the maximum shearing stress are determined by using Eq. 9-7.

$$\tau_{\max} = \sqrt{[(3-1)/2]^2 + 2^2} = \sqrt{5} = 2.24 \text{ ksi}$$

$$\tan 2\theta_2 = -\frac{(3-1)/2}{2} = -0.500$$

$$2\theta_2 = 153°26' \quad \text{or} \quad 153°26' + 180° = 333°26'$$

Hence $\quad \theta_2' = 76°43' \quad$ and $\quad \theta_2'' = 166°43'$

These planes are shown in Figs. 9-6(g) and (h). Then, using $2\theta_2' = 153°26'$ in Eq. 9-2,

$$\tau_{x'y'} = -\frac{3-1}{2} \sin 153°26' + 2 \cos 153°26' = -2.24 \text{ ksi}$$

293

which means that the shear along the plane EF has an opposite sense to that in Fig. 9-4(b). From Eq. 9-7

$$\sigma' = \frac{3+1}{2} = 2 \text{ ksi}$$

The complete results are shown in Fig. 9-6(i).

The description of the state of stress now can be exhibited in three alternative forms: as the originally given data, and in terms of the stresses found in parts (b) and (c) of this problem. In matrix representation of the stress tensors this yields

$$\begin{pmatrix} 3 & 2 \\ 2 & 1 \end{pmatrix} \text{ or } \begin{pmatrix} 4.24 & 0 \\ 0 & -0.24 \end{pmatrix} \text{ or } \begin{pmatrix} 2 & -2.24 \\ -2.24 & 2 \end{pmatrix} \text{ksi}$$

All these descriptions of the state of stress at the given point are equivalent. Note that in one of the stated forms the matrix is diagonal.

9-6. AN IMPORTANT TRANSFORMATION OF STRESS

A significant transformation of one description of a state of stress at a point to another occurs when pure shearing stress is converted into principal stresses. For this purpose consider an element subjected only to shearing stresses τ_{xy} as in Fig. 9-7(a). Then from Eq. 9-4 the principal stresses $\sigma_{1 \text{ or } 2} = \pm \tau_{xy}$, i.e., numerically σ_1, σ_2, and τ_{xy} are all equal, although σ_1 is a tensile stress and σ_2 is a compressive stress. In this case, from Eq. 9-3 the principal planes are given by tan $2\theta_1 = \infty$, i.e., $2\theta_1 = 90°$ or 270°. Hence $\theta_1' = 45°$ and $\theta_1'' = 135°$; the planes corresponding to these angles are in Fig. 9-7(b). To determine on which plane the tensile stress σ_1 acts, a substitution into Eq. 9-1 is made with $2\theta_1' = 90°$. This computation

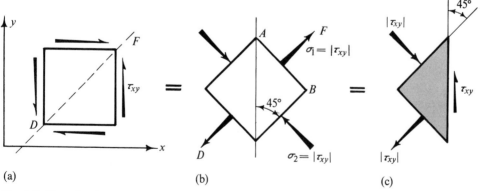

Fig. 9-7. Pure shearing stress is equivalent to tension-compression stresses acting on inclined planes at 45° to the shearing planes.

shows that $\sigma_1 = +\tau_{xy}$, hence the tensile stress acts perpendicular to the plane AB. Both principal stresses which are equivalent to the pure shearing stress are shown in Figs. 9-7(b) and (c). Therefore, whenever pure shearing stress is acting on an element it may be thought of as causing tension along one of the diagonals and compression along the other. The diagonal such as DF in Fig. 9-7(a), along which a tensile stress acts, is referred to as the *positive shear diagonal.*

From the physical point of view, the transformation of stress found completely agrees with intuition. The material "does not know" the manner in which its state of stress is described, and a little imagination should convince one that the tangential shearing stresses combine to cause pull along the positive shear diagonal and compression along the other diagonal.

9-7. MOHR'S CIRCLE OF STRESS

In this article the basic Eqs. 9-1 and 9-2 for the stress transformation at a point will be re-examined in order to interpret them graphically. In doing this, two objectives will be pursued. First, by graphically interpreting these equations a greater insight into the general problem of stress transformation will be achieved. This is the main purpose of this article. Second, with the aid of graphical construction, a quicker solution of stress transformation problems can often be obtained. This will be discussed in the following article.

A careful study of Eqs. 9-1 and 9-2 shows that they represent a circle written in parametric form. That they do represent a circle is made clearer by first rewriting them as

$$\sigma_{x'} - \frac{\sigma_x + \sigma_y}{2} = \frac{\sigma_x - \sigma_y}{2} \cos 2\theta + \tau_{xy} \sin 2\theta \tag{9-9}$$

$$\tau_{x'y'} = -\frac{\sigma_x - \sigma_y}{2} \sin 2\theta + \tau_{xy} \cos 2\theta \tag{9-10}$$

Then by squaring both these equations, adding, and simplifying

$$\left(\sigma_{x'} - \frac{\sigma_x + \sigma_y}{2}\right)^2 + \tau_{x'y'}^2 = \left(\frac{\sigma_x - \sigma_y}{2}\right)^2 + \tau_{xy}^2 \tag{9-11}$$

In every given problem σ_x, σ_y, and τ_{xy} are the three known constants, and $\sigma_{x'}$ and $\tau_{x'y'}$ are the variables. Hence Eq. 9-11 may be written in more compact form as

$$(\sigma_{x'} - a)^2 + \tau_{x'y'}^2 = b^2 \tag{9-12}$$

where $a = (\sigma_x + \sigma_y)/2$ and $b^2 = [(\sigma_x - \sigma_y)/2]^2 + \tau_{xy}^2$ are constants.

This equation is the familiar expression of analytical geometry

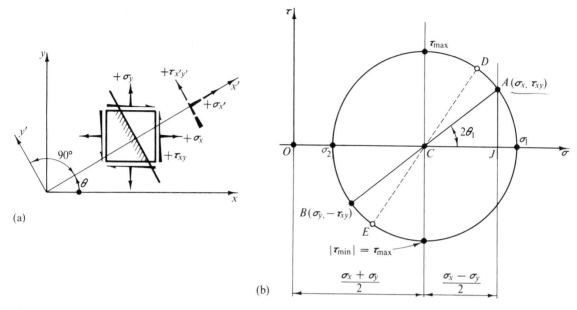

Fig. 9-8. Mohr's circle of stress.

$(x - a)^2 + y^2 = b^2$ for a circle of radius b with its center at $(+a, 0)$. Hence if a circle satisfying this equation is plotted, the simultaneous values of a point (x, y) on this circle correspond to $\sigma_{x'}$ and $\tau_{x'y'}$ for a particular orientation of an inclined plane. The ordinate of a point on the circle is the shearing stress $\tau_{x'y'}$, the abscissa is the normal stress $\sigma_{x'}$. The circle so constructed is called a *circle of stress* or *Mohr's circle of stress*.*

A Mohr's circle based on the information for the given stresses in Fig. 9-8(a) is plotted in Fig. 9-8(b) with σ and τ as the coordinate axes. The center is located at $(a, 0)$, and the radius equals b. Point A on the circle corresponds to the stresses on the right face of the given element when $\theta = 0°$. For this point, $\sigma_{x'} = \sigma_x$, and $\tau_{x'y'} = \tau_{xy}$. As $AJ/CJ = \tau_{xy}/[(\sigma_x - \sigma_y)/2]$, according to Eq. 9-3, the angle ACJ is equal to $2\theta_1$.

With $\theta = 90°$ the x' axis is directed upward and the y' axis points to the left. From this orientation of the axes, the coordinates for point B on the circle are $\sigma_{x'} = \sigma_y$, and $\tau_{x'y'} = -\tau_{xy}$. The coordinates of points B and A satisfy Eq. 9-11. The same reasoning can be applied to any other pair of points, such as D or E, on the circle. The coordinates of such points give the stresses associated with a particular orientation of the $x'y'$ axes which define a plane passing through an element. All the possible ways of describing the stresses for an element for different θ's are represented

---

* It is so named in honor of Professor Otto Mohr of Germany, who in 1895 suggested its use in stress analysis problems.

by points on the Mohr's circle of stress. Therefore the following important conclusions regarding the state of stress at a point can be drawn:

1. The largest possible normal stress is σ_1; the smallest is σ_2. No shearing stresses exist together with either one of these principal stresses.

2. The largest shearing stress τ_{max} is numerically equal to the radius of the circle, also to $(\sigma_1 - \sigma_2)/2$. A normal stress equal to $(\sigma_1 + \sigma_2)/2$ acts on each of the planes of maximum shearing stress.

3. If $\sigma_1 = \sigma_2$, Mohr's circle degenerates into a point, and no shearing stresses at all develop in the xy plane.

4. If $\sigma_x + \sigma_y = 0$, the center of Mohr's circle coincides with the origin of the σ-τ coordinates, and the state of pure shear exists.

5. The sum of the normal stresses on any two mutually perpendicular planes is invariant, i.e.

$$\sigma_x + \sigma_y = \sigma_1 + \sigma_2 = \sigma_{x'} + \sigma_{y'} = \text{constant}$$

9-8. CONSTRUCTION OF MOHR'S CIRCLE OF STRESS

Mohr's circle of stress is widely used in practice for stress transformation. To be of value, the procedure must be rapid and simple. As an aid in application, the recommended procedure is outlined below. All the steps in constructing the circle can be justified on the basis of the previously developed relations. A typical Mohr's circle is in Fig. 9-9.

1. Make a sketch of the element for which the normal and the shearing stresses are known and indicate on this element the proper sense of these stresses. In an actual problem the faces of this element must have a precise relationship to the axes of a member being analyzed.

2. Set up a rectangular coordinate system of axes where the horizontal axis is the normal stress axis and the vertical axis is the shearing stress axis. Directions of positive axes are taken as usual, upward and to the right.

3. Locate the center of the circle, which is on the horizontal axis at a distance of $(\sigma_x + \sigma_y)/2$ from the origin. Tensile stresses are positive, compressive are negative.

4. From the right face of the element prepared in (1), read the values for σ_x and τ_{xy} and plot the controlling point A on the circle. The coordinate distances to this point are measured from the origin. The sign of σ_x is positive if tensile, negative if compressive; that of τ_{xy} is positive if upward on the right face of the element, negative if downward.

5. Connect the center of the circle found in (3) with the point plotted in (4) and determine this distance, which is the radius of the circle.

6. Draw the circle using the radius found in (5). If only magnitudes and signs of stresses are of interest, this step completes the solution of the

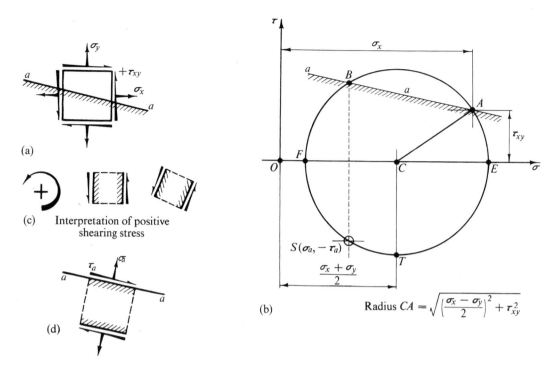

Fig. 9-9. Construction of Mohr's circle of stress.

problem. The coordinates of points on the circle provide the required information.

7. To determine the direction and sense of the stresses acting on any inclined plane, draw through point A a line parallel to the inclined plane and locate point B on the circle. The coordinates of point S lying vertically on the opposite side of the circle from B give the stresses acting on the inclined plane. In Fig. 9-9(b) such stresses are identified as σ_a and $-\tau_a$. A positive value of σ indicates a tensile stress, and vice versa. The sense of the shearing stress can be determined using the interpretation in Fig. 9-9(c). A tendency of the shearing stresses on two opposite faces of an element to cause counterclockwise rotation of the element is associated with a positive shearing stress. On this basis the result $(+\sigma_a, -\tau_a)$ has the meaning shown in Fig. 9-9(d).

8. By proceeding in the reverse order, the plane on which the stresses associated with any point on the circle act can be found. Thus, drawing a line from A toward E or F, i.e., having the point corresponding to B coincide with one of these intercepts, determines the inclination of the plane on which the respective principal stresses act. For this special case the distance BS degenerates into a point. The principal stress given by the particular intercept (either E or F) acts normal to the line connecting this

intercept point with point *A*. As before, positive stresses indicate tension, and vice versa.

By commencing with the highest or the lowest point on the circle, the planes on which the maximum shearing stresses and the associated normal stresses act can be found. For example, by imagining that point *S* is moved to *T*, the plane on which the stresses at *T* act is given by the new position of the line *BA* with point *B* at the highest point on the circle.

To solve the problems of stress transformation using Mohr's circle, the foregoing procedures may be applied graphically. However, it is recommended that trigonometric computations of the critical values be used in conjunction with the graphical construction. Then the work may be carried out on a crude sketch without scaling off the distances or angles, and the results will be accurate. Using Mohr's circle in this manner is equivalent to applying the basic equations of stress transformation.

Section 9-8 Construction of Mohr's circle of stress

EXAMPLE 9-3

Given the state of stress shown in Fig. 9-10(a), transform it (*a*) into the principal stresses, and (*b*) into the maximum shearing stresses and the

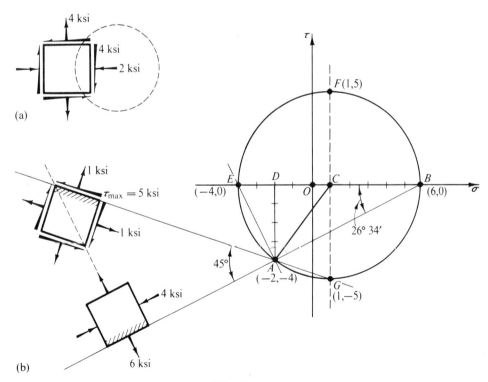

Fig. 9-10

associated normal stresses. Show the results for both cases on properly oriented elements.

SOLUTION

To construct Mohr's circle of stress the following quantities are required:

center of circle on the σ axis: $\quad (-2 + 4)/2 = +1$ ksi

point A on circle from data on the right face of element:

$$(-2, -4) \text{ ksi}$$

radius of circle: $\quad CA = \sqrt{CD^2 + DA^2} = 5$ ksi

After drawing the circle, one obtains $\sigma_1 = +6$ ksi, $\sigma_2 = -4$ ksi, and $\tau_{max} = 5$ ksi.

Drawing a line from σ_1 at B toward A locates the plane on which the stress σ_1 acts. Similarly, beginning at point E and drawing a line toward A gives the plane on which the stress σ_2 acts. The maximum shearing stress τ_{max} and the associated normal stress σ' are given by the coordinates of point F. Directly downward at point G, the inclined line AG locates the plane on which $\tau_{max} = +5$ ksi and $\sigma' = +1$ ksi act.

The complete results are shown on sketches in Fig. 9-10(b) on properly oriented elements. The angles shown are determined from suitable trigonometric relations. Thus since $\tan DBA = AD/DB = \frac{4}{8} = 0.5$, the angle $DBA = 26°34'$. The plane of maximum shear is located at 45° from the planes of principal stress. Of course the solution could have been made entirely by graphics.

It is significant to note that the approximate direction of the algebraically larger principal stress found in the above example might have been anticipated. Instead of thinking in terms of the normal and the shearing stresses as given in the original data, Fig. 9-11(a), an equivalent problem in Fig. 9-11(b) may be considered. Here the shearing stresses have been replaced by the equivalent tension-compression stresses acting along the proper shear diagonals. Then, for qualitative reasoning, the outline of the original element may be obliterated, and the tensile stresses may be singled out as in Fig. 9-11(c). From this new diagram it is apparent that regardless of the magnitudes of the particular stresses involved, the resultant maximum tensile stress must act somewhere between the given tensile stress and the positive shear diagonal. In other words, *the line of action of the algebraically larger principal stress is "straddled" by the algebraically larger given normal stress and the positive shear diagonal.* The use of the negative shear diagonal, located at 90° to the positive shear diagonal, is helpful in visualizing this effect for cases where both given normal stresses are compressive, Figs. 9-11(d) and (e). This procedure provides a qualitative check on the orientation of an element for the principal stresses.

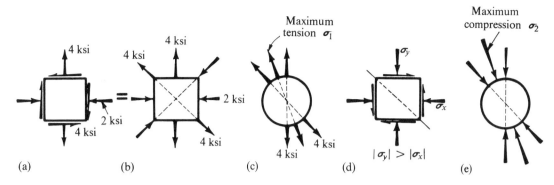

Fig. 9-11. A method for estimating the direction of the absolute maximum principal stresses.

EXAMPLE 9-4

Using Mohr's circle, transform the stresses shown in Fig. 9-12(a) into stresses acting on the plane at an angle of $22\frac{1}{2}°$ with the vertical axis.

SOLUTION

Here the center of Mohr's circle is at $(3 + 1)/2 = +2$ ksi on the σ axis. The stresses on the right face of the element give (3, 3) for the coordinates of point A on the circle. Therefore the radius of the circle is 3.16.

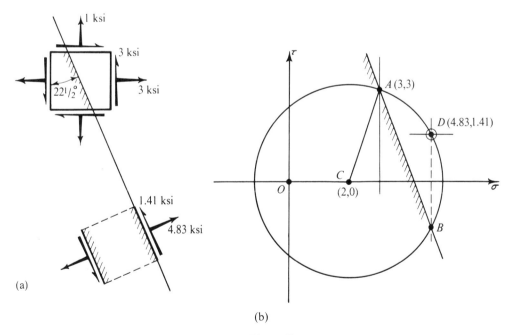

Fig. 9-12

*Chapter 9
Transformation of
stress and strain;
yield and fracture
criteria*

A line *AB* drawn parallel to the required inclined plane locates point *B*; directly above lies point *D*. The stresses acting on the required plane are given by the coordinates of point *D*. This solution is very easily accomplished by graphical construction; analytically the procedure is less direct. This type of graphical construction can be used effectively to provide a rapid qualitative check on analytical or experimental work.

For numerical work special trigonometric schemes may be devised in each particular case. However, this approach often proves cumbersome and direct application of Eqs. 9-1 and 9-2 is easier. Alternatively, one can always construct a wedge bounded by two axes and the inclined plane and solve the problem as illustrated in Example 9-1. In some instances the latter approach is least ambiguous.

9-9. MOHR'S CIRCLE OF STRESS FOR THE GENERAL STATE OF STRESS

So far in this chapter stress transformation and the associated Mohr's circle of stress for it were presented for a plane-stress problem. The treatment of the general three-dimensional stress transformation problem is beyond the scope of this book. However, some results of such an analysis are necessary for a more complete understanding of this subject. Therefore, several comments on transformation of the three-dimensional state of stress will be made now.

It is shown in books on elasticity and plasticity that any three-dimensional state of stress (see Fig. 3-2, or 9-2(a)) can be transformed into three principal stresses acting in three orthogonal directions. This is a direct generalization of the case discussed earlier where two principal stresses were shown to act in two orthogonal directions in the plane-stress problem. An element after the appropriate stress transformation with three principal stresses acting on it is in Fig. 9-13(a). This element can be viewed from three different directions as in Fig. 9-13(b).

Corresponding to each projection of the element in Fig. 9-13(b) a Mohr's circle can be drawn using the procedures developed earlier. For example, for an element situated in the 1–3 plane, the corresponding Mohr's circle to it passes through σ_1 and σ_3 as in Fig. 9-13(c). Analogous circles can be drawn for the 1–2 and the 2–3 planes. The three circles cluster together as in Fig. 9-13(c).

Next, suppose that instead of considering the planes on which principal stresses act, one considers an arbitrary plane such as the shaded plane *K* in Fig. 9-13(a). Then it can be shown* that the normal and shearing stresses acting on all such possible planes, when plotted as in Fig. 9-13(c), fall within the shaded part of the diagram. This means that the three circles already drawn give the limiting values of all possible stresses. This

---

* See O. Hoffman and G. Sachs, *Introduction to the Theory of Plasticity for Engineers* (New York: McGraw-Hill Book Company, 1953), p. 13.

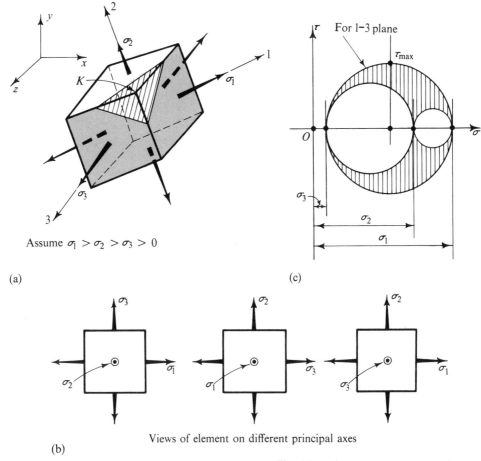

Fig. 9-13. Three-dimensional state of stress.

is an important fact and will be used in discussing material properties in a multiaxial state of stress.

In comparison with the above general problem, in the plane-stress problem $\sigma_3 = 0$. However, even in this less general problem the element is three-dimensional. Therefore, it is possible to study stresses on arbitrarily oriented planes corresponding to the plane K of Fig. 9-13(a). This has not been done earlier. With $\sigma_3 = 0$, three Mohr's circles are necessary to exhibit on a plot all the stresses on all the possible orientations of planes. For example, consider an element with $\sigma_1 = \sigma_2$, for which from a two-dimensional point of view Mohr's circle degenerates into a point. The same element, observed along different axes such as 1 and 3, with, for example, $\sigma_1 \neq 0$ and $\sigma_3 = 0$, generates a circle with a radius of $\sigma_1/2$. Thus, the direction from which an element is viewed is of the utmost importance.

PART B
TRANSFORMATION OF STRAIN

9-10. GENERAL REMARKS

In this part of the chapter the transformation of known strains associated with one set of axes or with known directions will be related to strains in any direction. It will be shown that the transformation of linear and shearing strains completely resembles the transformation of normal and shearing stresses presented earlier. Thus, after establishing the strain transformation equations, Mohr's circle of strain will be introduced. Attention will be confined to the two-dimensional case, or more precisely to the plane-strain case, which according to Eq. 4-9 means that $\varepsilon_z = \gamma_{zx} = \gamma_{zy} = 0$. The extension of the strain transformation to the general case involving Mohr's circle of strain for the three-dimensional problem will not be considered. Since the maximum strains usually occur on the free outer surfaces of a member, the two-dimensional problem is by far the most important one.

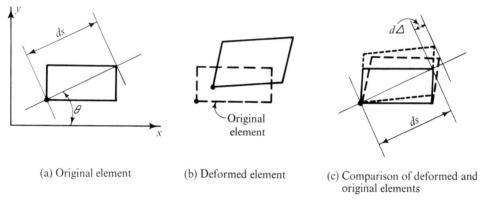

(a) Original element (b) Deformed element (c) Comparison of deformed and original elements

Fig. 9-14. Strains are determined on the basis of relative deformations.

In studying the strains at a point, only the relative displacement of the adjoining points is of importance. Translation and rotation of an element as a whole are of no consequence since these displacements are rigid-body displacements. For example, if the linear strain of a diagonal ds long of the element in Fig. 9-14(a) is being studied, the element in its deformed condition can be brought back for comparison purposes as in Fig. 9-14(c). It is immaterial whether the horizontal (dashed) or the vertical (dotted) sides of the deformed and the undeformed elements are matched to determine $d\Delta$. For the small strains considered throughout this text, the relevant quantity, elongation $d\Delta$ in the direction of the diagonal, is essentially the same regardless of the method of comparison employed.

In treating strains in the above manner only kinematic questions have relevance. The mechanical properties of material do not enter the problem. However, after the main features of strain transformation have been presented, some additional relations between stress and strain for linearly elastic material will be given at the end of this part of the chapter.

Section 9-11
Equations for the transformation of plane strain

9-11. EQUATIONS FOR THE TRANSFORMATION OF PLANE STRAIN

In establishing the equations for the transformation of strain, strict adherence to a sign convention is necessary. The sign convention used here was introduced in Chapter 4 and is related to the one chosen for the stresses. The linear strains ε_x and ε_y corresponding to elongations in the x and y directions, respectively, are taken positive. The shearing strain is considered positive if it elongates a diagonal having a positive slope in the xy coordinate system. For convenience in deriving the strain transformation equations, the element distorted by positive shearing strain will be taken as that shown in Fig. 9-15 (a). As noted in the preceding article, this leads to perfectly general results providing the strains are small.

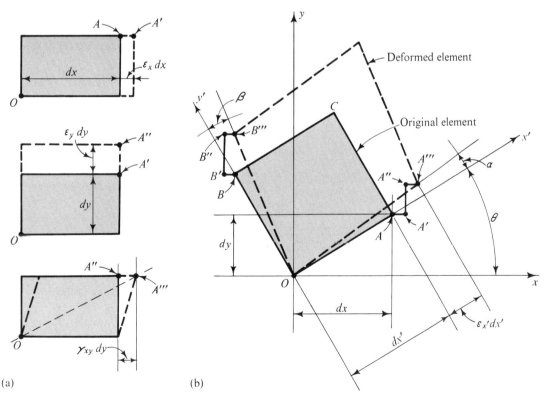

Fig. 9-15. Exaggerated deformations of elements for deriving strains along new axes.

Chapter 9
Transformation of stress and strain; yield and fracture criteria

Next, suppose that the strains ε_x, ε_y, and γ_{xy} associated with the xy axes are known and a linear strain along some new x' axis is required. The new $x'y'$ system of axes is related to the xy axes as in Fig. 9-15(b). In these new coordinates a length OA, which is dx' long, may be thought of as being a diagonal of a rectangular differential element dx by dy in the initial coordinates.

By considering the point O fixed in space, one can compute the displacements of point A caused by the imposed strains on a different basis in the two coordinate systems. The displacement in the x direction is $AA' = \varepsilon_x \, dx$; in the y direction, $A'A'' = \varepsilon_y \, dy$. For the shearing strain, assuming it causes the horizontal displacement shown in Fig. 9-15(a), $A''A''' = \gamma_{xy} \, dy$. The order in which these displacements occur is arbitrary. In Fig. 9-15(b), the displacement AA' is shown first, then $A'A''$, and finally $A''A'''$. By projecting these displacements onto the x' axis, one finds the displacement of point A along the x' axis. Then, recognizing that by definition $\varepsilon_{x'} \, dx'$ in the $x'y'$ coordinate system is also the elongation of OA, one has the following equality:

$$\varepsilon_{x'} \, dx' = AA' \cos \theta + A'A'' \sin \theta + A''A''' \cos \theta$$

On substituting the appropriate expressions for the displacements and dividing through by dx', one has

$$\varepsilon_{x'} = \varepsilon_x \frac{dx}{dx'} \cos \theta + \varepsilon_y \frac{dy}{dx'} \sin \theta + \gamma_{xy} \frac{dy}{dx'} \cos \theta$$

Since, however, $dx/dx' = \cos \theta$ and $dy/dx' = \sin \theta$

$$\varepsilon_{x'} = \varepsilon_x \cos^2 \theta + \varepsilon_y \sin^2 \theta + \gamma_{xy} \sin \theta \cos \theta \tag{9-13}$$

Equation 9-13 is the basic expression for the strain in an arbitrary direction defined by the x' axis. To apply this equation, ε_x, ε_y, and γ_{xy} must be known. By use of the trigonometric identities already encountered in deriving Eq. 9-1, Eq. 9-13 may be rewritten as

$$\varepsilon_{x'} = \frac{\varepsilon_x + \varepsilon_y}{2} + \frac{\varepsilon_x - \varepsilon_y}{2} \cos 2\theta + \frac{\gamma_{xy}}{2} \sin 2\theta \tag{9-14}$$

To complete the study of strain transformation at a point, shearing strain transformation must be also established. For this purpose consider an element $OACB$ with sides OA and OB directed along the x' and the y' axes as in Fig. 9-15(b). By definition the shearing strain for this element is the change in angle AOB. From the figure, the change of this angle is $\alpha + \beta$.

For small deformations the small angle α can be determined by projecting the displacements AA', $A'A''$, and $A''A'''$ onto a normal to OA and dividing this quantity by dx'. In applying this approach the tangent of the angle is assumed equal to the angle itself. This is acceptable if the strains are small. Thus

$$\alpha \approx \tan \alpha = \frac{-AA' \sin \theta + A'A'' \cos \theta - A''A''' \sin \theta}{dx'}$$

$$= -\varepsilon_x \frac{dx}{dx'} \sin \theta + \varepsilon_y \frac{dy}{dx'} \cos \theta - \gamma_{xy} \frac{dy}{dx'} \sin \theta$$

$$= -(\varepsilon_x - \varepsilon_y) \sin \theta \cos \theta - \gamma_{xy} \sin^2 \theta$$

By analogous reasoning

$$\beta \approx -(\varepsilon_x - \varepsilon_y) \sin \theta \cos \theta + \gamma_{xy} \cos^2 \theta$$

Therefore since the shearing strain $\gamma_{x'y'}$ of an angle included between the $x'y'$ axes is $\alpha + \beta$, one has

$$\gamma_{x'y'} = -2(\varepsilon_x - \varepsilon_y) \sin \theta \cos \theta + \gamma_{xy}(\cos^2 \theta - \sin^2 \theta)$$

or

$$\gamma_{x'y'} = -(\varepsilon_x - \varepsilon_y) \sin 2\theta + \gamma_{xy} \cos 2\theta \qquad (9\text{-}15)$$

This is the second fundamental expression for the transformation of strain. Note that when $\theta = 0°$, the shearing strain associated with the xy axes is recovered.

Equations 9-14 and 9-15 for strain transformation are analogous to Eqs. 9-1 and 9-2 for stress transformation. This feature will be emphasized further in discussing Mohr's circle of strain.

9-12. ALTERNATIVE DERIVATION OF EQUATION 9-13

Some readers may be interested in a different approach for deriving the strain transformation Eq. 9-13. This alternative derivation is more characteristic of the methods used in elasticity and plasticity and can be more readily generalized to three dimensions.

Consider an element AB initially ds long, Fig. 9-16. After straining, this element displaces to the position $A'B'$ and becomes ds^* long. The initial length $ds^2 = dx^2 + dy^2$; whereas the strained length of the element $(ds^*)^2 = (dx^*)^2 + (dy^*)^2$ with $dx^* = dx + du$ and $dy^* = dy + dv$.

The infinitesimal increments of strain du and dv can be found formally by applying the chain rule of differentiation to obtain total differentials, i.e.,

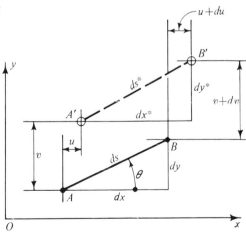

Fig. 9-16

$$du = \frac{\partial u}{\partial x} dx + \frac{\partial u}{\partial y} dy \quad \text{and} \quad dv = \frac{\partial v}{\partial x} dx + \frac{\partial v}{\partial y} dy$$

Using the above relations, the strain is most conveniently defined as the difference between $(ds^*)^2$ and ds^2. This difference is zero for unstrained bodies.

For the small deformation theory, in the expression for $(ds^*)^2$ the squares of small quantities can be neglected in comparison with the quantities themselves. Thus after some algebraic manipulations and simplifications

$$(ds^*)^2 = \left(1 + 2\frac{\partial u}{\partial x}\right)dx^2 + 2\frac{\partial u}{\partial y} dx\, dy + \left(1 + 2\frac{\partial v}{\partial y}\right)dy^2 + 2\frac{\partial v}{\partial x} dx\, dy$$

Hence
$$(ds^*)^2 - ds^2 = 2\frac{\partial u}{\partial x} dx^2 + 2\frac{\partial v}{\partial y} dy^2 + 2\left(\frac{\partial u}{\partial y} + \frac{\partial v}{\partial x}\right) dx\, dy$$

and, by recalling Eqs. 4-3 and 4-5, which define strains as derivatives of displacements, one has

$$(ds^*)^2 - ds^2 = 2\varepsilon_x\, dx^2 + 2\varepsilon_y\, dy^2 + 2\gamma_{xy}\, dy\, dx \qquad (9\text{-}16)$$

For small deformations, to a high degree of accuracy, one has

$$(ds^*)^2 - ds^2 = (ds^* + ds)\left(\frac{ds^* - ds}{ds}\right) ds \approx 2\varepsilon_\theta\, ds^2 \qquad (9\text{-}17)$$

where the linear strain $\varepsilon_\theta = (ds^* - ds)/ds$ from the classical definition of small strain, and, since ds^* differs very little from ds, $ds^* + ds \approx 2\, ds$.

On equating Eqs. 9-16 and 9-17, dividing through by ds^2, and recognizing that $\cos\theta = dx/ds$ and $\sin\theta = dy/ds$, one obtains

$$\varepsilon_\theta = \varepsilon_x \cos^2\theta + \varepsilon_y \sin^2\theta + \gamma_{xy} \sin\theta \cos\theta \qquad (9\text{-}18)$$

This equation for linear strain is identical to Eq. 9-13.

By taking two initially mutually perpendicular sides of an element and then forming the scalar product for the same two sides in the deformed state, Eq. 9-15 for the shearing strain can be obtained.

9-13. MOHR'S CIRCLE OF STRAIN

The two basic equations for the transformation of strains derived in the preceding article mathematically resemble the equations for the transformation of stresses derived in Art. 9-3. To achieve greater similarity between the appearances of the new equations and those of the earlier ones, Eq. 9-15 after division throughout by two is rewritten below as Eq. 9-19

$$\varepsilon_{x'} = \frac{\varepsilon_x + \varepsilon_y}{2} + \frac{\varepsilon_x - \varepsilon_y}{2} \cos 2\theta + \frac{\gamma_{xy}}{2} \sin 2\theta \qquad (9\text{-}14)$$

$$\frac{\gamma_{x'y'}}{2} = -\frac{\varepsilon_x - \varepsilon_y}{2} \sin 2\theta + \frac{\gamma_{xy}}{2} \cos 2\theta \qquad (9\text{-}19)$$

These equations in mathematical form are identical to the stress transformation Eqs. 9-1 and 9-2, except that the γ terms are divided by two. As these equations define the law of tensor transformation, elements of the strain tensor must be ε_x, ε_y, and $\gamma_{xy}/2$. This condition was anticipated in Eqs. 4-7 and 4-8 where not γ_{xy} but rather $\gamma_{xy}/2 \equiv \varepsilon_{xy}$ together with ε_{yz} and ε_{xz} were considered to be the elements of a strain tensor.

Since the strain transformation equations with the shearing strains divided by two are mathematically identical to the stress transformation equations, Mohr's circle of strain can be constructed. In this construction every point on the circle gives two values—one for the linear strain, the other for the shearing strain divided by two. Strains corresponding to elongation are positive; for contraction they are negative. Positive shearing strains distort the element as shown in Fig. 9-15(a). In plotting the circle the positive axes are taken in the usual manner, upward and to the right. The vertical axis is measured in terms of $\gamma/2$.

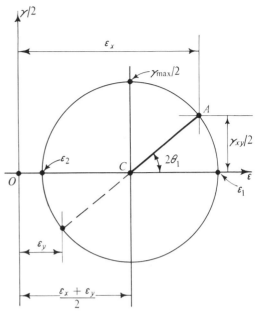

Fig. 9-17. Mohr's circle of strain.

As an illustration of Mohr's circle of strain, consider that ε_x, ε_y, and $+\gamma_{xy}$ are given. Then on the $\varepsilon - \tfrac{1}{2}\gamma$ axes in Fig. 9-17 the center of the circle C is at $[(\varepsilon_x + \varepsilon_y)/2, 0]$ and, from the given data, point A on the circle is at $(\varepsilon_x, \gamma_{xy}/2)$. An examination of this circle leads to conclusions analogous to those reached before for the circle of stress.

1. The maximum linear strain is ε_1; the minimum is ε_2. These are the principal strains, and no shearing strains are associated with them. The directions of the linear strains coincide with the directions of the principal stresses. As may be deduced from the circle, the analytical expression for the principal strains is

$$(\varepsilon_{x'})_{\substack{\max \\ \min}} = \varepsilon_{1 \text{ or } 2} = \frac{\varepsilon_x + \varepsilon_y}{2} \pm \sqrt{\left(\frac{\varepsilon_x - \varepsilon_y}{2}\right)^2 + \left(\frac{\gamma_{xy}}{2}\right)^2} \qquad (9\text{-}20)$$

where the positive sign in front of the radical is to be used for ε_1, the maximum principal strain in the algebraic sense. The negative sign is to be used for ε_2, the minimum principal strain. The planes on which the principal strains act may be defined analytically from Eq. 9-19 by setting it equal to zero. Thus

$$\tan 2\theta_1 = \frac{\gamma_{xy}}{\varepsilon_x - \varepsilon_y} \qquad (9\text{-}21)$$

since this equation has two roots, it is completely analogous to Eq. 9-3 and can be treated in the same manner.

*Chapter 9
Transformation of
stress and strain;
yield and fracture
criteria*

2. The largest shearing strain γ_{max} is equal to two times the radius of the circle. Linear strains of $(\varepsilon_1 + \varepsilon_2)/2$ in two mutually perpendicular directions are associated with the maximum shearing strain.

3. The sum of linear strains in any two mutually perpendicular directions is invariant, i.e., $\varepsilon_1 + \varepsilon_2 = \varepsilon_x + \varepsilon_y =$ constant. Other properties of strains at a point may be established by studying the circle further.

EXAMPLE 9-5

It is observed that an element of a body contracts 0.00050 in. per inch along the x-axis, elongates 0.00030 in. per inch in the y direction, and distorts through an angle* of 0.00060 radian as in Fig. 9-18(a). Find the principal strains and determine the directions in which these strains act. Use Mohr's circle of strain to obtain the solution.

SOLUTION

The data given indicate that $\varepsilon_x = -5 \times 10^{-4}$, $\varepsilon_y = +3 \times 10^{-4}$, and $\gamma_{xy} = -6 \times 10^{-4}$. Hence, on a $\varepsilon - \frac{1}{2}\gamma$ system of axes, Fig. 9-18, the center C of the circle is located at $(\varepsilon_x + \varepsilon_y)/2 = -1 \times 10^{-4}$ on the ε axis. Point A is at $(-5 \times 10^{-4}, -3 \times 10^{-4})$. The radius of the circle AC is equal to 5×10^{-4}. Hence $\varepsilon_1 = +4 \times 10^{-4}$ in. per inch takes

---

* This measurement may be made by scribing a small square on a body, straining the body, and then measuring the change in angle which takes place. Photographic enlargements of grids have been used for this purpose.

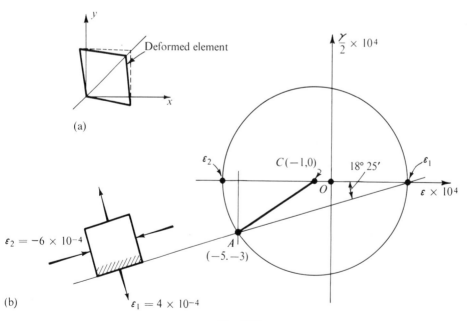

Fig. 9-18

place in the direction perpendicular to the line $A - \varepsilon_1$; and $\varepsilon_2 = -6 \times 10^{-4}$ in. per inch occurs in the direction perpendicular to the line $A - \varepsilon_2$. From the geometry of the figure, $|\theta| = \tan^{-1}(0.0003/0.0009) = 18°25'$.

9-14. STRAIN MEASUREMENTS; ROSETTES

Measurements of linear strain are particularly simple to make, and highly reliable techniques have been developed for this purpose. In such work linear strains are measured along several closely clustered gage lines, diagrammatically indicated in Fig. 9-19(a) by lines *a-a*, *b-b*, and *c-c*.

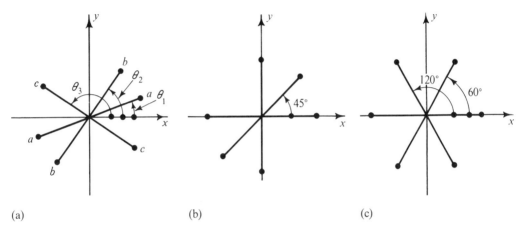

Fig. 9-19. (a) General strain rosette; (b) rectangular or 45° strain rosette; (c) equiangular or delta rosette.

These gage lines may be located on the member investigated with reference to some coordinate axes (as x and y) by the respective angles θ_1, θ_2, and θ_3. By comparing the initial distance between any two corresponding gage points with the distance in the stressed member, the elongation in the gage length is obtained. Dividing the elongation by the gage length gives the strain in the θ_1 direction, which will be designated ε_{θ_1}. By performing the same operation with the other gage lines, ε_{θ_2} and ε_{θ_3} are obtained. If the distances between the gage points are small, measurements approximating the strains at a point are obtained.

As an alternate to the foregoing experimental procedure, electric strain gages are unusually convenient to employ. These consist of very thin wires or foil which is glued to the member being investigated. As the forces are applied to a member, elongation or contraction of the wires or foil takes place concurrently with similar changes in the material. These changes in length alter the electrical resistance of the gage, which may be measured and calibrated to indicate the strain taking place.

Arrangements of gage lines at a point in a cluster, as shown in

Fig. 9-19, are known as *strain rosettes*. If three strain measurements are taken at a rosette, the information is sufficient to determine the complete state of plane strain at a point.

If the angles θ_1, θ_2, and θ_3, together with the corresponding strains ε_{θ_1}, ε_{θ_2}, and ε_{θ_3} are known from measurements, three simultaneous equations patterned after Eq. 9-13 may be written. In writing these equations it is convenient to employ the following notation: $\varepsilon_{x'} \equiv \varepsilon_{\theta_1}$, $\varepsilon_{x''} \equiv \varepsilon_{\theta_2}$, and $\varepsilon_{x'''} \equiv \varepsilon_{\theta_3}$.

$$\varepsilon_{\theta_1} = \varepsilon_x \cos^2 \theta_1 + \varepsilon_y \sin^2 \theta_1 + \gamma_{xy} \sin \theta_1 \cos \theta_1$$

$$\varepsilon_{\theta_2} = \varepsilon_x \cos^2 \theta_2 + \varepsilon_y \sin^2 \theta_2 + \gamma_{xy} \sin \theta_2 \cos \theta_2 \quad (9\text{-}22)$$

$$\varepsilon_{\theta_3} = \varepsilon_x \cos^2 \theta_3 + \varepsilon_y \sin^2 \theta_3 + \gamma_{xy} \sin \theta_3 \cos \theta_3$$

This set of equations may be solved for ε_x, ε_y, and γ_{xy} and the problem reverts back to the cases already considered.

To minimize computational work, the gages in a rosette are usually arranged in an orderly manner. For example, in Fig. 9-19(b), $\theta_1 = 0°$, $\theta_2 = 45°$, and $\theta_3 = 90°$. This arrangement of gage lines is known as the *rectangular* or the 45° *strain rosette*. By direct substitution into Eq. 9-22, it is found that for this rosette

$$\varepsilon_x = \varepsilon_{0°}, \quad \varepsilon_y = \varepsilon_{90°}, \quad \varepsilon_{45°} = \frac{\varepsilon_x}{2} + \frac{\varepsilon_y}{2} + \frac{\gamma_{xy}}{2}$$

or

$$\gamma_{xy} = 2\varepsilon_{45°} - (\varepsilon_{0°} + \varepsilon_{90°})$$

Thus ε_x, ε_y, and γ_{xy} become known.

Another arrangement of gage lines is shown in Fig. 9-19(c). This is known as the *equiangular*, or the *delta*, or the 60° *rosette*. Again, by substituting into Eq. 9-22 and simplifying, $\varepsilon_x = \varepsilon_{0°}$, $\varepsilon_y = (2\varepsilon_{60°} + 2\varepsilon_{120°} - \varepsilon_{0°})/3$, and $\gamma_{xy} = (2/\sqrt{3})(\varepsilon_{60°} - \varepsilon_{120°})$.

Other types of rosettes are occasionally used in experiments. The data from all rosettes may be analyzed by applying Eq. 9-22, solving for ε_x, ε_y, and γ_{xy}, and then applying Mohr's circle of strain.*

Sometimes rosettes with more than three lines are used. An additional gage line measurement provides a check on the experimental work. For these rosettes, the invariance of the strains in the mutually perpendicular directions may be used to check the data.

The application of the experimental rosette technique in complicated problems of stress analysis is almost indispensable.

---

* Convenient graphical solutions for principal strains from measured strains have been developed. See G. Murphy, "A Graphical Method for the Evaluation of Principal Strains from Normal Strains," *Journal of Applied Mechanics*, **12** (1945), A-209.

9-15. ADDITIONAL LINEAR RELATIONS BETWEEN STRESS AND STRAIN AND AMONG E, G, AND ν

Additional relations between stress and strain for linearly elastic, isotropic materials are discussed in this article. These relations are useful for obtaining stresses from planar strains and for finding volumetric changes in elastic materials subjected to uniform external pressure. The fundamental relation among the elastic constants E, G, and ν is also established.

Relation between principal stresses and strains

In many practical investigations strains on the surface of a member are determined by means of rosettes. By using Mohr's circle of strain or strain transformation equations, the principal strains can be found. From these it is possible to determine directly the principal stresses. To establish the appropriate equations, it should be noted that in a plane-stress problem $\sigma_z = 0$, and Eq. 4-11 written in terms of the principal stresses simplifies to

$$\varepsilon_1 = \frac{\sigma_1}{E} - \nu\frac{\sigma_2}{E} \quad \text{and} \quad \varepsilon_2 = \frac{\sigma_2}{E} - \nu\frac{\sigma_1}{E}$$

Solving these equations simultaneously for the principal stresses, one obtains the required relations:

$$\sigma_1 = \frac{E}{1 - \nu^2}(\varepsilon_1 + \nu\varepsilon_2) \quad \sigma_2 = \frac{E}{1 - \nu^2}(\varepsilon_2 + \nu\varepsilon_1) \quad (9\text{-}23)$$

The elastic constants E and ν must be determined from some appropriate experiments. With the aid of such experimental work very complicated problems may be successfully solved.*

EXAMPLE 9-6

At a certain point on a steel machine part measurements with an electric rectangular rosette indicate that $\varepsilon_{0°} = -0.00050$, $\varepsilon_{45°} = +0.0002$, and $\varepsilon_{90°} = +0.00030$. Assuming that $E = 30 \times 10^6$ psi and $\nu = 0.3$ are accurate enough, find the principal stresses at the point investigated.

SOLUTION

From the data given $\varepsilon_x = -0.00050$, $\varepsilon_y = +0.00030$, and

$$\gamma_{xy} = 2\varepsilon_{45°} - (\varepsilon_{0°} + \varepsilon_{90°})$$

$$= 2(+0.0002) - (-0.00050 + 0.00030) = +0.00060$$

---

* See M. Hetényi, editor-in-chief, *Handbook of Experimental Stress Analysis*, Society for Experimental Stress Analysis (New York: John Wiley & Sons, Inc., 1950).

The principal strains for these data were found in Example 9-5 and are $\varepsilon_1 = +0.00040$ and $\varepsilon_2 = -0.00060$. Hence by Eq. 9-23 the principal stresses are

$$\sigma_1 = \frac{30(10)^6}{1-(0.3)^2}[+0.00040 + 0.3(-0.00060)] = +7{,}250 \text{ psi}$$

$$\sigma_2 = \frac{30(10)^6}{1-(0.3)^2}[-0.00060 + 0.3(+0.00040)] = -15{,}830 \text{ psi}$$

The tensile stress σ_1 acts in the direction of ε_1, see Fig. 9-18. The compressive stress σ_2 acts in the direction of ε_2.

Relation among E, G, and ν

The methods of transforming one description of the state of stress or strain into another have been established. In Art. 9-6 particular emphasis was placed on the fact that pure shearing stresses can be transformed into purely normal stresses. Therefore, one must conclude that the deformations caused by pure shearing stresses must be related to the deformations caused by the normal stresses. Based on this assertion a fundamental relation among E, G, and ν for linearly elastic, isotropic materials can be established.

According to Eq. 9-13 with only $\gamma_{xy} \neq 0$, for an x' axis at $\theta = 45°$, the linear strain $\varepsilon_{x'} = \gamma_{xy}/2$. This linear strain $\varepsilon_{x'}$ can be related to the shearing stress τ_{xy} since according to Eq. 4-11 $\tau_{xy} = G\gamma_{xy}$. On this basis

$$\varepsilon_{x'} = \tau_{xy}/(2G)$$

On the other hand, according to Art. 9-6, pure shearing stress τ_{xy} can be expressed alternatively in terms of the principal stresses $\sigma_1 = \tau_{xy}$, and $\sigma_2 = -\tau_{xy}$, acting at 45° to the directions of shearing stresses (see Fig. 9-7). So, by using Eq. 4-11, one finds that the linear strain along the x' axis at $\theta = 45°$ in terms of the principal stresses becomes

$$\varepsilon_{x'} = \varepsilon_1 = \frac{\sigma_1}{E} - \nu\frac{\sigma_2}{E} = \frac{\tau_{xy}}{E}(1+\nu)$$

Equating the two alternative relations for the strain along the positive shear diagonal and simplifying

$$G = \frac{E}{2(1+\nu)} \tag{9-24}$$

This is the basic relation between E, G, and ν; it shows that these quantities are not independent of one another. If any two of these are determined experimentally, the third may be computed. Note that the shearing modulus G is always less than the elastic modulus E since the Poisson ratio ν is a positive quantity. For most materials ν is in the neighborhood of ¼.

Section 9-15 Additional linear relations between stress and strain and among E, G, and ν

Dilatation; bulk modulus

By extending some of the established concepts, one can derive an equation for volumetric changes in elastic materials subjected to stress. In the process of doing this two new terms are introduced and defined.

The sides dx, dy, and dz of an infinitesimal element after straining become $(1 + \varepsilon_x)\,dx$, $(1 + \varepsilon_y)\,dy$, and $(1 + \varepsilon_z)\,dz$, respectively. After subtracting the initial volume from the volume of the strained element, the change in volume is determined. This is

$$(1 + \varepsilon_x)\,dx(1 + \varepsilon_y)\,dy(1 + \varepsilon_z)\,dz - dx\,dy\,dz \approx (\varepsilon_x + \varepsilon_y + \varepsilon_z)\,dx\,dy\,dz$$

where the products of strain $\varepsilon_x\varepsilon_y + \varepsilon_y\varepsilon_z + \varepsilon_z\varepsilon_x + \varepsilon_x\varepsilon_y\varepsilon_z$, being small, are neglected. Therefore, in the infinitesimal (small) strain theory, e, the change in volume per unit volume, often referred to as *dilatation*, is defined as

$$e = \varepsilon_x + \varepsilon_y + \varepsilon_z = \varepsilon_1 + \varepsilon_2 + \varepsilon_3 \tag{9-25}$$

where the last equality follows from the fact that e is an invariant. A more restricted case of strain invariance was encountered in Art. 9-13 for the two-dimensional case, where it was shown that $\varepsilon_1 + \varepsilon_2 = \varepsilon_x + \varepsilon_y$. The shearing strains cause no change in volume.

Based on the generalized Hooke's law, the dilatation can be found in terms of stresses and material constants. For this purpose the first three parts of Eq. 4-11 must be added together. This yields

$$e = \varepsilon_x + \varepsilon_y + \varepsilon_z = \frac{1 - 2\nu}{E}(\sigma_x + \sigma_y + \sigma_z) \tag{9-26}$$

which means that dilatation is proportional to the algebraic sum of all normal stresses. As a direct counterpart to the strain invariant, the sum $(\sigma_x + \sigma_y + \sigma_z)$ is the stress invariant.

If an elastic body is subjected to hydrostatic pressure of uniform intensity p, so that $\sigma_x = \sigma_y = \sigma_z = -p$, then from Eq. 9-26

$$e = -\frac{3(1 - 2\nu)}{E}p \quad \text{or} \quad \frac{-p}{e} = k = \frac{E}{3(1 - 2\nu)} \tag{9-27}$$

The quantity k represents the ratio of the hydrostatic compressive stress to the decrease in volume and is called *modulus of compression* or *bulk modulus*.

PART C
YIELD AND FRACTURE CRITERIA

9-16. PRELIMINARY REMARKS

In the early chapters of the text when stresses due to axial loads or pure torsion were considered, the calculated stresses could be related to some analogous experiments for the same material. Based on such experimental evidence the behavior of members with respect to the onset of yielding and to probable fracture can be predicted with a reasonable degree of accuracy. The response of a material to uniaxial stress or pure shearing stress can be conveniently displayed on stress-strain diagrams. This was discussed in Chapters 4 and 5. Such a direct approach is not possible, however, for a complex state of stress which is characteristic of many elements in machines and structures. Therefore, it is important to establish criteria for behavior of materials under combined states of stress.

Unfortunately, at this date the quantitative criteria for yielding and fracture of materials under multiaxial states of stress are incomplete. A number of questions remain unsettled and are a part of an active area of materials research. As yet no complete answer can be given by any one theory. In this part of the chapter, the two widely accepted criteria for the onset of ductile behavior for combined stresses in ductile materials will be discussed first. This is followed by the presentation of a fracture criterion for brittle materials. A few additional fracture, or failure, criteria suitable for some materials are given at the end of the chapter. In classifying the materials in this manner, strictly speaking, one refers to the brittle or ductile state of the material as this characteristic is greatly affected by temperature as well as by the state of stress itself. More complete answers to such questions are beyond the scope of this book.

9-17. MAXIMUM SHEARING STRESS THEORY

The maximum shearing stress theory,* or simply the maximum shear theory, results from the observation that in a ductile material slipping occurs during yielding along critically oriented planes. This suggests that the maximum shearing stress plays the key role, and it is assumed that yielding of the material depends only on the maximum shearing stress which is attained within an element. Therefore, whenever

---

* This theory appears to have been originally proposed by C. A. Coulomb in 1773. In 1868 H. Tresca presented the results of his work on the flow of metals under great pressures to the French Academy. Now this theory often bears his name.

a certain critical value τ_{cr} is reached, yielding in an element commences.* For a given material this value usually is set equal to the shearing stress at yield in simple tension or compression. Thus, according to Eq. 9-6, if $\sigma_x = \pm\sigma_1 \neq 0$, and $\sigma_y = \tau_{xy} = 0$,

Section 9-17 Maximum shearing stress theory

$$\tau_{max} \equiv \tau_{cr} = \left|\pm\frac{\sigma_1}{2}\right| = \frac{\sigma_{yp}}{2} \qquad (9\text{-}28)$$

which means that if σ_{yp} is the yield-point stress found, for example, in a simple tension test, the corresponding maximum shearing stress is half as large. This conclusion also follows easily from Mohr's circle of stress.

To apply the maximum shearing stress criterion to a biaxial state of stress, the maximum shearing stress is determined and is equated to τ_{max} given by Eq. 9-28. In doing this for the principal stresses σ_1 and σ_2 with $\sigma_3 = 0$, two cases must be considered. In the one case, the signs of σ_1 and σ_2 are the same. The case when σ_1 and σ_2 are both tensile is illustrated in Fig. 9-20(a). An isometric view together with the three projections of the element onto the axes of principal stresses and the corresponding Mohr's circles for the three-dimensional state of stress are shown on the diagram. The alternative planes of the maximum shearing stress along which slip may occur are indicated in the figure.

The maximum shearing stress for the case in Fig. 9-20(a) is seen to be the same as that for a uniaxial stress. Therefore, if $|\sigma_1| > |\sigma_2|$, according to Eq. 9-28 $|\sigma_1|$ must not exceed σ_{yp}. Similarly, if $|\sigma_2| > |\sigma_1|$, $|\sigma_2|$ must not be greater than σ_{yp}. Therefore the criterion corresponding to this case is

$$|\sigma_1| \leq \sigma_{yp} \quad \text{and} \quad |\sigma_2| \leq \sigma_{yp} \qquad (9\text{-}29)$$

If the signs of σ_1 and σ_2 are opposite, the maximum shearing stress $\tau_{max} = [|\sigma_1| + |\sigma_2|]/2$. The planes of these stresses which correspond to the possible slip planes are in Fig. 9-20(b). As before, to obtain the yield criterion, τ_{max} must not exceed the maximum shearing stress at yield in uniaxial experiment. Expressed mathematically

$$\left|\pm\frac{\sigma_1 - \sigma_2}{2}\right| \leq \frac{\sigma_{yp}}{2}$$

or, for impending yield,

$$\frac{\sigma_1}{\sigma_{yp}} - \frac{\sigma_2}{\sigma_{yp}} = \pm 1 \qquad (9\text{-}30)$$

---

* In single crystals slip occurs along preferential planes and in preferential directions. In studies of this phenomenon the effective component of the shearing stress causing slip must be carefully determined. Here it is assumed that because of the random orientation of numerous crystals the material has isotropic properties, and so by determining τ_{max} one finds the critical shearing stress.

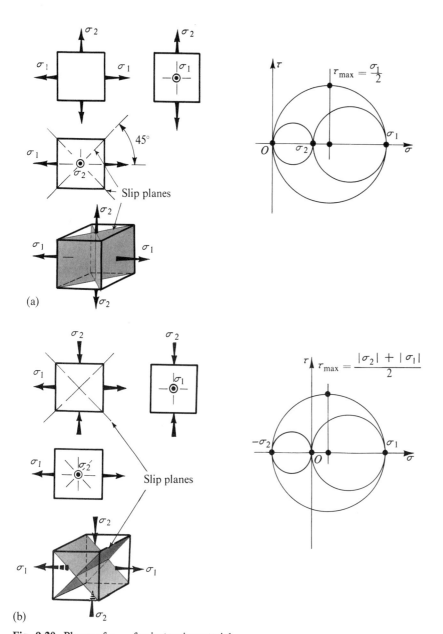

Fig. 9-20. Planes of τ_{max} for isotropic material.

Equation 9-30 can be plotted as in Fig. 9-21. Its results have relevance only in the second and the fourth quadrants. In the first and third quadrants the criterion expressed by Eq. 9-29 applies.

By considering σ_1 and σ_2 as the coordinates of a point, one may see that the stresses falling within the hexagon of Fig. 9-21 indicate that no yielding of the material has occurred, i.e., that the material behaves elastically. The state of stress corresponding to the points falling on the hexagon shows that the material is yielding. No points can lie outside the hexagon.

Note that, according to the maximum shear theory, if hydrostatic tensile or compressive stresses are added, i.e., stresses such that $\sigma_1' = \sigma_2' = \sigma_3'$, no change in the material response is predicted. Adding these stresses merely shifts the Mohr's circles of stress, such as in Fig. 9-20, along the σ axis and τ_{max} remains the same. This matter will be commented upon further in the next article.

The yield criterion just derived is often referred to as the *Tresca yield condition* and is one of the widely used laws of plasticity.

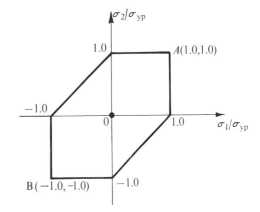

Fig. 9-21. Yield criterion based on maximum shearing stress.

9-18. MAXIMUM DISTORTION ENERGY THEORY

Another widely accepted criterion of yielding for ductile, isotropic materials is based on the energy concepts.* In this approach the total elastic energy is divided into two parts: one associated with the volumetric changes of the material, and the other causing shearing distortions. By equating the shearing distortion energy at yield point in simple tension to that under combined stress, the yield criterion for combined stress is established.

In order to derive the expression giving the yield condition for combined stress, the procedure of resolving the general state of stress must be employed. This is based on the concept of superposition. For example, it is possible to consider the stress tensor of the three principal stresses—σ_1, σ_2, and σ_3—to consist of two additive component tensors. The elements of one component tensor are defined as the mean "hydrostatic" stress

$$\bar{\sigma} = \frac{\sigma_1 + \sigma_2 + \sigma_3}{3} \qquad (9\text{-}31)$$

The elements of the other tensor are $(\sigma_1 - \bar{\sigma})$, $(\sigma_2 - \bar{\sigma})$, and $(\sigma_3 - \bar{\sigma})$. Writing this in matrix representation, one has

$$\begin{pmatrix} \sigma_1 & 0 & 0 \\ 0 & \sigma_2 & 0 \\ 0 & 0 & \sigma_3 \end{pmatrix} = \begin{pmatrix} \bar{\sigma} & 0 & 0 \\ 0 & \bar{\sigma} & 0 \\ 0 & 0 & \bar{\sigma} \end{pmatrix} + \begin{pmatrix} \sigma_1 - \bar{\sigma} & 0 & 0 \\ 0 & \sigma_2 - \bar{\sigma} & 0 \\ 0 & 0 & \sigma_3 - \bar{\sigma} \end{pmatrix} \qquad (9\text{-}32)$$

---

* The first attempt to use the total energy as the criterion of yielding was made by E. Beltrami of Italy in 1885. In its present form the theory was proposed by M. T. Huber of Poland in 1904 and was further developed and explained by R. von Mises (1913) and H. Hencky (1925), both of Germany and the United States.

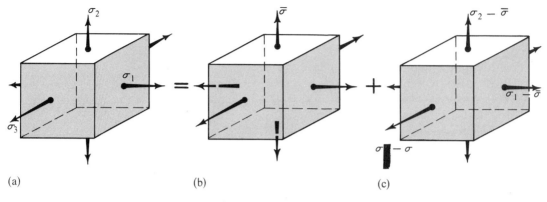

General state of stress

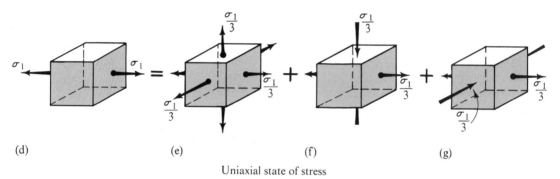

Uniaxial state of stress

Fig. 9-22. Resolution of principal stresses into spherical (dilatational) and deviatoric (distortional) stresses.

This resolution of the general state of stress is shown schematically in Fig. 9-22. The special case of resolving the uniaxial state of the stress in the figure has been carried a step further. The sum of the stresses in Figs. 9-22(f) and (g) corresponds to the last tensor of Eq. 9-32.

For the three-dimensional state of stress the Mohr's circle for the first tensor component of Eq. 9-32 degenerates into a point located at $\bar{\sigma}$ on the σ axis. Therefore the stresses associated with this tensor are the same in every possible direction. For this reason this tensor is called the *spherical stress tensor*. Alternatively, from Eq. 9-26, which states that dilatation of an elastic body is proportional to $\bar{\sigma}$, this tensor is also called the *dilatational stress tensor*.

The last tensor of Eq. 9-32 is called the *deviatoric or distortional stress tensor*. A good reason for the choice of these terms may be seen from Figs. 9-22(f) and (g). The state of stress consisting of tension and compression on the mutually perpendicular planes is equivalent to pure shearing stress. The latter system of stresses is known to cause no volumetric changes but instead distorts or deviates the element from its initial cubic shape.

Section 9-18
Maximum distortion energy theory

Having established the basis for resolving or decomposing the state of stress into dilatational and distortional components, one may find the strain energy due to distortion. For this purpose Eq. 4-21 is rewritten in terms of the principal stresses, i.e., with $\tau_{xy} = \tau_{yz} = \tau_{zx} = 0$. This gives a general expression for the total strain energy per unit volume:

$$U_{\text{total}} = \frac{1}{2E}(\sigma_1^2 + \sigma_2^2 + \sigma_3^2) - \frac{\nu}{E}(\sigma_1\sigma_2 + \sigma_2\sigma_3 + \sigma_3\sigma_1) \quad (9\text{-}33)$$

The strain energy per unit volume due to the dilatational stresses can be determined from Eq. 9-33 by first setting $\sigma_1 = \sigma_2 = \sigma_3 = p$, and then replacing p by $\bar{\sigma} = (\sigma_1 + \sigma_2 + \sigma_3)/3$. Thus

$$U_{\text{dilatation}} = \frac{3(1-2\nu)}{2E}p^2 = \frac{1-2\nu}{6E}(\sigma_1 + \sigma_2 + \sigma_3)^2 \quad (9\text{-}34)$$

By subtracting Eq. 9-34 from Eq. 9-33, simplifying, and noting from Eq. 9-24 that $G = E/2(1+\nu)$, one finds the distortion strain energy for combined stress:

$$U_{\text{distortion}} = \frac{1}{12G}[(\sigma_1 - \sigma_2)^2 + (\sigma_2 - \sigma_3)^2 + (\sigma_3 - \sigma_1)^2] \quad (9\text{-}35)$$

According to the basic assumption of the distortion energy theory, the expression of Eq. 9-35 must be equated to the maximum distortion energy in simple tension. The latter condition occurs when one of the principal stresses reaches the yield point σ_{yp} of the material. The distortion strain energy for this is $2\sigma_{yp}^2/12G$. Equating this to Eq. 9-35, after minor simplifications, one obtains the basic law for ideally plastic material:

$$(\sigma_1 - \sigma_2)^2 + (\sigma_2 - \sigma_3)^2 + (\sigma_3 - \sigma_1)^2 = 2\sigma_{yp}^2 \quad (9\text{-}36)$$

For plane stress $\sigma_3 = 0$, and Eq. 9-36 in dimensionless form becomes

$$\left(\frac{\sigma_1}{\sigma_{yp}}\right)^2 - \left(\frac{\sigma_1}{\sigma_{yp}}\frac{\sigma_2}{\sigma_{yp}}\right) + \left(\frac{\sigma_2}{\sigma_{yp}}\right)^2 = 1 \quad (9\text{-}37)$$

This is an equation of an ellipse, a plot of which is shown in Fig. 9-23. Any stress falling within the ellipse indicates that the material behaves elastically. Points on the ellipse indicate that the material is yielding. This is the same interpretation as that given earlier for Fig. 9-21. On unloading the material behaves elastically.

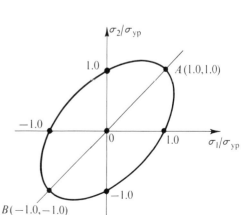

Fig. 9-23. Yield criterion based on maximum distortion energy.

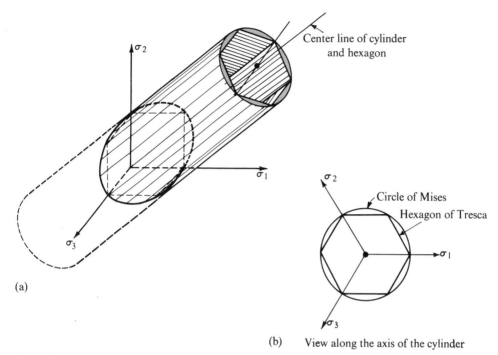

Fig. 9-24. Yield surfaces for three-dimensional state of stress.

It is important to note that this theory does not predict changes in the material response when hydrostatic tensile or compressive stresses are added. This follows from the fact that, since only differences of the stresses are involved in Eq. 9-36, adding a constant stress to each does not alter the yield condition. For this reason in the three-dimensional stress space, the yield surface becomes a cylinder with an axis having all three direction cosines equal to $1/\sqrt{3}$. Such a cylinder is in Fig. 9-24. The ellipse in Fig. 9-23 is simply the intersection of this cylinder with the σ_1-σ_2 plane. It can be shown also that the yield surface for the maximum shearing stress criterion is a hexagon which fits into the tube, Fig. 9-24.

The yield condition expressed by Eq. 9-36 can be shown to be another stress invariant. It is also a continuous function. These features make the use of this law of plastic yielding for combined stresses particularly attractive from the theoretical point of view. This widely used law is often referred to as the *Huber-Hencky-Mises* or simply the *von Mises yield condition.**

---

* In the past this condition has been also frequently referred to as the *octahedral shearing stress theory.* See A. Nadai, *Theory of Flow and Fracture of Solids* (New York, McGraw-Hill Book Company, 1950), p. 104, or, F. B. Seely and J. O. Smith, *Advanced Mechanics of Materials* (2nd. ed.) (New York: John Wiley & Sons, Inc., 1952), p. 61.

Both the maximum shearing stress and the distortion energy yield conditions have been used in the study of viscoelastic phenomena under combined stress. Extension of these ideas to strain hardening materials is also possible. Such topics, however, are beyond the scope of this text.

9-19. MAXIMUM NORMAL STRESS THEORY

The maximum normal stress theory or simply the maximum stress theory* asserts that failure or fracture of a material occurs when the maximum normal stress at a point reaches a critical value regardless of the other stresses. Only the largest principal stress must be determined to apply this criterion. The critical value of stress σ_{ult} is usually determined in a tensile experiment, where the failure of a specimen is defined to be either excessively large elongation or fracture. Usually the latter is implied.

Experimental evidence indicates that this theory applies well to brittle materials in all ranges of stresses providing a tensile principal stress exists. Failure is characterized by the separation, or the cleavage, fracture. This mechanism of failure differs drastically from the ductile fracture, which is accompanied by large deformations due to slip along the planes of maximum shearing stress.

The maximum stress theory can be interpreted on graphs as could the other theories. This is done in Fig. 9-25. Failure occurs if points fall on the surface. Unlike the previous theories, this stress criterion gives a bounded surface of the stress space.

---

* This theory is generally credited to W. J. M. Rankine, an eminent British educator (1820–72). An analogous theory based on the maximum strain, rather than stress, being the basic criterion of failure was proposed by the great French elastician, B. de Saint Venant (1797–1886). Experimental evidence does not corroborate the latter approach.

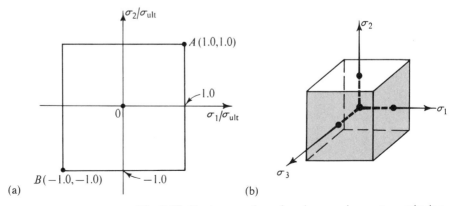

Fig. 9-25. Fracture envelopes based on maximum stress criterion.

9-20. COMPARISON OF THEORIES; OTHER THEORIES

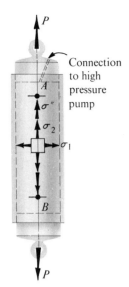

Fig. 9-26. Arrangement for obtaining controlled ratios of the principal stresses.

Most of the information on yielding and fracture of materials under the action of biaxial stresses comes from experiments on thin-walled cylinders. A typical arrangement for such an experiment is in Fig. 9-26. The ends of a thin-walled cylinder of the material being investigated are closed by substantial caps. This forms the hollow interior of a cylindrical pressure vessel. By pressurizing the available space* and simultaneously applying an additional tensile or compressive force P to the caps, one obtains different ratios of principal stresses. By maintaining a fixed ratio between the principal stresses until yielding or failure is reached, the desired data on a material are obtained. Analogous experiments with tubes simultaneously subjected to torque, axial force, and pressure are also used.

Comparison of some classical experimental results with the yield and fracture theories presented above is shown in Fig. 9-27.† Note the

---

* See Art. 10-6 for a discussion of pressure vessels.

† The experimental points shown on this figure are based on classical experiments by several investigators. The figure is adapted from a compilation made by G. Murphy, *Advanced Mechanics of Materials* (New York: McGraw-Hill Book Company, 1964), p. 83.

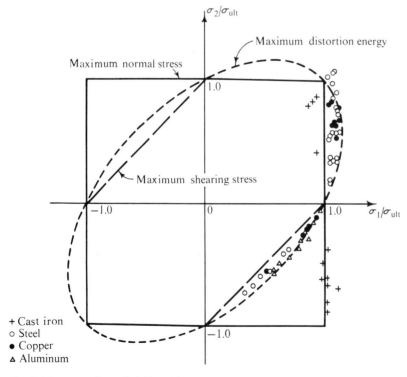

Fig. 9-27. Comparison of yield and fracture criteria with test data.

324

particularly good agreement between the maximum distortion energy theory and experimental results for ductile materials. However, the maximum normal stress theory appears to be best for brittle materials and can be unsafe for ductile materials.

Section 9-20 Comparison of theories; other theories

All the theories for uniaxial stress agree since the simple tension test is the standard of comparison. Therefore, if one of the principal stresses at a point is large in comparison with the other, all theories give practically the same results. The discrepancy between the theories is greatest in the second and fourth quadrants, when both principal stresses are numerically equal.

In the development of the theories discussed above it has been assumed that the properties of material in tension and compression are alike—the plots shown in Figs. 9-21, 9-23, and 9-25 have two axes of symmetry. On the other hand, it is known that some materials such as rocks, cast iron, concrete, and soils, have drastically different properties depending on the sense of the applied stress. An early modification of the maximum shear theory by C. Duguet in 1885 to achieve better agreement with experiment is shown in Fig. 9-28(a). This modification recognizes the high strength of some materials when subjected to biaxial compression. A. A. Griffith,* in a sense, refined the explanation for the above observation by introducing the idea of surface energy at microscopic cracks and showing the greater seriousness of tensile stresses compared with compressive ones with respect to failure. According to this theory an existing crack will rapidly propagate if the available elastic strain energy release rate is greater than the increase in the surface energy of the crack. The original Griffith concept has been considerably expanded by G. R. Irwin.†

Otto Mohr, in addition to showing the construction of the stress circle bearing his name, suggested another approach for predicting failure of a material. Different experiments such as one in simple tension, one in pure shear, and one in compression are performed first, see Fig. 9-28(b). Then an envelope to these circles defines the failure envelope. Circles drawn tangent to this envelope give the condition of failure at the point of tangency. This approach finds favor in soil mechanics.

Instead of studying the response of materials on stress-space plots as in Fig. 9-28, the stress invariant ($\sigma_x + \sigma_y + \sigma_z$) and the stress given by Eq. 9-36 may be used as the coordinate axes. Useful fracture criteria have been established from studies based on this approach.‡

---

* A. A. Griffith, "The Phenomena of Rupture and Flow of Solids," *Philosophical Transactions of the Royal Society of London*, Series A, **221** (1920), 163–98.

† G. R. Irwin, "Fracture Mechanics," *Proceedings, First Symposium on Naval Structural Mechanics* (Long Island City, N.Y.: Pergamon Press, 1958), p. 557. Also see *A Symposium on Fracture Toughness Testing and Its Applications*, American Society for Testing and Materials Special Technical Publication No. 381 (American Society for Testing and Materials and National Aeronautics and Space Administration, 1965).

‡ B. Bresler and K. Pister, "Failure of Plain Concrete Under Combined Stresses," *Transactions of the American Society of Civil Engineers* **122** (1957) 1049.

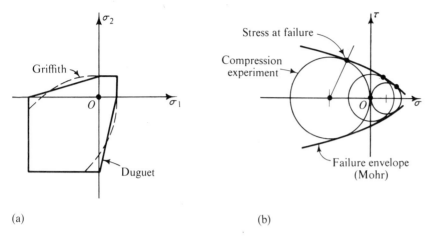

Fig. 9-28. Possible fracture criteria.

Sometimes the yield and fracture criteria discussed above are inconvenient to apply. In such cases interaction curves such as in Fig. 8-6 can be used to advantage. Experimentally determined curves of this type, unless complicated by a local or buckling phenomenon, are equivalent to the strength criteria discussed here.

In the next chapter, in the design of members departures will be made from strict adherence to the yield and fracture criteria established here, although unquestionably these theories provide the rational basis for design.

PROBLEMS FOR SOLUTION

9-1. Infinitesimal elements A, B, C, D, and E are shown on the figures for two different members. Draw each one of these elements separately, and indicate on the isolated elements the stress acting on it. For each stress clearly show its direction and sense by arrows, and state the formula one would use in its calculation. Neglect the weight of the members.

9-2 through 9-4. For the infinitesimal elements shown in the figures, find the normal and shearing stresses acting on the indicated inclined planes. Use the "wedge" method of analysis discussed in Example 9-1. *Ans. Prob. 9-3.* $\sigma = -400$ psi, $\tau = 4{,}970$ psi.

9-5. Derive Equation 9-2.

9-6. (a) For the data given in Prob. 9-2 plot

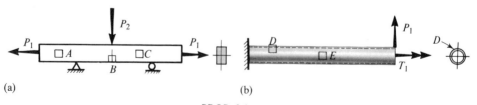

PROB. 9-1

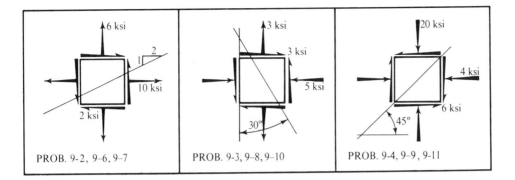

$\sigma_{x'}$ and $\tau_{x'y'}$ as ordinates with θ as abscissa for $0 \leq \theta \leq 2\pi$. (b) Generalize and discuss the results, especially with regard to the maxima and the minima of the functions.

9-7. Rework Prob. 9-2 using Eqs. 9-1 and 9-2.

9-8. Rework Prob. 9-3 using Eqs. 9-1 and 9-2.

9-9. For the data of Prob. 9-4 find the stresses on $\theta = 45°$ and $\theta = 135°$. Show the complete results on the newly oriented element.

9-10. For the data of Prob. 9-3, (a) find the principal stresses and show their directions and senses on a properly oriented element; (b) determine the maximum shearing stresses and the associated normal stresses. Show the results on a properly oriented element.

9-11. Same as preceding problem for data of Prob. 9-4.

9-12 through 9-15. Draw Mohr's circle of stress for the states of stress given in the figures. (a) Clearly show the planes on which the principal stresses act, and for each stress indicate with arrows its direction and sense. (b) Same as (a) for the maximum shearing stresses and the associated normal stresses, *Ans. Prob.* 9-15. (a) 6 ksi, −4 ksi; (b) 5 ksi. 1 ksi.

9-16. The state of two-dimensional stress at three different points is given in matrix representation as

(a) $\begin{pmatrix} 12 & 5 \\ 5 & 6 \end{pmatrix}$ ksi (b) $\begin{pmatrix} -6 & 6 \\ 6 & -8 \end{pmatrix}$ ksi

(c) $\begin{pmatrix} 3 & -9 \\ -9 & -12 \end{pmatrix}$ ksi

For each case draw Mohr's circle of stress, and then, using trigonometry, find the principal stresses and show their directions and senses on properly oriented elements. Also find the maximum shearing stresses with the associated normal stresses, and show the results on properly oriented elements. *Ans.* (a) 14.83 ksi, 3.17 ksi; 5.83 ksi, 9 ksi; (b) −0.9 ksi, −13.1 ksi; 6.1 ksi, −7 ksi; (c) 7.2 ksi, −16.2 ksi; 11.7 ksi, −4.5 ksi.

9-17. If $\sigma_x = \sigma_1 = 0$ and $\sigma_y = \sigma_2 = -4,000$ psi, using Mohr's circle of stress, find the

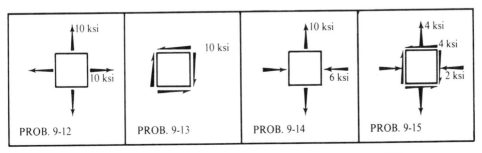

stresses acting on a plane defined by $\theta = +30°$. *Ans.* -1 ksi, -1.73 ksi.

9-18. Using Mohr's circle of stress, for the data of Prob. 9-15, find the stresses acting on $\theta = 30°$.

9-19. The magnitudes and directions of the stresses on two planes intersecting at a point are as shown in the figure. Determine the directions and magnitudes of the principal stresses at this point. Sketch the results on an element.

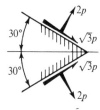

PROB. 9-19

9-20. At point A on an unloaded edge of an elastic body, oriented as shown in the figure with respect to the xy axes, the maximum shearing stress is 500 psi. (a) Find the principal stresses, and (b) determine the state of stress on an element oriented with its edges parallel to the xy axes. Show the results on a sketch of the element at A. (*Hint:* An effective solution may be obtained by constructing Mohr's circle of stress.) *Ans.* (b) $\sigma_x = 640$ psi.

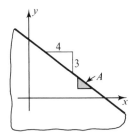

PROB. 9-20

9-21. A clevice transmits a force F to a bracket as shown in the figure. Stress analysis of this bracket gives the following stress components acting on the element A: 1,000 psi due to bending, 1,500 psi due to axial force, and

PROB. 9-21

600 psi due to shear. (Note that these are stress magnitudes only, their directions and senses must be determined by inspection.) (a) Indicate the resultant stresses on a sketch of the isolated element A. (b) Using Mohr's circle for the state of stress found in (a), determine the principal stresses and the maximum shearing stresses with the associated normal stresses. Show the results on properly oriented elements. *Ans.* (b) 900 psi, -400 psi; 650 psi, 250 psi.

9-22. The loading applied to a beam, such as shown in the figure, causes the following stresses on the element A: 1,000 psi shearing

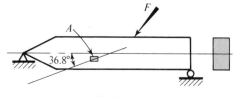

PROB. 9-22

stress due to $|V|$, 4,000 psi normal stress due to $|P|$, and 2,000 psi normal stress due to $|M|$. Determine the normal and shearing stresses acting on the element on a plane making an angle of 36.8° with the longitudinal axis as shown. *Ans.* 240 psi, 680 psi.

9-23. A solid circular shaft is loaded as

shown in the figure. At section $ABCD$ the stresses due to the 2,000-lb force, and the weight of the shaft and round drum are found to be as follows: maximum bending stress is 6,000 psi, maximum torsional stress is 4,000 psi, and maximum shearing stress due to V is 1,000 psi. (a) Set up elements at points A, B, C, and D and indicate the magnitudes and directions of the stresses acting on them. In each case state from which direction the element is observed. (b) Using Mohr's circle, find directions and magnitudes of the principal stresses and of the maximum shearing stress at point A. *Ans.* (b) 8 ksi, -2 ksi; 5 ksi.

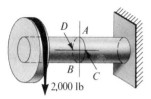

PROB. 9-23

9-24. Consider a semi-infinite linearly elastic body with a concentrated line load of P lb per inch, see figure. This approximates a long footing on a soil foundation or a knife edge pressing on a large, flat piece of metal (an idealization of roller-bearing reaction on the race). Using the methods of elasticity, one can show* that the application of the load causes

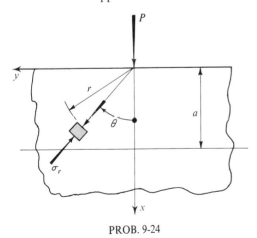

PROB. 9-24

* S. Timoshenko and J. N. Goodier, *Theory of Elasticity* (2nd ed.) (New York: McGraw-Hill Book Company, 1951), p. 85.

only radial stresses, which are given by

$$\sigma_r = -\frac{2P}{\pi}\frac{\cos\theta}{r}$$

Since this is the only stress, it is a principal stress. Also for infinitesimal elements no distinction need be made between the Cartesian and the polar elements. Transform σ_r into σ_x, σ_y, and τ_{xy} and plot the resulting stress distribution for σ_x and τ_{xy} at a constant depth a below the surface.

9-25. Consider an elastic wedge of unit thickness subjected to a concentrated line load P at the apex, as shown in the figure. According to the elasticity solution† this loading

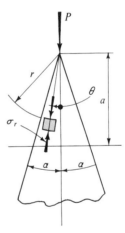

PROB. 9-25

causes only radial stress distribution, which is given by

$$\sigma_r = -\frac{P\cos\theta}{r[\alpha + \tfrac{1}{2}\sin 2\alpha]}$$

Based on this formula, determine the vertical stress distribution on a horizontal section at a distance a below the apex. Compare the maximum stress so found with the one given by Eq. 3-5 for $\alpha = 10°$ and $45°$.

9-26. An elastic wedge of unit thickness is subjected to a vertical force P as shown in the

† Timoshenko and Goodier, *Theory of Elasticity*, p. 97.

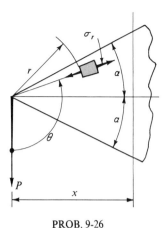

PROB. 9-26

figure. For such a wedge the elasticity solution shows that only radial stress distribution exists and is given* by

$$\sigma_r = -\frac{P \cos \theta}{r[\alpha - \tfrac{1}{2} \sin 2\alpha]}$$

Determine the normal and the shearing stresses on a vertical section at distance x from the applied force P and compare with the elementary solutions. If $\alpha = 30°$ find the percentage of discrepancy among the maximum stresses in the alternative solutions.

9-27. Using the stress transformation equations for a three-dimensional state of stress,† one may diagonalize any stress matrix. Suppose this were done and it yields

$$\begin{pmatrix} 12{,}000 & 0 & 0 \\ 0 & -6{,}000 & 0 \\ 0 & 0 & 8{,}000 \end{pmatrix} \text{psi}$$

For this state of stress what is the maximum shearing stress? Illustrate the plane or planes on which it acts in a sketch.

9-28. An investigation of stresses in the plate of a thin-walled pressure vessel indicates that the stress matrix is

---

* Timoshenko and Goodier, *Theory of Elasticity*, p. 97.
† See any book on elasticity or plasticity. For a brief discussion of this point see Art. 9-9.

$$\begin{pmatrix} 20 & 0 & 0 \\ 0 & 10 & 0 \\ 0 & 0 & 0 \end{pmatrix} \text{ksi}$$

where it is to be noted that $\sigma_3 \approx 0$. (This state of stress is analogous to that shown in Prob. 4-6.) Are there any shearing stresses in the material? Illustrate with a sketch.

9-29. Let l, m, and n define the direction cosines of a linear element. Using this notation, Eq. 9-18 can be rewritten as

$$\varepsilon_\theta = \varepsilon_x l^2 + \varepsilon_y m^2 + \gamma_{xy} lm$$

Show that for the three-dimensional case

$$\varepsilon_\theta = \varepsilon_x l^2 + \varepsilon_y m^2 + \varepsilon_z n^2 \\ + \gamma_{xy} lm + \gamma_{yz} mn + \gamma_{zx} nl$$

9-30. If the unit strains are $\varepsilon_x = -120 \times 10^{-6}$, $\varepsilon_y = +1{,}120 \times 10^{-6}$, and $\gamma_{xy} = -200 \times 10^{-6}$, what are the principal strains and in which direction do they occur? Use Eqs. 9-20 and 9-21 or Mohr's circle of strain, as directed. *Ans.* $1{,}130 \times 10^{-6}$, -130×10^{-6}.

9-31. If the unit strains are $\varepsilon_x = -800 \times 10^{-6}$, $\varepsilon_y = -200 \times 10^{-6}$, and $\gamma_{xy} = +800 \times 10^{-6}$, what are the principal strains and in which directions do they occur? Use Eqs. 9-20 and 9-21 or Mohr's circle, as directed. *Ans.* 0, $1{,}000 \times 10^{-6}$.

9-32. If the strain measurements given in the above problem were made on a steel member ($E = 29.5 \times 10^6$ psi and $\nu = 0.3$), what are the principal stresses and in which direction do they act?

9-33. The data for a rectangular rosette attached to a stressed steel member are $\varepsilon_{0°} = -220 \times 10^{-6}$, $\varepsilon_{45°} = +120 \times 10^{-6}$, $\varepsilon_{90°} = +220 \times 10^{-6}$. What are the principal stresses and in which directions do they act? $E = 30 \times 10^6$ psi and $\nu = 0.3$. *Ans.* ± 5.76 ksi, $14°18'$.

9-34. The data for an equiangular rosette, attached to a stressed, aluminum-alloy member, are $\varepsilon_{0°} = +400 \times 10^{-6}$, $\varepsilon_{60°} = +400 \times 10^{-6}$, and $\varepsilon_{120°} = -600 \times 10^{-6}$. What are the principal stresses and in which directions do they act? $E = 10^7$ psi and $\nu = \tfrac{1}{4}$. *Ans.* $+6.22$ ksi, -4.44 ksi, $30°$.

9-35. The data for a strain rosette with four gage lines attached to a stressed, aluminum-alloy member are $\varepsilon_{0°} = -120 \times 10^{-6}$, $\varepsilon_{45°} = +400 \times 10^{-6}$, $\varepsilon_{90°} = +1{,}120 \times 10^{-6}$, and $\varepsilon_{135°} = +600 \times 10^{-6}$. Check the consistency of the data. Then determine the principal stresses and the directions in which they act. Use the values of E and ν given in Prob. 9-34. *Ans.* $+11.7$ ksi, 1.6 ksi, $4°35'$.

9-36. At a point in a stressed elastic plate the following information is known: maximum shearing strain $\gamma_{max} = 5 \times 10^{-4}$, and the sum of the normal stresses on two perpendicular planes passing through the point is 4,000 psi. The elastic properties of the plate are $E = 30 \times 10^6$ psi, $G = 12 \times 10^6$ psi, $\nu = 1/4$. Calculate the magnitude of the principal stresses at the point. *Ans.* $\sigma_1 = 8$ ksi.

9-37. (a) Show that dilatation in shear is zero. (b) Using Eq. 9-27 establish an upper bound on Poisson's ratio for isotropic materials.

9-38. By commencing with Eqs. 9-33 and 9-34, derive in detail Eq. 9-35.

9-39. Establish the ratio of $U_\text{distortion}$ to U_total in the elastic range for a bar of mild steel subjected to tension. Let $E = 30 \times 10^3$ ksi, and $\nu = 0.25$.

9-40. The design stresses for an element of a thin plate are $\sigma_1 = 10$ ksi, $\sigma_2 = -10$ ksi (consider $\sigma_3 = 0$). Doubling this stress causes yielding in the material. If the design compressive stress σ_2 is reduced to one-half its former value, what tensile stress σ_1 may be applied? Maintain the same factor of safety in yielding in both cases and assume the Tresca flow law.

For additional stress analysis problems see end of next chapter.

10 Problems in stress analysis

10-1. INTRODUCTION

In the preceding chapter stress transformation laws as well as yield and fracture criteria have been established. Now the analysis of stresses in members can be considered from a more comprehensive point of view. In this connection two types of problems arise, and accordingly this chapter is divided into two parts. Part A discusses the analysis of stress in given members subjected to given loads. Part B shows how to select or design members according to the strength requirements for given loading conditions. In both parts of the chapter only statically determinate cases are treated and the analysis is confined principally to elastic cases.

In Part A first the stresses at a point in axially loaded members and shafts subjected to torque are re-examined from a broader point of view. This is followed by the analysis of stresses in flexural members which also transmit shear. A brief description of a related subject, the photoelastic method of experimental stress analysis, is then given. An introduction to the important problem of stress analysis of shells with applications to pressure vessels ends Part A.

In Part B of this chapter, strength requirements are considered as the design criterion, although, as pointed out in Chapter 1, the design of a member may depend on its strength, or its stiffness, or its stability. The main objective in this treatment is to establish simple and rapid procedures which may be used in practical design problems for selecting a member of adequate strength. Several formulas developed in the earlier chapters and the information on yield and fracture criteria discussed in the preceding chapter form the basis for the design of members. For these reasons, in many respects this chapter will serve as a review chapter.

*Section 10-2
Investigation of
stress at a point*

PART A
ANALYSIS OF STRESSES

10-2. INVESTIGATION OF STRESS AT A POINT

Based on the stress transformation equations or the representation of the state of stress on a Mohr's circle, it is evident that the state of stress at a point can be represented in infinite ways depending on the selected planes of stress. For quantitative solutions of problems, stresses on known planes are found first, using the equations derived previously.

In confining attention for the present to statically determinate cases, the familiar approach outlined in Art. 1-3 is employed. The reactions are found first. Then a segment of the body is isolated by passing a section perpendicular to its axis through the point to be investigated, and the system of forces necessary to maintain the equilibrium of the segment is determined. The magnitudes of the stresses are determined next by the conventional formulas. Then, on an element isolated from the member, the computed stresses are indicated. The sense of the computed stresses is noted on this element by arrows agreeing with the sense of the internal forces at the cut. Two sides of this element are parallel and two sides are perpendicular to the axis of the member being investigated. The definite relation of the sides of this element to the actual member must be clearly understood by the analyst. After the sketch of an element is prepared and stresses of the same kind are added algebraically, the stresses may be found on planes with any orientation through the same point. For this purpose, either analytical formulas or Mohr's circle of stress, discussed in the preceding chapter, is used. The principal stresses or the maximum shearing stress is usually the quantity sought.

In the following three examples, an axially loaded rod, a circular shaft in torsion, and a rectangular beam with transversely applied force will be examined for principal stresses and stresses acting on inclined planes.

*Chapter 10
Problems in stress analysis*

EXAMPLE 10-1

Find the stresses acting on an arbitrarily inclined plane in an axially loaded rod of constant cross-sectional area.

SOLUTION

Consider the prismatic bar subjected to axial tension in Fig. 10-1(a). By passing a section X-X perpendicular to the axis of the rod through a general point G and applying Eq. 3-5, one finds the stress $\sigma = P/A$, where A is the cross-sectional area of the rod. Moreover, since this normal stress is the only stress acting on the element, Fig. 10-1(b), it is the principal stress. Designating this stress as σ_y, and noting that $\sigma_x = 0$ and $\tau_{xy} = 0$, the normal and the shearing stresses acting on any inclined plane defined by the x' axis normal to this plane can be found using Eqs. 9-1 and 9-2:

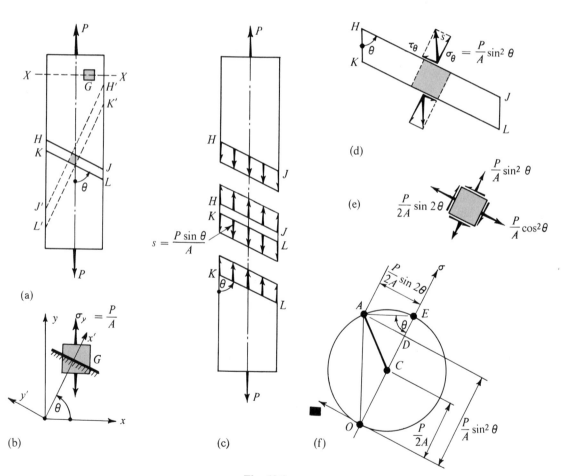

Fig. 10-1

$$\sigma_{x'} = \sigma \sin^2 \theta = \frac{\sigma}{2} - \frac{\sigma}{2} \cos 2\theta$$

$$\tau_{x'y'} = +\frac{\sigma}{2} \sin 2\theta \qquad (10\text{-}1)$$

Section 10-2
Investigation of
stress at a point

Equation 10-1 shows that the maximum or the principal stress occurs when $\theta = 90°$; the minimum principal stress occurs when $\theta = 0°$. When $\theta = 0°$ or $90°$ the shearing stress vanishes. The shearing stresses reach their maximum value of $\sigma/2$ when $\theta = \pm 45°$. For materials which are strong in tension or compression but weak in shearing strength, failures tend to be along the $45°$ planes.*

ALTERNATIVE SOLUTION

Instead of solving this problem by the formulas already developed, it is instructive to rework it from basic principles. Thus, consider the same bar, Fig. 10-1(a), and pass through it two parallel planes *HJ* and *KL* inclined at an angle θ with the vertical. Every vertical fiber in the block *HJLK*, shown isolated in Fig. 10-1(c), elongates the same amount. All these fibers are subjected to the same intensity of force. Therefore, although this has not been done before in this text, the stress s acting in a vertical direction on an inclined plane may be said to be $s = P/(A/\sin \theta)$ since $A/\sin \theta$ is the inclined cross-sectional area of the bar. This manner of expressing the stress is unusual; ordinarily it is resolved into the normal and the shearing stresses (Art. 3-2). This is done here by direct resolution of s into the components, Fig. 10-1(d), since s, σ_θ, and τ_θ all act on the same unit of area. Thus

$$\sigma_\theta = s \sin \theta = (P/A) \sin^2 \theta$$

and $$\tau_\theta = s \cos \theta = \frac{P}{A} \sin \theta \cos \theta = \frac{P}{2A} \sin 2\theta$$

These results agree with Eq. 10-1.

It is instructive to carry the above solution a step further. By isolating the block $H'J'L'K'$ (shown dashed in Fig. 10-1(a)), whose sides are perpendicular to those of the block *HJLK*, one obtains the element in Fig. 10-1(e). All normal and shearing stresses acting on this element may be determined by combining with the first solution an additional one for the block $H'J'L'K'$.

The principal stresses for the element of Fig. 10-1(e) can be obtained using Mohr's circle of stress as in Fig. 10-1(f). With the aid of trigonometry it can be shown that the maximum principal stress $\sigma_1 = P/A$ and acts normal to the line AE in the vertical direction, as to be expected. The minimum principal stress on the vertical side of the element is zero.

---

* In some materials such as concrete or duralumin failures do not take place precisely on the $45°$ planes since the normal stress existing simultaneously with the shearing stress also influences the breakdown of the material. This fact is rationalized in Mohr's theory of failure.

In matrix representation the above results mean that

$$\begin{pmatrix} 0 & 0 \\ 0 & \sigma_y \end{pmatrix} \text{ is equivalent to } \begin{pmatrix} \sigma_y \sin^2 \theta & \tfrac{1}{2} \sigma_y \sin 2\theta \\ \tfrac{1}{2} \sigma_y \sin 2\theta & \sigma_y \cos^2 \theta \end{pmatrix}$$

EXAMPLE 10-2

Determine the principal stresses occurring in a solid circular shaft subjected to a torque T, Fig. 10-2(a).

SOLUTION

In a circular shaft, the maximum shearing stress occurs in the outermost thin lamina (at but not on the outer surface) and, by Eq. 5-4, is $\tau_{max} = Tc/J$, where c is the radius of the shaft and J is its polar moment of inertia. This state of pure shearing stress is shown acting on an element in Fig. 10-2(a). However, according to Art. 9-6, a state of pure shearing stress transforms into tensile and compressive principal stresses, which are equal in magnitude to the shearing stress and act along the respective shear diagonals. Therefore the principal stresses are $\sigma_1 = +Tc/J$ and $\sigma_2 = -Tc/J$, acting in the direction shown in the figure.

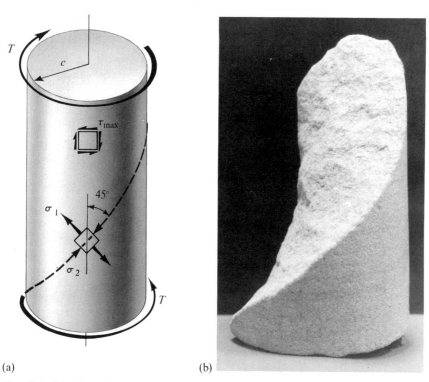

Fig. 10-2. (a) Alternative description of stresses for a shaft in torsion; (b) a core sample of sandstone after a torsion test. (Experiment by Professor D. Pirtz.)

The above stress transformation enables one to predict the type of failure that will take place in materials weak in tension. Such materials fail by tearing in a line perpendicular to the direction of σ_1. An example of such a failure for a sandstone sample is shown in Fig. 10-2(b); cast iron shafts fail in the same manner.* The failure takes place along a helix, shown in Fig. 10-2(a) by the dashed lines. Shafts made from materials weak in shearing strength, such as mild steel, break squarely across. In matrix representation the above stress transformation emphasizes that

$$\begin{pmatrix} 0 & \tau_{max} \\ \tau_{max} & 0 \end{pmatrix} \quad \text{is equivalent to} \quad \begin{pmatrix} \sigma_1 & 0 \\ 0 & \sigma_2 \end{pmatrix}$$

Section 10-2 Investigation of stress at a point

EXAMPLE 10-3

A weightless rectangular beam spans 40 in. and is loaded with a vertical downward force $P = 18.44$ kips at midspan, Fig. 10-3(a). Find the principal stresses at points A, B, C, B', and A'.

SOLUTION

At section AA' a shear of 9.22 kips and a bending moment of 92.2 kip-in. are necessary to maintain the equilibrium of the segment of the beam. These quantities with their proper sense are in Fig. 10-3(c).

The principal stress at points A and A' follows directly by applying the flexure formula, Eq. 6-4. Since the shearing stresses are distributed parabolically across the cross section of a rectangular beam, no shearing stresses act on these elements, Figs. 10-3(d) and (h).

$$\sigma_{A \text{ or } A'} = \frac{Mc}{I} = \frac{M}{S} = \frac{6M}{bh^2} = \frac{6(92.2)}{1.535(12)^2} = \pm 2.50 \text{ ksi}$$

The normal stresses acting on the elements B and B' shown in the first sketches of Figs. 10-3(e) and (g) are obtained by direct proportion from the normal stresses acting on the elements A and A' (or Eq. 6-3 could be used directly). The shearing stresses acting on both these elements are alike. Their sense on the right face of the elements agrees with the sense of the shear at the section AA' in Fig. 10-3(c). The magnitude of these shearing stresses is obtained by applying Eq. 7-6, $\tau_{xy} = VA_{fghj}\bar{y}/(It)$. For use in this equation, the area A_{fghj}, with the corresponding $\bar{y}$, is shown shaded in Fig. 10-3(b).

$$\sigma_{B \text{ or } B'} = (5.5/6)\sigma_A = \pm 2.292 \text{ ksi}$$

$$\tau_{B \text{ or } B'} = \frac{VA_{fghj}\bar{y}}{It} = -\frac{(9.22)(1.535)(0.5)(5.75)}{(1/12)(1.535)(12)^3(1.535)} = -0.12 \text{ ksi}$$

To obtain the principal stresses at B, Mohr's circle of stress is used. Its

---

* Ordinary chalk behaves similarly. This may be demonstrated in a classroom by twisting a piece of chalk to failure.

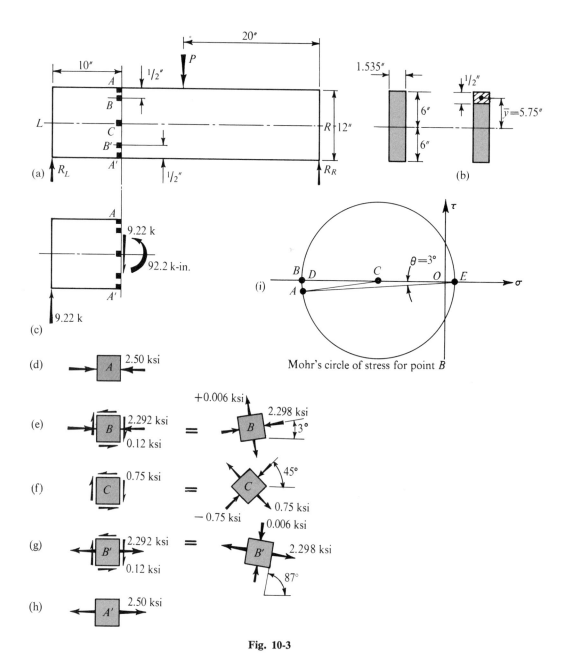

Fig. 10-3

construction is indicated in Fig. 10-3(i), and the results obtained are shown in the second sketch of Fig. 10-3(e). Note the invariance of the sum of the normal stresses, i.e., $\sigma_x + \sigma_y = \sigma_1 + \sigma_2$ or $-2.292 + 0 = -2.298 + 0.006$. A similar solution for the principal stresses at point B' yields the results indicated in the second sketch of Fig. 10-3(g).

Point C lies on the neutral axis of the beam, hence no flexural stress acts on the corresponding element, shown in the first sketch of Fig. 10-3(f). The shearing stress on the right face of the element at C acts in the same direction as the shear in Fig. 10-3(c). Its magnitude may be obtained by applying Eq. 7-6, or directly by using Eq. 7-7, i.e.

Section 10-3
Members in a state of two-dimensional stress

$$\tau_{max} = \frac{3V}{2A} = \frac{1.5(9.22)}{1.535(12)} = 0.75 \text{ ksi}$$

The pure shearing stress transformed into the principal stresses according to Art. 9-6 is shown in the second sketch of Fig. 10-3(f).

It is significant to further examine qualitatively the results obtained. For this purpose the computed principal stresses acting on the corresponding planes are shown in Figs. 10-4(a) and (b). By examining Fig. 10-4(a), the characteristic behavior of the algebraically larger (tensile) principal stress at a section of a rectangular beam may be seen. This stress progressively diminishes in magnitude from a maximum value at A' to zero at A. At the same time the corresponding directions of σ_1 gradually turn through 90°. A similar observation may be made regarding the algebraically smaller (compressive) principal stress σ_2, shown in Fig. 10-4(b).

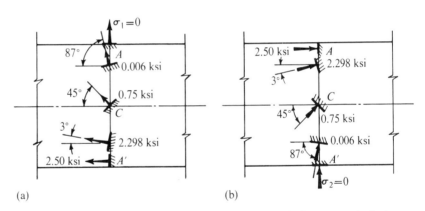

Fig. 10-4. (a) Behavior of the algebraically larger principal stress σ_1. (b) Behavior of the algebraically smaller principal stress σ_2.

10-3. MEMBERS IN A STATE OF TWO-DIMENSIONAL STRESS

Within the scope of the formulas developed in this text, bodies in a state of two-dimensional stress may be studied as was done in the preceding example. A great many points in a stressed body may be investigated for the magnitude and direction of the principal stresses. Then, to study the general behavior of the stresses, selected points may be interconnected to give a visual interpretation of the various aspects of the computed data. For example, the points of algebraically equal principal stresses, regardless of their sense, when connected, provide a "map" of stress contours. Any point lying on a stress contour has a principal stress of the same algebraic magnitude.

Chapter 10
Problems in stress analysis

Similarly, the points at which the directions of the minimum principal stresses form a constant angle with the x axis may be connected. Moreover, since the principal stresses are mutually perpendicular, the direction of the maximum principal stresses through the same points also forms a constant angle with the x axis. The line so connected is a locus of points along which the principal stresses have parallel directions. This line is called an *isoclinic line*. The adjective isoclinic is derived from two Greek words *isos* meaning equal and *klino* meaning slope or incline. Three isoclinic lines may be found by inspection in a rectangular prismatic beam subjected to a transverse load acting normal to its axis. The lines corresponding to the upper and the lower boundaries of a beam form two isoclinic lines as at the boundary the flexural stresses are the principal stresses and they act parallel to the boundaries. On the other hand, the flexural stress is zero at the neutral axis, and there only pure shearing stresses exist. These pure shearing stresses transform into principal stresses, all of which act at an angle of 45° with the axis of the beam. Hence another isoclinic line (the 45° isoclinic) is located on the axis of the beam. The other isoclinic lines are curved and are more difficult to determine.

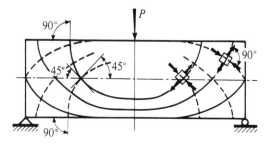

Fig. 10-5. Principal stress trajectories for a rectangular beam.

Another set of curves may be drawn for a stressed body for which the magnitude and the sense of the principal stresses are known at a great many points. A curve whose tangent is changing in direction to conform with the direction of the principal stresses is called a *principal stress trajectory* or *isostatic line*. Like the isoclinic lines, the principal stress trajectories do not connect the points of equal stresses but rather indicate the directions of the principal stresses. Since the principal stresses at any point are mutually perpendicular, the principal stress trajectories for the two principal stresses form a family of orthogonal curves.*
An example of stress trajectories for a rectangular beam loaded with a concentrated force at the midspan is in Fig. 10-5. The principal stress trajectories corresponding to the tensile stresses are shown in the figure by solid lines; those for the compressive stresses are shown dashed. The trajectory pattern (not shown) is severely disturbed at the supports and at the applied load P.

10-4. THE PHOTOELASTIC METHOD OF STRESS ANALYSIS

The state of stress in any two-dimensional stress problem may be expressed in terms of the stress contours, the isoclinic lines, and the principal stress trajectories discussed in the preceding article. Moreover, it is significant that the application of the same forces in the same manner to any two geometrically similar bodies made from different elastic materials causes the same stress distribution. The stress distribution is unaffected† by the elastic constants of a

---

* A somewhat analogous situation is found in fluid mechanics, where in "two-dimensional" fluid-flow problems the streamlines and the equipotential lines form an orthogonal system of curves—the *flow net*.
† For this to be true, strictly speaking, the bodies must be simply connected, i.e., without interior holes.

material. Therefore, to determine stresses experimentally, instead of finding the stresses in an actual member, the test specimen is prepared from any material suitable for the type of test to be performed. Glass, celluloid, and particularly certain grades of bakelite have the required optical properties for photoelastic work. In a stressed specimen made from one of these materials, the principal stresses temporarily change the optical properties of the material. This change in the optical properties may be detected and related to the principal stresses which cause it. The experimental and analytical technique necessary for the analysis of problems in this manner is known as *the photoelastic method of stress analysis*. Only a brief outline* of this method will be given here, commencing with some remarks on light.

Light travels through any given medium in a straight line at a constant velocity. For the purposes at hand, the behavior of light may be explained by considering its single ray as a series of chaotic waves which travel in a number of planes containing the ray. By restricting the vibration of the waves to a single plane, a plane polarized light is obtained. This is done by passing the light through a polarizer, which may be a suitable pile of plates, a Nicol prism, or a commercially manufactured Polaroid element. The plane of the polarizer through which the transverse vibrations of the light are allowed to pass is called the *plane of polarization*. A schematic diagram of the foregoing definitions is in Fig. 10-6(a), where a second polarizer, called the *analyzer*, is also shown. Note

Section 10-4
The photoelastic method of stress analysis

---

* For more details see M. M. Frocht, *Photoelasticity*, vols. I and II (New York: John Wiley & Sons, Inc., 1941 and 1948).

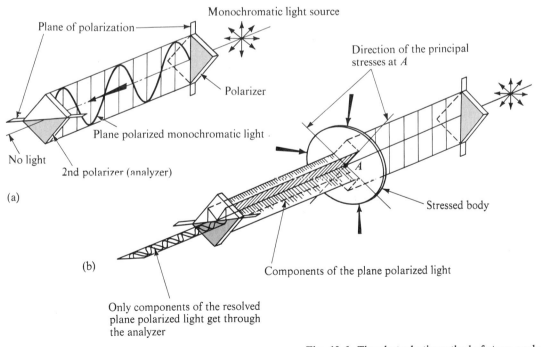

Fig. 10-6. The photoelastic method of stress analysis.

*Chapter 10
Problems in stress
analysis*

that if the planes of polarization of the two polarizers are at right angles to each other, no light gets through the analyzer. This arrangement of the analyzer with respect to the polarizer is termed *crossed*.

If the light source* used is monochromatic, i.e., of one color, the transverse vibrations of the plane polarized light are regular as the wavelength in a given medium for any one color is constant. When propagating through the same medium, these transverse vibrations are described by a sinusoidal wave of constant amplitude and frequency.

By inserting an annealed specimen made from a suitable transparent material between the polarizer and the analyzer in the arrangement shown in Fig. 10-6(a), no new phenomenon is observed. However, by stressing the specimen, the optical properties of the material change, and two phenomena take place:†

1. At each point of the stressed body the polarized light wave is resolved into two mutually perpendicular components lying in the planes of the principal stresses occurring at that point.

2. The linear velocity of each of the components of the light wave is retarded through the stressed specimen in direct proportion to the related principal stress.

These facts are the basis of the photoelastic method of stress analysis. A schematic representation of the behavior of a monochromatic, plane-polarized wave as it passes through a stressed body and an analyzer is in Fig. 10-6(b). In the stressed specimen, a plane-polarized wave is resolved into two components whose planes coincide with the planes of the principal stresses, as at point A. These components of the sinusoidal vibration leave the specimen with the same frequency but are out of phase. The latter effect is caused by the different amount of retardation of the light in the two principal planes of stress. Finally, the light waves emerging from the analyzer are again brought into the same plane since only certain of the components of the polarized light may go through the analyzer. The two monochromatic light waves which leave the analyzer vibrate out of phase in the same plane with the same frequency. Their phase difference, which is directly proportional to the difference in the principal stresses at a point such as A of the stressed body, indicates several possibilities, which may be observed on a screen placed after the analyzer. If the two waves are out of phase by a full wavelength of the light used, they reinforce each other and the brightest light is seen on the screen. For other conditions, some interference takes place between the two light waves. A complete elimination of the light occurs if the amplitudes of the two light waves are equal and are out of phase by one-half wavelength or its odd integer multiple. Therefore, since an infinite number of points in the stressed body affect the plane polarized light in a manner analogous to point A, alternate bright and dark bands become apparent on a screen. The dark bands are called *fringes*.

---

* Mercury-vapor lamps are commonly used for this purpose.
† The first phenomenon stated was discovered by Sir David Brewster in 1816. The quantitative relation was established by G. Wertheim in 1854. The modern development of photoelasticity and its engineering applications probably owes most to the two English professors, E. G. Coker and L. N. G. Filon, whose treatise on this subject was first published in 1930.

Section 10-4
The photoelastic method
of stress analysis

The greater the difference in the principal stresses, the greater the phase difference between the two light waves emerging from the analyzer. Hence, if forces are gradually applied to a specimen until the principal stresses differ sufficiently to cause a phase difference of one-half wavelength between the two light waves at some points, the first fringe appears on the screen. Then as the magnitude of the applied forces is increased, the first fringe shifts to a new position and another "higher order" fringe makes its appearance on the screen. The second fringe corresponds to the principal stresses which cause a phase difference of 1½ wavelengths. This process may be continued as long as the specimen behaves elastically; more and more fringes appear on the screen. A photograph with several fringes for a rectangular beam loaded at midspan is in Fig. 10-7. Fringes may be calibrated in a separate experiment with a bar in tension or a beam in pure flexure. The stresses for these simple members may be accurately computed. It is necessary to make the calibration specimens from the same material as that of the specimen to be investigated. With the calibration data, complex members subjected to complicated loading may be investigated. For each fringe order the difference of the principal stresses, $\sigma_1 - \sigma_2$, is known from calibration, hence fringes represent a map of the difference in the principal stresses. Moreover, since according to Eq. 9-8 the difference of the principal stresses divided by two is equal to the maximum shearing stress, the fringes also represent the loci of the principal shearing stresses.

Fig. 10-7. Fringe photograph of a rectangular beam. (Photograph by Professor R. W. Clough.)

A fringe photograph of the stressed body and calibration data are sufficient for determining the magnitude of the maximum or principal shearing stresses. The principal stress at any point of the unloaded boundary may also be obtained: At a free boundary, one of the principal stresses which acts normal to the boundary must be zero, and the fringe order is directly related to the other principal stress. Additional experimental work must be performed to determine normal stresses away from the boundaries.

One method of completing the problem consists of obtaining some very accurate measurements of the change in thickness of the stressed specimen at a number of points. These measurements, which may be designated by Δt, where t is the thickness of the specimen, are related to the principal stresses, i.e., from the generalized Hooke's law, Eq. 4-11, with $\sigma_z = 0$ one obtains

$$\Delta t = -\nu \left(\frac{\sigma_1 + \sigma_2}{E} \right) t \quad \text{or} \quad \sigma_1 + \sigma_2 = -\frac{E}{\nu t} \Delta t \quad (10\text{-}2)$$

Then from an additional experiment on the same material in simple tension where $\sigma_1 \neq 0$ and $\sigma_2 = 0$, a new calibration chart may be prepared which gives the sum of the principal stresses versus Δt. From the information obtained from these two experiments, a map of the sum of the principal stresses for the specimen investigated may be prepared. By superposing this map with the map of the differences of the principal stresses obtained from the fringe photograph, one may determine the magnitudes of the principal stresses at any point of the stressed specimen.

Additional information must be found in the picture of fringes to determine the direction of the principal stresses. This information is given by the isoclinic line—a black line corresponding to the locus of the points where the direction of one of the principal stresses in the stressed body coincides with the plane of the polarized light leaving the polarizer. Rays passing through these points in the stressed specimen are not resolved and are blacked out by the analyzer. By rotating the polarizer into several known positions and maintaining the analyzer crossed, the isoclinic lines may be determined. These lines may be difficult to distinguish from the fringes as both simultaneously appear on the screen. One method of differentiating the isoclinic lines from the fringes uses white instead of monochromatic light. Using white light makes isoclinics appear black, but the fringes are colored and contain all the visible spectral colors of the white light. On the other hand, to eliminate the isoclinic lines, which are undesirable for fringe photographs, two quarter-wave plates may be inserted into the optical system. Quarter-wave plates resolve the plane polarized light into two mutually perpendicular components; one of these is a quarter wave out of phase with the other. Combination of these components results in a "circularly polarized light." One of the quarter-wave plates is placed between the polarizer and the specimen and the other between the specimen and the analyzer. The fringe photograph in Fig. 10-7 was obtained by using this method.

With the aid of analytical methods, a sequence of isoclinic lines and fringe photographs is sufficient to solve the photoelastic problem without finding the sum of the principal stresses experimentally. These procedures are very detailed and laborious, and the reader is referred to books on photoelasticity for further information.

The photoelastic method of stress analysis is very versatile and has been used to solve numerous problems. Nearly all solutions for the stress-concentration factors have been established by photoelasticity. The inaccuracy of the elementary formulas at concentrated forces is clearly brought out by the fringe photographs. For example, in Fig. 10-7, according to the elementary formulas, the fringes in the upper half of the beam should be like those in the lower half. Also note the local disturbance of the stresses at the supports in the same photograph.

The photoelastic method is best adapted to two-dimensional stress problems. Three-dimensional problems have also been analyzed by specialized techniques. The extension of the method to inelastic or plastic problems remains for the present unsolved.

During the last decade the rapid development of another optical procedure, the Moiré-fringe method, has met with much success. This method will not be described here.*

10-5. THIN SHELLS OF REVOLUTION

In numerous applications thin curved sheets are used as structural components. Examples of such construction are pressure vessels, roof domes, containers for liquids, missiles, airplane wings, etc. These structures are usually called shells. A very large and highly developed

---

* See for example, P. S. Theocaris, "Moiré Fringes: A Powerful Measuring Device," *Applied Mechanics Review*, **15**, (May 1962), 333.

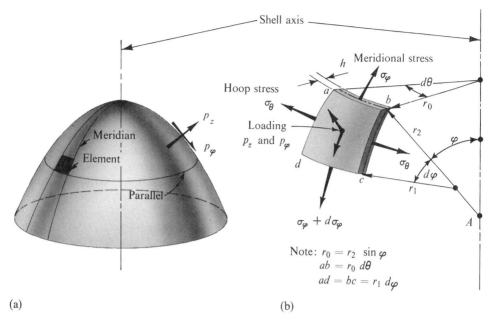

Fig. 10-8. Thin shell of revolution.

method evolved for the stress analysis of shells. Here attention will be confined to the analysis of shells of revolution subjected to distributed loading such as the weight of the shell or internal pressure.

An example of a shell of revolution is in Fig. 10-8(a); other examples are a sphere, a cylinder, and a cone. Such shells are generated by rotating a plane curve, called the meridian, about an axis lying in the plane of the curve. The meridional curve may have any arbitrarily varying radius of curvature r_1, Fig. 10-8(b). To completely define the geometry of a shell surface another radius of curvature r_2 is required. The center of curvature of this radius, as can be shown by the methods of differential geometry, is located on the shell axis. This radius generates the shell surface in the direction perpendicular to the direction of the tangent to the meridian. In most cases, however, it is more convenient to work with another radius of curvature r_o, which lies in a plane perpendicular to the shell axis. The two radii r_o and r_2 are related since $r_o = r_2 \sin \varphi$, see Fig. 10-8(b). With known radii r_o and r_1, and the subtended infinitesimal angles $d\theta$ and $d\varphi$, the infinitesimal arc lengths of the curvilinear shell element become $r_o\, d\theta$ and $r_1\, d\varphi$, respectively.

In this analysis it will be assumed that the shell thickness h is negligible compared with r_1 and r_2, i.e., no distinction will be made between the inner, mean, and outer radii of the shell. It will be further assumed that the shell is free to deform under load and that the deformations are small. These assumptions permit treating the shell as a membrane, where only uniform, in-plane stresses exist. In membranes no bending moments and transverse shears of significant magnitudes are said to develop. A membrane is a two-dimensional analog of a flexible string, but it can resist compressive stresses. These assumptions are known to be quite reasonable for thin shells in regions away from external constraints.

Here the analysis is confined to stresses in symmetrically loaded shells of revolution. For this case the loading per unit of surface area can consist of loading p_z acting normal to the shell surface and loading p_φ applied tangentially to the meridian. For a given angle φ these quantities remain constant along a parallel, see Fig. 10-8.

For axisymmetrical loading conditions, by virtue of symmetry, the hoop or the tangential stresses σ_θ are constant on either side of an infinitesimal element, as in Fig. 10-8(b). This is not generally true for the meridional stress σ_φ. A possible in-plane shearing stress $\tau_{\theta\varphi}$ (not shown) vanishes because of the symmetry of the problem.

10-6. EQUILIBRIUM EQUATIONS FOR THIN SHELLS OF REVOLUTION

In axisymmetrical problems of shells of revolution there are only two unknown stresses, σ_φ and σ_θ, and the governing equations for these are established from two equilibrium conditions. One of the equilibrium equations is obtained by summing the forces in the direction normal to the tangent plane of the infinitesimal element. In this summation the components of forces acting on the edges of the shell element and the force caused by the applied loading p_z are involved. These forces are shown in three consecutive diagrams of Fig. 10-9.

Since the cross-sectional area along each of the two vertical edges of an infinitesimal element is $hr_1\,d\varphi$ and the hoop stress acting on these areas is σ_θ, the horizontal forces acting on the infinitesimal hoop are $\sigma_\theta hr_1\,d\varphi$ as in Fig. 10-9(a). These two hoop forces each inclined at an angle of $d\theta/2$ with the tangent plane produce a horizontal component of $2\sigma_\theta hr_1\,d\varphi\,(d\theta/2)$ acting toward the shell axis. This horizontal force must be multiplied by $\sin\varphi$ to determine the normal force component acting toward point A as in Fig. 10-9(a).

The normal force component caused by the meridional stresses is found similarly, Fig. 10-9(b). Although the meridional stresses as well as the length of the edges of the element may change from top to bottom, in this projection these changes are infinitesimal quantities of higher order than the other quantities involved and can be neglected. Finally, by taking the surface area of the infinitesimal element as $r_o\,d\theta\,r_1\,d\varphi$ the resultant due to p_z acting on this surface can be found as in Fig. 10-9(c). The tangential load p_φ gives no force component in the direction considered, and for this reason it is not included in these diagrams.

For the forces considered above, from $\Sigma F_n = 0$, one has

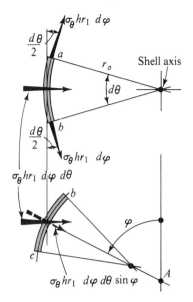

(a) Normal component of hoop force

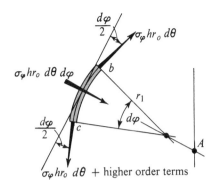

(b) Normal component of meridional force

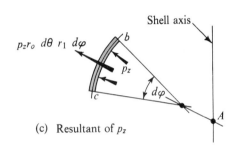

(c) Resultant of p_z

Fig. 10-9

$$\sigma_\theta h r_1\, d\varphi\, d\theta \sin\varphi + \sigma_\varphi h r_o\, d\theta\, d\varphi - p_z r_o\, d\theta r_1\, d\varphi = 0$$

Recalling that $r_o = r_2 \sin\varphi$ and simplifying, one obtains one of the basic relations for membrane shells:

$$\frac{\sigma_\varphi}{r_1} + \frac{\sigma_\theta}{r_2} = \frac{p_z}{h} \qquad (10\text{-}3)$$

Here, as before, h is the thickness of the shell.

A second equation for determining the unknown membrane stresses can be found by considering a free body of the whole shell above a parallel circle as in Fig. 10-10(a). Here the vertical resultant R due to loading p_z and p_φ is maintained in equilibrium by the vertical component of the force developed by the meridional stress σ_φ.

Keeping in mind that the circumference of the circle at the section through the shell is $2\pi r_o$, from $\Sigma F_y = 0$ one has

$$2\pi r_o h \sigma_\varphi \sin\varphi - R = 0$$

$$\sigma_\varphi = \frac{R}{2\pi r_o h \sin\varphi} \qquad (10\text{-}4)$$

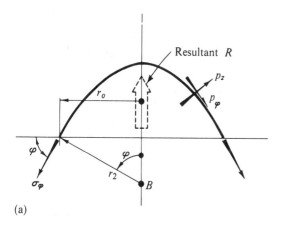
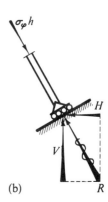

Fig. 10-10. Equilibrium diagrams: (a) segment of a shell, (b) compressive force at a boundary.

Equations 10-3 and 10-4 are sufficient for determining the membrane stresses in axisymmetrically loaded shells of revolution. Negative answers in the results indicated compressive stresses. Note that since no properties of materials were used in establishing these relations, they are not restricted to elastic materials.

The supporting conditions for membrane shells must be as in Fig. 10-10(b) providing the support for the tangential force of $\sigma_\varphi h$ lb per inch. A vertical component alone does not fulfill this condition. Unless a ring to resist the horizontal component of $\sigma_\varphi h$ is provided, severe bending stresses develop in the shell. The introduction of such a ring, although usually desirable, unfortunately causes significant local bending stresses. The detailed treatment of this problem is beyond the scope of the book.

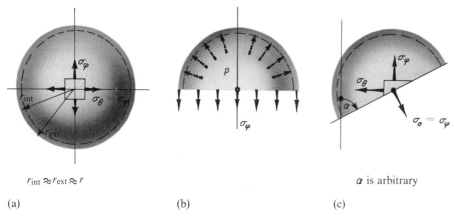

Fig. 10-11. Spherical pressure vessel.

EXAMPLE 10-4

Determine the stresses caused by the internal pressure p in a thin-walled spherical shell.

Section 10-6 Equilibrium equations for thin shells of revolution

SOLUTION

For a thin-walled vessel the internal radius r_{int} and the external radius r_{ext} may be taken as equal. For a sphere $r_1 = r_2$; therefore

$$r_{\text{int}} = r_{\text{ext}} = r_1 = r_2 = r$$

In order to apply Eq. 10-4, the hemisphere shown in Fig. 10-11(b) is isolated from the sphere by passing a section through the center of the sphere. The applied load $p_z = p$, and for $\varphi = 90°$, one has $r_o = r$; the resultant $R = \pi r^2 p$. Since for a sphere any section through the center yields the same kind of free body, $\sigma_\varphi = \sigma_\theta = \sigma_\alpha$ (see Fig. 10-11(c)). Therefore

$$\sigma_\varphi = \sigma_\theta = pr/(2h) \qquad (10\text{-}5)$$

After determining σ_φ, since $r_1 = r_2 = r$, the equality of σ_φ with σ_θ could also be established from Eq. 10-3.

The stresses σ_φ and σ_θ are the principal stresses. For an element of a sphere viewed from the outside, Mohr's circle of stress degenerates into a point. The maximum shearing stress is located by looking at the element from the side. (See Art. 9-9 and Fig. 9-13 for a three-dimensional state of stress.)

If contents are of negligible weight, a sphere is an ideal shape for a closed pressure vessel.

EXAMPLE 10-5

Determine the stresses caused by the internal pressure p in a cylindrical vessel, Fig. 10-12.

SOLUTION

For a cylindrical vessel $r_1 = \infty$ and $r_2 = r$. Hence, with $p_z = p$, from Eq. 10-3,

$$\sigma_\theta = pr/h \qquad (10\text{-}6)$$

which is twice as large as the stress found for a comparable sphere. This fact may be understood better by considering the cylindrical element in Fig. 10-12(b). Here, in resisting the internal pressure p, unlike the doubly curved shell, the longitudinal stress σ_φ develops no inward force component.

The magnitude of the longitudinal stress σ_φ follows by applying Eq. 10-4. From Fig. 10-12(c) it is seen that the situation is identical to that of a sphere; hence for a cylinder

$$\sigma_\varphi = pr/(2h) \qquad (10\text{-}7)$$

Both σ_θ and σ_φ are the principal stresses as conditions of symmetry

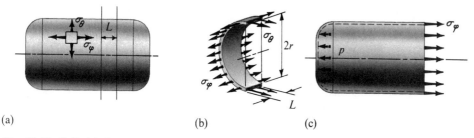

Fig. 10-12. Cylindrical pressure vessel.

exclude the existence of shearing stresses in the planes of the sections considered.

EXAMPLE 10-6

Determine the membrane stresses in a spherical dome of radius a caused by its own weight of q lb per square inch, Fig. 10-13.

SOLUTION

The surface area of the dome above an arbitrary angle φ is found first. Multiplying this area by q yields the resultant R which acts downward—i.e., R must be taken as negative. Then from Eq. 10-4, the meridional stress σ_φ can be determined. Upon substituting the known σ_φ into Eq. 10-3 and simplifying, one may evaluate the remaining stress σ_θ. In making these calculations recall that $r_o = a \sin \varphi$, and note that $p_z = -q \cos \varphi$.

$$R = -q \int_0^\varphi (2\pi r_o) a \, d\varphi = -2\pi a^2 q \int_0^\varphi \sin \varphi \, d\varphi = -2\pi a^2 (1 - \cos \varphi) q$$

$$\sigma_\varphi = \frac{R}{2\pi r_o h \sin \varphi} = -\frac{aq}{h(1 + \cos \varphi)} \tag{10-8}$$

$$\sigma_\theta = \frac{p_z a}{h} - \sigma_\varphi = \frac{aq}{h}\left(\frac{1}{1 + \cos \varphi} - \cos \varphi\right) \tag{10-9}$$

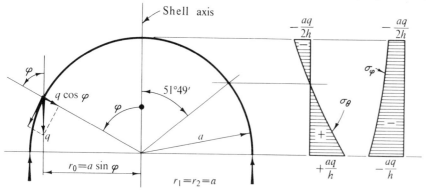

Fig. 10-13

These results are plotted in Fig. 10-13. It is interesting to note the change in sign of σ_θ, which means that up to $\varphi = 51°49'$ no hoop tension develops in the dome. This is an important fact for materials weak in tension and accounts for the successful use of domes in roof construction during the medieval era.

*Section 10-7
Remarks on
thin-walled pressure
vessels*

10-7. REMARKS ON THIN-WALLED PRESSURE VESSELS

The state of stress for an element of a thin-walled pressure vessel as given by Eqs. 10-3 through 10-7 is considered to be biaxial, although the internal pressure acting normal to the wall causes a local compressive stress equal to the internal pressure. Actually a state of triaxial stress exists on the inside of the vessel. However, for thin-walled pressure vessels the third stress is much smaller than σ_φ and σ_θ and for this reason can be neglected. The extent of the error committed in making this approximation can be rigorously determined from the general theory which applies to thick as well as thin cylinders. It can be shown* from solutions based on the theory of elasticity that the wall thickness may reach one-tenth the internal radius and the error in applying the formulas for thin-walled vessels will still be small. For internal pressure, if in Eq. 10-6 the internal radius is used for r, the average hoop stress is always correct.

A discontinuity of the membrane action of a shell occurs at all points of external restraint or at junctures of shell elements possessing different stiffness characteristics. For example, this situation occurs at the juncture of the cylindrical portion of a pressure vessel with the ends. Under the action of the internal pressure, the cylinder tends to expand as shown diagrammatically by the dashed lines in Fig. 10-14, while the ends tend to expand a different amount because of differences in stress. This incompatibility of deformations causes local bending and shearing stresses in the neighborhood of the joint since there must be physical continuity between the ends and the cylindrical wall. For this reason, properly curved ends must be used for pressure vessels. Flat ends are undesirable. The ASME Unfired Pressure Vessel Code gives practical information on the design of ends; the necessary theory is beyond the scope of this text.

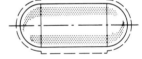

Fig. 10-14. Dashed lines showing the tendency (exaggerated) for the cylinder and the ends to expand a different amount under the action of internal pressure.

A majority of pressure vessels are manufactured from separate curved sheets which are joined. A common method of accomplishing this is to arc-weld the abutting material. Grooves into which the welding metal is deposited are prepared in a number of different ways, depending on the thickness of the plates. Calculations for the joints are made from an allowable weld tensile stress. These are usually expressed as a certain percentage of the strength of the original solid plate of the parent material. This percentage factor varies greatly, depending on the workmanship.

---

* S. Timoshenko and J. N. Goodier, *Theory of Elasticity* (2nd. ed.) (New York: McGraw-Hill Book Company, 1951), p. 60.

*Chapter 10
Problems in stress
analysis*

For ordinary work, a 20 per cent reduction from the allowable stress for the solid plate to that for the weld may be used. For this factor the efficiency of the joint is said to be 80 per cent. On high-grade work, some of the specifications allow 100 per cent efficiency for the welded joint.

In conclusion it must be emphasized that the formulas derived for thin-walled pressure vessels should be used only for cases of internal pressure. If a vessel, such as a vacuum tank or a submarine, is to be designed for external pressure, instability (buckling) of the walls may occur and stress calculations based on the above formulas are meaningless.

PART B
DESIGN OF MEMBERS TO MEET STRENGTH REQUIREMENTS

10-8. GENERAL REMARKS

The design of structural systems is based on many considerations. Functional requirements certainly play the key role in the selection and arrangement of structural members and machine parts. The discussion of these questions is beyond the scope of this text. If, however, the geometrical configuration for a member is given and the applied loads are specified, the procedures developed in the previous chapters enable one to select members of adequate size to meet strength requirements in numerous situations. Since by this time the reader has become familiar with stress transformations and yield and fracture criteria, he should be able to develop a deeper appreciation of the design of members to meet strength requirements. Some of the discussion presented below is in the nature of a review.

10-9. DESIGN OF AXIALLY LOADED MEMBERS

Axially loaded tensile members and short compression blocks* are designed by using Eq. 3-7, i.e., $A = P/\sigma_{\text{allow}}$. The critical section for an axially loaded member occurs at a section of minimum cross-sectional area, where the stress is a maximum. If an abrupt discontinuity in the cross-sectional area is imposed by the design requirements, the use of Eq. 4-34, $\sigma_{\max} = KP/A$, is appropriate. The use of the latter formula is necessary in the design of machine parts to account for the local stress concentrations where fatigue failure may start.

Beside the normal stresses, given by the above equations, shearing stresses act on inclined planes, even in a state of uniaxial stress. Hence, if

---

* Slender compression members are discussed in Chapter 14.

a material is weak in shearing strength in comparison with its strength in tension or compression, it will fail along planes approximating the planes of the maximum shearing stress. For example, concrete or cast iron members in uniaxial compression and duralumin members in uniaxial tension fail on planes inclined to the direction of the load. This observation is accounted for by Mohr's theory of failure, which was briefly discussed in Art. 9-20.

Regardless of the type of failure that may actually take place, the allowable stress for design of axially loaded members is customarily based on the normal stress. This design procedure is consistent. The maximum normal stress which a material can withstand at the failure point is directly related to the ultimate strength of the material. Hence, although the actual break may occur on an inclined plane, the maximum normal stress may be considered as the ultimate normal stress.

10-10. DESIGN OF TORSION MEMBERS

The pertinent formulas for the design of torsion members were established in Chapter 5. For circular shafts, the solution of Eq. 5-8, $J/c = T/\tau_{max}$, at a critical section gives the parameter J/c required to provide a member of adequate strength. As shafts are mainly used as parts of machines, Eq. 5-17, $\tau_{max} = KTc/J$ should be used in most cases. This equation, 5-17, with the stress-concentration factor K, takes care of the high local shearing stress at the changes of the cross-sectional area.

Most torsion members are designed by selecting an allowable shearing stress, which is substituted for τ_{max} in Eqs. 5-8 or 5-17. This amounts to a direct use of the maximum shear theory of failure. However, it is well to bear in mind that a state of pure shearing stress, which occurs in torsion, may be transformed into the principal stresses by rotating an element through 45° (Art. 9-6). In some materials, failure may be caused by one of these principal stresses. For example, a member made of cast iron, a material strong in compression but weaker in tension than in shear, fails in tension.

10-11. DESIGN CRITERIA FOR PRISMATIC BEAMS

Section 10-11 Design criteria for prismatic beams

In a beam subjected to pure bending, the critical section is the section where the greatest bending moment occurs. By assigning an allowable stress, one may determine the section modulus of such a beam using Eq. 6-8, $S = M/\sigma_{max}$. If the required section modulus is known, a beam of correct proportions to be of adequate strength may be selected. However, if a beam resists shear in addition to bending, its design becomes slightly more involved.

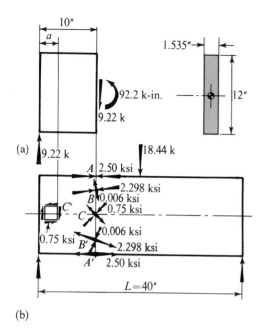

Fig. 10-15

Consider the prismatic rectangular beam of Example 10-3 at a section 10 in. from the left support, where the beam transmits a bending moment and a shear, Fig. 10-15(a). The principal stresses at points A, B, C, B', and A' at this section were found before and are reproduced in Fig. 10-15(b). If this section were the critical section, the design of this beam based on the maximum normal stress theory would be governed by the stresses at the extreme fibers as no other stresses exceed these stresses. For a prismatic beam, these stresses depend only on the magnitude of the bending moment and are largest at a section where the maximum bending moment occurs. Therefore, in ordinary design it is not necessary to perform the combined stress analysis for interior points. In the example considered, the maximum bending moment is at the middle of the span. The foregoing may be generalized into a basic rule for the design of beams: A critical section for a prismatic beam carrying transverse forces acting normal to its axis occurs where the bending moment reaches its absolute maximum* value.

The above criterion for the design of prismatic beams is incomplete. In some cases, the shearing stresses caused by the shear at a section may control the design. In the example considered, Fig. 10-15, the magnitude of the shear remains constant at every section through the beam. At a small distance a from the right support, the maximum shear is still 9.22 kips, and the bending moment, $9.22a$ kip-in., is small. The maximum shearing stress at the neutral axis corresponding to the shear of 9.22 kips is the same at point C' as it is at point C.† Therefore, since in a general problem the bending stresses may be small, they may not control the selection of a beam, and another critical section for any prismatic beam occurs where the shear is a maximum. In applying this criterion it is customary to work directly with the maximum shearing stress that may be obtained from Eq. 7-6, $\tau = VQ/(It)$, and not to transform $\tau_{\max}$ so found into the principal stresses. For rectangular and I beams, the maximum shearing stress given by Eq. 7-6 reduces to Eqs. 7-7 and 7-9, $\tau_{\max} = 3V/2A$

---

* For cross sections without two axes of symmetry, such as T beams, made from material which has different properties in tension and in compression, the largest moments of both senses (positive and negative) must be examined. Under some circumstances, a smaller bending moment of one sense may cause a more critical stress than a larger moment of another sense. The section at which the extreme fiber stress of either sign in relation to the respective allowable stress is highest is the critical section.

† At point C, the maximum shearing stresses are shown transformed into the principal stresses.

and $(\tau_{max})_{approx} = V/A_{web}$, respectively. With the exception of brittle materials, the allowable shearing stress is less than the allowable bending stress. This is consistent with the known yield and fracture criteria (see Art. 9-20).

Usually the bending stresses control the selection of a beam. Only in beams spanning a short distance does shear control the design. For small lengths of beams, the applied forces and reactions have small moment arms, and the required resisting bending moments are small. On the other hand, the shearing forces may be large if the applied forces are large.

Section 10-11
Design criteria for prismatic beams

The two criteria for the design of beams are accurate if the two critical sections are in different locations. However, in some instances the maximum bending moment and the maximum shear occur at the same section through the beam. In such situations, higher combined stresses than σ_{max} and τ_{max}, as given by Eqs. 6-8 and 7-6, may exist at the interior points. For example, consider an I beam of negligible weight which carries a force P at the middle of the span, Fig. 10-16(a). The maximum bending moment occurs at the midspan. Except for sign, the shear is the same on either side of the applied force. At a section just to the right or just to the left of the applied force, the maximum moment and the maximum shear occur simultaneously. A section just to the left of P, with the corresponding system of forces acting on it, is shown in Fig. 10-16(a). For this section, it can be shown that the stresses at the extreme fibers are 2.50 ksi, and the principal stresses at the juncture of the web with the flanges, neglecting stress concentrations, are ±2.81 ksi and ±0.51 ksi, acting as shown in Fig.

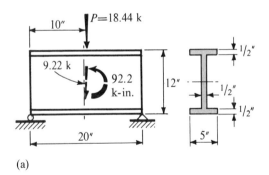
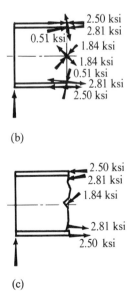

Fig. 10-16

10-16(b) and (c). As usual, local disturbance of stresses in the neighborhood of the applied force P is neglected. From this example it is seen that the maximum normal stress does not always occur at the extreme fibers. Nevertheless, only the extreme fiber stresses and the shearing stresses at the neutral axis are investigated in ordinary design. In the design codes, the allowable stresses are presumably set low enough so that an adequate factor of safety remains even if the higher combined stresses are disregarded. Also note that, for the same applied force, by increasing the span, the flexural stresses rapidly increase, and the shearing stresses remain constant. In most cases, the flexural stresses are dominant, and the extreme fiber stress is the maximum normal stress. Only for very short beams and unusual arrangements need the combined stress analysis be performed.

10-12. DESIGN OF PRISMATIC BEAMS

The design of prismatic members is controlled by the maximum stresses developed at the critical sections. As pointed out in the preceding article, one critical section occurs where the bending moment is a maximum, the other where the shear is a maximum. For determining the location of these critical sections, shear and moment diagrams are very useful. The absolute* maximum value of the moment is used in design, whether positive or negative. Likewise, the absolute maximum shear ordinate is the significant ordinate. For example, consider a simple beam with a concentrated load, as in Fig. 10-17. The shear diagram, neglecting the weight of the beam, is shown in Fig. 10-17(a) as it is ordinarily constructed by assuming the applied force concentrated at a point. The shear diagram as it more nearly exists is shown in Fig. 10-17(b). Here an allowance is made for the width of the applied force and reactions, assuming them to be uniformly distributed. The assumption of concentrated forces merely straightens the oblique shear lines. In either case, the design shear value is the greater of the positive or negative ordinates and is not the full value of the applied force.

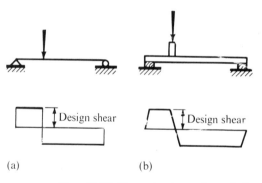

Fig. 10-17. Determination of a design ordinate from the shear diagram.

The allowable stresses to be used in design are prescribed by various authorities. In most cases the designer must follow a code depending on the location of the installation. In different codes even for the same material and the same use the allowable stresses differ. The allowable stresses in bending and in shear are different, with the allowable shearing stresses usually being lower.

---

* This is not always true for materials which have different properties in tension and compression. See footnote, p. 354.

Sometimes the design of beams is based on their ultimate (plastic) moment capacity. (See Art. 6-8 and Eq. 6-13, which defines the plastic section modulus of a section.) In such problems the assumed design loads are multiplied by a load factor, which defines the ultimate load the beam must carry. For compact, statically determinate beams AISC Code (1963) specifies a factor of 1.70. This means that the collapse of a beam would occur after the design loads are increased by a factor of 1.70. Therefore, the load factor is analogous to the factor of safety in elastic stress analysis. Plastic or limit analysis of beams will be considered further in Chapter 12.

Section 10-12
Design of prismatic beams

In the elastic design, after the critical values of moment and shear are determined and the allowable stresses are selected, the beam is usually chosen first to resist a maximum moment. Then the beam is checked for shearing stress. As most beams are governed by flexural stresses, this procedure is convenient. However, in some cases, particularly in timber (and concrete) design, the shearing stress frequently controls the dimensions of the cross section.

Usually there are several types or sizes of commercially available members that may be used for a given beam. Unless specific size limitations are placed on the beam, the lightest member is used for economy. The procedure of selecting a member is a trial-and-error process.

It should also be noted that some beams must be selected on the basis of allowable deflections. This topic will be treated in the next chapter.

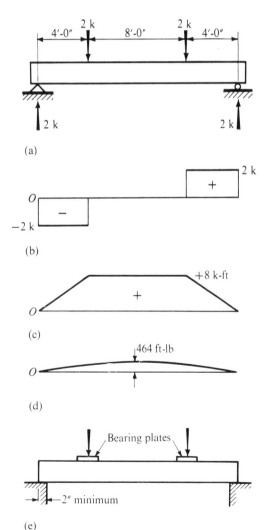

Fig. 10-18

EXAMPLE 10-7

Select a Douglas Fir beam of rectangular cross section to carry two concentrated forces as shown in Fig. 10-18(a). The allowable stress in bending is 1,200 psi, in shear 100 psi, and in bearing perpendicular to the grain of the wood 200 psi.

SOLUTION

Shear and moment diagrams for the applied forces are prepared first and are shown, respectively, in Figs. 10-18(b) and (c). From Fig. 10-18(c) it is seen that $M_{max} = 8$ kip-ft.

From Eq. 6-8

$$S = \frac{M}{\sigma_{allow}} = \frac{8{,}000(12)}{1{,}200} = 80 \text{ in.}^3$$

By arbitrarily assuming that the depth h of the

beam is to be two times greater than its width b, from Eq. 6-9,

$$S = \frac{bh^2}{6} = \frac{h^3}{12} = 80 \qquad \text{hence} \qquad h = 9.86 \text{ in. and } b = 4.93 \text{ in.}$$

From Table 10 in the Appendix, a surfaced 6-in.-by-10-in. beam is seen to fulfill this requirement. The actual size of this beam is $5\frac{1}{2}$ in. by $9\frac{1}{2}$ in., and its section modulus is $S = 82.7$ in.³ For this beam, from Eq. 7-7

$$\tau_{max} = \frac{3V}{2A} = \frac{3(2,000)}{2(5.5)(9.5)} = 57.3 \text{ psi}$$

This stress is within the allowable limit. Hence the beam is satisfactory. Note that other proportions of the beam can be used; a more direct method of design is to find a beam of size corresponding to that of the wanted section modulus directly from Table 10.

The above analysis was made without regard for the weight of the beam, which initially was unknown. (Experienced designers usually make an allowance for the weight of the beam at the outset.) However, this may be accounted for now. Assuming that wood weighs 40 lb per cubic foot the beam selected weighs 14.5 lb per lineal foot. This uniformly distributed load causes a parabolic bending-moment diagram, shown in Fig. 10-18(d), where the maximum ordinate is $p_o L^2/8 = 14.5(16)^2/8 = 464$ ft-lb (see Example 2-6). This bending-moment diagram should be added to the moment diagram caused by the applied forces. Inspection of these diagrams shows that the maximum bending moment due to both causes is $464 + 8,000 = 8,464$ ft-lb. Hence the required section modulus actually is $S = M/\sigma_{allow} = 8,464(12)/(1,200) = 84.64$ in.³ The surfaced 6-in.-by-10-in. beam originally selected provides an S of 82.7 in.³, which is about $2\frac{1}{2}$ per cent below the required value. Under most circumstances this would be considered satisfactory.

In actual construction, beams are not supported as in Fig. 10-18(a). Wood may be crushed by the supports or the applied concentrated forces. For this reason an adequate bearing area must be provided at the supports and at the applied forces. Assuming that both reactions and the applied forces are 2 kips each, i.e., by neglecting the weight of the beam, it is found, by Eq. 3-7, that the required bearing area at each concentrated force is

$$A = \frac{P}{\sigma_{allow}} = \frac{2,000}{200} = 10 \text{ in.}^2$$

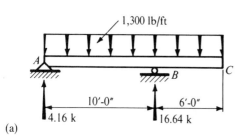

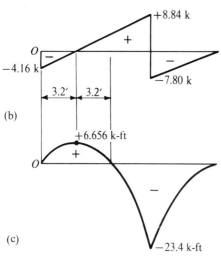

Fig. 10-19

These areas may be provided by specifying that the ends of the beam rest on at least 2-in.-by-5.5-in. (11-in.²) pads; and at the concentrated forces 3.5-in.-by-3.5-in. (12.2-in.²) steel washers can be used.

Section 10-12 Design of prismatic beams

EXAMPLE 10-8

Select an *I* beam or a wide-flange steel beam to support the load in Fig. 10-19(a). Given, $\sigma_{\text{allow}} = 24{,}000$ psi, $\tau_{\text{allow}} = 14{,}500$ psi.

SOLUTION

The shear and the bending-moment diagrams for the loaded beam are shown in Fig. 10-19(b) and (c), respectively. The maximum moment is 23.4 kip-ft.

From Eq. 6-8
$$S = \frac{(23.4)12}{24} = 11.6 \text{ in.}^3$$

Examination of Tables 3 and 4 in the Appendix shows that this requirement for the section modulus is met by a 7-in. *I* beam weighing 20.0 lb per foot. ($S = 12.0$ in.³) However, lighter members, such as an 8-in. *I* beam weighing 18.4 lb per foot ($S = 14.2$ in.³) and an 8-in. wide-flange section weighing 17 lb per foot ($S = 14.1$ in.³) can also be used. For weight economy the 8 WF 17 section will be used. The weight of this beam is very small in comparison with the applied load and so is neglected.

From Fig. 10-19(b), $V_{\max} = 8.84$ kips. Hence, from Eq. 7-9,

$$(\tau_{\max})_{\text{approx}} = \frac{V}{A_{\text{web}}} = \frac{8{,}840}{(0.23)8} = 4{,}800 \text{ psi}$$

This stress is within the allowable value, and the beam selected is satisfactory.

At the supports or at concentrated loads, *I* and wide-flange beams should be checked for crippling of the webs. This phenomenon is illustrated at the bottom of Fig. 10-20(a). Crippling of the webs is more critical for members with thin webs than direct bearing of the flanges, which may be investigated as in the preceding problem. To preclude crippling, a design rule is specified by the AISC. It states that the direct stress on area, $(a + k)t$ at the ends or $(a_1 + 2k)t$ at the interior points, must not exceed $0.75\sigma_{\text{yp}}$. In these expressions, a and a_1 are the respective lengths of bearing of the applied forces at exterior or interior portions of a beam, Fig. 10-20(b); t is the thickness of the web; and k is the distance from the outer face of the flange to the toe of the web fillet. The values of k and t are tabulated in manufacturers' catalogues.

For the above problem assuming $\sigma_{\text{yp}} = 36$ ksi, the minimum widths of the supports, according to the above rule, are as follows:

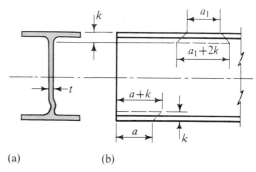

Fig. 10-20

At support A:

$27(a + k)t = 4.16$ or $27(a + 5/8)(0.23) = 4.16$ $a = 0.05$ in.

At support B:

$27(a_1 + 2k)t = 16.64$ or $27(a_1 + 5/4)(0.23) = 16.64$ $a_1 = 1.43$ in.

The preceding two examples illustrate the design of beams whose cross sections have two axes of symmetry. In both cases the bending moments controlled the design, and, since this is usually true, it is significant to note which members are efficient in flexure. A concentration of as much material as possible away from the neutral axis results in the best sections for resisting flexure, Fig. 10-21(a). Material concentrated near the outside fibers works at a high stress. For this reason, I sections, which approximate this requirement, are widely used in practice.

The above statements apply for materials which have nearly equal properties in tension and compression. If this is not the case, a deliberate shift of the neutral axis from the mid-height position is desirable. This accounts for the wide use of T and channel sections for cast iron beams (see Example 6-5).

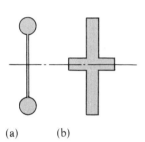

Fig. 10-21. (a) Efficient (b) inefficient sections in flexure.

Finally, two other items warrant particular attention in the design of beams. In many cases, the loads for which a beam is designed are transient in character. They may be placed on the beam all at once, piecemeal, or in different locations. The loads, which are not a part of the "dead weight" of the structure itself, are called *live loads*. Live loads must be so placed as to cause the highest possible stresses in a beam. In many cases the placement may be determined by inspection. For example, in a simple beam with a single moving load, the placement of the load at mid-span causes the largest bending moment, but placing the same load at the support causes the greatest shear. For most building work, the live load, which supposedly provides for the most severe expected loading condition is specified in building codes on the basis of so many pounds per square foot of floor area. Multiplying this live load by the spacing of parallel beams gives the *uniformly distributed live load* per unit length of the beam. For design purposes, this load is added to the dead weight of the construction.

The second item pertains to the lateral instability of beams. The flanges of a beam, if not held, may be so narrow in relation to the span that the beam may buckle sideways and collapse. The qualitative aspect of this problem was discussed in Art. 6-2. Analytical treatment of such problems is beyond the scope of this book.

10-13. DESIGN OF NONPRISMATIC BEAMS

It should be apparent from the preceding discussion that the selection of a prismatic beam is based only on the stresses at the critical sections. At all other sections of the beam the stresses will be below the allowable level. Therefore the potential capacity of a given material is not fully utilized. This situation may be improved by designing a beam of variable cross section, i.e., by making the beam nonprismatic. Since flexural stresses control the design of most beams as has been shown, the cross sections may everywhere be made just strong enough to resist the corresponding moment. Such beams are called *beams of constant strength*. Shear governs the design at sections through these beams where the bending moment is small.

*Section 10-13
Design of
nonprismatic beams*

EXAMPLE 10-9

Design a cantilever of constant strength for resisting a concentrated force applied at the end. Neglect the weight of the beam.

SOLUTION

A cantilever with a concentrated force applied at the end is in Fig. 10-22(a); the corresponding moment diagram is plotted in Fig. 10-22(b). Basing

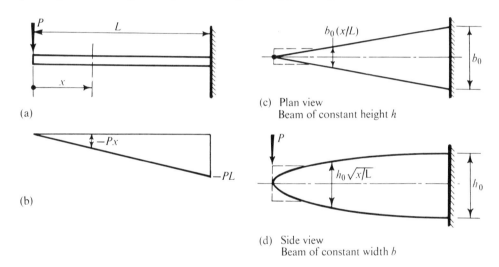

Fig. 10-22

the design on the bending moment, the required section modulus at an arbitrary section is given by Eq. 6-8:

$$S = \frac{M}{\sigma_{\text{allow}}} = \frac{Px}{\sigma_{\text{allow}}}$$

A great many cross-sectional areas satisfy this requirement; so, first, it will be assumed that the beam will be of rectangular cross section and of constant height h. The section modulus for this beam is given by Eq. 6-9 as $bh^2/6 = S$; hence

$$\frac{bh^2}{6} = \frac{Px}{\sigma_{\text{allow}}} \quad \text{or} \quad b = \left[\frac{6P}{h^2 \sigma_{\text{allow}}}\right] x = \frac{b_o}{L} x$$

where the bracketed expression is a constant and is set equal to b_o/L, so that when $x = L$ the width is b_o. A beam of constant strength with a constant depth in a plan view looks like the wedge* in Fig. 10-22(c). Near the free end this wedge must be modified to be of adequate strength to resist the shearing force.

If the width or breadth b of the beam is constant

$$\frac{bh^2}{6} = \frac{Px}{\sigma_{\text{allow}}} \quad \text{or} \quad h = \sqrt{\frac{6Px}{b\sigma_{\text{allow}}}} = h_o\sqrt{\frac{x}{L}}$$

This expression indicates that a cantilever of constant width loaded at the end is also of constant strength if its height varies parabolically from the free end, Fig. 10-22(d).

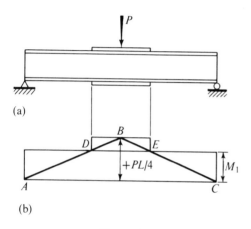

Fig. 10-23

Beams of constant strength are used in leaf springs and in many machine parts which are cast or forged. In structural work an approximation to a beam of constant strength is frequently made. For example, the moment diagram for the beam loaded as in Fig. 10-23(a) is given by the lines AB and BC in Fig. 10-23(b). By selecting a beam of flexural capacity equal only to M_1, the middle portion of the beam is overstressed. However, cover plates may be provided near the middle of the beam to boost the flexural capacity of the composite beam to the required value of the maximum moment. For the case shown, the cover plates must extend at least over the length DE of the beam, and in practice they are made somewhat longer.

10-14. DESIGN OF COMPLEX MEMBERS

In many instances the design of complex members cannot be accomplished in the routine manner of the preceding examples. Sometimes the size of a member must be assumed and a complete stress analysis performed at sections where the stresses appear critical. Designs of this type may require several revisions and much labor. Even experimental methods of stress analysis must be occasionally used since elementary formulas may not be sufficiently accurate. In accurate analyses of manufactured machine parts, the yield and fracture criteria discussed in Chapter 9 are frequently used.

---

* Since this beam is not of constant cross-sectional area, the use of the elementary flexure formula is not entirely correct. When the angle included by the sides of the wedge is small, little error is involved. As this angle becomes large, the error may be considerable. An exact solution shows that when the total included angle is 40° the solution is in error by nearly 10 per cent.

As a last example in this chapter, a transmission shaft problem will be analyzed. A direct analytical procedure is possible in this problem, which is of great importance in the design of power equipment.

Section 10-14
Design of complex members

EXAMPLE 10-10

Select the size of a solid steel shaft to drive the two sprockets shown in Fig. 10-24(a). These sprockets drive 1¾-in. pitch roller chains* as shown in Figs. 10-24(b) and (c). Pitch diameters of the sprockets in the figures are from a manufacturer's catalogue. A 20-hp speed-reducer unit is coupled directly to the shaft and drives it at 63 rpm. At each sprocket 10 hp is taken off. Assume the maximum shear theory of failure, and let $\tau_{\text{allow}} = 6{,}000$ psi.

SOLUTION

According to Eq. 5-1 the torque delivered to the shaft segment CD is $T = 63{,}000(\text{hp})/N = (63{,}000)20/63 = 20{,}000$ in-lb. Hence the torques

---

* Similar sprockets and roller chains are commonly used on bicycles.

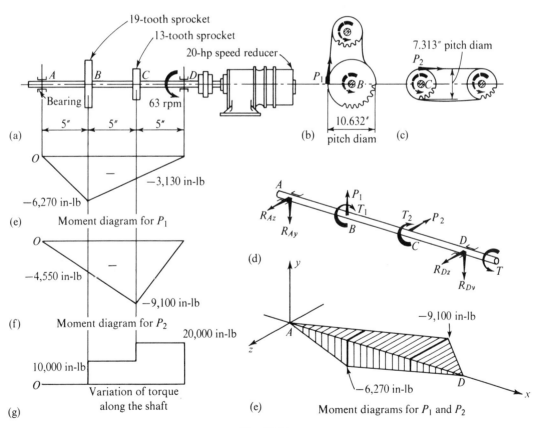

Fig. 10-24

T_1 and T_2 delivered to the sprockets are $T/2 = 10{,}000$ in-lb each. Since the chains are arranged as in Figs. 10-24(b) and (c), the pull in the chain at sprocket B is $P_1 = T_1/(D_1/2) = 10{,}000/(10.632/2) = 1{,}880$ lb. Similarly, $P_2 = 10{,}000/(7.313/2) = 2{,}730$ lb. The pull P_1 on the chain is equivalent to a torque T_1 and a vertical force at B as shown in Fig. 10-24(d). At C the force P_2 acts horizontally and exerts a torque T_2. A complete free-body diagram for the shaft AD is in Fig. 10-24(d).

It is seen from the free-body diagram that this shaft is simultaneously subjected to bending and torque. These effects on the member are best studied with the aid of appropriate diagrams, Figs. 10-24(e), (f), and (g). Next note that, although bending takes place in two planes, a vectorial resultant of the moments may be used in the flexure formula since the beam has a circular cross section. Bearing the last statement in mind, one will see that the general Eq. 9-6, which gives the principal shearing stress at the surface of the shaft, reduces in this problem of bending and torsion to

$$\tau_{max} = \sqrt{\left(\frac{\sigma_{bending}}{2}\right)^2 + \tau_{torsion}^2} \quad \text{or} \quad \tau_{max} = \sqrt{\left(\frac{Mc}{2I}\right)^2 + \left(\frac{Tc}{J}\right)^2}$$

However, since for a circular cross-section, $J = 2I$ (Eq. 6-7), $J = \pi d^4/32$ (Eq. 5-3), and $c = d/2$, the last expression simplifies to

$$\tau_{max} = \frac{16}{\pi d^3}\sqrt{M^2 + T^2}$$

Then, by assigning the allowable shearing stress to τ_{max}, one finds that a design formula, based on the maximum shear theory* of failure, for a shaft subjected to bending and torsion is

$$d = \sqrt[3]{\frac{16}{\pi \tau_{allow}}\sqrt{M^2 + T^2}} \tag{10-10}$$

This formula may be used to select the diameter of a shaft simultaneously subjected to bending and torque. In the problem investigated, a few trials should convince the reader that the $\sqrt{M^2 + T^2}$ is largest at the sprocket C; hence the critical section is there. Thus

$$M^2 + T^2 = (M_{vert})^2 + (M_{horiz})^2 + T^2$$
$$= (3{,}130)^2 + (9{,}100)^2 + (20{,}000)^2$$
$$= 492{,}600{,}000 \text{ in}^2\text{-lb}^2$$

$$d = \sqrt[3]{\frac{16}{6{,}000\pi}\sqrt{492{,}600{,}000}} = 2.68 \text{ in.}$$

A $2^{11}/_{16}$-in. diameter shaft, which is a commercial size, should be used.

---

* A formula based on the maximum normal stress theory of failure is also occasionally used in practice.

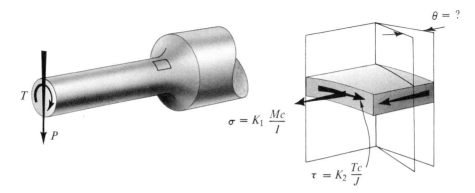

Fig. 10-25. Analysis of a shaft with stress concentrations.

The effect of a shock load on the shaft has been neglected in the foregoing analysis. For some equipment, where operation is jerky, this condition requires special consideration. The initially assumed allowable stress presumably allows for keyways and fatigue of the material.

Although Eq. 10-10 and similar ones based on other failure criteria are widely used in practice, the reader is cautioned in applying them. In many machines shaft diameters change abruptly giving rise to stress concentrations. In stress analysis this requires the use of stress concentration factors in bending which are usually different from those in torsion. Therefore the problem must be analyzed by considering the actual stresses at the critical section. (See Fig. 10-25.) Then, an appropriate procedure, such as Mohr's circle of stress, must be used to determine the significant stress depending on the selected fracture criteria.

PROBLEMS FOR SOLUTION

10-1. On a flat polished tensile specimen of mild steel it is possible to observe slip lines at approximately 45° to the direction of the pull. Such lines are called Lüder's lines. Explain the reason for their appearance. Consult your materials science text.

10-2. A concrete cylinder tested in a vertical position failed at a compressive stress of 4,000 psi. The failure occurred on a plane of 30° with the vertical. On a clear sketch show the normal and the shearing stresses which acted on the plane of failure. Which point on the Mohr's failure envelope (Fig. 9-28(b)) does this experiment establish? *Ans.* $\sigma = -1$ ksi, $\tau = 1.73$ ksi.

10-3. A cast iron beam is loaded as shown in the figure. Determine the principal stresses at

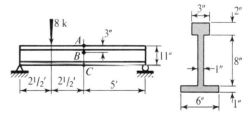

PROB. 10-3

the three points A, B, and C caused by the applied force. The moment of inertia of the cross-sectional area around the neutral axis is 316.2 in.[4] *Ans.* At A, 0, $-2,220$ psi; at B, $+37$ psi, $-1,117$ psi; at C, $+1,960$, 0.

10-4. A 4-in.-by-18-in. rectangular wooden beam supports an 8.1-kip load as in the figure.

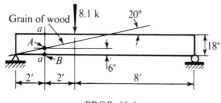

PROB. 10-4

At section *a–a* the grain of the wood makes an angle of 20° with the axis of the beam. Find the shearing stress along the grain of the wood at points *A* and *B* caused by the applied concentrated force. *Ans.* 140.6 psi at *A*.

10-5. A very short *I*-beam cantilever is loaded as shown in the figure. Find the principal stresses and their direction at points *A*, *B*, and *C*. Point *B* is in the web at the juncture with the flange. Neglect the weight of the beam and ignore the effect of stress concentrations. *I* around the neutral axis for the whole section is 221 in.^{4} Use the accurate formula to determine the shearing stresses. *Ans.* See Fig. 10-16.

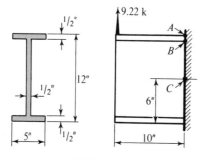

PROB. 10-5

10-6. The maximum shearing stress at point *A* in a beam, see figure, is 120 psi. Determine the magnitude of the force *P*. Assume the beam to be weightless.

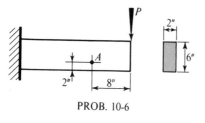

PROB. 10-6

10-7. A special hoist is loaded with a 15-kip load suspended by a cable as in the figure. Determine the state of stress at point *A* caused by this load. Show the results on an element with horizontal and vertical faces. *I* of the cross-sectional area around the neutral axis is 165 in.^{4} *Ans.* −4.66 ksi, 1.17 ksi.

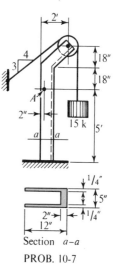

PROB. 10-7

10-8. A *T* bracket for a machine has the dimensions in the figure ($I = 136$ in.^{4}). When a concentrated force *F*, causing no torsion, is applied to the bracket, longitudinal strain of 184×10^{-6} in. per inch is observed at gage *A*. (a) What is the magnitude of the force *F*? Let $E = 29 \times 10^6$ psi and $G = 12 \times 10^6$ psi. (b) Set up a differential element showing the complete state of stress at *A*. *Ans.* (a) 10 kips.

10-9. After the erection of a heavy structure, it is estimated that the state of stress in the rock foundation will be essentially two-dimensional and as shown in the figure. If the rock is stratified, the strata making an angle of 30° with the vertical, is the anticipated state of stress permissible? Assume that the static coefficient of friction of rock on rock is 0.50, and along the planes of stratification cohesion amounts to 12 psi.

10-10. A 2-in.-diameter vertical shaft in a turbine is to be designed to carry an axial tensile force of 16 kips and simultaneously transmit a torque *T*. If the maximum permissible shearing stress is 8 ksi, what is the allowable torque on the shaft?

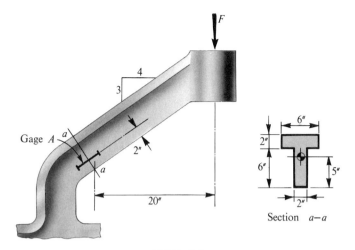

PROB. 10-8

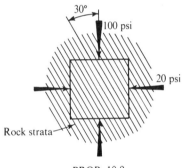

PROB. 10-9

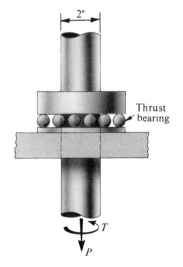

PROB. 10-10

10-11. The principal features of a stabilizer bar for the rear suspension of an automobile can be idealized as in the figure. The bar is 1 in. in diameter with 90° bends and is used in a horizontal position. If equal and opposite forces P of 150 lb each act on this bar, what is the state of stress in the bar just outside the bearings? Confine the investigation to a point on the top of the bar and to a point on the near side at the right bearing. Show the results on differential elements clearly related to the points considered.

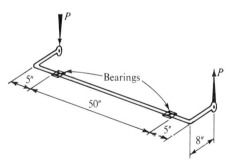

PROB. 10-11

10-12. A machine bracket is loaded with an inclined force of 444 lb as in the figure. Find the principal stresses at point A. Show the results on a properly oriented element. Neglect the weight of the member. *Ans.* $+570$ psi, $-4{,}470$ psi, $19°40'$.

367

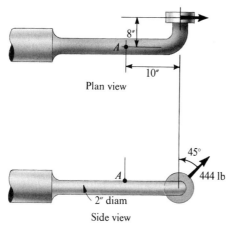

Plan view

Side view

PROB. 10-12

10-13. A crank has the dimensions shown in the figure, with the main shaft being 2 in. in diameter. Determine the magnitude of the force P which may be applied as governed by the combined stress at point A at the centerline of the bearing. The maximum shearing stress may not exceed 12 ksi, and the maximum normal stress cannot be greater than 24 ksi. Assume concentrated reaction at the bearing. Comment on this assumption.

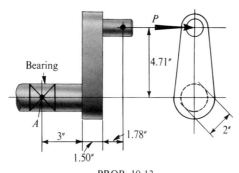

PROB. 10-13

10-14. A 400-lb sign is supported by a 2½-in. standard-weight steel pipe as in the figure. The maximum horizontal wind force acting on this sign is estimated to be 90 lb. Determine the state of stress caused by this loading at points A and B at the built-in end. Principal stresses are not required. Indicate results on sketches of elements cut out from the pipe at these points. These elements are

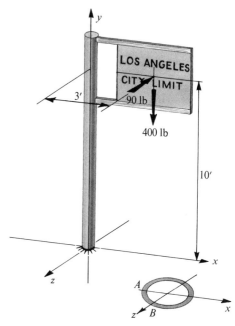

PROB. 10-14

to be viewed from outside the pipe. *Ans.* For A: 13,350 psi, 1,422 psi; for B: 9,945 psi, 1,523 psi.

10-15. A bent pipe 10 in. in diameter is used to support an inclined force $F = 6$ kips as in

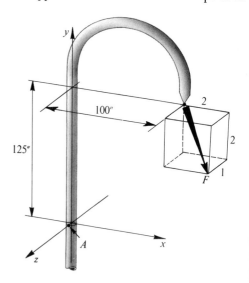

PROB. 10-15

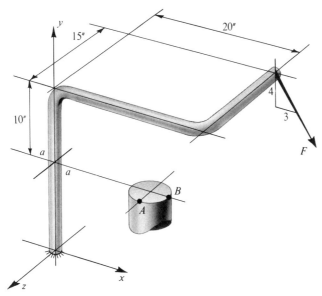

PROB. 10-16

the figure. Determine the state of stress at A caused by the applied force. Show the results on an element. Principal stresses are not required. The pipe wall thickness is 0.318 in. For simplicity assume that the outside and the mean diameters are equal. Hence, $A = 10$ in.^{2}, $I = 125$ in.^{4}, and $J = 250$ in.^{4} *Ans.* $\sigma_y = -10.4$ ksi.

10-16. A 2-in.-diameter rod is subjected at its free end to an inclined force $F = 50\pi$ lb as in the figure. (The force F in plan view acts in the direction of the x axis.) Determine the magnitude and directions of the stresses due to F on the elements A and B at section a–a. Show the results on elements clearly related to the points on the rod. Principal stresses are

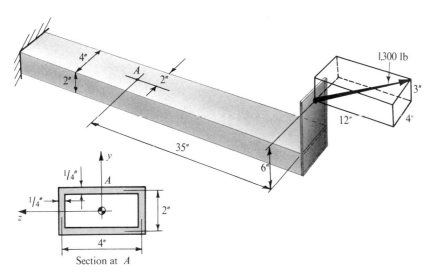

PROB. 10-17

not required. *Ans.* At A, $\sigma_y = 2{,}360$ psi, $\tau_{xy} = -860$ psi.

For additional data for problems of this type see Chapter 2.

10-17. A hollow square tube is attached to a machine and is loaded by an inclined force of 1,300 lb as in the figure. Determine the state of stress at point A caused by the applied force. The tube may be considered "thin," i.e., do not differentiate among the stresses at the outside, inside, or centerline of the tube. Thus, using centerline dimensions, $A = 3.00$ in.^{4}, $I_{yy} = 6.67$ in.^{4}, $I_{zz} = 2.33$ in.^{4}, and $J = 9.00$ in.^{4} *Ans.* $\sigma_x = 500$ psi, $\tau_{xy} = 940$ psi.

10-18. The containment vessel for the San Onofre, California nuclear generating station is made in the form of a hemisphere with a diameter of 140 ft. The steel shell is 1 in. thick weighing 40.8 lb per square foot. (a) Determine the maximum tensile and compressive stresses in this shell caused by its own weight. (b) What additional internal pressure can be developed within the vessel before a stress of 40 ksi will be reached?

10-19. In the design of roof structures it is customary to assume a uniformly distributed load applied over the entire projected area in addition to the weight of the structure. Such a load is called a live load. Show that the stresses in a hemispherical shell, such as shown in Fig. 10-13, due to a live load of p_o lb per square foot of the horizontal projected area are

$$\sigma_\varphi = -p_o a/(2h)$$

and $$\sigma_\theta = -[p_o a \cos 2\varphi]/(2h)$$

10-20. A "penstock," a pipe for conveying water to a hydroelectric turbine, operates at a head of 300 ft. If the diameter of the penstock is 30 in. and the allowable stress is 8,000 psi, what wall thickness is required? (The allowable stress is set low to provide for corrosion and inefficiency of welded joints.)

10-21. A cylindrical vessel for storing ammonia (NH$_3$) at the maximum temperature of 50°C is to be 8 ft in diameter. Determine the required wall thickness for the cylinder. The steel to be used has a yield strength of 42 ksi. Assume a factor of safety of 6 on the yield but make no further reduction in the stress for possible imperfections in the welds as these will be inspected with X-rays. The vapor pressure of NH$_3$ at 50°C is 20 atm.

10-22. A piece of 10-in.-diameter tubing of 0.1-in. wall thickness was closed off at the ends as shown in Fig. 9-26. Then this assembly was put into a testing machine and subjected simultaneously to an axial pull P and an internal pressure of 240 psi. What was the magnitude of the applied force P if the gage points A and B, initially precisely 8 in. apart, were found to be 8.0016 in. apart after all the forces were applied? $E = 30 \times 10^6$ psi and $\nu = 0.25$. *Ans.* 9.43 kips.

10-23. A cylindrical pressure vessel of 120-in. diameter outside, used for processing rubber, is 36 ft long. If the cylindrical portion of the vessel is made from 1 in. thick steel plate and the vessel operates at 120 psi internal pressure determine the total elongation of the circumference and the increase in the diameter's dimension caused by the operating pressure. $E = 29 \times 10^6$ psi and $\nu = 0.3$. *Ans.* 0.0778 in. 0.0247 in.

10-24. It can be shown* by the methods of elasticity that in a thick-walled cylinder subjected to internal pressure p_i the radial and the tangential stresses respectively are

$$\sigma_r = \frac{p_i a^2}{b^2 - a^2}\left(1 - \frac{b^2}{r^2}\right)$$

and $$\sigma_\theta = \frac{p_i a^2}{b^2 - a^2}\left(1 + \frac{b^2}{r^2}\right)$$

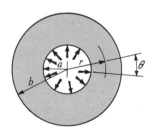

PROB. 10-24

---

* See, for example, E. P. Popov, *Mechanics of Materials* (Englewood Cliffs, N.J.: Prentice-Hall, Inc., 1952), p. 419, or any text on elasticity.

where the radii a, b, and r are defined in the figure. Make a comparison of the tangential stress distribution caused by p_i given by the above exact formulas with that given by the approximate formula for thin-walled cylinders. Investigate cases when $b = 1.1a$, and $b = 4a$. For the latter case also study the variation of σ_r. Plot the results on appropriate diagrams for section $\theta = 0°$ or $180°$.

10-25. Exceptionally light-weight pressure vessels have been developed by employing glass filaments for resisting the tensile forces and using epoxy resin as a binder. A diagram of a filament-wound cylinder is in the figure. If the winding is needed to resist only hoop stresses, the helix angle $\alpha = 90°$. If, however, the cylinder is closed, both hoop and longitudinal forces develop, and the required helix angle of the filaments $\alpha \approx 55°$ ($\tan^2 \alpha = 2$). Verify this result. (*Hint:* Isolate an element of unit width and a developed length of $\tan \alpha$ as in the figure. For such an element the same number of filaments is cut by each section. Therefore, if F is a force in a filament and n is the number of filaments at a section, $P_y = Fn \sin \alpha$. The force P_x can be found similarly. An equation based on the known ratio between the longitudinal and the hoop stress leads to the required result.)

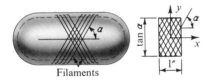

PROB. 10-25

10-26. A thin-walled conical shell is filled to the top with liquid of unit weight γ as in the figure. Show that $\sigma_\varphi = \gamma y(3a - 2y) \sin \alpha / (6h \cos^2 \alpha)$, and $\sigma_\theta = \gamma y(a - y) \sin \alpha / (h \cos^2 \alpha)$.

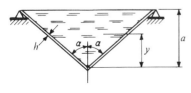

PROB. 10-26

10-27. A closed, cylindrical, steel pressure vessel of 100-in. mean diameter with walls ½ in. thick has a butt-welded seam along a helix angle $\alpha = 30°$ (see figure for Prob. 10-25). During pressurization the strain measurement across the weld, i.e., on a gage line of $\alpha + 90°$, is 430×10^{-6} psi. (a) What was the pressure in the vessel? (b) What was the shearing stress along the seam? Let $E = 29 \times 10^6$ psi, and $G = 12 \times 10^6$ psi. *Ans.* (a) 200 psi; (b) 4,330 psi.

10-28. A vertical fractionating column 30 ft high is made of 12-in. standard steel pipe weighing 49.56 lb per foot (see Table 8 in the Appendix). If the pipe is pressurized to 250 psi, and a horizontal wind force of 20 lb per linear foot of pipe acts on the column, what is the state of stress in the column 5 ft above the bottom on the windward side? Use actual internal pressurized areas in computations rather than the mean diameter. Show the results on an element. (The outward appearance of the fractionating column resembles that of the chimney in Prob. 8-28.) *Ans.* 4,000 psi, 3,630 psi.

10-29. In a certain research investigation on the creep of lead, it was necessary to control the state of stress for the element of a tube. In one such case, a long cylindrical tube with closed ends was pressurized and simultaneously subjected to a torque. The tube was 4 in. in outside diameter with ¼-in. walls. What were the principal stresses at the outside surface of the wall of the cylinder if the chamber was pressurized to 200 psi and the externally applied torque was 2,000 in-lb? Use the actual pressurized areas in computations rather than the mean diameter. *Ans.* 1,563 psi, 489 psi.

10-30. A compressed-air, cylindrical tank 10 ft long is 2 ft in mean diameter and has a wall

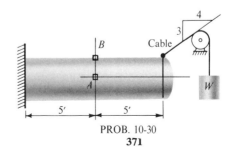

PROB. 10-30

thickness of ¼ in. In addition to pressurization, this tank is subjected at the end to a pull from the cable, as illustrated in the figure. The cable is in the vertical plane containing the vertical axis of the tank. If $W = 5,000$ lb and the pressure in the tank is 40 psi, what is the state of stress at points A and B? Neglect the weight of the tank. Show the results on sketches of infinitesimal elements viewed from outside.

10-31. A steel pressure vessel 20 in. in diameter and of 0.25-in. wall thickness acts also as an eccentrically loaded cantilever as in the figure. If the internal pressure is 250 psi and the applied weight $W = 31.4$ kips, determine the state of stress at point A. Show the results on an infinitesimal element. Principal stresses are not required. Neglect the weight of the vessel. *Ans.* $\sigma_x = 5$ ksi, $\sigma_y = 10$ ksi, $\tau = 6$ ksi.

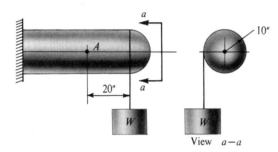

PROB. 10-31

10-32. Select four alternative beams of different materials to resist the same moment of 20 k-ft and compare the weights.* One beam is to be of wood, based on an allowable stress of 1,500 psi. The cross section of this beam is to be the nearest commercial size for a rectangular member with its depth twice the width. The other three beams are to have I sections using the following allowable stresses: 24,000 psi for steel as well as for an aluminum alloy, and 12,000 psi for fiber-glass-reinforced polyester plastic which weighs 5.5×10^{-2} lb per cubic inch. The cross-sectional properties of all I beams are as given in Table 3 of the Appendix.

10-33. A full-sized 4-in.-by-6-in. wooden

---
* If costs are readily available, also make a cost comparison.

---

beam (6 in. high) acts as a simple beam. What may the span L be, and what uniformly distributed load p (including the weight of the beam) may this beam support if the allowable stresses of $\sigma = 1,200$ psi and $\tau = 75$ psi are to be reached simultaneously? *Ans.* 8 ft, 300 lb per foot.

10-34. A 4-in.-by-6-in. (actual size) wooden beam is to be symmetrically loaded with two equal loads P as in the figure. Determine the position of these loads and their magnitude when a bending stress of 1,600 psi and a shearing stress of 100 psi are just reached. Neglect the weight of the beam.

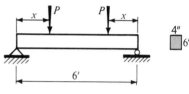

PROB. 10-34

10-35. A four-wheel car running on rails is to be used in light industrial service. When loaded this car will weigh a total of 4 tons. If the bearings are located with respect to the rails as in the figure, what size round axle should be used? Assume the allowable bending stress to be 10 ksi and the allowable shearing stress to be 6 ksi.

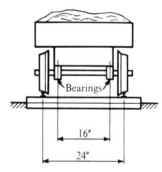

PROB. 10-35

10-36. A wall is to be temporarily supported as indicated in the figure to permit construction of a new foundation. The weight of the wall is 5 tons per foot horizontally. If beams A are to be used every 10 ft, what must their

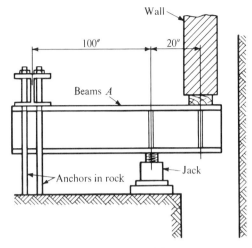

PROB. 10-36

size be? Use steel WF or I beams, whichever are lighter. The allowable bending stress is 20 ksi, and the allowable shearing stress is 10 ksi.

10-37. Select the required cross section for a rectangular wooden beam that is to carry a load of $p_o = 1.0$ k per foot, including its own weight, for the span in the figure. The allowable bending stress is 1,300 psi and the allowable shearing stress is 150 psi. The beam is to be twice as high as it is wide. Let $a = 10$ ft and $b = 4$ ft.

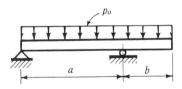

PROB. 10-37

10-38. A T beam is supported in the same manner as the beam in the preceding problem; however, $a = 16$ ft and $b = 8$ ft. The cross-sectional dimensions of the T are in the figure; the moment of inertia around the centroidal axis I is 820 in.[4] If the allowable stresses are 1,200 psi in bending and 100 psi in shear, what is the largest load p_o in lb per foot that this beam can carry?

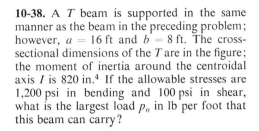

PROB. 10-38

10-39. A plastic beam is to be made from two 1-in.-by-3-in. pieces to span 24 in. and to carry an intermittently applied, uniformly distributed load p. The pieces can be arranged in two alternative ways as in the figure. The allowable stresses are 500 psi in flexure, 100 psi shear in plastic, and 75 psi shear in glue. Which arrangement of pieces should be used, and what load p may be applied?

PROB. 10-39

10-40. A box beam is fabricated from two pieces of $3/4$-in. plywood and two $4\frac{1}{2}$-in.-by-3-in. (actual size) solid wood pieces as shown

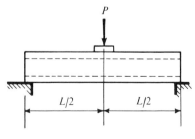

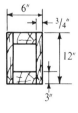

PROB. 10-40

373

in the cross-sectional view. If this beam is to be used to carry a concentrated force in the middle of a simple span, (a) what may the magnitude of the maximum applied load *P* be; (b) how long may the span be; and (c) what size bearing plate should be provided under the concentrated force? Neglect the weight of the beam and assume that there is no danger of lateral buckling. The allowable stresses are: 1,500 psi in bending, 120 psi for shear in plywood, 60 psi for shear in the glued joint, and 400 psi in bearing perpendicular to the grain. *Ans.* 3,210 lb, 244 in., 8 in.²

10-41. Determine the size required for an *I* beam rail of an overhead traveling crane of 4-ton capacity. The *I* beam is to be attached to the wall at one end and hung from a bracket as in the figure. Assume pinned connection at the wall, and in computations neglect the weight of the beam. Let the allowable bending stress be 12 ksi and the allowable shearing stress be 7 ksi. *Ans.* 15 *I* 42.9.

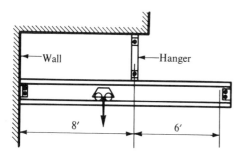

PROB. 10-41

10-42. A portion of the floor-framing plan for an office building is shown in the figure. Wooden joists spanning 12 ft are spaced 16 in. apart and support a wooden floor above and a plastered ceiling below. Assume that the floor may be loaded by the occupants everywhere by as much as 75 lb per square foot of floor area (live load). Assume further that floor, joists, and ceiling weigh 25 lb per square foot of the floor area (dead load). (a) Determine the depth required for standard commercial joists nominally 2 in. thick. For wood the allowable bending stress is 1,200 psi and the shearing stress is 100 psi. (b) Select the size required for the steel beam *A*. Since the joists delivering the load to this beam are

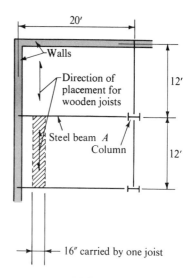

PROB. 10-42

spaced closely, assume that the beam is loaded by a uniformly distributed load. The allowable stresses for steel are 20,000 psi and 13,000 psi for bending and shear, respectively. Use a *WF* or an *I* beam, whichever is lighter. Neglect the width of the column. *Ans.* (a) 2 in. × 10 in. (nominal), (b) 14 WF 30.

10-43. In many engineering design problems it is very difficult to determine the magnitudes of the loads that will act on a structure or a machine part. Satisfactory performance in an

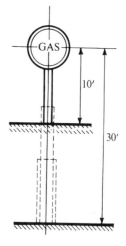

PROB. 10-43

existing installation may provide the basis for extrapolation. With this in mind, suppose that a certain sign, such as shown in the figure, has performed satisfactorily on a 4-in. standard steel pipe when its centroid was 10 ft above the ground. What should the size of pipe be if the sign were raised to 30 ft above the ground. Assume that the wind pressure on the sign at the greater height will be 50 per cent greater than it was in the original installation. Vary the size of the pipe along the length as required; however, for ease in fabrication, the successive pipe segments must fit into each other. In arranging the pipe segments also give some thought to aesthetic considerations. For simplicity in calculations, neglect the weight of the pipes and the wind pressure on the pipes themselves.

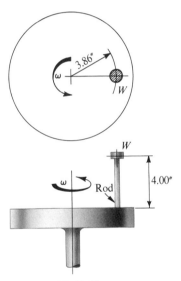

PROB. 10-44

10-44. A rigid plate rotates in a horizontal plane as shown in the figure. A vertical steel rod attached to the plate supports a concentrated mass at its free end. When the plate is in motion, (a) at what angular velocity ω will yielding commence in the rod, and (b) at what angular velocity ω will the ultimate bending capacity of the rod be reached? The rod is 0.20 in. in diameter and weighs $3\pi \times 10^{-3}$ lb per inch. The weight W at the top of the rod is $1.9\pi \times 10^{-2}$ lb. Assume that the rod behaves like a linearly elastic-perfectly plastic material

with $\sigma_{yp} = 40{,}000$ psi. Neglect vertical forces. (*Note:* This is a direct stress analysis problem. Design would be lengthier as trial and error solutions would be required.) *Ans.* (a) 100 radians per second.

10-45. The middle half of a simple beam of 10-ft total length is 6 in. wide by 12 in. deep; the end quarters are 6 in. wide by 8 in. deep,

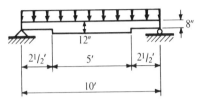

PROB. 10-45

see figure. Determine the safe, uniformly distributed load this beam may carry if the allowable bending stress is 1,500 psi and the allowable shearing stress is 150 psi. Neglect stress concentrations at the change in cross section.

10-46. A 10 in., 25.4 lb I beam is coverplated with two $\frac{1}{2}$-in.-by-6-in. plates as shown in Fig. 10-23a (I of the composite section is 287.1 in.^{4}), and it spans 20 ft. (a) What concentrated force may be applied at the center of the span if the allowable stress in bending is 16,000 psi? (b) For the above load, where are the theoretical points beyond which the cover plates need not extend? Neglect the weight of the beam, and assume that the beam is braced laterally. *Ans.* (a) 13.9 kips; (b) 4.67 ft from ends.

10-47. Design a cantilever beam of constant strength for resisting a uniformly distributed load. Assume that the width of the beam is constant.

10-48. (a) Show that the larger principal stress for a circular shaft simultaneously subjected to a torque and a bending moment is $\sigma_1 = (c/J)(M + \sqrt{M^2 + T^2})$.

(b) Show that the design formula for shafts, on the basis of the maximum stress theory, is $d = \sqrt[3]{[16/(\pi \tau_{\text{allow}})](M + \sqrt{M^2 + T^2})}$.

10-49. The head shaft of an inclined bucket elevator is arranged as in the figure. It is driven at A at 11 rpm and requires 60 hp for

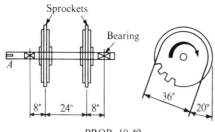

PROB. 10-49

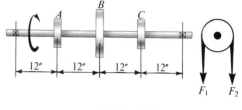

PROB. 10-50

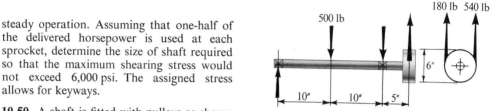

PROB. 10-51

steady operation. Assuming that one-half of the delivered horsepower is used at each sprocket, determine the size of shaft required so that the maximum shearing stress would not exceed 6,000 psi. The assigned stress allows for keyways.

10-50. A shaft is fitted with pulleys as shown in the figure. The end bearings are self-aligning, i.e., they do not introduce moment into the shaft at the supports. Pulley B is the driving pulley. Pulley A and C are the driven pulleys and take off 9,000 in-lb and 3,000 in-lb of torque, respectively. The resultant of the pulls at each pulley is 400 lb acting downward. Determine the size of shaft required so that the principal shearing stress would not exceed 6,000 psi. *Ans.* 2.24 in.

10-51. A motor armature weighing 500 lb is supported by a shaft in bearings 20 in. apart. The pulley on the end of the shaft is 6 in. in diameter and overhangs the bearing on the right end by 5 in., see figure. Power is taken off upward through a vertical belt drive at the pulley; the total tension on the tight side is 540 lb, on the slack side 180 lb. Compute the required shaft diameter on the basis of the maximum shear stress theory if $\tau_{allow} = 7,000$ psi. *Ans.* 1.48 in.

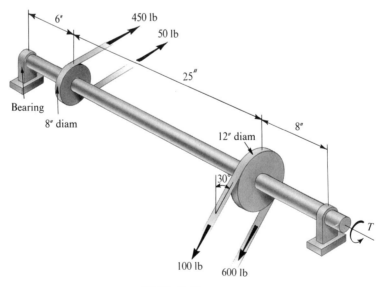

PROB. 10-52

10-52. A drive shaft for two pulleys is arranged as shown in the figure. The belt tensions are known. Determine the required size of the shaft. Assume from the ASME code that $\tau_{allow} = 6,000$ psi for shafts with keyways. Moreover, since the shaft will operate under conditions of suddenly applied load, multiply the given loads by a shock and fatigue factor of $1\frac{1}{2}$.

10-53. A low-speed shaft is acted upon by an eccentrically applied load P caused by a force developed between the gears. Determine the allowable magnitude of the force P on the basis of the maximum shearing stress theory if $\tau_{allow} = 6,500$ psi. The small diameter of the overhung shaft is 3 in. Consider the critical section to be where the shaft changes diameter, and that $M = 3P$ in-lb and $T = 6P$ in-lb. Note that since the diameter size changes abruptly, the following stress concentration factors must be considered: $K_1 = 1.6$ in bending, and $K_2 = 1.2$ in torsion. *Ans.* 4,000 lb.

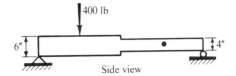

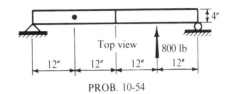

PROB. 10-54

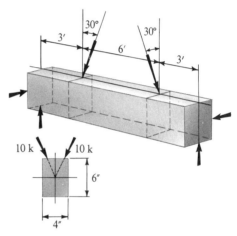

PROB. 10-55

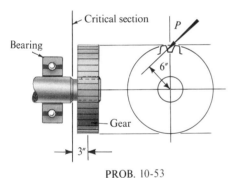

PROB. 10-53

10-54. Neglecting the weight of the beam and stress concentrations at the change in cross section, find the largest bending stress for the beam loaded as shown in the figure.

10-55. A 12 ft long beam is loaded as shown in the figure. The two applied forces act perpendicularly to the long axis of the beam and are inclined 30° with the vertical. If these forces act through the centroid of the cross-sectional area, find the location and magnitude of the maximum bending stress. Neglect the weight of the beam. *Ans.* ±18,625 psi.

11 Deflection of beams

11-1. INTRODUCTION

Under the action of applied forces the axis of a beam deflects from its initial position. Accurate values for beam deflections are sought in many practical cases. Elements of machines must be sufficiently rigid to prevent misalignment and to maintain dimensional accuracy under load. In buildings the floor beams cannot deflect excessively to obviate the undesirable psychological effect on the occupants and to minimize or prevent distress in brittle finish materials. Likewise, information on deformation characteristics of members is essential in the study of vibrations of machines as well as of stationary and flight structures.

Basic differential equations for the deflection of beams will be developed in this chapter. Solution of these equations is illustrated in detail. Only deflections caused by forces acting perpendicularly to the axis of a beam are considered. Situations in which axial forces occur simultaneously are discussed in Chapter 14.

As will become apparent from the derivation, the basic theory developed in this chapter is limited to deflections which are small in relation to span length. An idea of the accuracy involved may be gained by noting, for example, that there is approximately a 1 per cent error from the exact solution, if deflections of a simple span are on the order of one-twentieth of its length. By doubling the deflection to one-tenth of the span length, which ordinarily would be considered an intolerably large deflection, the error is raised to approximately 4 per cent. As stiff flexural members are required in most engineering applications, this limitation of the theory is not serious. For clarity, however, the deflections of beams will be shown greatly exaggerated on all diagrams.

Section 11-2 Strain-curvature and moment-curvature relations

Only deflections caused by bending are considered in this chapter. Those due to shear are discussed in Chapter 13 (see especially Example 13-4).

Both the elastic and the inelastic deflections of beams are considered in this chapter. Since, however, calculations for inelastic deflections of beams are tedious to perform, illustrations are drawn principally from elastic cases. As the solution of some statically indeterminate elastic beam problems presents no additional mathematical difficulties in comparison with the determinate cases, the solution of such problems is discussed in this chapter. A more complete treatment of statically indeterminate structural systems is given in the next chapter.

After deriving the basic differential equations for beam deflections and exhibiting the boundary conditions, the remainder of the chapter is divided for convenience into two parts. In Part A, direct integration procedures are discussed; in Part B, a special approach, the so-called *moment-area method*, is presented.

11-2. STRAIN-CURVATURE AND MOMENT-CURVATURE RELATIONS

To develop the theory of beam deflection, the geometry or kinematics of deformation of a beam element must be considered. The fundamental kinematic hypothesis that plane sections remain plane during deformation, first introduced in Art. 6-3, provides the basis for the theory. This treatment neglects shear deformation of a beam. Fortunately the deflections due to shear usually are very small. (See Example 13-4.)

A segment of an initially straight beam is shown in a deformed state in Fig. 11-1(a). This diagram is completely analogous to Fig. 6-1, used in establishing the stress distribution in beams due to bending. The deflected axis of the beam, i.e., its elastic curve, is shown bent into a radius ρ. The center of curvature O for the radius of any element can be found by extending to intersection any two adjoining sections such as $A'B'$ and $D'C'$. For the present it will be assumed that bending is taking place around one of the principal axes of the cross section.

In the enlarged view of the element $A'B'C'D'$ in Fig. 11-1(b), it can

379

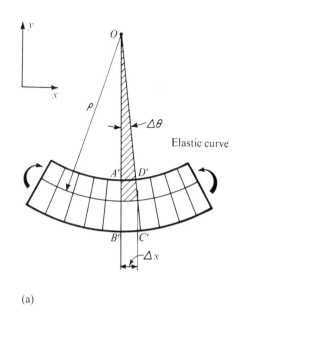
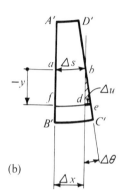

(a)

(b)

Fig. 11-1. Deformation of a beam segment in bending.

be seen that in a bent beam the included angle between two adjoining sections is $\Delta\theta$. If the distances y from the neutral surface to the strained fibers are measured in the usual manner as being positive upwards, the deformation Δu of any fiber can be expressed as

$$\Delta u = -y\,\Delta\theta \tag{11-1}$$

For negative y's this yields elongation which is consistent with the deformation shown in the figure.

The fibers lying in the curved neutral surface of the deformed beam, characterized in Fig. 11-1(b) by the fiber ab, are not strained at all. Therefore the arc length Δs corresponds to the initial length of all fibers between sections $A'B'$ and $D'C'$.ABearing this in mind, upon dividing Eq. 11-1 by Δs, one can form the following relations:

$$\lim_{\Delta s \to 0} \frac{\Delta u}{\Delta s} = -y \lim_{\Delta s \to 0} \frac{\Delta\theta}{\Delta s} \quad \text{or} \quad \frac{du}{ds} = -y \frac{d\theta}{ds} \tag{11-2}$$

One can recognize that du/ds is the linear strain in a beam fiber at a distance y from the neutral axis. Hence

$$du/ds = \varepsilon \tag{11-3}$$

The term $d\theta/ds$ in Eq. 11-2 has a clear geometrical meaning. With the aid of Fig. 11-1(a) it is seen that since $\Delta s = \rho \, \Delta \theta$

$$\lim_{\Delta s \to 0} \frac{\Delta \theta}{\Delta s} = \frac{d\theta}{ds} = \frac{1}{\rho} = \kappa \qquad (11\text{-}4)$$

Section 11-2 Strain-curvature and moment-curvature relations

which is the definition of *curvature* κ (kappa).

On the above basis, upon substituting Eqs. 11-3 and 11-4 into Eq. 11-2, one may express the fundamental relation between curvature of the elastic curve and the linear strain as

$$\frac{1}{\rho} = \kappa = -\frac{\varepsilon}{y} \qquad (11\text{-}5)$$

It is important to note that as no material properties were used in deriving Eq. 11-5, this relation can be used for inelastic as well as for elastic problems. In the latter case, it is expedient to note that, since $\varepsilon = \varepsilon_x = \sigma_x/E$, and $\sigma_x = -My/I$,

$$\frac{1}{\rho} = \frac{M}{EI} \qquad (11\text{-}6)$$

This equation relates the bending moment M at a given section of an elastic beam having a moment of inertia I around the neutral axis to the curvature $1/\rho$ of the elastic curve.

EXAMPLE 11-1

For cutting metal a band saw $\frac{1}{2}$ in. wide and 0.025 in. thick runs over two pulleys of 16-in. diameter. What maximum bending stress is developed in the saw as it goes over a pulley? Let $E = 30 \times 10^6$ psi.

SOLUTION

In this application the material must behave elastically. As the thin saw blade goes over the pulley it conforms to the radius of the pulley; hence $\rho \approx 8$ in.

Equation 6-3, $\sigma = -My/I$, together with Eq. 11-6, after some minor simplifications, yields a generally useful relation:

$$\sigma = -Ey/\rho \qquad (11\text{-}7)$$

With $y = \pm c$, the maximum bending stress in the saw is determined:

$$\sigma_{\max} = \frac{Ec}{\rho} = \frac{(30)10^6(0.0125)}{8} = 46{,}800 \text{ psi}$$

The high stress developed in the band saw necessitates superior materials for this application.

---

* Note that both θ and s must increase in the same direction.

11-3. THE GOVERNING DIFFERENTIAL EQUATION FOR DEFLECTION OF ELASTIC BEAMS

In texts on analytic geometry it is shown that in Cartesian coordinates the curvature of a line is defined as

$$\frac{1}{\rho} = \frac{\dfrac{d^2v}{dx^2}}{\left[1 + \left(\dfrac{dv}{dx}\right)^2\right]^{3/2}} = \frac{v''}{[1 + (v')^2]^{3/2}} \tag{11-8}$$

where x and v are the coordinates of a point on a curve. In terms of the problem being considered, the distance x locates a point on the elastic curve of a deflected beam, and v gives the deflection of the same point from its initial position.

If Eq. 11-8 were substituted into Eq. 11-5 or 11-6, the exact differential equation of the elastic curve would result. In general, the solution of such an equation is very difficult to achieve. Since, however, the deflections tolerated in the vast majority of engineering structures are very small, the slope dv/dx of the elastic curve is also very small. Therefore the square of the slope v' is a negligible quantity in comparison with unity, and Eq. 11-8 simplifies to

$$\frac{1}{\rho} \approx \frac{d^2v}{dx^2} \tag{11-9}$$

On this basis the governing differential equation for the deflection of an elastic beam* follows from Eq. 11-6 and is

$$\frac{d^2v}{dx^2} = \frac{M}{EI} \tag{11-10}$$

where it is understood that $M = M_{zz}$, and $I = I_{zz}$.

Note that in Eq. 11-10 the xyz coordinate system is employed to locate the material points in a beam for calculating the moment of inertia I. On the other hand, in the planar problem, it is the xv system of axes that is used to locate points on the elastic curve.

The positive direction of the v axis is taken to have the same sense as that of the positive y axis and the positive direction of the applied load p, Fig. 11-2(a). Note especially that if the positive slope dv/dx of the elastic curve becomes more positive as x increases, the curvature $1/\rho \approx d^2v/dx^2$ is positive. This sense of curvature agrees with the induced curvature caused

---

* The equation of the elastic curve was formulated by James Bernoulli, a Swiss mathematician, in 1694. Leonhard Euler (1707–83) greatly extended its applications.

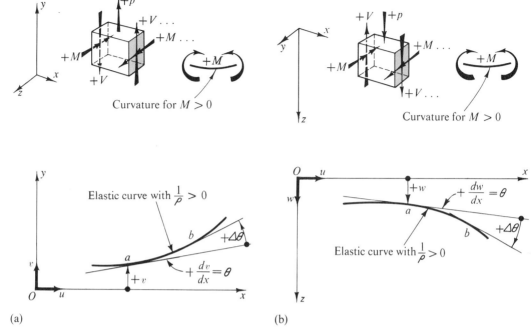

Fig. 11-2. Moment and its relation to curvature in different coordinate planes.

by the applied positive moments M. For this reason on both sides of Eq. 11-10 the signs are positive.

In contrast, if bending takes place in the xz plane, a different situation is found. To illustrate, consider the initial axes rotated 90° with the pertinent quantities having positive senses as shown in Fig. 11-2(b). Here the curvature induced by the positive moments M is opposite to that associated with the positive curvature $1/\rho \approx d^2w/dx^2$ of the elastic curve. Therefore

$$\frac{d^2w}{dx^2} = -\frac{M}{EI} \qquad (11\text{-}11)$$

where $M = M_{yy}$ and $I = I_{yy}$; except for notation and sign, the equation is the same as Eq. 11-10.

In some books and technical papers the axes in Fig. 11-2(b) are preferred, which leads to the use of Eq. 11-11.* In this text the detailed study is confined to Eq. 11-10.

---

*In conjunction with this equation it should be noted that the differential equations of equilibrium are

$$dV/dx = -p \quad \text{and} \quad dM/dx = V$$

The reader should compare these relations with Eqs. 2-4 and 2-5.

It is important and interesting to note that for the elastic curve, at the level of accuracy of Eq. 11-10, one has $ds = dx$. This follows from the fact that, as before, the square of the slope dv/dx is negligibly small compared with unity, and

$$ds = \sqrt{dx^2 + dv^2} = \sqrt{1 + (v')^2}\, dx \approx dx \qquad (11\text{-}12)$$

Therefore in the small deflection theory no difference in length is said to exist between the initial length of the beam axis and the arc of the elastic curve. Stated alternatively, there is no horizontal displacement u of the points lying on the neutral surface, i.e., at $y = 0$. This approximation can be made the basis of an alternative derivation of Eq. 11-10, which follows.

11-4. AN ALTERNATIVE DERIVATION OF EQUATION 11-10

In the classical theories of plates and shells which deal with small deflections, equations analogous to Eq. 11-10 must be established. The characteristic approach can be illustrated on the beam problem.

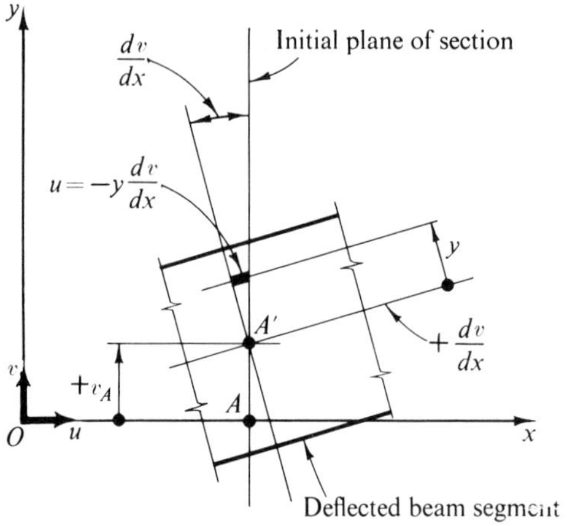

In a deformed condition a point A on the axis of an unloaded beam, Fig. 11-3, according to Eq. 11-12 is directly above its initial position. The tangent to the elastic curve at the same point rotates through an angle dv/dx. A plane section with the centroid at A' also rotates through the same angle dv/dx as during bending deformation sections remain normal to the bent axis of a beam. Therefore, the displacement u of a material point at a distance* y from the elastic curve is

$$u = -y \frac{dv}{dx} \qquad (11\text{-}13)$$

Fig. 11-3. Longitudinal displacements in a beam due to rotation of a plane section.

where the negative sign shows that for positive y and v' the displacement u is toward the origin. For $y = 0$, there is no displacement u, as required by Eq. 11-12.

Next, recall Eq. 4-3, which states that $\varepsilon_x = \partial u/\partial x$. Therefore, from Eq. 11-13, $\varepsilon_x = -y\, d^2v/dx^2$ since v is only a function of x.

The same linear strain also can be found from Eqs. 4-11 and 6-3 yielding $\varepsilon_x = -My/EI$. On equating the two alternative expressions for ε_x and eliminating y from both sides of the equation, one has

$$\frac{d^2v}{dx^2} = \frac{M}{EI}$$

which is the previously derived Eq. 11-10.

---

* As the angle dv/dx is small its cosine can be taken as unity.

11-5. ALTERNATIVE DIFFERENTIAL EQUATIONS OF ELASTIC BEAMS

In Chapter 2 a number of differential relations were shown among shear, moment, and the applied load (Eqs. 2-4, 2-5, and 2-6). These can be combined with Eq. 11-10 to yield the following useful sequence of equations:

$v = $ deflection of the elastic curve

$$\theta = \frac{dv}{dx} = v' = \text{slope of the elastic curve}$$

$$M = EI\frac{d^2v}{dx^2} = EIv'' \quad\quad (11\text{-}14)$$

$$V = -\frac{dM}{dx} = -\frac{d}{dx}\left(EI\frac{d^2v}{dx^2}\right) = -(EIv'')'$$

$$p = -\frac{dV}{dx} = \frac{d^2}{dx^2}\left(EI\frac{d^2v}{dx^2}\right) = (EIv'')''$$

For beams with constant flexural rigidity EI, Eqs. 11-14 simplifies into three alternative equations for determining the deflection of a loaded beam:

$$EI\frac{d^2v}{dx^2} = M(x) \quad\quad (11\text{-}15)$$

$$EI\frac{d^3v}{dx^3} = -V(x) \quad\quad (11\text{-}16)$$

$$EI\frac{d^4v}{dx^4} = p(x) \quad\quad (11\text{-}17)^*$$

The choice of equation for a given case depends on the ease with which an expression for load, shear, or moment can be formulated. Fewer constants of integration are needed in the lower-order equations.

11-6. BOUNDARY CONDITIONS

For the solution of beam deflection problems, in addition to the differential equations, boundary conditions must be prescribed. Several types of homogeneous boundary conditions are as follows:

(A) CLAMPED OR FIXED SUPPORT: In this case the displacement v

---

* If in Eq. 11-17 in accordance with the d'Alembert principle one sets $p = -m\ddot{v}$, where m is the mass of the beam per unit length and $\ddot{v} = \partial^2 v/\partial t^2$, the basic equation for the free lateral vibration of a beam is obtained. This equation is $EI\,\partial^4 v/\partial x^4 + m\,\partial^2 v/\partial t^2 = 0$.

and the slope dv/dx must vanish. Hence at the end considered, where $x = a$,

$$v(a) = 0, \qquad v'(a) = 0 \qquad \text{(11-18a)}$$

(B) ROLLER OR PINNED SUPPORT: At the end considered no deflection v nor moment M can exist. Hence

$$v(a) = 0, \qquad M(a) = EIv''(a) = 0 \qquad \text{(11-18b)}$$

Here the physically evident condition for M is related to the derivative of v with respect to x from one part of Eq. 11-14.

(C) FREE END: Such an end is free of moment and shear. Hence

$$M(a) = EIv''(a) = 0, \qquad V(a) = -(EIv'')'_{x=a} = 0 \qquad \text{(11-18c)}$$

(D) GUIDED SUPPORT: In this case free vertical movement is permitted, but the rotation of the end is prevented. The support is not capable of resisting any shear. Therefore

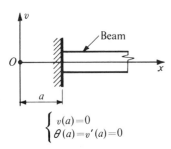

$$\begin{cases} v(a)=0 \\ \theta(a)=v'(a)=0 \end{cases}$$

(a) Clamped support

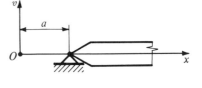

$$\begin{cases} v(a)=0 \\ M(a)=EIv''(a)=0 \end{cases}$$

(b) Simple support

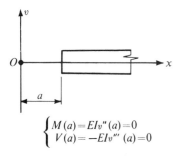

$$\begin{cases} M(a)=EIv''(a)=0 \\ V(a)=-EIv'''(a)=0 \end{cases}$$

(c) Free end

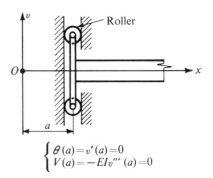

$$\begin{cases} \theta(a)=v'(a)=0 \\ V(a)=-EIv'''(a)=0 \end{cases}$$

(d) Guided support

Fig. 11-4. Homogeneous boundary conditions for beams with constant EI. In (a) both conditions are *kinematic*; in (c) both are *static*; in (b) and (d) conditions are mixed.

$$v'(a) = 0, \quad V(a) = -(EIv'')'_{x=a} = 0 \qquad (11\text{-}18d)$$

The same boundary conditions for constant EI are summarized in Fig. 11-4. Note the two basically different types of boundary conditions. Some pertain to the force quantities and are said to be *static boundary conditions*. Others describe geometrical or deformational behavior of an end; these are *kinematic boundary conditions*.

Nonhomogeneous boundary conditions, where a given shear, moment, rotation, or displacement are prescribed at the boundary, also occur in applications. In such cases the zeros in the appropriate Eqs. 11-18a through 11-18d are replaced by the specified quantity.

In some solutions the physical requirements of continuity of the elastic curve must be brought in to supplement the boundary conditions. This means that at any juncture of two zones of a beam the deflection and the tangent to the elastic curve must be the same, regardless of the direction from which the common point is approached. The situations in Fig. 11-5 are impossible. The requirements of force and moment equilibrium are contained implicitly in the conditions of continuity.

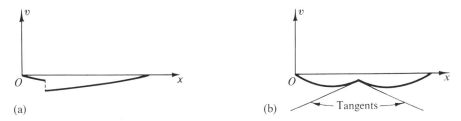

Fig. 11-5. Impossible situations in a continuous elastic curve.

11-7. DEFLECTION OF VISCOELASTIC BEAMS

If a loaded beam is viscoelastic, deflections change with time. As an introduction to this important topic, linear, viscoelastic beams will be considered in this article. In the analysis it will be assumed that the load $p(x)$ is applied at time $t = 0$ and remains constant thereafter.*

For elastic and plastic beams, according to Eq. 11-5, the fundamental kinematic hypothesis of the flexure theory relating the curvature of the elastic curve to the linear strain can be stated as

$$\varepsilon(x, y) = -\kappa(x)y$$

For a viscoelastic material the same hypothesis gives

---

* As noted in Art. 5-10, this loading can be expressed as $p(x, t) = p(x)H(t)$, where $H(t)$ is the Heaviside operator having the value of unity for $t > 0$.

$$\varepsilon(x, y, t) = -\kappa(x, t)y \qquad (11\text{-}19\text{a})$$

i.e., now the curvature κ is a function not only of the position x on the beam but also of the time t.

For a *linear*, viscoelastic material an alternative formulation for this strain can be based on the constitutive relation given by Eq. 4-27. Thus, for a uniaxial stress σ_o, which is a function of its position x and y, Eq. 4-27 can be generalized as

$$\varepsilon(x, y, t) = \sigma_o(x, y)J_c(t) \qquad (11\text{-}19\text{b})$$

where $J_c(t)$ is creep compliance.

By eliminating the strain ε from Eqs. 11-19a and 11-19b, one obtains an expression for the bending stress σ_o in a beam:

$$\sigma_o(x, y) = -[\kappa(x, t)/J_c(t)]y \qquad (11\text{-}19\text{c})$$

This relation shows that the bending stress varies linearly with y. Moreover, it is independent of time. Therefore, the stress distribution in a beam of a linearly viscoelastic material is identical to that in a linearly elastic beam.

In Eq. 11-19c the expression in brackets plays the same role as the constant B in Eq. 6-2, used in deriving the elastic flexure formula. Therefore, since an equilibrium condition requires that $M = -BI$ (see Art. 6-4), with $B = -[\kappa(x, t)/J_c(t)]$, one now has

$$\kappa(x, t) = \frac{M(x)}{I} J_c(t) \qquad (11\text{-}19\text{d})$$

This expression for beam curvature provides the basis for determining the deflection of linear, viscoelastic beams under sustained loadings.

Since the right side of Eq. 11-19d is a product only of a function of x and a function of t, the time-dependent deflection of a beam must be also a product of two functions in the same two variables. Therefore the beam deflection can be assumed to be

$$v(x, t) = \mathrm{v}(x)EJ_c(t) \qquad (11\text{-}19\text{e})$$

In a few steps it will become apparent that this is the correct choice for the deflection function.

The second derivative of Eq. 11-19e with respect to x gives an expression for the curvature of the elastic curve:

$$\kappa(x, t) = \frac{\partial^2 v(x, t)}{\partial x^2} = \frac{\partial^2 \mathrm{v}(x)}{\partial x^2} EJ_c(t) \qquad (11\text{-}19\text{f})$$

This relation together with Eq. 11-19d yields

$$\frac{\partial^2 v(x)}{\partial x^2} EJ_c(t) = \frac{M(x)}{I} J_c(t) \qquad (11\text{-}20a)$$

Differentiating this equation twice with respect to x and recalling that, according to Eq. 2-6, $p = d^2M/dx^2$, one has

$$\frac{\partial^4 v(x)}{\partial x^4} EJ_c(t) = \frac{p(x)}{I} J_c(t) \qquad (11\text{-}20b)$$

The term $J_c(t)$ appears on both sides of both of the above two equations and can be canceled. This reduces these equations to functions of x only, which makes them identical to Eqs. 11-15 and 11-17 for the deflection of linearly elastic beams. Therefore, it can be concluded that $v(x) = v_{el}(x)$, where $v_{el}(x)$ is the deflection of a linearly elastic beam with the same loading and boundary conditions as those of the original viscoelastic beam. On this basis, using Eq. 11-19e, one can formalize the deflection for linearly viscoelastic beams as

$$v(x, t) = v_{el}(x) EJ_c(t) \qquad (11\text{-}21)$$

This important equation states that in order to determine the time-dependent deflection of a linear, viscoelastic beam, one simply multiplies the elastic deflection $v_{el}(x)$ by the E and the creep compliance $J_c(t)$ for the material of the beam. The material properties E and J_c must be known from experiments.

The solution expressed by Eq. 11-21 is applicable to both statically determinate and statically indeterminate beams. However, as stated, the solution is limited to a step-function input of the applied load at $t = 0$ which then remains constant. The boundary conditions also must not change with time. If a sequence of load applications occurs, it is necessary to employ the Boltzmann superposition principle (Art. 4-16).

PART A
DIRECT INTEGRATION METHODS

11-8. SOLUTION OF BEAM DEFLECTION PROBLEMS BY DIRECT INTEGRATION

As a general example of calculating beam deflection consider Eq. 11-17, $EIv^{iv} = p(x)$. By successively integrating this expression four times, the formal solution for v is obtained. Thus

**Chapter 11
Deflection of beams**

$$EIv^{iv} = EI\frac{d^4v}{dx^4} = EI\frac{d}{dx}(v''') = p(x)$$

$$EIv''' = \int_0^x p\,dx + C_1$$

$$EIv'' = \int_0^x dx \int_0^x p\,dx + C_1 x + C_2 \qquad (11\text{-}22)$$

$$EIv' = \int_0^x dx \int_0^x dx \int_0^x p\,dx + C_1 x^2/2 + C_2 x + C_3$$

$$EIv = \int_0^x dx \int_0^x dx \int_0^x dx \int_0^x p\,dx + C_1 x^3/3! + C_2 x^2/2! + C_3 x + C_4$$

If, instead, one commenced with Eq. 11-15, $EIv'' = M(x)$, after two integrations, the solution is

$$EIv = \int_0^x dx \int_0^x M\,dx + C_3 x + C_4 \qquad (11\text{-}23)$$

In both equations the constants C_1, C_2, C_3, and C_4, corresponding to the homogeneous solution of the differential equations, must be determined from the conditions at the boundaries. The constants C_1 and C_2 were encountered in Chapter 2 in the solution of the differential equations of equilibrium (Art. 2-13). In Eq. 11-23 the constants C_1 and C_2 are incorporated into the expression for M. The constants $-C_1$, C_2, $C_3/(EI)$, and $C_4/(EI)$, respectively, are usually* the initial values of V, M, θ, and v at the origin.

The first term on the right hand of the last part of Eq. 11-22 and the corresponding one in Eq. 11-23 are the particular solutions of the respective differential equations. The one in Eq. 11-22 is especially interesting as it depends only on the loading condition of the beam. This term remains the same regardless of the prescribed boundary conditions. The latter are brought into the problem from the homogeneous solution of the differential equation.

If the loading, shear, and moment functions are continuous and the flexural rigidity EI is constant, the evaluation of the particular integrals is very direct. When discontinuities occur, the singularity functions introduced in Chapter 2 can be used to advantage. This, however, is not essential. Solutions can be found for each segment of a beam in which the functions are continuous; the complete solution is then achieved by enforcing continuity conditions at the common boundaries of the beam

---

* In certain cases where transcendental functions are used, these constants do not have this meaning. Basically, the whole function, which includes the constants of integration, must satisfy the conditions at the boundary.

segments. Alternatively, graphical or numerical procedures* of successive integrations can be used very effectively in the solution of practical problems.

The procedures discussed above are quite general and are applicable to both statically determinate and indeterminate elastic beams. In the next four examples, however, alternative solutions for determinate cases only will be illustrated. In one of the examples the case of a variable I is treated. Statically indeterminate beams will be considered in the following article.

Section 11-8
Solution of beam deflection problems by direct integration

EXAMPLE 11-2

A bending moment M_1 is applied at the free end of a cantilever of length

---

* Such procedures are of great importance in complicated problems. For example, see N. M. Newmark, "Numerical Procedure for Computing Deflections, Moments, and Buckling Loads," *Trans. ASCE*, **108** (1943), 1161.

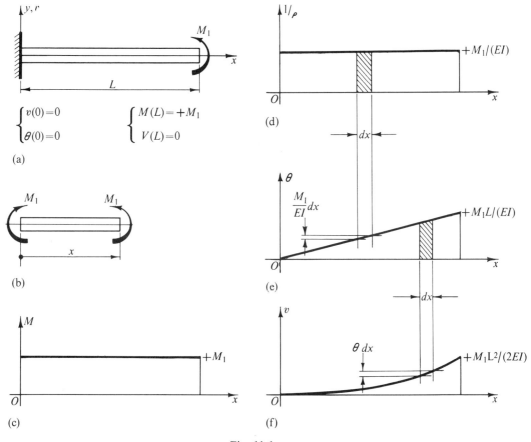

Fig. 11-6

L and of constant flexural rigidity EI, Fig. 11-6(a). Find the equation of the elastic curve.

SOLUTION

The boundary conditions are recorded near the figure from inspection of the conditions at the ends. At $x = L$, $M(L) = +M_1$, a nonhomogeneous condition.

From a free-body diagram of Fig. 11-6(b) it can be observed that throughout the beam the bending moment is $+M_1$. By applying Eq. 11-15, integrating successively, and making use of the boundary conditions, one obtains the solution for v:

$$EI \frac{d^2v}{dx^2} = M = M_1$$

$$EI \frac{dv}{dx} = M_1 x + C_3$$

But $\theta(0) = 0$; hence at $x = 0$ one has $EIv'(0) = C_3 = 0$ and

$$EI \frac{dv}{dx} = M_1 x$$

$$EIv = \tfrac{1}{2} M_1 x^2 + C_4$$

But $v(0) = 0$; hence $EIv(0) = C_4 = 0$ and

$$v = M_1 x^2/(2EI) \tag{11-24}$$

The positive sign of the result indicates that the deflection due to M_1 is upward. The largest value of v occurs at $x = L$. The slope of the elastic curve at the free end is $+M_1 L/(EI)$ radians.

Equation 11-24 shows that the elastic curve is a parabola. However, every element of the beam experiences equal moments and deforms alike. Therefore the elastic curve should be a part of a circle. The inconsistency results from the use of an approximate relation for the curvature $1/\rho$. It can be shown that the error committed is in the ratio of $(\rho - v)^3$ to ρ^3. As the deflection v is much smaller than ρ, the error is not serious.

It is important to associate the above successive integration procedure with a graphical solution or interpretation. This is shown in the sequence of Figs. 11-6(c) through (f). First the conventional moment diagram is shown. Then from Eqs. 11-9 and 11-10, $1/\rho \approx d^2v/dx^2 = M/(EI)$, the curvature diagram is plotted in Fig. 11-6(d). For the elastic case this is simply a plot of $M/(EI)$. By integrating the curvature diagram one obtains the θ diagram. In the next integration the elastic curve is obtained. In this problem since the beam is fixed at the origin, the conditions $\theta(0) = 0$, and $v(0) = 0$ are used in constructing the diagrams. This graphical approach or its numerical equivalents are very useful in the solution of problems with variable EI.

EXAMPLE 11-3

A simple beam supports a uniformly distributed downward load p_o. The flexural rigidity EI is constant. Find the elastic curve by the following three methods: (a) Use the second-order differential equation to obtain the deflection of the beam. (b) Use the fourth-order equation instead of the one in (a). (c) Illustrate a graphical solution of the problem.

*Section 11-8
Solution of beam deflection problems by direct integration*

SOLUTION

Case (a). A diagram of the beam together with the implied boundary conditions is in Fig. 11-7(a). The expression for M for use in the second-order differential equation has been found in Example 2-6. From Fig. 2-22

$$M = \frac{p_o L x}{2} - \frac{p_o x^2}{2}$$

Substituting this relation into Eq. 11-15, integrating it twice in succession, and making use of the boundary conditions, one finds the equation of the elastic curve:

$$EI \frac{d^2 v}{dx^2} = M = \frac{p_o L x}{2} - \frac{p_o x^2}{2}$$

$$EI \frac{dv}{dx} = \frac{p_o L x^2}{4} - \frac{p_o x^3}{6} + C_3$$

$$EIv = \frac{p_o L x^3}{12} - \frac{p_o x^4}{24} + C_3 x + C_4$$

But $v(0) = 0$; hence $EIv(0) = 0 = C_4$; and, since also $v(L) = 0$,

$$EIv(L) = 0 = \frac{p_o L^4}{24} + C_3 L \quad \text{and} \quad C_3 = -\frac{p_o L^3}{24}$$

$$v = -\frac{p_o}{24 EI}(L^3 x - 2Lx^3 + x^4) \tag{11-25}$$

By virtue of symmetry, the largest deflection occurs at $x = L/2$. On substituting this value of x into Eq. 11-25 one obtains

$$|v|_{\max} = 5 p_o L^4 / (384 EI) \tag{11-26}$$

The condition of symmetry could also have been used to determine the constant C_3. As it is known that $v'(L/2) = 0$, one has

$$EIv'(L/2) = \frac{p_o L(L/2)^2}{4} - \frac{p_o (L/2)^3}{6} + C_3 = 0$$

and, as before, $C_3 = -(1/24) p_o L^3$.

Case (b). Application of Eq. 11-17 to the solution of this problem is direct. The constants are found from the boundary conditions.

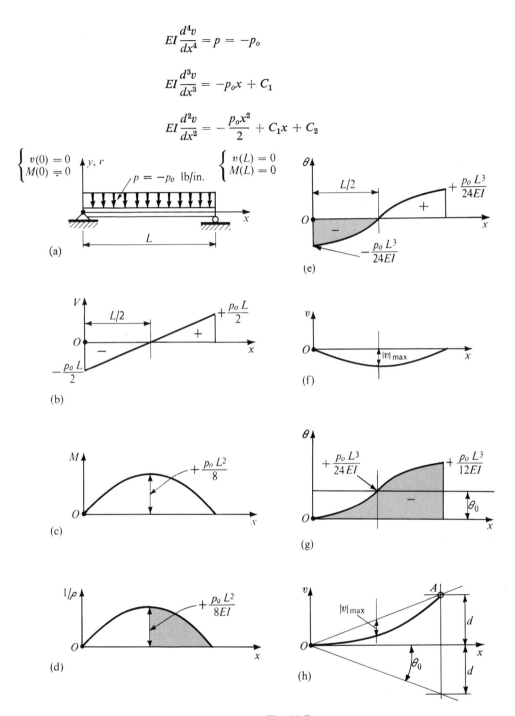

Fig. 11-7

But $M(0) = 0$; hence $EIv''(0) = 0 = C_2$; and, since also $M(L) = 0$,

$$EIv''(L) = 0 = -\frac{p_o L^2}{2} + C_1 L \quad \text{or} \quad C_1 = \frac{p_o L}{2}$$

hence
$$EI\frac{d^2v}{dx^2} = \frac{p_o L x}{2} - \frac{p_o x^2}{2}$$

Section 11-8 Solution of beam deflection problems by direct integration

The remainder of the problem is the same as in Case (a). In this approach no preliminary calculation of reactions is required. As will be shown later, this is advantageous in some statically indeterminate problems.

Case (c). The steps needed for a graphical solution of the complete problem are in Figs. 11-7(b) through (f). In Figs. 11-7(b) and (c) the conventional shear and moment diagrams are shown. The curvature diagram is obtained by plotting $M/(EI)$, as in Fig. 11-7(d).

Since by virtue of symmetry the slope to the elastic curve at $x = L/2$ is horizontal, $\theta(L/2) = 0$. Therefore, the construction of the θ diagram can be begun from the center. In this procedure, the right ordinate in Fig. 11-7(e) must equal the shaded area of Fig. 11-7(d), and vice versa. By summing the θ diagram, one finds the elastic deflection v. The shaded area of Fig. 11-7(e) is equal numerically to the maximum deflection. In the above, the condition of symmetry was employed. A generally applicable procedure follows.

After the curvature diagram is established as in Fig. 11-7(d), the θ diagram can be constructed with an assumed initial value of θ at the origin. For example, let $\theta(0) = 0$ and sum the curvature diagram to obtain the θ diagram, Fig. 11-7(g). Note that the shape of the curve so found is identical to that of Fig. 11-7(e). Summing the area of the θ diagram gives the elastic curve. In Fig. 11-7(h) this curve extends from O to A. This violates the boundary condition at A, where the deflection must be zero. Correct deflections are given, however, by measuring them vertically from a straight line passing through O and A. This inclined line corrects the deflection ordinates caused by the incorrectly assumed $\theta(0)$. In fact, after constructing Fig. 11-7(h), one knows that $\theta(0) = -d/L = -p_o L^3/(24EI)$. When this value of $\theta(0)$ is used, the problem reverts to the preceding solution (Figs. 11-7(e) and (f)). In Fig. 11-7(h) inclined measurements have no meaning. The procedure described is applicable for beams with overhangs. In such cases the base line for measuring deflections must pass through the support points.

EXAMPLE 11-4

A simple beam supports a concentrated downward force P at a distance a from the left support, Fig. 11-8(a). The flexural rigidity EI is constant. (a) Find the equation of the elastic curve without the use of operational notation. (b) Rework the problem using singularity functions.

SOLUTION

Case (a). The solution will be made by using the second-order differential equation. The reactions and boundary conditions are noted in

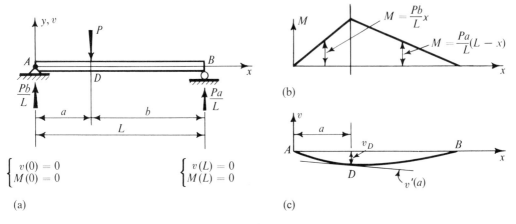

Fig. 11-8

Fig. 11-8(a). The moment diagram plotted in Fig. 11-8(b) clearly shows that a discontinuity at $x = a$ exists in $M(x)$, requiring two different functions for it. At first the solution proceeds independently for each segment of the beam.

For segment AD:

$$\frac{d^2v}{dx^2} = \frac{M}{EI} = \frac{Pb}{EIL} x$$

$$\frac{dv}{dx} = \frac{Pb}{EIL} \frac{x^2}{2} + A_1$$

$$v = \frac{Pb}{EIL} \frac{x^3}{6} + A_1 x + A_2$$

For segment DB:

$$\frac{d^2v}{dx^2} = \frac{M}{EI} = \frac{Pa}{EI} - \frac{Pa}{EIL} x$$

$$\frac{dv}{dx} = \frac{Pa}{EI} x - \frac{Pa}{EIL} \frac{x^2}{2} + B_1$$

$$v = \frac{Pa}{EI} \frac{x^2}{2} - \frac{Pa}{EIL} \frac{x^3}{6} + B_1 x + B_2$$

To determine the four constants A_1, A_2, B_1, and B_2, two boundary and two continuity conditions must be used.

For segment AD: $\qquad v(0) = 0 = A_2$

For segment DB: $\qquad v(L) = 0 = \dfrac{PaL^2}{3EI} + B_1 L + B_2$

Equating deflections for both segments at $x = a$:

$$v_D = v(a) = \frac{Pa^3 b}{6EIL} + A_1 a = \frac{Pa^3}{2EI} - \frac{Pa^4}{6EIL} + B_1 a + B_2$$

Equating slopes for both segments at $x = a$:

$$\theta_D = v'(a) = \frac{Pa^2 b}{2EIL} + A_1 = \frac{Pa^2}{EI} - \frac{Pa^3}{2EIL} + B_1$$

Upon solving the above four equations simultaneously, one finds

$$A_1 = -\frac{Pb}{6EIL}(L^2 - b^2) \qquad A_2 = 0$$

$$B_1 = -\frac{Pa}{6EIL}(2L^2 + a^2) \qquad B_2 = \frac{Pa^3}{6EI}$$

With these constants, for example, the elastic curve for the left segment AD of the beam becomes

$$v = [(Pb/(6EIL)][x^3 - (L^2 - b^2)x] \qquad (11\text{-}27)$$

The largest deflection occurs in the longer segment of the beam. If $a > b$, the point of maximum deflection is at $x = \sqrt{a(a + 2b)/3}$, which follows from setting the expression for the slope equal to zero. The deflection at this point is

$$|v|_{\max} = \frac{Pb(L^2 - b^2)^{3/2}}{9\sqrt{3}EIL} \qquad (11\text{-}28)$$

Usually the deflection at the center of the span is very nearly equal to the numerically largest deflection. Such a deflection is much simpler to determine, recommending its use in practice. If the force P is applied at the middle of the span, i.e., $a = b = L/2$, it can be shown by direct substitution into Eq. 11-27 or 11-28 that at $x = L/2$

$$|v|_{\max} = PL^3/(48EI) \qquad (11\text{-}29)$$

Case (b). The solution of the same problem using singularity functions is very direct and follows the procedure used previously:

$$EI\frac{d^4v}{dx^4} = p = -P\langle x - a \rangle_*^{-1}$$

$$EI\frac{d^3v}{dx^3} = -P\langle x - a \rangle^0 + C_1$$

$$EI\frac{d^2v}{dx^2} = -P\langle x - a \rangle^1 + C_1 x + C_2$$

But $M(0) = 0$; hence $EIv''(0) = 0 = C_2$; and, since also $M(L) = 0$,

$$EIv''(L) = -Pb + C_1 L = 0 \qquad \text{or} \qquad C_1 = Pb/L$$

$$EI\frac{dv}{dx} = -\frac{P}{2}\langle x - a \rangle^2 + \frac{Pb}{2L}x^2 + C_3$$

$$EIv = -\frac{P}{6}\langle x - a \rangle^3 + \frac{Pb}{6L}x^3 + C_3 x + C_4$$

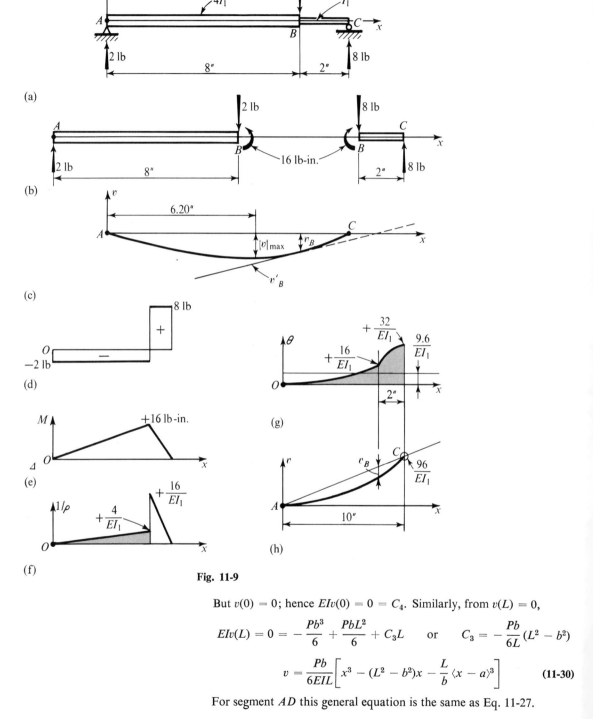

Fig. 11-9

But $v(0) = 0$; hence $EIv(0) = 0 = C_4$. Similarly, from $v(L) = 0$,

$$EIv(L) = 0 = -\frac{Pb^3}{6} + \frac{PbL^2}{6} + C_3 L \quad \text{or} \quad C_3 = -\frac{Pb}{6L}(L^2 - b^2)$$

$$v = \frac{Pb}{6EIL}\left[x^3 - (L^2 - b^2)x - \frac{L}{b}\langle x - a \rangle^3\right] \quad \text{(11-30)}$$

For segment AD this general equation is the same as Eq. 11-27.

EXAMPLE 11-5

A simply supported beam 10 in. long is loaded with a 10-lb downward force 8 in. from the left support, Fig. 11-9(a). The cross section of the beam is such that in the segment AB the moment of inertia is $4I_1$; in the remainder of the beam it is I_1. Determine the elastic curve.

Section 11-8
Solution of beam deflection problems by direct integration

SOLUTION

An analytical solution of this problem can be achieved by either one of the two methods used in the previous example. Since, however, a special procedure must be used with singularity functions, this approach will be illustrated here.*

The beam is separated at the point of discontinuity in I and the forces necessary for the equilibrium of segments are computed, Fig. 11-9(b). Then the solution is commenced independently for each segment of the beam with the same location of the origin at A. The successive integrations are carried out until the moment-curvature equations are found. No constants of integration appear in the first two integrations as the reactive forces are computed beforehand. For segment AB:

$$\frac{d^4v}{dx^4} = \frac{p}{EI} = \frac{1}{4EI_1}[+2\langle x\rangle_*^{-1} - 2\langle x-8\rangle_*^{-1} - 16\langle x-8\rangle_*^{-2}]$$

$$\frac{d^3v}{dx^3} = -\frac{V}{EI} = \frac{1}{2EI_1}\langle x\rangle^0 - \frac{1}{2EI_1}\langle x-8\rangle^0 - \frac{4}{EI_1}\langle x-8\rangle_*^{-1}$$

$$\frac{d^2v}{dx^2} = \frac{M}{EI} = \frac{1}{2EI_1}\langle x\rangle^1 - \frac{1}{2EI_1}\langle x-8\rangle^1 - \frac{4}{EI_1}\langle x-8\rangle^0$$

$$\frac{dv}{dx} = \theta = \frac{1}{4EI_1}\langle x\rangle^2 - \frac{1}{4EI_1}\langle x-8\rangle^2 - \frac{4}{EI_1}\langle x-8\rangle^1 + \theta_o$$

where θ_o is an unknown constant of integration. For segment BC:

$$\frac{d^4v}{dx^4} = \frac{p}{EI} = \frac{1}{EI_1}[+16\langle x-8\rangle_*^{-2} - 8\langle x-8\rangle_*^{-1} + 8\langle x-10\rangle_*^{-1}]$$

$$\frac{d^3v}{dx^3} = -\frac{V}{EI} = \frac{16}{EI_1}\langle x-8\rangle_*^{-1} - \frac{8}{EI_1}\langle x-8\rangle^0 + \frac{8}{EI_1}\langle x-10\rangle^0$$

$$\frac{d^2v}{dx^2} = \frac{M}{EI} = \frac{16}{EI_1}\langle x-8\rangle^0 - \frac{8}{EI_1}\langle x-8\rangle^1 + \frac{8}{EI}\langle x-10\rangle^1$$

At this stage of integration it must be recognized that by virtue of the continuity requirements, the terminal value of θ for the segment AB is the initial one for the segment BC. Moreover, this expression of θ for

---

* Several schemes for using singularity functions for problems with variable I can be devised. For the procedure used here the author is indebted to Professor E. L. Wilson.

the segment AB for $x \geq 8$ remains constant. Therefore it is possible to integrate the last expression above for the segment BC and add to it the θ for segment AB. This yields a complete continuous function of θ for the whole beam AC. On subsequent integrations the boundary conditions can be used for determining the constants. For segment BC:

$$\frac{dv}{dx} = \theta = \frac{16}{EI_1} \langle x - 8 \rangle^1 - \frac{4}{EI_1} \langle x - 8 \rangle^2$$

Here the last term of the earlier expression, having no relevance, has been dropped. Adding this expression to the one found earlier for the segment AB, one has for the entire beam AC:

$$\frac{dv}{dx} = \theta = \frac{1}{4EI_1} \langle x \rangle^2 - \frac{4.25}{EI_1} \langle x - 8 \rangle^2 + \frac{12}{EI_1} \langle x - 8 \rangle^1 + \theta_o$$

and $\quad v = \dfrac{1}{12EI_1} \langle x \rangle^3 - \dfrac{4.25}{3EI_1} \langle x - 8 \rangle^3 + \dfrac{6}{EI_1} \langle x - 8 \rangle^2 + \theta_o x + v_o$

But since $v(0) = 0$, one has $v_o = 0$. The condition that $v(10) = 0$ yields $\theta_o = -9.6/(EI_1)$. This completes the solution of the problem.

The equation for the slope in segment AB is $\theta = x^2/(4EI_1) - 48/(5EI_1)$. Upon setting this quantity equal to zero, x is found to be 6.20 in. The largest deflection occurs at this value of x, and $|v|_{\max} = 39.7/(EI_1)$. Characteristically, the deflection at the center of the span—i.e., at $x = 5$ in.—is nearly the same, being $37.6/(EI_1)$.

A self-explanatory graphical procedure is in Figs. 11-9(d) through (h). Variation in I causes virtually no complications in the graphical solution, a great advantage in complex problems.

11-9. STATICALLY INDETERMINATE ELASTIC BEAM PROBLEMS

In a large and important class of beam problems reactions cannot be determined using the conventional procedures of statics. For example, for the beam shown in Fig. 11-10(a), four reaction components are unknown. The three vertical components cannot be found from equations of static equilibrium. Further examination of Fig. 11-10(a) shows that any one of the vertical reactions can be removed and the beam would remain in equilibrium. Therefore any one of these reactions may be said to be *superflous*, or *redundant*, for maintaining equilibrium. Problems with extra or redundant reactive forces and/or moments are called (*externally*) *statically indeterminate*.

When the number of unknown reactions exceeds by one that which may be determined by statics, the member is said to be indeterminate to the *first degree*. As the number of unknowns increases, the degree of indeterminacy also increases. For example, by providing one more support than shown for the beam in Fig. 11-10(a), the beam would become

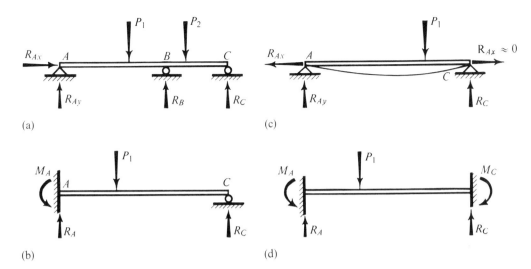

Fig. 11-10. Illustrations of statical indeterminacy of beams. In (a) and (b) the beams are indeterminate to the first degree. If it is assumed that the horizontal components of the reactions are negligible, the beam in (c) is determinate, and in (d) indeterminate to the second degree.

indeterminate to the second degree. The beam of Fig. 11-10(b) is indeterminate to the first degree since either M_A or R_C can be considered redundant.

As according to Eq. 11-12 for small deflections $ds \approx dx$, no significant axial strain can develop in a transversely loaded beam.* Therefore, the horizontal components of the reactions in situations with immovable supports, such as shown in Figs. 11-10(c) and (d), are negligible. On this basis, the beam shown in Fig. 11-10(c), with pins at both ends, is a determinate beam. The beam of Fig. 11-10(d) is indeterminate to the second degree.

To determine the elastic deflection of statically indeterminate beams, the procedure of solving the differential equations is practically the same as that discussed above for determinate beams. The only difference is that kinematic boundary conditions replace some of the static ones. As the degree of indeterminacy increases, as in continuous members, the number of simultaneous equations for determining the constants increases. In such problems the number of constants to be found is no longer limited to a maximum of four.

A simple example of a statically indeterminate beam follows. After solving the problem using the fourth-order differential equation, an approach for applying the second-order equation is given.

---

* The horizontal force becomes important in thin plates. See S. Timoshenko and S. Woinowsky-Krieger, *Theory of Plates and Shells* (2nd. ed.) (New York: McGraw-Hill Book Company, 1959), p. 6.

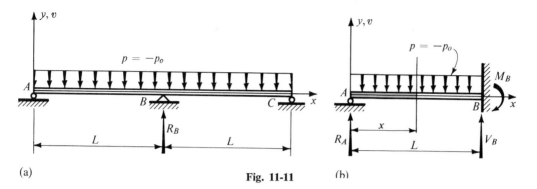

Fig. 11-11

EXAMPLE 11-6

Find the equation of the elastic curve for the uniformly loaded, two-span continuous beam shown in Fig. 11-11(a). The EI is constant.

SOLUTION

Equation 11-17, $EIv^{iv} = p$, can be used here. In addition to the boundary conditions of zero moment and zero deflection at the ends A and C, the deflection at B is also zero. The reaction R_B at B must be treated as an unknown force. On this basis

$$EI\frac{d^4v}{dx^4} = p = -p_o + R_B \langle x - L \rangle_*^{-1}$$

$$EI\frac{d^3v}{dx^3} = -p_o x + R_B \langle x - L \rangle^0 + C_1$$

$$EI\frac{d^2v}{dx^2} = -\frac{p_o x^2}{2} + R_B \langle x - L \rangle^1 + C_1 x + C_2$$

$$EI\frac{dv}{dx} = -\frac{p_o x^3}{6} + \frac{R_B}{2} \langle x - L \rangle^2 + \frac{C_1 x^2}{2} + C_2 x + C_3$$

$$EIv = -\frac{p_o x^4}{24} + \frac{R_B}{6} \langle x - L \rangle^3 + \frac{C_1 x^3}{6} + \frac{C_2 x^2}{2} + C_3 x + C_4$$

The five constants R_B, C_1, C_2, C_3, and C_4 are found from three deflection and two moment conditions:

$$v_A = v(0) = 0, \qquad v_B = v(L) = 0, \qquad v_C = v(2L) = 0$$
$$M_A = EIv''(0) = 0 \quad \text{and} \quad M_C = EIv''(2L) = 0$$

The boundary conditions at $x = 0$ yield directly $C_4 = 0$ and $C_2 = 0$. The remaining three conditions $v_B = v_C = M_C = 0$ give the following three simultaneous equations:

$$+\tfrac{1}{6} L^3 C_1 + L C_3 = \tfrac{1}{24} p_o L^4$$
$$+\tfrac{1}{6} L^3 R_B + \tfrac{4}{3} L^3 C_1 + 2 L C_3 = \tfrac{2}{3} p_o L^4$$
$$+L R_B + 2 L C_1 = 2 p_o L^2$$

from which $R_B = \tfrac{5}{4} p_o L$, $C_1 = \tfrac{3}{8} p_o L$, and $C_3 = -\tfrac{1}{48} p_o L^3$. Therefore

$$v = -[p_o/(48 EI)][2x^4 - 3 L x^3 + L^3 x - 10 L \langle x - L \rangle^3] \quad (11\text{-}31)$$

ALTERNATE SOLUTIONS

The stated problem is symmetrical around the support at B. Therefore the tangent to the elastic curve at B is horizontal, and an equivalent problem involving one-half of the original beam shown in Fig. 11-11(b) can be analyzed instead. This new problem can be solved using the fourth-order differential equation with the following four boundary conditions:

$$v_A = 0, \quad M_A = EI v''(0) = 0, \quad v_B = 0, \quad \text{and} \quad v'_B = 0$$

Alternatively, on designating the unknown reaction at A as R_A, one may state the bending moment within the span as

$$M = R_A x - p_o x^2/2$$

By substituting this relation into Eq. 11-15, integrating it twice, and making use of three of the kinematic boundary conditions stated above, one finds the unknown constants R_A, C_3, and C_4.

11-10. TWO ADDITIONAL SINGULARITY FUNCTIONS

In some assemblies of beams it is necessary to introduce connections which permit movement. One commonly used type, a hinge, cannot resist bending moment, but can transmit shear. Thus it is a shear connection. Another, basically different connection, a moment connection, can transmit moment but not shear. The elastic curve at such connections is not continuous. At a hinge a local concentrated angle change between the tangents to the elastic curve takes place. In a moment connection the adjoining ends of the elastic curve undergo relative translations. The characteristic discontinuities in the elastic curves at such connections are shown in Fig. 11-12. An abrupt change of slope at a hinge is shown as $\Delta \theta_c$; an abrupt change in deflection at a moment connection as Δv_b.

For the analysis of beams having the above connections it is convenient to introduce two new singularity functions expressing a fictitious load acting on a beam as

$$p = \Delta \theta_a EI \langle x - a \rangle_*^{-3} \quad [\text{lb/in.}] \quad (11\text{-}32)$$

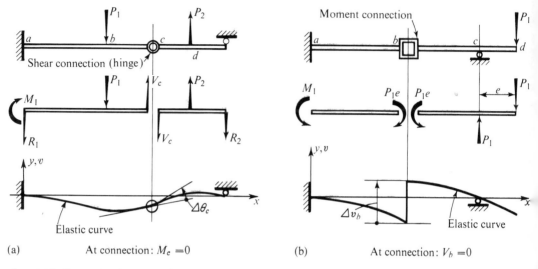

Fig. 11-12. Discontinuities in the elastic curves at connections.

for a concentrated change of slope $\Delta\theta_a$ at $x = a$; and as

$$p = \Delta v_a EI \langle x - a \rangle_*^{-4} \quad \text{[lb/in.]} \tag{11-33}$$

for a concentrated or sudden change in deflection Δv_a at $x = a$.

These functions together with the previously defined ones, $P\langle x - a \rangle_*^{-1}$ for a concentrated force and $M_a \langle x - a \rangle_*^{-2}$ for a concentrated moment, integrate according to the following rule:

$$\int_0^x \langle x - a \rangle_*^n \, dx = \langle x - a \rangle_*^{n+1} \quad \text{for} \quad n < 0 \tag{11-34}$$

As before, Eq. 2-16 applies for $n \geq 0$.

The negative exponents in Eqs. 11-32 and 11-33 are so taken that on successive integration of $EIv^{\text{iv}} = p$, one obtains the correct quantities for slope and deflections.

Having available the set of four singularity functions permits remarkable versatility in the analysis of beams. The beams can be either determinate or indeterminate. In either case the boundary conditions at both ends must be used in the solution of problems. If connections exist, then an additional condition of $M = 0$ holds true at each hinge and of $V = 0$ at every moment connection. In indeterminate beams, for every constraint caused by a redundant reaction, a kinematic condition on the deflection becomes available. The solution of beams with the aid of

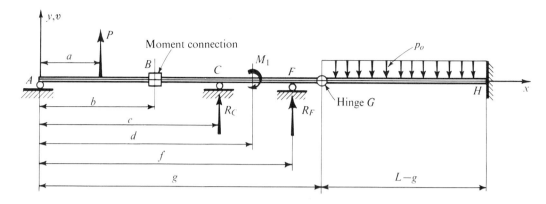

Fig. 11-13

singularity functions is completely general.* The method becomes, however, very cumbersome for stepped beams.

As an example of general approach, consider the beam shown in Fig. 11-13. The pertinent equation here is

$$EI\frac{d^4v}{dx^4} = p = P\langle x-a\rangle_*^{-1} + \Delta v_B EI\langle x-b\rangle_*^{-4} + R_C\langle x-c\rangle_*^{-1}$$
$$+ M_1\langle x-d\rangle_*^{-2} + R_F\langle x-f\rangle_*^{-1} + \Delta\theta_G EI\langle x-g\rangle_*^{-3} - p_o\langle x-g\rangle^0$$

where the loads P, M_1, and p_o are given. The boundary conditions are

$$v_A = v(0) = 0, \qquad M(0) = EIv''(0) = 0,$$
$$v_H = v(L) = 0, \quad \text{and} \quad v'(L) = 0$$

The conditions at the connections are

$$V_B = V(b) = 0 \quad \text{and} \quad M_G = M(g) = 0$$

And the two restraining conditions are

$$v_C = v(c) = 0 \quad \text{and} \quad v_F = v(f) = 0$$

The above information is sufficient for solving this complex problem, which, however, is indeterminate only to the first degree.

---

* Some of the presentation in this article follows R. J. Brungraber, "Singularity Functions in the Solution of Beam-Deflection Problems," *Journal of Engineering Education*, **55**, no. 9 (May 1965), 278–80. Although no cases can be found in literature, these functions can be used very effectively in constructing the influence lines for beams. This topic, however, is beyond the scope of this book.

11-11. REMARKS ON THE ELASTIC DEFLECTION OF BEAMS

The integration procedures discussed above for obtaining the elastic deflections of loaded beams are generally applicable. The reader must realize, however, that numerous problems with different loadings have been solved and are readily available.* Nearly all the tabulated solutions are made for simple loading conditions. Therefore in practice the deflections of beams subjected to several or complicated loading conditions are usually synthesized from the simpler loadings, using the principle of superposition. For example, the problem in Fig. 11-14 may be separated into three different cases as shown. The algebraic sum of the three separate deflections caused by the separate loads for the same point gives the total deflection.

Note that in applying superposition, the solution of Example 11-4 for a concentrated force P at an arbitrary location may be used for determining deflections of beams with the same boundary conditions for any loading. For distributed loads, P must be replaced by $p\,dx$ and integrated over the loaded range.

The procedures discussed above for determining elastic deflection of straight beams can be extended to structural systems consisting of several flexural members. For example, consider the simple frame shown in Fig. 11-15(a), for which the deflection of point C due to the applied force P is sought. The deflection of the vertical leg BC alone can be found by treating it as a cantilever fixed at B. However, due to the applied load, joint B deflects and rotates. This is determined by studying the behavior of the member AB.

A free-body diagram for the member AB is in Fig. 11-15(b). This member is seen to resist the axial force P and a moment $M_1 = Pa$. Usually the effect of the axial force P on deflections due to bending can be neglected.† The axial elongation of a member usually is also very small in

---

* See any civil or mechanical engineering handbook.
† Recall discussion in connection with Fig. 8-1 and see Art. 14-3 on beam columns.

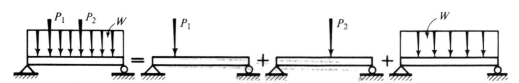

Fig. 11-14. Resolution of a complex problem into several simpler problems in computing deflections.

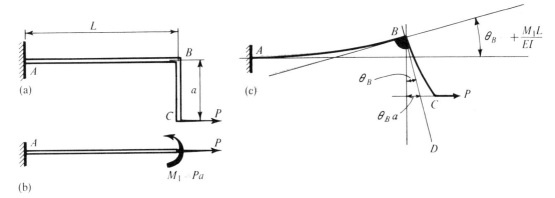

Fig. 11-15. A method of analyzing deflections of frames.

comparison with the bending deflections. Therefore here the problem can be reduced to that of determining the deflection and rotation of B caused by an end moment M_1. This has been done in Example 11-2; from it the angle θ_B is noted on Fig. 11-15(c). By multiplying this angle θ_B by the length a of the vertical member, the deflection of point C due to rotation of joint B is found. The cantilever deflection of the member BC when treated alone is augmented by the amount $\theta_B a$. The vertical deflection of C is equal to the vertical deflection of point B.

In interpreting the shape of deformed structures such as shown in Fig. 11-15(c) it must be kept clearly in mind that the deformations are greatly exaggerated. In the small deformation theory discussed here, the cosines of all small angles such as θ_B are taken to equal unity. Both the deflections and the rotations of the elastic curve are small.

Instead of using the singularity functions, occasionally it is desirable to analyze beams with overhangs in the manner described above. For this purpose, for example, the portion of a beam between the supports, as AB in Fig. 11-16(a), is isolated* and rotation of the tangent at B is found. The remainder of the problem is analogous to the case discussed before.

---

* The effect of the overhang on the beam segment AB must be included by introducing at the support B a bending moment $-Pa$.

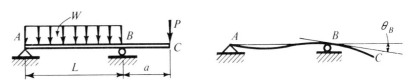

Fig. 11-16. A method of analyzing deflections of an overhang.

11-12. ELASTIC DEFLECTION OF BEAMS IN SKEW BENDING

In the preceding discussion the deflection of beams was assumed to take place in one plane. More precisely, the foregoing theory applies to deflections of beams when the applied moments act around one of the principal axes of the cross section and when deflection takes place in a plane normal to such an axis. Unless this is the case, another moment develops tending to bend the beam around the other principal axis (see Art. 6-5). If unsymmetrical bending occurs, the elastic deflection problem may be solved by superposition. The elastic curve in the plane containing one of the principal axes is determined by considering only the effect of the components of forces acting parallel to this axis. The elastic curve in the plane containing the other principal axis is found similarly. A vectorial addition of the deflections so found at a particular point of a beam gives the total displacement of the beam at that point. For example, if a certain beam made of a Z section is subjected to unsymmetrical bending and the deflection at a particular point is v_1 in the y direction, and w_1 in the z direction, Fig. 11-17, the total deflection is $v_1 \leftrightarrow w_1$, i.e., the distance AA'.

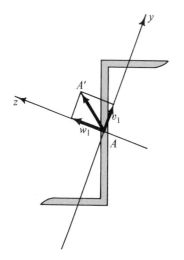

Fig. 11-17. Deflection of a beam subjected to unsymmetrical bending.

Such deflections, without torsion, occur only if the applied forces pass through the shear center (see Art. 7-7).

11-13. INELASTIC DEFLECTION OF BEAMS

All the preceding solutions for beam deflections apply only if the material behaves elastically. This limitation is the result of introducing Hooke's law into the strain-curvature relation, Eq. 11-5, to yield the moment-curvature equation, Eq. 11-6. The subsequent procedures of approximating the curvature as d^2v/dx^2 and the integration schemes have nothing to do with the material properties.

If attention is limited to statically determinate beams, the bending moments throughout a member can be determined regardless of the material properties of the beam. Then further, if for a given cross section a relationship between the bending moment and curvature is available, the curvature diagram or function for the given beam can be established. Upon two successive integrations of the curvature relation, with adjustments for the boundary conditions, the inelastic deflection of a given beam can be found. This will be illustrated in the next two examples.

Superposition does not apply in inelastic problems since deflections are not linearly related to the applied forces. As a consequence of this, in the indeterminate beams time-consuming trial-and-error solutions are often required to calculate deflections. The bending moments depend on the reactions, and the latter depend on the nonlinear response of the beam to deformations. This will not be pursued in this book. An approach for the plastic strength analysis of statically indeterminate beams will be given, however, in the next chapter.

Section 11-13
Inelastic deflection of beams

EXAMPLE 11-7

Determine and plot the moment-curvature relationship for an elastic-ideally-plastic rectangular beam.

SOLUTION

In a rectangular elastic-plastic beam at y_0, see Fig. 6-17, where the juncture of the elastic and plastic zones occurs, the linear strain $\varepsilon_x = \pm\varepsilon_{yp}$. Therefore, according to Eq. 11-5, with the curvature $1/\rho = \kappa$,

$$\frac{1}{\rho} = \kappa = -\frac{\varepsilon_{yp}}{y_0} \quad \text{and} \quad \kappa_{yp} = -\frac{\varepsilon_{yp}}{h/2}$$

where the last expression gives the curvature of the member at impending yielding when $y_o = h/2$. From the above relations

$$\frac{y_o}{h/2} = \frac{\kappa_{yp}}{\kappa}$$

On substituting this expression into Eq. 6-14, one obtains the required

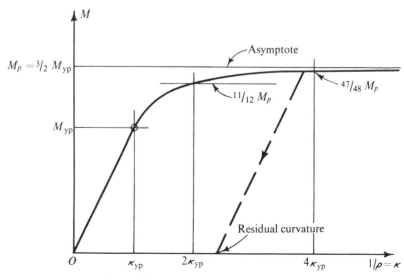

Fig. 11-18. Moment-curvature relation for a rectangular beam.

moment-curvature relationship:

$$M = M_p\left[1 - \tfrac{1}{3}\left(\tfrac{y_o}{h/2}\right)^2\right] = \tfrac{3}{2} M_{yp}\left[1 - \tfrac{1}{3}\left(\tfrac{\kappa_{yp}}{\kappa}\right)^2\right] \quad (11\text{-}35)$$

This function is plotted in Fig. 11-18. Note how rapidly it approaches the asymptote. At curvature just double that of the impending yielding, eleven-twelfths or 91.6 per cent of the ultimate plastic moment M_p is already reached. At this point the middle half of the beam remains elastic.

On releasing an applied moment the beam rebounds elastically as shown in the figure. On this basis residual curvature can be determined.

The reader should recall that the ratio of M_p to M_{yp} varies for different cross sections.

EXAMPLE 11-8

A 3-in.-wide mild-steel cantilever beam has the other dimensions as shown in Fig. 11-19(a). Determine the tip deflection caused by applying the two loads of 5 kips each. Assume $E = 30 \times 10^3$ ksi, and $\sigma_{yp} = \pm 40$ ksi.

SOLUTION

The moment diagram is in Fig. 11-19(b). From $\sigma_{max} = Mc/I$ it is found that the largest stress in the beam segment ab is 24.4 ksi, which indicates

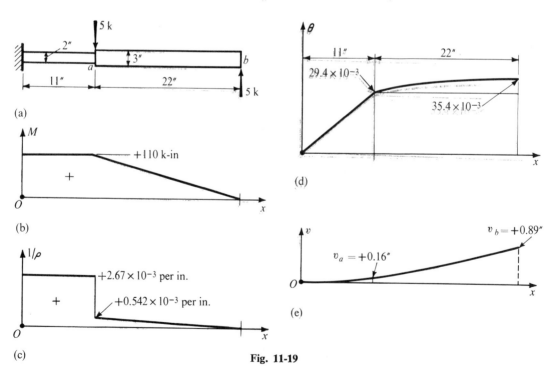

Fig. 11-19

elastic behavior. An analogous calculation for the shallow section of the beam gives a stress of 55 ksi, which is not possible as the material yields at 40 ksi.

*Section 11-14
Introduction to the moment-area method*

A check of the ultimate capacity for the 2-in.-deep section based on Eq. 6-12 gives

$$M_p = M_{ult} = \sigma_{yp}\frac{bh^2}{4} = \frac{40 \times 3 \times 2^2}{4} = 120 \text{ k-in.}$$

This calculation shows that although the beam yields partially, it can carry the applied moment. The applied moment is $1\tfrac{1}{12} M_p$. According to the results found in the preceding example, this means that the curvature in the 2-in.-deep section of the beam is twice that at the beginning of yielding. Therefore the curvature in the 11-in. segment of the beam adjoining the support is

$$\frac{1}{\rho} = 2\kappa_{yp} = 2\frac{\varepsilon_{yp}}{h/2} = 2\frac{\sigma_{yp}}{Eh/2} = \frac{2 \times 40}{30 \times 10^3 \times 1} = 2.67 \times 10^{-3} \text{ per in.}$$

The maximum curvature for segment ab is

$$\frac{1}{\rho} = \frac{M_{max}}{EI} = \frac{\sigma_{max}}{Ec} = \frac{24.4}{30 \times 10^3 \times 1.5} = 0.542 \times 10^{-3} \text{ per in.}$$

These data on curvatures are plotted in Fig. 11-19(c). On integrating this twice with $\theta(0) = 0$ and $v(0) = 0$, the deflected curve, Fig. 11-19(e), is obtained. The tip deflection is 0.89 in. upward.

If the applied loads were released, the beam would rebound elastically. This amounts to 0.64 in. at the tip and a residual tip deflection of 0.25 in. would remain. The residual curvature would be confined to the 2-in.-deep segment of the beam.

If the end load were applied alone, the beam would collapse. Superposition cannot be used to solve this problem.

PART B
MOMENT-AREA METHOD*

11-14. INTRODUCTION TO THE MOMENT-AREA METHOD†

In numerous engineering applications where deflections of beams must be determined, the loading is complex, and the cross-sectional areas of the beam vary. This is the usual situation in shafts of machines, where

---

* In a short course, the remainder of this chapter may be omitted.
† The development of the moment-area method for finding deflections of beams is due to Charles E. Greene, of the University of Michigan, who taught it to his classes in 1873. Somewhat earlier, in 1868, Otto Mohr, of Dresden, Germany, developed a similar method which appears to have been unknown to Professor Greene.

gradual or stepwise variations in the shaft diameter are made to accommodate rotors, bearings, collars, retainers, etc. Likewise, haunched or tapered beams are frequently employed in aircraft as well as in bridge construction. By interpreting semigraphically the mathematical operations of solving the governing differential equation, an effective procedure for obtaining deflections in complicated situations has been developed. Using this alternative procedure, one finds that problems with load discontinuities and arbitrary variations of inertia of the cross-sectional area of a beam cause no complications and require only a little more arithmetical work for their solution. The solution of such problems is the objective in this part of the chapter on the moment-area method.

The method to be developed usually is used to obtain only the displacement and rotation at a single point on a beam. It can be used to determine the equation of the elastic curve, but no advantage is gained in comparison with the direct solution of the differential equation. Often, however, it is the deflection or the angular rotation of the elastic curve at a particular point of a beam or both that are of greatest interest in the solution of practical problems.

The method of moment areas is just an alternative method for solving the deflection problem. It possesses the same approximations and limitations discussed earlier in connection with the solution of the differential equation of the elastic curve. By applying it, one determines only the deflection due to the flexure of the beam; deflection due to shear is neglected. Here the application of the method will be limited to statically determinate beams. Statically indeterminate situations will be considered in the next chapter.

11-15. DERIVATION OF THE MOMENT-AREA THEOREMS

The necessary theorems are based on the geometry of the elastic curve and the associated $M/(EI)$ diagram. Boundary conditions do not enter into the derivation of the theorems since the theorems are based only on the interpretation of definite integrals. As will be shown later, further geometrical considerations are necessary to solve a complete problem.

For deriving the theorems, Eq. 11-10, $d^2v/dx^2 = M/(EI)$, can be rewritten in the following alternative forms:

$$\frac{d^2v}{dx^2} = \frac{d}{dx}\left(\frac{dv}{dx}\right) = \frac{d\theta}{dx} = \frac{M}{EI} \quad \text{or} \quad d\theta = \frac{M}{EI}dx \qquad (11\text{-}36)$$

As may be seen from Fig. 11-20(a), the quantity $[M/(EI)]\,dx$ corresponds to an infinitesimal area of the $M/(EI)$ diagram. According to Eq. 11-36 this area is equal to the change in angle between two adjoining tangents. The contribution of an angle change in one element to the deformation of the elastic curve is shown in Fig. 11-20(b).

If the small angle change $d\theta$ for an element is multiplied by a distance x from an arbitrary origin to the same element, a vertical distance dt is obtained, see Fig. 11-20(b). As only small deflections are considered, no distinction between the arc AA' and the vertical distance dt need be made. Based on this geometrical reasoning, one has

$$dt = x\, d\theta = \frac{M}{EI} x\, dx \qquad (11\text{-}37)$$

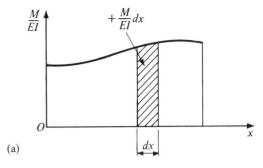

(a)

Formally integrating Eqs. 11-36 and 11-37 between any two points such as A and B on a beam, see Fig. 11-21, yields the two moment-area theorems. The first is

$$\int_A^B d\theta = \theta_B - \theta_A = \Delta\theta_{BA} = \int_A^B \frac{M}{EI} dx \qquad (11\text{-}38)$$

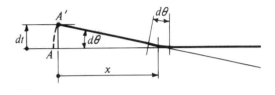

(b)

This states that the change in angle measured in radians between the two tangents at any two points A and B on the elastic curve is equal to the $M/(EI)$ area bounded by the ordinates through A and B.

Fig. 11-20. Interpretation of a small angle change in an element.

Therefore, if the slope of the elastic curve at one point, as at A, is known, the slope at another point on the right, as at B, can be determined:

$$\theta_B = \theta_A + \Delta\theta_{BA} \qquad (11\text{-}39)$$

The first theorem shows that a numerical evaluation of the $M/(EI)$ area bounded between the ordinates through any two points on the elastic curve gives the angular rotation between the corresponding tangents. In performing this summation, areas corresponding to the positive bending moments are taken positive, and those corresponding to the negative moments are taken negative. If the sum of the areas between any two points, such as A and B, is positive, the tangent on the right rotates in the counterclockwise direction; if negative, the tangent on the right rotates in a clockwise direction (see Fig. 11-21(b)). If the net area is zero, the two tangents are parallel.

The quantity dt in Fig. 11-21(b) is due to the effect of curvature of an element. By summing this effect for all elements from A to B, the vertical distance AF is obtained. Geometrically this distance represents the displacement or deviation of a point A from a tangent to the elastic curve at B. Henceforth, it will be termed the *tangential deviation* of a point A from a tangent at B and will be designated t_{AB}. The foregoing, in mathematical form, gives the second moment-area theorem:

$$t_{AB} = \int_A^B d\theta\, x = \int_A^B \frac{M}{EI} x\, dx \qquad (11\text{-}40)$$

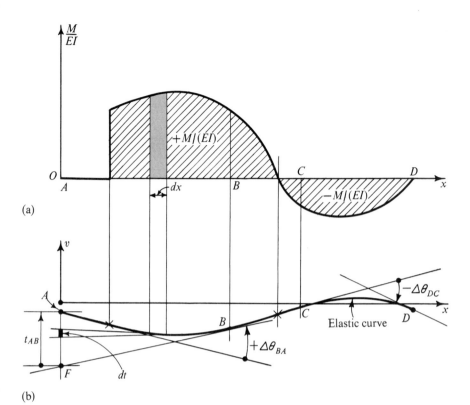

Fig. 11-21. Relation between the $M/(EI)$ diagram and the elastic curve.

This states that the tangential deviation of a point A on the elastic curve from a tangent through another point B also on the elastic curve is equal to the statical (or first) moment of the bounded section of the $M/(EI)$ diagram around a vertical line through A. In most cases, the tangential deviation is not in itself the desired deflection of a beam.

Using the definition of the center of gravity of an area, one may for convenience restate Eq. 11-40 in numerical applications in a simpler form as

$$t_{AB} = \Phi \bar{x} \qquad (11\text{-}41)$$

where Φ is the total area of the $M/(EI)$ diagram between the two points considered and $\bar{x}$ is the horizontal distance to the centroid of this area from A.

By analogous reasoning the deviation of a point B from a tangent through A is

$$t_{BA} = \Phi \bar{x}_1 \qquad (11\text{-}42)$$

where the same $M/(EI)$ area is used, but $\bar{x}_1$ is measured from the vertical line through point B, Fig. 11-22. Note carefully the order of the subscript letters for *t* in these two equations. The point whose deviation is being determined is written first.

In the above equations, the distances $\bar{x}$ or $\bar{x}_1$ are always taken positive, and as E and I are also positive quantities, the sign of the tangential deviation depends on the sign of the bending moments. A positive value for the tangential deviation indicates that a given point lies above a tangent to the elastic curve drawn through the other point and vice versa, Fig. 11-22.

The above two theorems are applicable between any two points on a continuous elastic curve of any beam for any loading. They apply between and beyond the reactions for overhanging and continuous beams. However, it must be emphasized that only relative rotation of the tangents and only tangential deviations are obtained directly. A further con-

*Section 11-15
Derivation of the moment-area theorems*

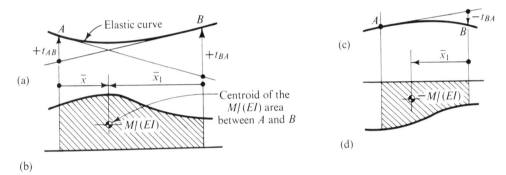

Fig. 11-22. The meaning of signs for tangential deviation.

sideration of the geometry of the elastic curve at the supports to include the boundary conditions is necessary in every case to determine deflections. This will be illustrated in the examples that follow.

In applying the moment-area method a carefully prepared sketch of the elastic curve is always necessary. Since no deflection is possible at a pinned or a roller support, the elastic curve is drawn passing through such supports. At a fixed support, neither displacement nor rotation of the tangent to the elastic curve is permitted, so the elastic curve must be drawn tangent to the direction of the unloaded axis of the beam. In preparing a sketch of the elastic curve in the above manner, it is customary to exaggerate the anticipated deflections. On such a sketch the deflection of a point on a beam is usually referred to as being above or below its initial position, without much emphasis on the signs. To aid in the application of the method, useful properties of areas enclosed by curves and centroids are assembled in Table 2 of the Appendix.

EXAMPLE 11-9

Consider an aluminum cantilever beam 16 in. long with a 1,000-lb force applied 4 in. from the free end, as shown in Fig. 11-23(a). For a distance

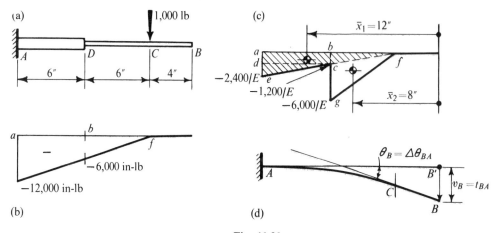

Fig. 11-23

of 6 in. from the fixed end, the beam is of greater depth than it is beyond, having $I_1 = 5$ in.^{4} For the remaining 10 in. of the beam, $I_2 = 1$ in.^{4} Find the deflection and the angular rotation of the free end. Neglect the weight of the beam, and assume E for aluminum at 10^7 psi.

SOLUTION

The bending-moment diagram is in Fig. 11-23(b). By dividing all ordinates of the M diagrams by EI, the $M/(EI)$ diagram in Fig. 11-23(c) is obtained. Two ordinates appear at point D. One, $-1,200/E$, is applicable just to the left of D, the other, $-6,000/E$, applies just to the right of D. Since the bending moment is negative from A to C, the elastic curve throughout this distance is concave down, Fig. 11-23(d). At the fixed support A, the elastic curve must start out tangent to the initial direction AB' of the unloaded beam. The unloaded straight segment CB of the beam is tangent to the elastic curve at C.

After the foregoing preparatory steps, from the geometry of the sketch of the elastic curve it may be seen that the distance BB' represents the desired deflection of the free end. However, BB' is also the tangential deviation of the point B from the tangent at A. Therefore the second moment-area theorem may be used to obtain t_{BA}, which in this special case represents the deflection of the free end. Also, from the geometry of the elastic curve it is seen that the angle included between the lines BC and AB' is the angular rotation of the segment CB. This angle is the same as the one included between the tangents to the elastic curve at the points A and B, and the first moment-area theorem may be used to compute this quantity.

It is convenient to extend the line ec in Fig. 11-23(c) to the point f for computing the area of the $M/(EI)$ diagram. This gives two triangles, the areas of which are easily calculated.*

---

* A little ingenuity in such cases saves arithmetical work. Of course it is perfectly correct in this example to use the two triangular areas dce and bfg, and a rectangle $abcd$.

Section 11-15
Derivation of the
moment-area
theorems

The area of triangle *afe*:

$$\Phi_1 = -\tfrac{1}{2}(12)(2,400)/E = -14,400/E$$

The area of triangle *fcg*:

$$\Phi_2 = -\tfrac{1}{2}(6)(4,800)/E = -14,400/E$$

$$\theta_B = \Delta\theta_{BA} = \int_A^B \frac{M}{EI}\,dx = \Phi_1 + \Phi_2 = -\frac{28,800}{10^7} = -0.00288 \text{ radian}$$

$$v_B = t_{BA} = \Phi_1 \bar{x}_1 + \Phi_2 \bar{x}_2 = (-14,400/E)(12) + (-14,400/E)(8)$$

$$= -0.0288 \text{ in.}$$

Note the numerical smallness of both the above values. The negative sign of $\Delta\theta$ indicates clockwise rotation of the tangent at B in relation to the tangent at A. The negative sign of t_{BA} means that point B is below a tangent through A.

EXAMPLE 11-10

Find the deflection due to the concentrated force P applied as shown in Fig. 11-24(a) at the center of a simply supported beam. The flexural rigidity EI is constant.

SOLUTION

The bending-moment diagram is in Fig. 11-24(b). Since EI is constant, the $M/(EI)$ diagram need not be made, as the areas of the bending-

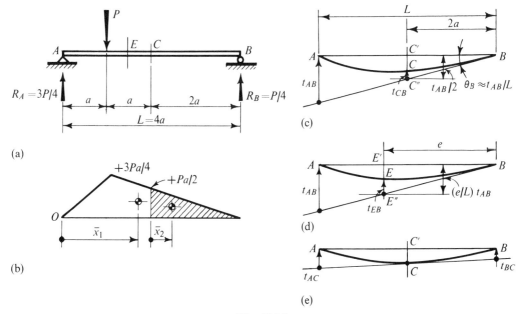

Fig. 11-24

moment diagram divided by EI give the necessary quantities for use in the moment-area theorems. The elastic curve is in Fig. 11-24(c). It is concave upward throughout its length as the bending moments are positive. This curve must pass through the points of the support at A and B.

It is apparent from the sketch of the elastic curve that the desired quantity is represented by the distance CC'. Moreover, from purely geometrical or kinematic considerations, $CC' = C'C'' - C''C$, where the distance $C''C$ is measured from a tangent to the elastic curve passing through the point of support B. However, since the deviation of a support point from a tangent to the elastic curve at the other support may always be computed by the second moment-area theorem, a distance such as $C'C''$ may be found by proportion from the geometry of the figure. In this case, t_{AB} follows by taking the whole $M/(EI)$ area between A and B and multiplying it by* its $\bar{x}$ measured from a vertical through A, whence $C'C'' = \frac{1}{2} t_{AB}$. By another application of the second theorem, t_{CB}, which is equal to $C''C$, is determined. For this case, the $M/(EI)$ area is shaded in Fig. 11-24(b), and, for it, the $\bar{x}$ is measured from C. Since the right reaction is $P/4$ and the distance $CB = 2a$, the maximum ordinate for the shaded triangle is $+Pa/2$.

$$v_C = C'C'' - C''C = (t_{AB}/2) - t_{CB}$$

$$t_{AB} = \Phi_1 \bar{x}_1 = \frac{1}{EI}\left(\frac{4a}{2}\frac{3Pa}{4}\right)\frac{(a+4a)}{3} = +\frac{5Pa^3}{2EI}$$

$$t_{CB} = \Phi_2 \bar{x}_2 = \frac{1}{EI}\left(\frac{2a}{2}\frac{Pa}{2}\right)\frac{(2a)}{3} = +\frac{Pa^3}{3EI}$$

$$v_C = \frac{t_{AB}}{2} - t_{CB} = \frac{5Pa^3}{4EI} - \frac{Pa^3}{3EI} = \frac{11Pa^3}{12EI}$$

The positive signs of t_{AB} and t_{CB} indicate that points A and C lie above the tangent through B. As may be seen from Fig. 11-24(c), the deflection at the center of the beam is in a downward direction.

The slope of the elastic curve at C can be found from the slope at one of the ends and from Eq. 11-39. For point B on the right

$$\theta_B = \theta_C + \Delta\theta_{BC} \quad \text{or} \quad \theta_C = \theta_B - \Delta\theta_{BC}$$

$$\theta_C = \frac{t_{AB}}{L} - \Phi_2 = \frac{5Pa^2}{8EI} - \frac{Pa^2}{2EI} = \frac{Pa^2}{8EI} \quad \text{radians counterclockwise}$$

The above procedure for finding the deflection of a point on the elastic curve is generally applicable. For example, if the deflection of

---

* See Table 2 in the Appendix for the centroid of the whole triangular area. Alternatively, by treating the whole $M/(EI)$ area as two triangles,

$$t_{AB} = \frac{1}{EI}\left(\frac{a}{2}\frac{3Pa}{4}\right)\frac{2a}{3} + \frac{1}{EI}\left(\frac{3a}{2}\frac{3Pa}{4}\right)\left(a + \frac{3a}{3}\right) = +\frac{5Pa^3}{2EI}$$

the point E, Fig. 11-24(d), at a distance e from B is wanted, the solution may be formulated as

$$v_E = E'E'' - E''E = (e/L)t_{AB} - t_{EB}$$

By locating the point E at a variable distance x from one of the supports, the equation of the elastic curve may be obtained.

To simplify the arithmetical work, some care in selecting the tangent at a support must be exercised. Thus, although $v_C = t_{BA}/2 - t_{CA}$ (not shown in the diagram), this solution would involve the use of the unshaded portion of the bending-moment diagram to obtain t_{CA}, which is more tedious.

ALTERNATE SOLUTION

The solution of the foregoing problem may be based on a different geometrical concept. This is illustrated in Fig. 11-24(e), where a tangent to the elastic curve is drawn at C. Then, since the distances AC and CB are equal,

$$v_C = CC' = (t_{AC} + t_{BC})/2$$

i.e., the distance CC' is an average of t_{AC} and t_{BC}. The tangential deviation t_{AC} is obtained by taking the first moment of the unshaded $M/(EI)$ area in Fig. 11-24(b) about A, and t_{BC} is given by the first moment of the shaded $M/(EI)$ area about B. The numerical details of this solution are left for completion by the reader. This procedure is usually longer than the first.

Note particularly that if the elastic curve is not symmetrical, the tangent at the center of the beam is not horizontal.

EXAMPLE 11-11

For a prismatic beam loaded as in the preceding example, find the maximum deflection caused by the applied force P, Fig. 11-25(a).

SOLUTION

The bending-moment diagram and the elastic curve are in Figs. 11-25(b) and (c) respectively. The elastic curve is concave up throughout its length, and the maximum deflection occurs where the tangent to the elastic curve is horizontal. This point of tangency is designated in the figure by D and is located by the unknown horizontal distance d measured from the right support B. Then, by drawing a tangent to the elastic curve through point B at the support, one sees that $\Delta\theta_{BD} = \theta_B$ since the line passing through the supports is horizontal. However, the slope θ_B of the elastic curve at B may be determined by obtaining t_{AB} and dividing it by the length of the span. On the other hand, by using the first moment-area theorem, $\Delta\theta_{BD}$ may be expressed in terms of the shaded area in Fig. 11-25(b). Equating $\Delta\theta_{BD}$ to θ_B and solving for d locates the horizontal tangent at D. Then, again from geometrical considerations, it is seen that the maximum deflection represented by DD' is equal to the tangential deviation of B from a horizontal tangent through D, i.e., t_{BD}.

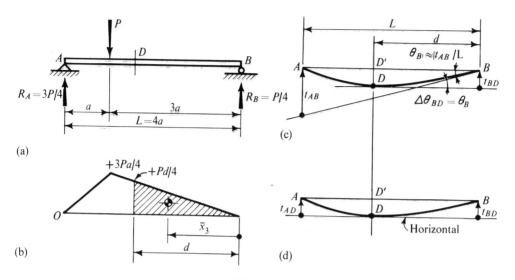

Fig. 11-25

$$t_{AB} = \Phi_1 \bar{x}_1 = +\frac{5Pa^3}{2EI} \quad \text{(see Example 11-10)}$$

$$\theta_B = \frac{t_{AB}}{L} = \frac{t_{AB}}{4a} = \frac{5Pa^2}{8EI}$$

$$\Delta\theta_{BD} = \frac{1}{EI}\left(\frac{d}{2}\frac{Pd}{4}\right) = \frac{Pd^2}{8EI} \quad \text{(area between } D \text{ and } B\text{)}$$

Since $\theta_B = \theta_D + \Delta\theta_{BD}$ and it is required that $\theta_D = 0$,

$$\Delta\theta_{BD} = \theta_B, \quad \frac{Pd^2}{8EI} = \frac{5Pa^2}{8EI} \quad \text{hence } d = \sqrt{5}\,a$$

$$v_{\max} = v_D = DD' = t_{BD} = \Phi_3 \bar{x}_3$$

$$= \frac{1}{EI}\left(\frac{d}{2}\frac{Pd}{4}\right)\frac{2d}{3} = \frac{5\sqrt{5}\,Pa^3}{12EI} = \frac{11.2\,Pa^3}{12EI}$$

After the distance d is found, the maximum deflection may also be obtained, as $v_{\max} = t_{AD}$, or $v_{\max} = (d/L)t_{AB} - t_{DB}$ (not shown). Also note that using the condition $t_{AD} = t_{BD}$, Fig. 11-25(d), an equation may be set up for d.

It should be apparent from the above solution that it is easier to calculate the deflection at the center of the beam, which was illustrated in Example 11-10, than to determine the maximum deflection. Yet, by examining the end results, one sees that numerically the two deflections differ little: $v_{center} = 11Pa^3/(12EI)$ vs. $v_{max} = 11.2Pa^3/(12EI)$. For this reason, in many practical problems of simply supported beams where all the applied forces act in the same direction, it is often sufficiently accurate to calculate the deflection at the center instead of attempting to obtain the true maximum.

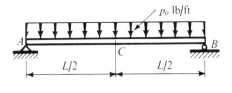

(a)

EXAMPLE 11-12

In a simply supported beam, find the maximum deflection and rotation of the elastic curve at the ends caused by the application of a uniformly distributed load of p_o lb per foot, Fig. 11-26(a). The flexural rigidity EI is constant.

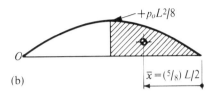

(b)

SOLUTION

The bending-moment diagram is in Fig. 11-26(b). As established in Example 2-6, it is a second-degree parabola with a maximum value at the vertex of $p_o L^2/8$. The elastic curve passing through the points of the support A and B is shown in Fig. 11-26(c).

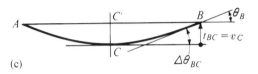

(c)

Fig. 11-26

In this case, the $M/(EI)$ diagram is symmetrical about a vertical line passing through the center. Therefore the elastic curve must be symmetrical, and the tangent to this curve at the center of the beam is horizontal. From the figure, it is seen that $\Delta\theta_{BC}$ is equal to θ_B, and the rotation of the end B is equal to one-half the area* of the whole $M/(EI)$ diagram. The distance CC' is the desired deflection, and from the geometry of the figure it is seen to be equal to t_{BC} (or t_{AC}, not shown).

$$\Phi = \frac{1}{EI}\left(\frac{2}{3}\frac{L}{2}\frac{p_o L^2}{8}\right) = \frac{p_o L^3}{24EI}$$

$$\theta_B = \Delta\theta_{BC} = \Phi = +\frac{p_o L^3}{24EI}$$

$$v_C = v_{max} = t_{BC} = \Phi\bar{x} = \frac{p_o L^3}{24EI}\frac{5L}{16} = \frac{5p_o L^4}{384EI}$$

The value of the deflection agrees with Eq. 11-26, which expresses the same quantity derived by the integration method. Since the point B is above the tangent through C, the sign of v_C is positive.

EXAMPLE 11-13

Find the deflection of the free end A of the beam shown in Fig. 11-27(a) caused by the applied forces. The EI is constant.

---

* See Table 2 in the Appendix for a formula giving an area enclosed by a parabola as well as for $\bar{x}$.

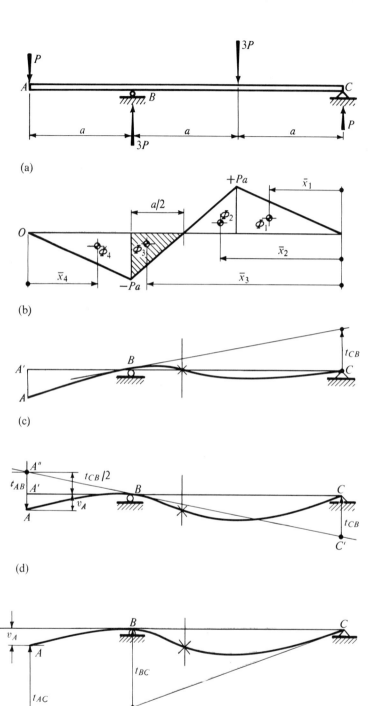

Fig. 11-27

SOLUTION

The bending-moment diagram for the applied forces is in Fig. 11-27(b). The bending moment changes sign at $a/2$ from the left support. At this point an inflection in the elastic curve takes place. Corresponding to the positive moment, the curve is concave up, and vice versa. The elastic curve is so drawn and passes over the supports at B and C, Fig. 11-27(c). To begin, the inclination of the tangent to the elastic curve at the support B is determined by finding t_{CB} as the statical moment of the areas with the proper signs of the $M/(EI)$ diagram between the verticals through C and B about C.

Section 11-15 Derivation of the moment-area theorems

$$t_{CB} = \Phi_1 \bar{x}_1 + \Phi_2 \bar{x}_2 + \Phi_3 \bar{x}_3$$

$$= \frac{1}{EI} \left[\frac{a}{2}(+Pa)\frac{2a}{3} + \frac{1}{2}\frac{a}{2}(+Pa)\left(a + \frac{1}{3}\frac{a}{2}\right) + \frac{1}{2}\frac{a}{2}(-Pa)\left(\frac{3a}{2} + \frac{2}{3}\frac{a}{2}\right) \right]$$

$$= +\frac{Pa^3}{6EI}$$

The positive sign of t_{CB} indicates that the point C is above the tangent through B. Hence a corrected sketch of the elastic curve is made, Fig. 11-27(d), where it is seen that the deflection sought is given by the distance AA' and is equal to $AA'' - A'A''$. Further, since the triangles $A'A''B$ and $CC'B$ are similar, the distance $A'A'' = t_{CB}/2$. On the other hand, the distance AA'' is the deviation of the point A from the tangent to the elastic curve at the support B. Hence

$$v_A = AA' = AA'' - A'A'' = t_{AB} - (t_{CB}/2)$$

$$t_{AB} = \frac{1}{EI}(\Phi_4 \bar{x}_4) = \frac{1}{EI}\left[\frac{a}{2}(-Pa)\frac{2a}{3}\right] = -\frac{Pa^3}{3EI}$$

where the negative sign means that point A is below the tangent through B. This sign is not used henceforth as the geometry of the elastic curve indicates the direction of the actual displacements. Thus the deflection of point A below the line passing through the supports is

$$v_A = \frac{Pa^3}{3EI} - \frac{1}{2}\frac{Pa^3}{6EI} = \frac{Pa^3}{4EI}$$

This example illustrates the necessity of watching the signs of the quantities computed in the applications of the moment-area method, although usually less difficulty is encountered than in the above example. For instance, if the deflection of the end A is established by first finding the rotation of the elastic curve at C, no ambiguity in the direction of tangents occurs. This scheme of analysis is shown in Fig. 11-27(e), where $v_A = \tfrac{3}{2} t_{BC} - t_{AC}$.

The foregoing examples illustrate the manner in which the moment-area method may be used to obtain the deflection of any statically determinate beam. No matter how complex the $M/(EI)$ diagrams may become, the above procedures are applicable. In practice, any $M/(EI)$ diagram whatsoever may be approximated by a number of rectangles and triangles. It is also possible to introduce concentrated angle changes at hinges to account for discontinuities in the directions of the tangents at such points. The magnitudes of the concentrations can be found from kinematic requirements.*

For complicated loading conditions, deflections of elastic beams determined by the moment-area method are often best found by superposition. In this manner the areas of the separate $M/(EI)$ diagrams may become simple geometrical shapes. In the next chapter superposition will be used in solving statically indeterminate problems.

The method described here can be used very effectively in determining the inelastic deflection of beams, providing the $M/(EI)$ diagrams are replaced by the appropriate curvature diagrams.

PROBLEMS FOR SOLUTION

11-1. A long, flat, rectangular bar of aluminum alloy is $\frac{1}{2}$ in. thick by 3 in. wide. (a) Determine the smallest diameter of the cylinder around which this bar could be wrapped so that the elastic limit of the material would not be exceeded. Let $\sigma_{yp} = 24{,}000$ psi, and $E = 10 \times 10^6$ psi. (b) What bending moment would develop in the bar for the above condition?

11-2. Assume that a straight, rectangular bar after severe cold working has a residual stress distribution such as was found in Example 6-7, see Fig. 6-16. (a) If one-sixth of the thickness of this bar is machined off on the top and on the bottom, reducing the bar to two-thirds of its original thickness, what will the curvature ρ of the machined bar be? Assign the necessary parameters to solve this problem in general terms. (b) For the above conditions, if the bar is 1 in.^{2} and 40 in. long, what will the deflection of the bar at the center from the chord through the end be? Let $\sigma_{yp} = 54$ ksi, and $E = 27 \times 10^6$ psi. Note that for small deflections the maximum deflection from a chord L long of a curve bent into a circle of radius R is approximately† $L^2/(8R)$. (*Hint:* The machining operation removes the internal microresidual stresses.)

11-3. Consider a pipe of a linearly viscoelastic material which spans horizontally across a distance L. This pipe is empty for 16 hr a day, and is filled 8 hr. The ratio of the weight of the filled pipe to its empty weight is 5. (a) Assuming that the material is Maxwellian, sketch a diagram showing the center deflection as a function of time for two typical days. (b) For the same conditions of input, sketch another diagram for a material having the properties of a Standard Solid (see Prob. 4-17).

11-4. For many materials the assumption of linear viscoelasticity is not satisfactory. To treat such cases a number of empirical relations for steady-state creep have been proposed. One such widely used relation is

---

* For a systematic treatment of more complex problems see for example A. C. Scordelis and C. M. Smith, "An Analytical Procedure for Calculating Truss Displacements," *Proceedings of the American Society of Civil Engineers*, paper no. 732, **81** (July 1955).
† This follows by retaining the first term of the expansion of $R(1 - \cos\theta)$ where θ is one-half the included angle.

$\dot{\varepsilon} = B\sigma^n$ where B is a constant and the experimentally determined exponent $n > 1$. Show that, using this relation, the maximum stress in a rectangular beam is $\sigma_{max} = (Mc/I)[(2n + 1)/(3n)]$ and that at a distance y from the centroidal axis $\sigma = (y/c)^{1/n}\sigma_{max}$. Sketch the resulting stress distribution for an $n = 6$.

11-5. Using the exact differential equation, Eq. 11-8, show that the equation of the elastic curve in Example 11-2 is $x^2 + (v - \rho)^2 = \rho^2$, where ρ is a constant. Compare the second derivative of this exact solution with the approximate one, Eq. 11-9. (*Hint:* Let $dv/dx = \tan\theta$ and integrate.)

11-6 through 11-14. For the statically determinate beams loaded as shown in the figures, solve one of the following alternatives as directed:

A. Using the second-order differential equation, obtain the equation for the elastic curve and make a careful sketch of it. Use the singularity functions wherever necessary. For all beams EI is constant.

B. Same as above, but use the fourth-order differential equation for beam deflection.

C. Same as **B** above, and illustrate the solution with sketches showing the integration steps graphically.

D. Same as **B** above, and, in addition, after completing two integrations, compute the reactions directly and check the expressions found for $V(x)$ and $M(x)$ before completing the problem.

Ans. Probs. 11-6 *and* 11-9. See Table 11 in the Appendix; *Prob.* 11-12. $EIv = -\frac{1}{6}\langle x - 4\rangle^4 + \frac{1}{6}\langle x - 6\rangle^4 + \frac{1}{2}x^3 - 27x$; *Prob.* 11-14. $V(x) = +24\langle x - 2\rangle_*^{-1} + 3\langle x - 4\rangle^1 - \frac{1}{4}\langle x - 4\rangle^2 - 9$. (For additional data for problems of this type see Chapter 2.)

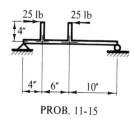

PROB. 11-15

11-15. Using a semigraphical procedure such as shown in Figs. 11-7 and 11-9, find the deflection of the beam at midspan, see figure. Let $EI = 23,200$ lb-in.^{2} Neglect the effect of the axial force on the deflection. *Ans.* 0.0905 in.

11-16. Using a semigraphical procedure such as shown in Figs. 11-7 and 11-9, find the deflection of the beam at the point of the applied load, see figure. Let $I_1 = 400$ in.^{4}, $I_2 = 300$ in.^{4}, and $E = 30 \times 10^6$ psi. *Ans.* 0.210 in.

11-17. Using singularity functions, find the equation of the elastic curve and calculate the tip deflection due to the applied force P, see figure. Assume E is constant. *Ans.* $v_{\max} = -743P/(EI)$.

PROB. 11-16

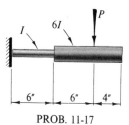

PROB. 11-17

PROB. 11-18	PROB. 11-19	PROB. 11-20
PROB. 11-21	PROB. 11-22	PROB. 11-23
PROB. 11-24	PROB. 11-25	

11-18 through 11-25. For the beams loaded as shown in the figures solve one of the following alternatives as directed:

A. Using the fourth-order differential equation, obtain the equation of the elastic curve and make a careful sketch of it. Use singularity functions wherever necessary. For all beams EI is constant.

B. Same as above, and plot the shear and moment diagrams.

C. Same as **A**, but using the second-order differential equation for beam deflection.

Ans. Prob. 11-18. $24 EIv = -p_o x^2 \times (L - x)^2$; Prob. 11-20. $EIv = -(3/32)PLx^2 + (11/96)Px^3 - (1/6) P \langle x - L/2 \rangle^3$; Prob. 11-21. $EIv = (M_1/2)\langle x - L/2 \rangle^2 - [M_1/(4L)] x^3 + (M_1/8)x^2$; Prob. 11-24. $EIv = -3x^2 + (1/6) x^3 + 6\langle x - 6 \rangle^1 - (1/3)\langle x - 12 \rangle^3$; Prob. 11-25. Deflection at P is $Pa^3/(3EI)$.

11-26. For the beam loaded as shown in the figure, (a) determine the ratio of the moment at the built-in end to the applied moment M_a; (b) determine the rotation of the free end. The EI is constant. *Ans.* $-1/2$; $-M_a L/(4EI)$.

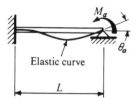

PROB. 11-26

11-27. One end of an elastic beam is displaced an amount Δ relative to the other end as shown in the figure. No rotation of the ends is permitted to occur. Derive the expression for the elastic curve, and plot the shear and moment diagrams. The EI is constant.

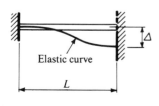

PROB. 11-27

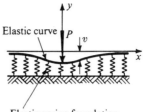

PROB. 11-28

11-28. Consider an infinitely long beam of constant EI supported on an elastic foundation which is capable of exerting a force normal to it proportional to the displacement v. This is the classical problem of a beam on an elastic foundation and is usually represented as shown in the figure. The linear, elastic springs in the foundation are capable of resisting both tensile and compressive forces. The foundation modulus is k lb per square inch. The homogeneous differential equation for this problem is $EIv^{iv} = -kv$. (a) Show that the solution* of the governing equation for a beam of an elastic foundation is

$$v = e^{\beta x}(C_1 \cos \beta x + C_2 \sin \beta x) + e^{-\beta x}(C_3 \cos \beta x + C_4 \sin \beta x)$$

where C_1, C_2, C_3, C_4 are constants, and $\beta = \sqrt[4]{k/(4EI)}$. (b) Show that for a singular force P at $x = 0$, for $x > 0$,

$$v = [P\beta/(2k)] e^{-\beta x}(\cos \beta x + \sin \beta x)$$

(*Hint*: If $x \to \infty$, then $v \to 0$. This condition eliminates two of the unknown constants.)

11-29. Consider the beam shown in Fig. 11-16. (a) Using the equations given in Table 11 of the Appendix and applying superposition, determine the deflection of the end C. (b) How large must the force P be to have zero deflection at C?

11-30. Consider the structure loaded with the

---

* For further details see S. P. Timoshenko, *Strength of Materials* (3rd ed.) Part II, Advanced Theory and Problems (Princeton, N.J.: D. Van Nostrand Co., Inc., 1956), p. 2.

force P as shown in Fig. 11-15a. What uniformly distributed load p_o must be applied to the horizontal beam AB so that the horizontal displacement of point C will be back to its unloaded condition?

11-31. Using the results found in Example 11-4 for deflection of a beam due to a concentrated force P, determine the deflection at the center of the beam caused by a uniformly distributed downward load p_o. (Treat $p_o\, dx$ as an infinitesimal concentrated force, and integrate.)

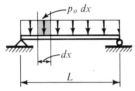

PROB. 11-31

11-32. An 8 WF 40 beam is loaded as shown in the figure. Using the equations given in the Appendix and the method of superposition, calculate the deflection at the center of the span. Let $E = 29 \times 10^6$ psi. *Ans.* 0.39 in.

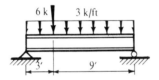

PROB. 11-32

11-33. A steel beam is supported and loaded as shown in the figure. The force $P = 10$ kips, and $M_A = 1,200$ k-in. With the horizontal line AC as a reference, determine the slope and the vertical deflection at end C. For this beam $E = 30 \times 10^6$ psi and $I = 1,000$ in.[4] The spring at B, when isolated, requires a force of 20 kips to shorten it 1 in. Use any method you wish to compute the required slope and

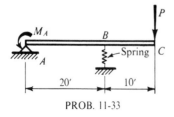

PROB. 11-33

deflection. *Ans.* -9.3×10^{-3} radian, -1.51 in.

11-34. The maximum deflection for a simple beam spanning 24 ft and carrying a uniformly distributed load of 40 kips total, including its weight, is limited to 0.5 in.[*] (a) Specify the required steel I beam. Let $E = 30 \times 10^6$ psi. (b) What size aluminum-alloy beam would be needed for the same requirements? Let $E = 10 \times 10^6$ psi, and use Table 11 in the Appendix for section properties. (c) Determine the maximum stresses in both cases. *Ans.* (a) 18 I 70, (b) 24 I.

11-35. A uniformly loaded 6-in.-by-12-in. (nominal size) wooden beam spans 10 ft and is considered to have satisfactory deflection characteristics. Select an aluminum-alloy I beam, a steel I beam, and a polyester-plastic I beam having the same deflection characteristics. In making the beam selections neglect the differences in their own weights. Let $E = 1.5 \times 10^6$ psi for wood and polyester plastic, $E = 10 \times 10^6$ psi for aluminum, and $E = 30 \times 10^6$ psi for steel. For section properties of all I beams use Table 2 in the Appendix. *Ans.* 10 in., 7 in., 18 in.

11-36. A 5-ft-long cantilever is loaded at the end with a force $P = 1,000$ lb forming an angle α with the vertical. The member is an 8-in., 18.4-lb steel I beam. Determine the total tip deflection for $\alpha = 0°$, $10°$, $45°$, and $90°$, caused by the applied force. Let $E = 29 \times 10^6$ psi.

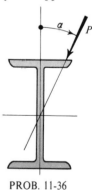

PROB. 11-36

---

[*] This is a more stringent requirement than the one commonly used in building design. In the latter applications, the deflection is commonly limited to $1/360$-th of the span length.

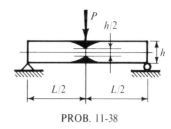

PROB. 11-38

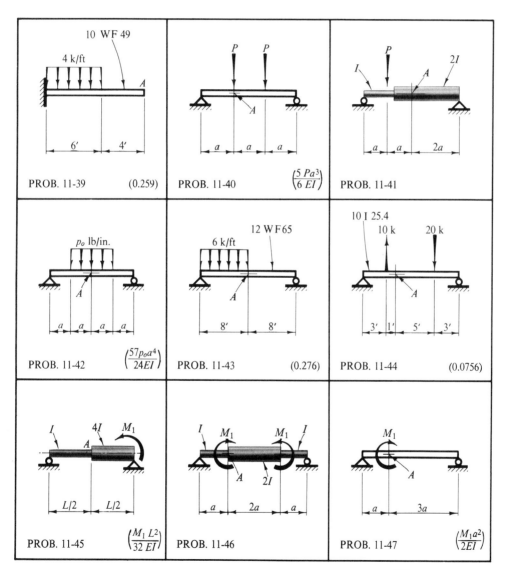

PROB. 11-39   (0.259)

PROB. 11-40 $\left(\dfrac{5\,Pa^3}{6\,EI}\right)$

PROB. 11-41

PROB. 11-42 $\left(\dfrac{57 p_o a^4}{24 EI}\right)$

PROB. 11-43   (0.276)

PROB. 11-44   (0.0756)

PROB. 11-45 $\left(\dfrac{M_1 L^2}{32\,EI}\right)$

PROB. 11-46

PROB. 11-47 $\left(\dfrac{M_1 a^2}{2EI}\right)$

11-37. A 1-in. square bar of a linearly elastic-plastic material is to be wrapped around a round mandrel as shown in the figure. (a) What mandrel diameter D is required so that the outer thirds of the cross sections become plastic, i.e., the elastic core is $\frac{1}{3}$ in. deep by 1 in. wide? Assume the material to be initially stress-free with $\sigma_{yp} = 40$ ksi. Let $E = 30 \times 10^6$ psi. The pitch of the helix angle is so small that only the bending of the bar in a plane need be considered. (b) What will be the diameter of the coil after the release of the forces used in forming it? Stated alternatively, determine the coil diameter after the elastic spring-back.

11-38. A rectangular, weightless, simple beam of linearly elastic-plastic material is loaded in the middle by the force P as shown in the figure. (a) Determine the magnitude of the force P that would cause the plastic zone to penetrate $\frac{1}{4}$ of the beam depth from each side. (b) For the above loading condition sketch the moment-curvature diagram clearly showing it for the plastic zone. (c) Describe how the deflection of this beam can be found. Numerical calculations of the deflections are not required.

11-39 through 11-47. Using the moment-area method for the members loaded as shown in the figures, determine the deflection and the slope of the elastic curve at points A. Specify whether the deflection is upward or downward. If the size of the member is not given, assume that EI is constant over the entire length. Neglect the weight of the members. Wherever needed, assume $E = 30 \times 10^6$ psi. Wherever the answer is expressed in terms of EI, no adjustment for units need be made. *Ans.* The deflection sought is noted in the lower right corner.

11-48 through 11-50. Using the moment-area method for the members loaded as shown in the figures, determine the location and magnitude of the maximum deflection between the supports. Disregard the effect of axial forces on deflections wherever this condition

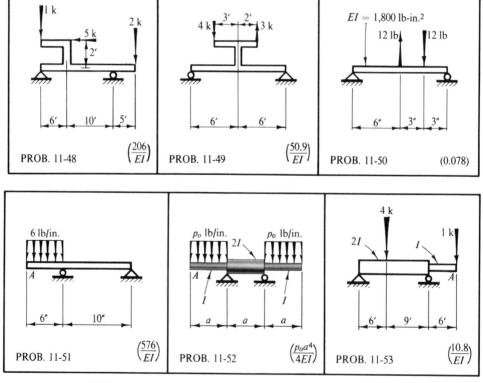

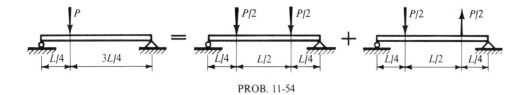

PROB. 11-54

occurs. Other conditions are the same as in Probs. 11-39 through 11-47. *Ans.* Lower right corner of each figure.

11-51 through 11-53. Using the moment-area method, determine the deflection of the overhang at *A* for the beams loaded as shown. Other conditions are the same as in Probs. 11-48 through 11-50. *Ans.* Lower right corner of each figure.

11-54. Determine the deflection at the midspan of a simple beam, loaded as shown in the figure, by solving the two separate problems indicated and superposing the results. Use the moment-area method. The EI is constant. *Ans.* $11PL^3/(768EI)$.

11-55. A light pointer is attached only at *A* to a 6-in.-by-6-in. (actual) wooden beam as shown in the figure. Determine the position of the end of the pointer after a concentrated force of 1,200 lb is applied. Let $E = 1.2 \times 10^6$ psi. *Ans.* 0.036 in.

11-56. Beam *AB* is subjected to an end moment at *A* and an unknown concentrated moment M_c as shown in the figure. Using the moment-area method, determine the magnitude of the bending moment M_c so that the deflection at point *B* will be equal to zero. The EI is constant. *Ans.* 16 k-ft.

11-57. The beam *ABCD* is initially horizontal. A load *P* is then applied at *C* as shown in the

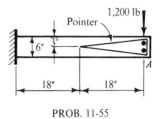

PROB. 11-55

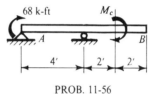

PROB. 11-56

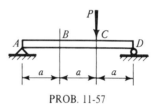

PROB. 11-57

figure. It is desired to place a vertical force at *B* to bring the position of the beam at *B* back to the original level *ABCD*. What force is required at *B*? *Ans.* $7P/8$.

12 Statically indeterminate problems

12-1. INTRODUCTION

The simplest problems in mechanics of solids are externally statically determinate. In such cases the reactions and the internal system of stress resultants at a section can be determined from statics without considering deformations. In the preceding chapter, only on introducing the differential equations for the deflection of beams did methods for the analysis of statically indeterminate elastic beams become possible. In this chapter procedures for the solution of statically indeterminate problems will be extended to include additional situations.

In Part A of this chapter procedures for the analysis of statically indeterminate systems applicable to both linear and nonlinear material response will be discussed. It will be shown how the equations of static equilibrium can be supplemented by additional ones based on considerations of geometry of deformation. The required additional equations will be formulated using the conditions of compatible displacements. In the inelastic analysis of the indeterminate systems such procedures can become very complex. The same is true of many statically indeterminate viscoelastic problems.

Section 12-2
A general approach

A very effective general method for the solution of highly indeterminate elastic problems is discussed in Part B of this chapter. To achieve the solution of problems, a statically indeterminate system is first reduced to a determinate one by removing the redundant reactions. Then the redundant reactions are re-applied and are so adjusted in their magnitude that the prescribed deformations at the points of their application are obtained. This widely used method of analysis is based on the superposition technique and is limited to the solution of linearly elastic problems. A useful recurrence formula for the analysis of continuous elastic beams is also given in this part of the chapter.

A procedure for determining the limit or the ultimate carrying capacity of determinate and indeterminate beams made of ductile material is treated in Part C of this chapter.

In Parts A and B particular attention will be directed toward finding the magnitude of the redundant reactions. After the redundant reactions are known, a member becomes statically determinate and may be analyzed for strength or stiffness characteristics by the methods introduced earlier.

PART A
ANALYSIS WITH THE AID OF DISPLACEMENT RELATIONS

12-2. A GENERAL APPROACH

In all statically indeterminate problems the equations of static equilibrium remain valid. These equations are necessary but not sufficient to solve the indeterminate problems. The supplementary equations are established from considerations of the geometry of deformation. In structural systems, of physical necessity, certain elements or parts must deflect together, twist together, expand together, etc., or remain stationary. Formulating such observations quantitatively provides one with the required additional equations. For example, a statement of a common displacement of several members of a joint can give the required relation. Such kinematic equations are independent of the mechanical properties of materials and thus are not limited to the linear elastic response.

The necessary procedures for determining the linear deformation of axially loaded rods, the angular twist of shafts, and the deflection of beams were developed earlier. Here, except for designating the forces acting on such members as unknowns by some appropriate algebraic symbols, the same procedures apply. As before the smallness of the deformations in comparison with the linear dimensions of the body is tacitly assumed.

Several examples illustrating the method of supplementing the equilibrium equations with the displacement relationships follow.

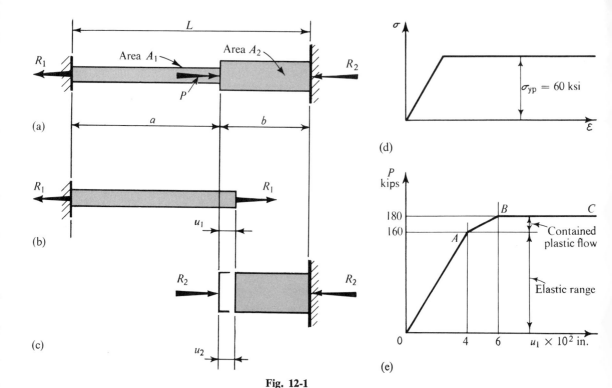

Fig. 12-1

EXAMPLE 12-1

A stepped bar is built in at both ends to immovable supports, Fig. 12-1(a). The left part of the bar has a cross-sectional area A_1; the area of the right part is A_2. (a) If the material of the bar is elastic with an elastic modulus E, what are the reactions R_1 and R_2 caused by the application of an axial force P at the point of discontinuity of the section? (b) If $A_1 = 1$ in.^{2}, $A_2 = 2$ in.^{2}, $a = 30$ in., $b = 20$ in., and the material is linearly elastic–perfectly plastic as shown in Fig. 12-1(d), determine the displacement u_1 of the step as a function of the applied force P. Let $E = 30 \times 10^3$ ksi.

SOLUTION

Case (a). The point on the rod where the force P is applied deflects the same amount whether the right or the left part of the bar is considered. By separating the bar at P, the two free-body diagrams in Figs. 12-1(b) and (c) are obtained. The left part of the rod is subjected throughout its length to a tensile force R_1 and elongates an amount u_1. The right part contracts u_2 under the action of a compressive force R_2. Of physical necessity, the absolute values of the two deflections must be the same:

From statics: $\qquad\qquad R_1 + R_2 = P$

From geometry:* $\qquad |u_1| = |u_2|$

On applying Eq. 4-33, $u = PL/(AE)$, the above relation yields

$$\frac{R_1 a}{A_1 E} = \frac{R_2 b}{A_2 E}$$

Solving the two equations simultaneously gives

$$R_1 = \frac{P}{1 + aA_2/(bA_1)} \quad \text{and} \quad R_2 = \frac{P}{1 + bA_1/(aA_2)} \quad (12\text{-}1)$$

Case (b). By direct substitution of the given data into Eq. 12-1, one finds

$$R_1 = \frac{P}{1 + 30(2)/20} = \frac{P}{4} \quad \text{and} \quad R_2 = \frac{3P}{4}$$

Hence the normal stresses are

$$\sigma_1 = R_1/A_1 = P/4 \quad \text{and} \quad \sigma_2 = R_2/A_2 = 3P/8 \quad \text{(compression)}$$

As $|\sigma_2| > \sigma_1$, the load at impending yield is found by setting $|\sigma_2| = 60$ ksi. At this load the right part of the bar just reaches yield and the strain attains the magnitude of $\varepsilon_{yp} = \sigma_{yp}/E$. Therefore

$$P_{yp} = 8\sigma_{yp}/3 = 160 \text{ kips} \quad \text{and} \quad |u_2| = |u_1| = \varepsilon_{yp} b = 4 \times 10^{-2} \text{ in.}$$

These quantities locate point A in Fig. 12-1(e).

On increasing P above 160 kips, the right part of the bar continues to yield carrying a compressive force $R_2 = \sigma_{yp} A_2 = 120$ kips. At the point of impending yield for the whole bar the left part just reaches yield. This occurs when $R_1 = \sigma_{yp} A_1 = 60$ kips, and the strain in the left part just reaches $\varepsilon_{yp} = \sigma_{yp}/E$. Therefore

$$P = R_1 + R_2 = 180 \text{ kips} \quad \text{and} \quad u_1 = \varepsilon_{yp} a = 6 \times 10^{-2} \text{ in.}$$

These quantities locate point B in Fig. 12-1(e). Beyond this point the plastic flow is uncontained, and $P = 180$ kips is the ultimate or limit load of the rod.

Note the simplicity of calculating the limit load, which, however, provides no information on the deflection characteristics of the system. In general, plastic limit analysis is simpler than elastic analysis, which in turn is simpler than tracing the elastic-plastic load-deflection relationship.

EXAMPLE 12-2

An elastic bar of constant cross-sectional area A is built in at the ends to

---

* By considering elongation of the bar positive and contraction negative, alternatively one has $u_1 + u_2 = 0$, which means that the total deformation of the bar from end to end is zero.

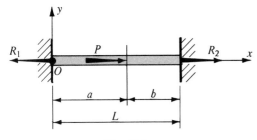

Fig. 12-2

immovable supports, Fig. 12-2. Using singularity functions, determine the reactions at the supports due to the application of the axial force P.

SOLUTION

Equation 4-32 is the appropriate differential equation for the solution of this problem. The expression for the singular force is given by Eq. 2-12. From the statement of the problem, the boundary conditions are $u(0) = u(L) = 0$. Note that both these conditions are kinematic, which is characteristic of statically indeterminate problems. On this basis

$$AE\frac{d^2u}{dx^2} = -p_x = -P\langle x - a\rangle_*^{-1}$$

$$AE\frac{du}{dx} = -P\langle x - a\rangle^0 + C_1$$

$$AEu = -P\langle x - a\rangle^1 + C_1 x + C_2$$

$$AEu(0) = 0 = C_2$$

$$AEu(L) = 0 = -Pb + C_1 L \quad \text{or} \quad C_1 = Pb/L$$

$$R_1 = P(0) = \left(AE\frac{du}{dx}\right)_{x=0} = +Pb/L$$

and $\quad R_2 = P(L) = \left(AE\dfrac{du}{dx}\right)_{x=L} = -P + C_1 = -Pa/L$

With $A_1 = A_2$, and $a + b = L$, Eq. 12-1 yields the same results. It is interesting to note that the closer force P is applied to a given support, the more of it that is carried by that support, i.e., the force follows the stiffest path.

If the cross section of the bar varies, the above procedure is not as good as the one used in the preceding example.

Shafts build in at the ends and subjected to torque are statically indeterminate. The solution of such problems is completely analogous to the above two examples.

EXAMPLE 12-3

A steel rod 2 in.² in cross-sectional area and 15.0025 in. long is loosely inserted into a copper tube as in Fig. 12-3. The copper tube has a

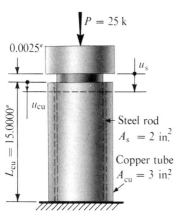

Fig. 12-3

cross-sectional area of 3 in.² and is 15.0000 in. long. If an axial force $P = 25$ kips is applied through a rigid cap, what stresses will develop in the two materials? Assume that the elastic moduli of steel and copper are $E_s = 30 \times 10^6$ psi and $E_{cu} = 17 \times 10^6$ psi, respectively.

*Section 12-2
A general approach*

SOLUTION

If the applied force P is sufficiently large to close the small gap, a force P_s will be developed in the steel rod and a force P_{cu} in the copper tube. Moreover, upon loading, the steel rod will compress axially u_s, which is as much as the axial deformation u_{cu} of the copper tube plus the initial gap. Hence

From statics: $\qquad P_s + P_{cu} = 25,000$ lb

From geometry: $\qquad u_s = u_{cu} + 0.0025$

On applying Eq. 4-33, $u = PL/(AE)$,

$$\frac{P_s L_s}{A_s E_s} = \frac{P_{cu} L_{cu}}{A_{cu} E_{cu}} + 0.0025$$

or $\qquad \dfrac{15.0025}{2(30)10^6} P_s - \dfrac{15}{3(17)10^6} P_{cu} = 0.0025$

or $\qquad P_s - 1.176 P_{cu} = 10,000$ lb

Solving these equations simultaneously gives

$$P_{cu} = 6,900 \text{ lb} \quad \text{and} \quad P_s = 18,100 \text{ lb}$$

and dividing these forces by the respective cross-sectional areas gives

$$\sigma_{cu} = 6,900/3 = 2,300 \text{ psi} \quad \text{and} \quad \sigma_s = 18,100/2 = 9,050 \text{ psi}$$

If either of the above stresses were above the proportional limit of its material or if the applied force were too small to close the gap, the above solution would not be valid. Also note that since the deformations considered are small, it is sufficiently accurate to use $L_s = L_{cu}$.

ALTERNATE SOLUTION

The force F necessary to close the gap may be found first, using Eq. 4-33. In developing this force the rod acts as a "spring" and resists a part of the applied force. The remaining force P' causes equal deflections u'_s and u'_{cu} in the two materials.

$$F = \frac{u A_s E_s}{L_s} = \frac{(0.0025)2(30)10^6}{15.0025} = 10,000 \text{ lb} = 10 \text{ kips}$$

$$P' = P - F = 25 - 10 = 15 \text{ kips}$$

*Chapter 12
Statically
indeterminate
problems*

Then if P'_s is the force resisted by the steel rod in addition to the force F and P'_{cu} is the force carried by the copper tube,

From statics: $\qquad P'_s + P'_{cu} = P' = 15$

From geometry: $\quad u'_s = u'_{cu} \quad$ or $\quad \dfrac{P'_s L_s}{A_s E_s} = \dfrac{P'_{cu} L_{cu}}{A_{cu} E_{cu}}$

$$\dfrac{15.0025}{2(30)10^6} P'_s = \dfrac{15}{3(17)10^6} P'_{cu}, \qquad P'_{cu} = \dfrac{17}{20} P'_s$$

By solving the two appropriate equations simultaneously, it is found that $P'_{cu} = 6.9$ kips and $P'_s = 8.1$ kips, or $P_s = P'_s + F = 18.1$ kips.

If $(\sigma_{yp})_s = 40$ ksi and $(\sigma_{yp})_{cu} = 10$ ksi, the limit load for this assembly can be determined as follows:

$$P_{ult} = (\sigma_{yp})_s A_s + (\sigma_{yp})_{cu} A_{cu} = 110 \text{ kips}$$

At the ultimate load both materials yield, therefore the small discrepancy in the initial lengths of the parts is of no consequence.

EXAMPLE 12-4

Three bars of elastic–perfectly plastic material are symmetrically arranged in a plane to form the system shown in Fig. 12-4(a). Investigate the load-deflection characteristics of joint C. The cross-sectional area A of each bar is the same.

SOLUTION

A free-body diagram of joint C is in Fig. 12-4(b), from which for small deformations an equation of equilibrium is

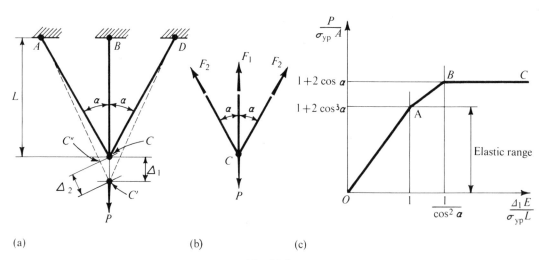

Fig. 12-4

$$F_1 + 2F_2 \cos \alpha = P \qquad (12\text{-}2)$$

Section 12-2
A general approach

This relation holds true regardless of the material response. The latter, however, depends on the attained magnitude of strain.

The deformed structure is shown in Fig. 12-4(a) by the dashed lines AC', BC', and DC'. The elongation of the bar BC' is Δ_1. The elongation of the inclined bars is Δ_2. For compatible displacements

$$\Delta_2 = \Delta_1 \cos \alpha \qquad (12\text{-}3)$$

where it is assumed that because of the smallness of the deformations being considered, the arc CC'' with the center at A can be replaced by a normal to AC'.

Equation 12-3 applies both in the elastic and the inelastic strain ranges provided the deformation remains small. For the elastic range, by noting that the inclined bars are $L/\cos \alpha$ long, on applying Eq. 4-33, one has

$$\frac{F_2(L/\cos \alpha)}{AE} = \frac{F_1 L}{AE} \cos \alpha, \quad \text{i.e.,} \quad F_2 = F_1 \cos^2 \alpha \qquad (12\text{-}4)$$

Solving Eqs. 12-2 and 12-4 simultaneously yields

$$F_1 = \frac{P}{1 + 2\cos^3 \alpha} \quad \text{and} \quad F_2 = \frac{P}{1 + 2\cos^3 \alpha} \cos^2 \alpha \qquad (12\text{-}5)$$

From this it is seen that the maximum force and stress occur in the vertical bar. At impending yield $F_1 = \sigma_{yp}A$ and $\Delta_1 = (\sigma_{yp}/E)L$. With F_1 known, the maximum force P which can be carried elastically follows from Eqs. 12-2 and 12-4. This condition with $P = \sigma_{yp}A(1 + 2\cos^3 \alpha)$ corresponds to point A in Fig. 12-4(c).

On increasing the force P above the impending yield in the vertical bar, the force $F_1 = \sigma_{yp}A$ remains constant, and Eq. 12-2 alone becomes sufficient for determining the force F_2. The inclined bars behave elastically until their stress reaches σ_{yp}. This occurs when $F_2 = \sigma_{yp}A$. At impending yield in the inclined bars, using Eq. 12-2, $P = \sigma_{yp}A(1 + 2\cos \alpha)$. This condition corresponds to the limit load for the system.

At impending yield $\Delta_2 = (\sigma_{yp}/E)(L/\cos \alpha)$. Whence from Eq. 12-3, $\Delta_1 = (\sigma_{yp}/E)L/\cos^2 \alpha$. This value locates the abscissa for point B in Fig. 12-4(c). Beyond this point uncontained plastic flow takes place.

Note again the simplicity in finding the limit load as one works directly with a statically determinate system.

EXAMPLE 12-5

Two cantilever beams AD and BE of equal flexural rigidity $EI = 9 \times 10^9$ lb-in.^{2}, shown in Fig. 12-5(a), are interconnected by a taut steel rod DC ($E_s = 30 \times 10^6$ psi). The rod DC is 12 ft 6 in. long and has a cross section of 0.5 in.^{2} Find the deflection of the cantilever AD at D due to a force $P = 10$ kips applied at E.

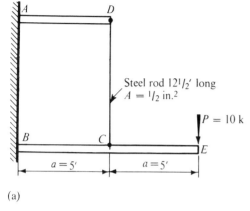
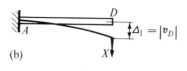
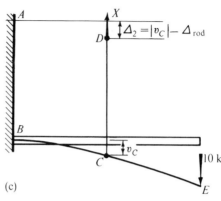

Fig. 12-5

SOLUTION

By separating the structure at D, the two free-body diagrams in Figs. 12-5(b) and (c) are obtained. In both diagrams the same unknown force X is shown acting (a condition of statics). The deflection of the point D is the same whether the beam AD at D or the top of the rod DC is considered. The deflection Δ_1 of the point D in Fig. 12-5(b) is caused by X. The deflection Δ_2 of the point D on the rod is equal to the deflection v_C of the beam BE caused by the forces P and X less the elastic stretch of the rod DC.

From statics: $\qquad X_{\text{pull on } AD} = X_{\text{pull on } DC} = X$

From geometry: $\qquad \Delta_1 = \Delta_2 \quad$ or $\quad |v_D| = |v_C| - \Delta_{\text{rod}}$

Beam deflections can be found using any one of the methods discussed in the preceding chapter. From Table 11 of the Appendix, in terms of the notation of this problem, one has

$$v_D = -\frac{Xa^3}{3EI} = -\frac{X \times 60^3}{3 \times 9 \times 10^9} = -8 \times 10^{-6} X \qquad \text{(down)}$$

$$v_{C \text{ due to } X} = +8 \times 10^{-6} X \qquad \text{(up)}$$

$$v_{C \text{ due to } P} = -\frac{P}{6EI}[2(2a)^3 - 3(2a)^2 a + a^3] = -0.200 \text{ in.} \qquad \text{(down)},$$

and using Eq. 4-33

$$\Delta_{\text{rod}} = \frac{XL_{CD}}{A_{CD}E} = \frac{X(12.5)12}{0.5(30)10^6} = 10 \times 10^{-6} X$$

Then $8 \times 10^{-6}X = (0.200 - 8 \times 10^{-6}X) - 10 \times 10^{-6}X$

and $X = +7,690$ lb

and $v_D = -8 \times 10^{-6} \times 7,690 = -0.0615$ in.   (down)

Note particularly that the deflection of point C is caused by the applied force P at the end of the cantilever as well as by the unknown force X.

12-3. STRESSES CAUSED BY TEMPERATURE

It was possible to disregard the deformations caused by temperature in statically determinate systems since in such situations the members are free to expand or contract. However, in statically indeterminate systems, expansion or contraction of a body may be inhibited or entirely prevented in certain directions. This may cause significant stresses and must be investigated.

The determination of free deformations caused by a change in temperature follows from Eq. 4-14. For a body of length L having a uniform thermal strain, the linear deformation Δ due to a change in temperature of δT degrees is

$$\Delta = \alpha(\delta T)L \qquad (12\text{-}6)$$

where α is the coefficient of thermal expansion.

The solution of indeterminate problems involving temperature deformations follows the concepts discussed in the preceding article. Two examples follow to illustrate some of the details of solution.

EXAMPLE 12-6

A copper tube 12 in. long and having a cross-sectional area of 3 in.² is placed between two very rigid caps made of Invar,* Fig. 12-6(a). Four ¾-in. steel bolts are symmetrically arranged parallel to the axis of the tube and are lightly tightened. Find the stress in the tube if the temperature of the assembly is raised from 60°F to 160°F. Let $E_{cu} = 17 \times 10^6$ psi, $E_s = 30 \times 10^6$ psi, $\alpha_{cu} = 0.0000091$ per °F, and $\alpha_s = 0.0000065$ per °F.

SOLUTION

If the copper tube and the steel bolts were free to expand, the axial thermal elongations

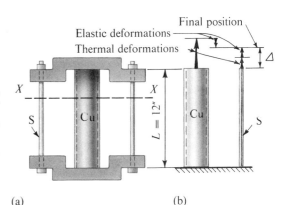

Fig. 12-6

---

* Invar is a steel alloy which at ordinary temperatures has an $\alpha \approx 0$ and for this reason is used in the best grades of surveyor's tapes and watch springs.

shown in Fig. 12-6(b) would take place. However, since the axial deformation of the tube must be the same as that of the bolts, the copper tube will be pushed back and the bolts will be pulled out, so that the net deformations will be the same. Moreover, as may be established by considering a free body of the assembly above some arbitrary plane, as X-X in Fig. 12-6(a), the compressive force P_{cu} in the copper tube and the tensile force P_s in the steel bolts are equal. Hence

From statics: $\quad\quad\quad\quad P_{cu} = P_s = P$

From geometry: $\quad\quad\quad \Delta_{cu} = \Delta_s = \Delta$

This kinematic relation, on the basis of Fig. 12-6(b) with the aid of Eqs. 12-6 and 4-33, becomes:

$$\alpha_{cu}\,\delta T L_{cu} - \frac{P_{cu} L_{cu}}{A_{cu} E_{cu}} = \alpha_s\,\delta T L_s + \frac{P_s L_s}{A_s E_s}$$

or, since $\delta T = 100°$ and 0.442 in.² is the cross section of one bolt,

$$(0.0000091)100 - \frac{P_{cu}}{3(17)10^6} = (0.0000065)100 + \frac{P_s}{4(0.442)30(10)^6}$$

Solving the two equations simultaneously, $P = 6{,}750$ lb. Therefore the stress in the copper tube is $\sigma_{cu} = 6{,}750/3 = 2{,}250$ psi.

The kinematic expression used above may also be set up on the basis of the following statement: The differential expansion of the two materials due to the change in temperature is accommodated by or is equal to the elastic deformations which take place in the two materials.

EXAMPLE 12-7

A steel bolt having a cross-sectional area $A_1 = 1$ in.² is used to grip two steel washers of total thickness L, each having the cross-sectional area

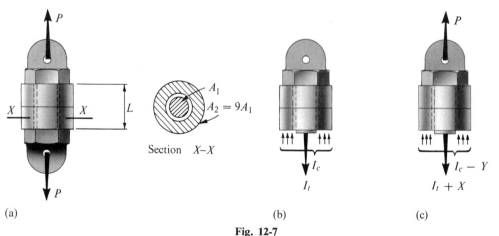

Fig. 12-7

$A_2 = 9$ in.², Fig. 12-7(a). If the bolt in this assembly is initially so tightened that its stress is 20 ksi, what will be the final stress in this bolt after a force $P = 15$ kips is applied to the assembly?

*Section 12-3
Stresses caused by temperature*

SOLUTION

A free body corresponding to the initial conditions of the assembly is in Fig. 12-7(b), where I_t is the initial tensile force in the bolt and I_c is the initial compressive force in the washers. From statics, $I_t = I_c$. A free body of the assembly after the force P is applied is shown in Fig. 12-7(c), where X designates the increase in the tensile force in the bolt, and Y is the decrease in the compressive force on the washers due to P. As a result of these forces X and Y, if the adjacent parts remain in contact, the bolt elongates the same amount as the washers expand elastically. Hence the final conditions are the following:

From statics: $\qquad P + (I_c - Y) = (I_t + X)$

or since $\qquad\qquad I_c = I_t$

$$X + Y = P$$

From geometry: $\qquad \Delta_{\text{bolt}} = \Delta_{\text{washers}}$

On applying Eq. 4-33

$$\frac{XL}{A_1 E} = \frac{YL}{A_2 E}, \quad \text{i.e.,} \quad Y = \frac{A_2}{A_1} X$$

Solving the two equations simultaneously,

$$X = \frac{P}{1 + (A_2/A_1)} = \frac{P}{1 + 9} = 0.1P = 1{,}500 \text{ lb}$$

Therefore the increase of the stress in the bolt is $X/A_1 = 1{,}500$ psi, and the stress in the bolt after the application of the force P becomes 21,500 psi. This remarkable result indicates that most of the applied force is carried by decreasing the initial compressive force on the assembled washers since $Y = 0.9P$.

The solution is not valid if one of the materials ceases to behave elastically or if the applied force is such that the initial precompression of the assembled parts is destroyed.

Situations approximating the above idealized problem are found in many practical applications. A hot rivet used in the assembly of plates, upon cooling, develops within it enormous tensile stresses. Thoroughly tightened bolts as in a head of an automobile engine or in a flange of a pressure vessel have high initial tensile stresses; so do the steel tendons in a prestressed concrete beam. It is crucially important that on applying the working loads, only a small increase occur in the initial tensile stresses.

PART B
ANALYSIS BY THE METHOD OF SUPERPOSITION

12-4. METHOD OF ANALYSIS

Structural systems made of linearly elastic materials which experience small deformations are linear structural systems. For such structures the method of superposition is applicable, and an effective procedure for the analysis of statically indeterminate systems is available.

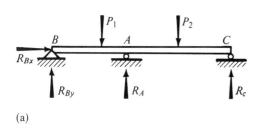

(a)

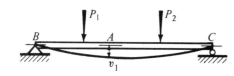

(b)

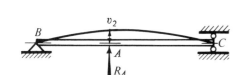

(c)

Fig. 12-8. Illustration of the superposition technique.

In analyzing the structural systems in this manner, it is necessary to remove temporarily the redundant reactions, which renders the structure statically determinate. Then, in this structure artificially reduced to statical determinacy, it is possible to find any desired deflection by the methods previously discussed. For example, by removing the redundant reaction* at A from the indeterminate beam shown in Fig. 12-8(a), the deflection v_1 at A, Fig. 12-8(b), may be found. By reapplying the removed redundant reaction R_A to the same determinate beam, Fig. 12-8(c), the deflection v_2, found as a function of R_A, may also be determined. Then, superposing (adding) the two deflections, since $v_1 + v_2 = 0$ one finds a solution for R_A. The effect of this superposition is that under the action of the applied forces and the redundant reaction, point A actually does not move.

The method of superposition is suitable for the analysis of systems with a large degree of indeterminacy and is widely used in practice. As in this approach the redundant forces are the unknowns, this procedure is often referred to as the *force method*.† Note especially that, after the redundant reactions are determined, a problem becomes statically determinate, and further analysis of stresses and deformations proceeds in the usual manner.

---

* In the analysis of beams, bending moments at the supports are often treated as redundants. Rotations of tangents at the supports are considered in such cases instead of deflections.
† A completely parallel approach based on making the displacements the unknowns is called the *displacement method*. A special procedure based on this concept is discussed in Art. 12-6. For more details see for example, H. C. Martin, *Introduction to Matrix Methods of Structural Analysis* (New York: McGraw-Hill Book Company, 1966).

EXAMPLE 12-8

Rework Example 12-1, using the method of superposition, Fig. 12-9(a).

SOLUTION

The rod is imagined cut at the left support. Then, in this determinate member, the cut end deflects an amount u_1 because of the applied force P, Fig. 12-9(b). Reapplication of the unknown force R_1, Fig. 12-9(c), causes a deflection u_2. Superposing these deflections in order to have no movement of the left end of the rod, as required by the conditions of the problem, gives

$$u_1 + u_2 = 0$$

Then, using Eq. 4-33, $u = PL/(AE)$, and taking deflections to the right as positive,

$$\frac{Pb}{A_2 E} - \left(\frac{R_1 a}{A_1 E} + \frac{R_1 b}{A_2 E} \right) = 0$$

and

$$R_1 = \frac{P}{1 + (aA_2/bA_1)}$$

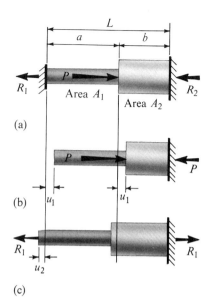

Fig. 12-9

which is the same result as that obtained in Example 12-1. The right reaction may be found from the condition of statics: $R_1 + R_2 = P$.

The analysis of the three-bar system of Example 12-4 can be made in the same manner as was done in the above example. First, the system is imagined cut at B and the displacement u_1 with zero stress in the member BC is determined, Fig. 12-10. Then, the force applied at B necessary to close the gap is found. The sum of the two solutions is the result sought. In using this approach, deformations caused by the changes of temperature can be readily included.

The superposition technique is also very convenient in the analysis of elastic statically indeterminate torsional problems. In such cases, the torsion member is imagined cut at one of the supports, and the rotation of the released end is computed. The magnitude of the redundant torque is determined by making it of sufficient magnitude to restore the cut end of the member to its true position.

EXAMPLE 12-9

Plot shear and moment diagrams for a uniformly loaded beam fixed at one end and simply supported at the other, Fig. 12-11(a). The EI is constant.

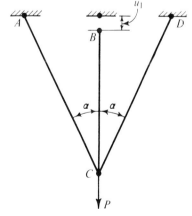

Fig. 12-10

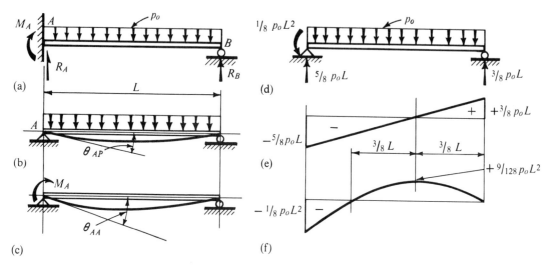

Fig. 12-11

SOLUTION

This beam is indeterminate to the first degree, but it can be reduced to determinacy by removing M_A as in Fig. 12-11(b). A positive moment M_A acting at A on the same structure is shown in Fig. 12-11(c). The rotations at A for the two determinate cases can be found from Table 11 in the Appendix. (Also see Example 11-3.) The requirement of zero rotation at A in the original structure provides the necessary equation for determining M_A.

and
$$\theta_{AP} = p_0 L^3/(24EI) \quad \text{(clockwise)}$$
$$\theta_{AA} = M_A L/(3EI) \quad \text{(clockwise)}$$
$$\theta_A = \theta_{AP} + \theta_{AA} = 0$$

Taking clockwise rotations as positive,

$$\frac{p_0 L^3}{24EI} + \frac{M_A L}{3EI} = 0 \quad \text{and} \quad M_A = -\frac{p_0 L^2}{8}$$

The negative sign of the result indicates that M_A acts in the direction opposite to that assumed. Its correct sense is shown in Fig. 12-11(d).

The remainder of the problem may be solved with the aid of statics. Reactions, shear diagram, and moment diagram are in Figs. 12-1(d), (e), and (f), respectively.

This problem may also be analyzed by treating R_B as the redundant.

As stated earlier, the superposition procedure is applicable to linear systems which are indeterminate to a high degree. In such cases, it is essential to remember that the displacement of every point on a

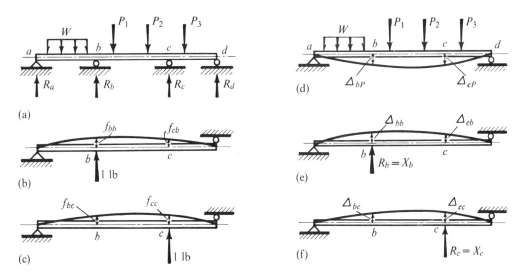

Fig. 12-12. Superposition method for a continuous beam.

structure reduced to statical determinacy is affected by each reapplied redundant force. As an example consider the beam in Fig. 12-12(a).

On removing the redundant reactions* R_b and R_c the beam becomes determinate and the deflections at b and c can be computed, Fig. 12-12(d). These deflections are designated Δ_{bP} and Δ_{cP}, respectively, where the first subscript indicates the point where the deflection occurs, and the second the cause of the deflection. By reapplying R_b to the same beam, Fig. 12-12(e), the deflections at b and c due to R_b at b can be found. These deflections are designated Δ_{bb} and Δ_{cb}, respectively. Similarly, Δ_{bc} and Δ_{cc}, Fig. 12-12(f), due to R_c, may be established. Superposing the deflections at each support and setting the sum equal to zero, since points b and c actually do not deflect one obtains two equations:

$$\Delta_b = \Delta_{bP} + \Delta_{bb} + \Delta_{bc} = 0$$
$$\Delta_c = \Delta_{cP} + \Delta_{cb} + \Delta_{cc} = 0 \qquad (12\text{-}7)$$

These can be rewritten in a more meaningful form using *flexibility coefficients* f_{bb}, f_{bc}, f_{cb}, and f_{cc}, which are defined as the deflections shown in Figs. 12-12(b) and (c) due to unit forces applied in the direction of the redundants. Then, since a linear structural system is being considered, the deflection at point b due to the redundants can be expressed as

$$\Delta_{bb} = f_{bb} X_b \quad \text{and} \quad \Delta_{bc} = f_{bc} X_c \qquad (12\text{-}8)$$

and similarly at point c as

$$\Delta_{cb} = f_{cb} X_b \quad \text{and} \quad \Delta_{cc} = f_{cc} X_c \qquad (12\text{-}9)$$

---
* The choice of redundant reactions is arbitrary.

where X_b and X_c are dimensionless factors which on being multiplied by the respective unit forces acquire the units of the redundant quantities. Using this notation Eq. 12-7 becomes

$$f_{bb}X_b + f_{bc}X_c + \Delta_{bP} = 0$$
$$f_{cb}X_b + f_{cc}X_c + \Delta_{cP} = 0 \qquad (12\text{-}10)$$

where the only unknown quantities are X_b and X_c; simultaneous solutions of these equations constitute the solution of the problem.

The canonical form of superposition equations* of the force method for a system with n unknown redundants reads:

$$f_{aa}X_a + f_{ab}X_b + \cdots + f_{an}X_n + \Delta_{aP} = \Delta_a$$
$$f_{ba}X_a + f_{bb}X_b + \cdots + f_{bn}X_n + \Delta_{bP} = \Delta_b \qquad (12\text{-}11)$$
$$\cdots\cdots\cdots\cdots\cdots\cdots\cdots\cdots\cdots\cdots\cdots$$
$$f_{na}X_a + f_{nb}X_b + \cdots + f_{nn}X_n + \Delta_{nP} = \Delta_n$$

In general, the deflections of various points designated in Eq. 12-11 as $\Delta_a, \Delta_b, \ldots \Delta_n$ need not necessarily be zero. In these equations the quantities f_{ij}, Δ_{iP}, and Δ_i represent either linear or angular deflections depending on whether they are associated with a force or a couple.

It is important to note that the matrix of the flexibility coefficients f_{ij} is symmetric, i.e., $f_{ij} = f_{ji}$. This follows from the law of reciprocal deflections. For proof see Art. 13-5.

---

* Sometimes these are referred to as the Maxwell-Mohr equations.

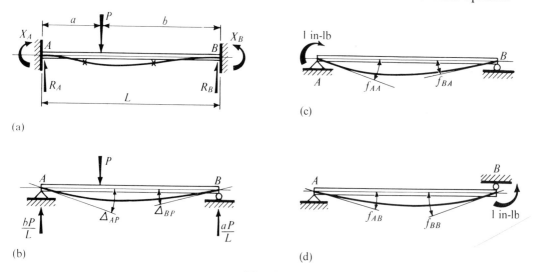

Fig. 12-13

EXAMPLE 12-10

For the fixed-ended beam shown in Fig. 12-13(a), using the force method, find the moments developed at the supports due to the applied force *P*. The *EI* is constant.

*Section 12-4
Method of analysis*

SOLUTION

This beam, indeterminate to the second degree, is reduced to determinacy by removing the end moments, Fig. 12-13(b). In this determinate beam the rotations of the tangents at the supports due to the applied load can be taken from the solution in Example 11-4. This yields

$$|\Delta_{AP}| = \left|\left(\frac{dv}{dx}\right)_{x=0}\right| = \frac{Pab}{6EIL}(a + 2b)$$

$$|\Delta_{BP}| = \left|\left(\frac{dv}{dx}\right)_{x=L}\right| = \frac{Pab}{6EIL}(b + 2a)$$

The rotations of the beam ends due to the unit couples shown in Figs. 12-13(c) and (d) can be found from Table 11 in the Appendix. With $M_o = 1$ this gives

$$|f_{AA}| = |f_{BB}| = \frac{L}{3EI} \quad \text{and} \quad |f_{AB}| = |f_{BA}| = \frac{L}{6EI}$$

The sense of all of the above rotations is in Fig. 12-13. This must be carefully noted in setting up the superposition relations formally stated by Eq. 12-11. In each equation the positive displacements are measured in the direction of the displacement caused by the corresponding redundant quantity. On this basis two equations are obtained:

$$\circlearrowleft + \quad \Delta_A \equiv \theta_A = +\frac{L}{3EI}X_A + \frac{L}{6EI}X_B + \frac{Pab}{6EIL}(a + 2b) = 0$$

$$\circlearrowleft + \quad \Delta_B \equiv \theta_B = +\frac{L}{6EI}X_A + \frac{L}{3EI}X_B + \frac{Pab}{6EIL}(b + 2a) = 0$$

Solving these equations simultaneously

$$X_A = -\frac{Pab^2}{L^2} \quad \text{and} \quad X_B = -\frac{Pa^2b}{L^2}$$

where the negative signs of the bending moments indicate that the assumed directions were chosen incorrectly.

This problem may also be solved by treating R_B and X_B as the redundants since their temporary removal makes the structure determinate. This procedure is particularly convenient to apply if it is specified that one of the supports moves vertically. In such a case the deflection caused by the applied forces and the redundants is equated to the movement of the support.

12-5. MOMENT-AREA METHOD FOR STATICALLY INDETERMINATE BEAMS*

The application of the moment-area method with the superposition technique to indeterminate beams may be greatly accelerated, according to the following reasoning: Restrained† and continuous beams differ from simply supported beams mainly by the presence of redundant moments at the supports. Therefore the bending-moment diagrams for these beams may be considered to consist of two independent parts—one part for the moment caused by all of the applied loading on a beam assumed simply supported, the other part for the redundant moments. Thus the effect of redundant end moments is superposed on a beam assumed simply supported. Physically this notion may be clarified by imagining an indeterminate beam cut through at the supports while the vertical reactions are maintained. The continuity of the elastic curve of the beam is preserved by the redundant moments.

Although the critical ordinates of the bending-moment diagrams caused by the redundant moments are not known, their shape is known. Application of a redundant moment at an end of a simple beam results in a triangular-shaped moment diagram with a maximum at the applied moment and a zero ordinate at the other end. Likewise, when end moments are present at both ends of a simple beam, two triangular moment diagrams superpose into a trapezoidal-shaped diagram (verify these statements).

The known and the unknown parts of the bending-moment diagram together give a complete bending-moment diagram. This whole diagram may then be used in applying the moment-area theorems to the continuous elastic curve of a beam. The geometrical conditions of a problem, such as the continuity of the elastic curve at the support or the tangents at built-in ends which cannot rotate, permit a rapid formulation of equations for the unknown values of the redundant moments at the supports.

For beams of variable flexural rigidity, $M/(EI)$ diagrams must be used.

EXAMPLE 12-11

Find the maximum downward deflection due to an applied force $P = 100$ lb for the small aluminum beam shown in Fig. 12-14(a). The beam's constant flexural rigidity $EI = 25,000$ lb-in.^{2}

SOLUTION

The solution of this problem consists of two parts. First, a redundant reaction is determined to establish the numerical values for the bending-

---

* The remainder of this part of the chapter may be omitted.
† Indeterminate beams with one or more ends fixed are called restrained beams.

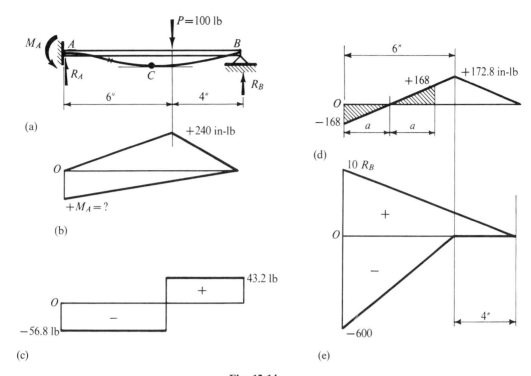

Fig. 12-14

moment diagram; then the usual moment area procedure is applied to find the deflection.

Imagining the beam released from the redundant end moment, one can construct the simple beam moment diagram above the baseline in Fig. 12-14(b). The moment diagram of known shape due to the unknown redundant moment M_A is shown on the same diagram below the base line. One assumes M_A to be positive since in this manner its correct sign according to the beam convention is automatically obtained. The composite diagram represents a complete bending-moment diagram.

The tangent at the built-in end remains horizontal after the application of the force P. Hence the geometrical condition is $t_{BA} = 0$. An equation formulated on this basis yields a solution for M_A.* The equations of static equilibrium are used to compute the reactions. The final bending-moment diagram, Fig. 12-14(d), is obtained in the usual manner after the reactions are known.

Thus, since $t_{BA} = 0$

$$\frac{1}{EI}\left[\frac{(10)(+240)}{2}\frac{(10+4)}{3} + \frac{(10)(+M_A)}{2}\tfrac{2}{3}(10)\right] = 0$$

A solution of this equation yields $M_A = -168$ in-lb.

---

* See Table 2 of the Appendix for the centroidal distance of a whole triangle.

$$\sum M_A = 0 \circlearrowright +, \qquad 100(6) - R_B(10) - 168 = 0, \qquad R_B = 43.2 \text{ lb}$$

$$\sum M_B = 0 \circlearrowleft +, \qquad 100(4) + 168 - R_A(10) = 0, \qquad R_A = 56.8 \text{ lb}$$

Check: $\qquad \sum F_y = 0 \uparrow +, \qquad 43.2 + 65.8 - 100 = 0$

The maximum deflection occurs where the tangent to the elastic curve is horizontal, point C in Fig. 12-14(a). Hence, by noting that the tangent at A is also horizontal and using the first moment-area theorem, one locates point C. This occurs when the shaded areas in Fig. 12-14(d) having opposite signs are equal, i.e., at a distance $2a = 2(168/56.8) = 5.92$ in. from A. The tangential deviation t_{AC} (or t_{CA}) gives the deflection of point C.

$$v_{\max} = v_C = t_{AC}$$

$$= \frac{1}{EI} \left\{ \frac{(2.96)}{2}(+168)\left[2.96 + \frac{2(2.96)}{3}\right] + \frac{(2.96)}{2}(-168)\frac{(2.96)}{3} \right\}$$

$$= (982/EI) = 0.0393 \text{ in. down}$$

ALTERNATE SOLUTION

A rapid solution may also be obtained by plotting the moment diagram as for a cantilever. This is shown in Fig. 12-14(e). Note that one of the ordinates is in terms of the redundant reaction R_B. Again using the geometrical condition $t_{BA} = 0$, one obtains an equation yielding R_B. Other reactions follow by statics.

From the condition $t_{BA} = 0$, one has

$$(1/EI)\{\tfrac{1}{2}(10)(+10R_B)\tfrac{2}{3}(10) + \tfrac{1}{2}(6)(-600)[4 + \tfrac{2}{3}(6)]\} = 0$$

Hence, $R_B = 43.2$ lb, up as assumed.

$$\sum M_A = 0 \circlearrowright +, \qquad M_A + 43.2(10) - 100(6) = 0,$$

$$M_A = 168 \text{ in-lb} \circlearrowleft$$

The remainder of the work is the same as in the preceding solution.

EXAMPLE 12-12

Find the moments at the supports for a fixed-ended beam loaded with a uniformly distributed load of p_o lb per unit length, Fig. 12-15(a).

SOLUTION

The moments at the supports are called *fixed-end moments*, and their determination is of great importance in structural theory. Due to symmetry in this problem, the fixed-end moments are equal, as are the vertical reactions, which are $p_o L/2$ each. The moment diagram for this beam considered simply supported is a parabola, Fig. 12-15(b), while the fixed-end moments give the rectangular diagram in the same figure.

Although this beam is indeterminate to the second degree, because of symmetry a single equation based on a geometrical condition is sufficient to yield the redundant moments. From the geometry of the elastic curve, several conditions may be used such as, $\Delta\theta_{AB} = 0$, $t_{BA} = 0$, or $t_{AB} = 0$. From the condition $\Delta\theta_{AB} = 0$,

$$\frac{1}{EI}\left[\frac{2L}{3}\left(+\frac{p_oL^2}{8}\right) + L(+M_A)\right] = 0$$

then $\qquad M_A = M_B = -p_oL^2/12$

The composite moment diagram is in Fig. 12-15(c). In comparison with the maximum bending moment of a simple beam, a considerable reduction in the magnitude of the critical moments occurs.

Section 12-5
Moment-area method for statically indeterminate beams

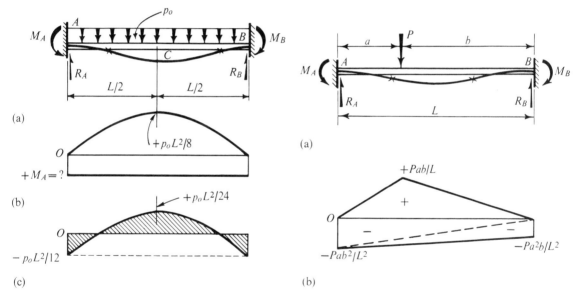

Fig. 12-15

Fig. 12-16

EXAMPLE 12-13

Rework Example 12-10, using the moment-area method, Fig. 12-16(a).

SOLUTION

Treating the beam AB as a simple beam gives the moment diagram due to P as shown above the baseline in Fig. 12-16(b). The fixed-end moments are not equal and result in the trapezoidal diagram. Three geometrical conditions for the elastic curve are available to solve this problem, indeterminate to the second degree: (a) $\Delta\theta_{AB} = 0$ since the tangents at A and B are parallel. (b) $t_{BA} = 0$ since the support B does not deviate from a fixed tangent at A. (c) Similarly, $t_{AB} = 0$.

Any two of the above conditions may be used; arithmetical simplicity of the resulting equations governs the choice. Thus, using condition (a), which is always the simplest, and condition (b), two equations are

$$\Delta\theta_{AB} = \frac{1}{EI}\left(\frac{L}{2}\frac{Pab}{L} + \frac{LM_A}{2} + \frac{LM_B}{2}\right) = 0$$

or

$$M_A + M_B = -Pab/L$$

$$t_{BA} = \frac{1}{EI}\left[\frac{L}{2}\frac{Pab}{L}\frac{(L+b)}{3} + \frac{L}{2}M_A\frac{2L}{3} + \frac{LM_B}{2}\frac{L}{3}\right] = 0$$

or

$$2M_A + M_B = -(Pab/L^2)(L+b)$$

Solving the two reduced equations simultaneously

$$M_A = -\frac{Pab^2}{L^2} \quad \text{and} \quad M_B = -\frac{Pa^2b}{L^2}$$

These results agree with those found in Example 12-10.

EXAMPLE 12-14

Plot moment and shear diagrams for the continuous beam loaded as shown in Fig. 12-17(a). The EI is constant for the whole beam.

SOLUTION

This beam is indeterminate to the second degree. By treating each span as a simple beam with the redundant moments, Fig. 12-17(b), one obtains the moment diagram in Fig. 12-17(c). No end moments exist at A as this end is on a roller. The clue to the solution is contained in two geometrical conditions for the elastic curve for the whole beam, Fig. 12-17(d):

(a) $\theta_B = \theta'_B$. Since the beam is physically continuous, there is a line at the support B which is tangent to the elastic curve in either span.

(b) $t_{BC} = 0$ since the support B does not deviate from a fixed tangent at C.

To apply condition (a), t_{AB} and t_{CB} are determined, and, by dividing these quantities by the respective span lengths, the two angles θ_B and θ'_B are obtained. These angles are equal. However, although t_{CB} is algebraically expressed as a positive quantity, the tangent through point B is above point C. Therefore this deviation must be considered negative. Whence, by using condition (a), one equation with the redundant moments is obtained:

$$t_{AB} = \frac{1}{EI}\left[\frac{2(10)}{3}(+30)\frac{(10)}{2} + \frac{(10)}{2}(+M_B)\frac{2(10)}{3}\right]$$

$$= \frac{1}{EI}\left[1{,}000 + \frac{(100)M_B}{3}\right]$$

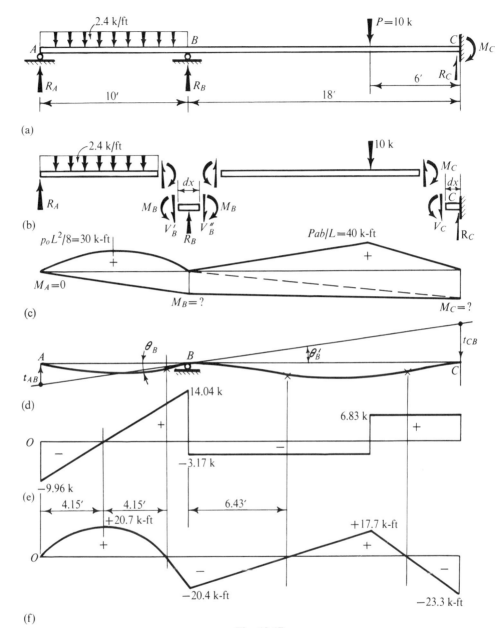

Fig. 12-17

$$t_{CB} = \frac{1}{EI}\left[\frac{(18)}{2}(+40)\frac{(18+6)}{3} + \frac{(18)(+M_B)}{2}\frac{2(18)}{3} + \frac{(18)(+M_C)}{2}\frac{(18)}{3}\right]$$

$$= (1/EI)(2{,}880 + 108M_B + 54M_C)$$

Since $\theta_B = \theta'_B$ or $(t_{AB}/L_{AB}) = -(t_{CB}/L_{CB})$,

455

$$\frac{1}{EI}\left(\frac{1{,}000 + \frac{1}{3}(100)M_B}{10}\right) = -\frac{1}{EI}\left(\frac{2{,}880 + 108M_B + 54M_C}{18}\right)$$

or
$$(^{28}\!/_{3})M_B + 3M_C = -260$$

Using condition (b) for the span BC provides another equation, $t_{BC} = 0$, or

$$\frac{1}{EI}\left[\frac{(18)}{2}(+40)\frac{(18+12)}{3} + \frac{(18)(+M_B)}{2}\frac{(18)}{3} + \frac{(18)(+M_C)}{2}\frac{2(18)}{3}\right] = 0$$

or
$$3M_B + 6M_C = -200$$

Solving the two reduced equations simultaneously,

$$M_B = -20.4 \text{ ft-lb} \quad \text{and} \quad M_C = -23.3 \text{ ft-lb}$$

where the signs agree with the convention of signs used for beams. These moments with their proper sense are shown in Fig. 12-17(b).

After the redundant moments M_A and M_C are found, no new techniques are necessary to construct the moment and shear diagrams. However, particular care must be exercised to include the moments at the supports while computing shears and reactions. Usually, isolated beams as shown in Fig. 12-17(b) are the most convenient free bodies for determining shears. Reactions follow by adding the shears on the adjoining beams. In units of kips and feet, for free body AB:

$\sum M_B = 0 \circlearrowleft +$, $\quad 2.4(10)5 - 20.4 - 10R_A = 0$, $\quad R_A = 9.96$ kips ↑

$\sum M_A = 0 \circlearrowright +$, $\quad 2.4(10)5 + 20.4 - 10V'_B = 0$, $\quad V'_B = 14.04$ kips ↑

For free body BC:

$$\sum M_C = 0 \circlearrowleft +, \quad 10(6) + 20.4 - 23.3 - 18V''_B = 0,$$
$$V''_B = 3.17 \text{ kips} \uparrow$$

$$\sum M_B = 0 \circlearrowright +, \quad 10(12) - 20.4 + 23.3 - 18V_C = 0,$$
$$V_C = R_C = 6.83 \text{ kips} \uparrow$$

Check:
$$R_A + V'_B = 24 \text{ kips} \uparrow \quad \text{and} \quad V''_B + R_C = 10 \text{ kips} \uparrow$$

From above, $R_B = V'_B + V''_B = 17.21$ kips ↑.

The complete shear and moment diagrams are in Figs. 12-17(e) and (f), respectively.

12-6. THE THREE-MOMENT EQUATION

Generalizing the procedure used in the preceding example, a recurrence formula, i.e., an equation which may be repeatedly applied for every two adjoining spans, may be derived for continuous beams. For any

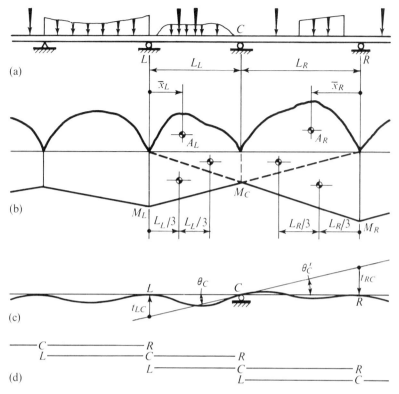

Fig. 12-18. Derivation of the three-moment equation.

n number of spans, $n - 1$ such equations may be written. This gives enough simultaneous equations for the solution of redundant moments over the supports. This recurrence formula is called the *three-moment equation* because three unknown moments appear in it.

Consider a continuous beam, such as shown in Fig. 12-18(a), subjected to any transverse loading. For any two adjoining spans, as LC and CR, the bending-moment diagram is considered to consist of two parts. The areas A_L and A_R to the left and to the right of the center support C, Fig. 12-18(b), correspond to the bending-moment diagrams in the respective spans if these spans are treated as being simply supported. These moment diagrams depend entirely upon the nature of the known forces applied within each span. The other part of the moment diagram of known shape is due to the unknown moments M_L at the left support, M_C at the center support, and M_R at the right support.

Next, the elastic curve in Fig. 12-18(c) must be considered. This curve is continuous for any continuous beam. Hence the angles θ_C and θ'_C, which define, from the respective sides, the inclination of the same tangent to the elastic curve at C, are equal. By using the second moment-area theorem to obtain t_{LC} and t_{RC}, these angles are defined as $\theta_C = t_{LC}/L_L$ and $\theta'_C = -t_{RC}/L_R$, where L_L and L_R are span lengths on the left and on the right of C, respectively. The negative sign for the second angle is necessary since the tangent from point C is above the support R and a positive deviation of t_{RC} locates a tangent below the same support. Hence, following the steps outlined,

$$\theta_C = \theta'_C \quad \text{or} \quad t_{LC}/L_L = -t_{RC}/L_R$$

and

$$\frac{1}{L_L}\frac{1}{EI_L}\left(A_L\bar{x}_L + \frac{L_L M_L}{2}\frac{L_L}{3} + \frac{L_L M_C}{2}\frac{2L_L}{3}\right)$$

$$= -\frac{1}{L_R}\frac{1}{EI_R}\left(A_R\bar{x}_R + \frac{L_R M_R}{2}\frac{L_R}{3} + \frac{L_R M_C}{2}\frac{2L_R}{3}\right)$$

where I_L and I_R are the respective moments of inertia of the cross-sectional area of the beam in the left and the right spans. Throughout each span, I_L and I_R are assumed constant. The term $\bar{x}_L$ is the distance from the left support L to the centroid of the area A_L, and $\bar{x}_R$ is a similar distance for A_R measured from the right support R. The terms M_L, M_C, and M_R denote the unknown moments at the supports.

Simplifying the above expression, the three-moment equation* is

$$L_L M_L + 2\left(L_L + \frac{I_L}{I_R}L_R\right)M_C + \frac{I_L}{I_R}L_R M_R$$

$$= -\frac{6A_L\bar{x}_L}{L_L} - \frac{6A_R\bar{x}_R}{L_R}\frac{I_L}{I_R} \quad (12\text{-}12)$$

This equation 12-12 applies to continuous beams on unyielding supports, with the beam in each span of constant I. In a particular problem, all terms, with the exception of the redundant moments at the supports, are constant. A sufficient number of simultaneous equations for the unknown moments is obtained by successively imagining the supports of the adjoining spans as L, C, and R as shown in Fig. 12-18(d). However, in these equations the subscripts of the M's must correspond to the actual designation of the supports, such as A, B, C, etc. Also note that at pinned ends of beams the moments are known to be zero. Likewise, if a continuous beam has an overhang, the moment at the first support is known from statics. Fixed supports will be discussed in Example 12-16. For symmetrical beams symmetrically loaded, work may be minimized by noting that moments at symmetrically placed supports are equal.

In deriving the three-moment equation, the moments at the supports were assumed positive. Hence an algebraic solution of simultaneous equations automatically gives the correct sign of moments according to the convention for beams.

---

* The three-moment equation was originally derived by E. Clapeyron, a French engineer, in 1857, and sometimes is referred to as Clapeyron's equation.

12-7. SPECIAL CASES

As a specific example of the evaluation of the constant terms on the right side of the three-moment equation, consider two adjoining spans loaded with the concentrated forces P_L and P_R, Fig. 12-19. Considering these spans simply supported, since the maximum moment in the left span is $+P_L ab/L_L$, and $\bar{x}_L = (L_L + a)/3$ one writes

$$-6A_L \frac{\bar{x}_L}{L_L} = -6\left(\frac{L_L}{2}\right)\frac{P_L ab}{L_L}\frac{(L_L + a)}{3L_L}$$

$$= -P_L ab\left(1 + \frac{a}{L_L}\right) \qquad (12\text{-}13)$$

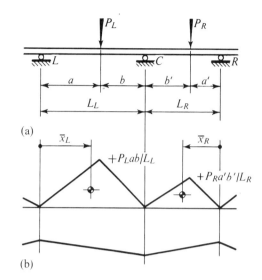

Fig. 12-19. Establishing the constants on the right hand of the three-moment equation for concentrated loads.

Similarly, by interchanging the role of the dimensions a and b in the right span, i.e., by always measuring a's from the outside support toward the force,

$$-6A_R \frac{I_L}{I_R}\frac{\bar{x}_R}{L_R} = -P_R a'b'\left(1 + \frac{a'}{L_R}\right)\frac{I_L}{I_R} \qquad (12\text{-}14)$$

If a number of concentrated forces occurs within a span, the contribution of each one of them to the above constant may be treated separately. Hence a constant term for the right side of the three-moment equation applicable for any number of concentrated forces applied within the spans is

$$-\sum P_L ab\left(1 + \frac{a}{L_L}\right) - \sum P_R a'b'\left(1 + \frac{a'}{L_R}\right)\frac{I_L}{I_R} \qquad (12\text{-}15)$$

where the summation sign designates the fact that a separate term appears for every concentrated force P_L in the left span, and similarly, for every force P_R in the right span. In both cases, a or a' is the distance from the outside support to the particular concentrated force, and b or b' is the distance to the force from the center support. If any one of these forces acts upward, the term contributed to the constant by such a force is of opposite sign.

The constant for the right side of the three-moment equation, when uniformly distributed loads are applied to a beam, is determined similarly. Thus, using the diagram in Fig. 12-20,

$$-6A_L \frac{\bar{x}_L}{L_L} = -6\left(\frac{2L_L}{3}\right)\left(\frac{p_L L_L^2}{8}\frac{L_L}{2L_L}\right) = -\frac{p_L L_L^3}{4}$$

$$(12\text{-}16)$$

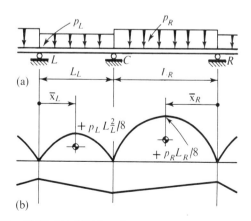

Fig. 12-20. Establishing the constants on the right hand of the three-moment equation for uniformly distributed loads.

and similarly

$$-6A_R \frac{I_L \bar{x}_R}{I_R L_R} = -\frac{p_R L_R^3}{4} \frac{I_L}{I_R} \tag{12-17}$$

Constants for other types of loading may be determined by using the same procedure as above.

EXAMPLE 12-15

Find the moments at all supports and the reactions at C and D for the continuous beam loaded as shown in Fig. 12-21(a). The flexural rigidity EI is constant.

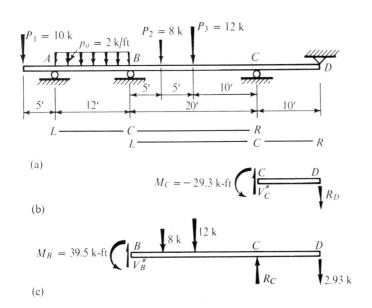

Fig. 12-21

SOLUTION

By using Eq. 12-12 and treating the span AB as the left span and BC as the right, one equation is written. From statics, the beam convention being used for signs, $M_A = -10(5) = -50$ kip-ft. Equations 12-15 and 12-16 are used to obtain the right-hand terms. The moments of inertia I_L and I_R are equal.

$$12M_A + 2(12 + 20)M_B + 20M_C$$

$$= -\frac{2(12)^3}{4} - 8(15)5\left(1 + \frac{15}{20}\right) - 12(10)10\left(1 + \frac{10}{20}\right)$$

Substituting $M_A = -50$ kip-ft and simplifying gives

$$64M_B + 20M_C = -3,114$$

Section 12-7
Special cases

Next, Eq. 12-12 is again applied for the spans BC and CD. No constant terms are contributed to the right side of the three-moment equation by the unloaded span CD. At the pinned end, $M_D = 0$.

$$20M_B + 2(20 + 10)M_C + 10M_D$$
$$= -8(5)15\left(1 + \frac{5}{20}\right) - 12(10)10\left(1 + \frac{10}{20}\right)$$

or
$$20M_B + 60M_C = -2,550$$

Solving the reduced equations simultaneously gives

$$M_B = -39.5 \text{ kip-ft} \quad \text{and} \quad M_C = -29.3 \text{ kip-ft}$$

Isolating the span CD as in Fig. 12-21(b), one obtains the reaction R_D from statics. Instead of isolating the span BC and computing V_C' to add to V_C'' to find R_C, as was done in Example 12-14, the free body shown in Fig. 12-21(c) is used. For free body CD:

$$\Sigma M_C = 0 \circlearrowright +, \quad 29.3 - 10R_D = 0, \quad R_D = 2.93 \text{ kips} \downarrow$$

For free body BD:

$$\Sigma M_B = 0, \circlearrowleft +,$$

$$8(5) + 12(10) - R_C(20) + 2.93(30) - 39.5 = 0, \quad R_C = 10.42 \text{ kips} \uparrow$$

EXAMPLE 12-16

Rework Example 12-14 using the three-moment equation, Fig. 12-22.

SOLUTION

No difficulty is encountered in setting up a three-moment equation for the spans AB and BC. This is done in a manner analogous to that in the preceding example. Note that an unknown moment does exist at the built-in end, and, since the end A is on a roller, $M_A = 0$.

$$10M_A + 2(10 + 18)M_B + 18M_C$$
$$= -\frac{2.4(10)^3}{4} - 10(6)12\left(1 + \frac{6}{18}\right)$$

$$56M_B + 18M_C = -1,560$$

To set up the next equation, an artifice is introduced. An imaginary span of zero length is added at the fixed end, and the

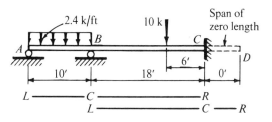

Fig. 12-22

three-moment equation is set up in the usual manner:

$$18M_B + 2(18 + 0)M_C + 0(M_D) = -10(12)6\left(1 + \frac{12}{18}\right)$$

$$18M_B + 36M_C = -1{,}200$$

Solving the reduced equations simultaneously

$$M_B = -20.4 \text{ kip-ft} \quad \text{and} \quad M_C = -23.3 \text{ kip-ft}$$

The remainder of the problem is the same as before.

The use of a zero-length span at the fixed ends of beams is justified by the moment-area procedure. This expedient is equivalent to the requirement of a zero deviation of a support nearest the fixed end from the tangent at the fixed end. For example, if the second of the above reduced equations is divided through by 6, the corresponding equation in Example 12-14 is obtained; there the latter condition was used directly.

PART C
LIMIT ANALYSIS OF BEAMS

12-8. ELASTIC-PLASTIC BENDING OF BEAMS

Examples of ultimate or limit load calculations for axially loaded bar systems made of elastic-plastic material were given at the beginning of this chapter (see Examples 12-1 and 12-4). It is important to note that in such problems there are three stages of loading. First, there is the range of linear elastic response (see Figs. 12-1(e) and 12-4(c)). Then a portion of a structure yields as the remainder continues to deform elastically. This is the range of contained plastic flow. Finally, the structure continues to yield at no increase in load. At this stage the plastic deformation of the structure becomes unbounded.* This condition corresponds to the limit load for the structure.

Since the same general behavior is exhibited by elastic-plastic beams, the objective now is to develop a procedure for determining the limit loads for them. By bypassing the earlier stages of loading and going directly to the determination of the limit load, the procedure becomes relatively simple. For background, some of the results previously established will be re-examined.

Typical moment-curvature relationships of elastic-plastic beams having several different cross sections are shown in Fig. 12-23. Such results were established in Example 11-7 for a rectangular beam. (See

---

* In reality a structure cannot be permitted to deform excessively as only small deformations are assumed to occur.

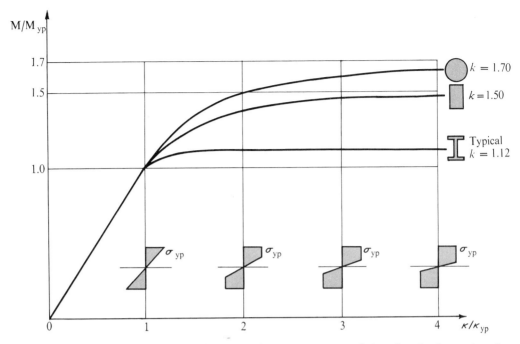

Fig. 12-23. Moment-curvature relations for circular, rectangular, and I cross sections. $M_p/M_{yp} = k$, the shape factor.

Fig. 11-18.) Note especially the rapid ascent of the curves toward their respective asymptotes as the cross sections plastify. This means that very soon after exhausting the elastic capacity of a beam, a rather constant moment is both achieved and maintained. This condition is likened to a plastic hinge. In contrast to a frictionless hinge capable of permitting large rotations at no moment, the plastic hinge allows large rotations to occur at a constant moment. This constant moment is approximately M_p, the ultimate or plastic moment for a cross section.

Using plastic hinges, a sufficient number may be inserted into a structure at the points of maximum moments to create a kinematically admissible collapse mechanism. Such a mechanism, permitting unbounded movement of a system, enables one to determine the ultimate or limit carrying capacity of a beam or of a frame. This approach will now be illustrated by several examples, confining the discussion to beams.

When the limit analysis approach is used for the selection of members, the working loads are multiplied by a load factor larger than unity to obtain the limit loads for which the calculations are performed. This is analogous to the use of the factor of safety in elastic analyses. In structural steel work the term *plastic design* is commonly applied to this approach.

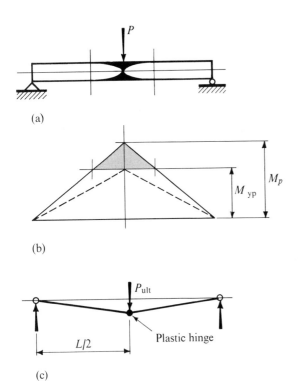

Fig. 12-24

EXAMPLE 12-17

A force P is applied at the middle of a simply supported beam, Fig. 12-24(a). If the beam is made of a ductile material, what is the limit load P_{ult}? Neglect the weight of the beam.

SOLUTION

The shape of the moment diagram is the same regardless of the load magnitude. For any value of P, the maximum moment $M = PL/4$, and if $M \leq M_{yp}$, the beam behaves elastically. Once M_{yp} is exceeded, contained yielding of the beam commences and continues until the maximum plastic moment M_p is reached. At that instant a plastic hinge is formed in the middle of the span forming the collapse mechanism shown in Fig. 12-24(c). By setting the plastic moment M_p equal to $PL/4$ with $P = P_{ult}$, one obtains the result sought:

$$P_{ult} = 4M_p/L$$

Note that consideration of the actual plastic zone indicated shaded in Fig. 12-24(a) is unnecessary in this calculation.

EXAMPLE 12-18

A restrained beam of ductile material is loaded as shown in Fig. 12-25(a). Find the limit load P_{ult}. Neglect the weight of the beam.

SOLUTION

The results of an elastic analysis are shown in Fig. 12-25(b) in the usual manner. The same results are replotted in Fig. 12-25(c) from a horizontal baseline AB. In both diagrams the values of the moment ordinates are the same, and the shaded portions of the diagrams represent the final results. Note that the auxiliary ordinate $PL/4$ has precisely the value of the maximum moment in a simple beam with a concentrated force in the middle.

By setting the maximum elastic moment equal to M_{yp}, one obtains the load P_{yp} at impending yield:

$$P_{yp} = (16/3)M_{yp}/L$$

When the load is increased above P_{yp}, the moment at the built-in end can reach but cannot exceed M_p. This is also true of the moment at the middle of the span. These limiting conditions are shown in Fig. 12-25(d). The sequence in which M_p's occur is unimportant. In determining the limit load it is necessary to have only a kinematically admissible mechanism. With the two plastic hinges and a roller on the right this condition is assured, Fig. 12-25(e).

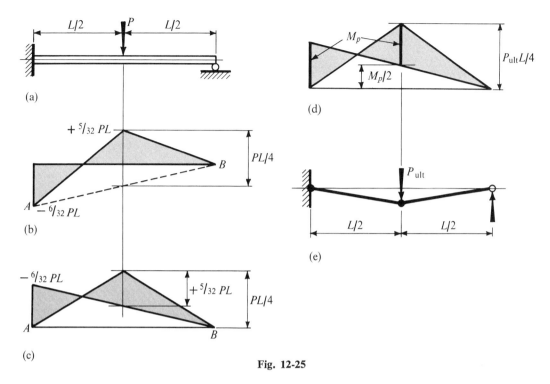

Fig. 12-25

From Fig. 12-25(d) it can be seen that by proportions the end moment M_p gives an ordinate of $M_\text{p}/2$ at the middle of the span. Therefore, the simple beam ordinate $P_\text{ult}L/4$ in the middle of the span must be equated to $3M_\text{p}/2$ to obtain the limit load. This gives

$$P_\text{ult} = 6M_\text{p}/L$$

Comparing this result with P_yp, one has

$$P_\text{ult} = \frac{9M_\text{p}}{8M_\text{yp}} P_\text{yp} = \tfrac{9}{8} k P_\text{yp}$$

which shows that the increase in P_ult over P_yp is due to two causes: $M_\text{p} > M_\text{yp}$, and the maximum moments are distributed more advantageously in the plastic case. (Compare the moment diagrams in Figs. 12-25(c) and (d).)

EXAMPLE 12-19

A restrained beam of ductile material carries a uniformly distributed load as shown in Fig. 12-26(a). Find the limit load p_ult.

SOLUTION

In this problem two plastic hinges are required to create a collapse mechanism. One of these hinges will be at the built-in end. The location

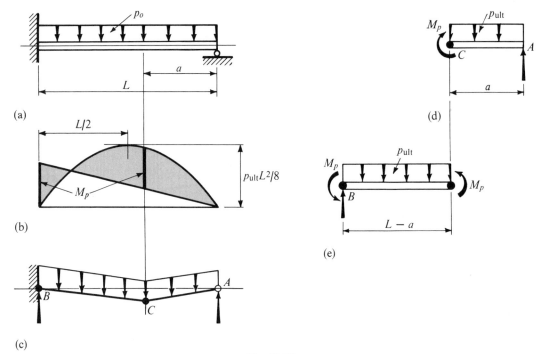

Fig. 12-26

of the hinge associated with the other maximum moment is not immediately known since the moment near the middle of the span changes very gradually. However, one can assume the mechanism to be as shown in Fig. 12-26(c) since this would be in agreement with the moment diagram of Fig. 12-26(b).

For purposes of analysis, the beam with the plastic hinges is separated into two parts as in Figs. 12-26(d) and (e). Then by noting that no shear is possible at C since it is the point of maximum moment on a continuous curve, one may write two equations of static equilibrium:

$$\sum M_A = 0 \;\circlearrowright +, \qquad M_p - p_{\text{ult}} a^2/2 = 0$$

$$\sum M_B = 0 \;\circlearrowleft +, \qquad 2M_p - p_{\text{ult}}(L-a)^2/2 = 0$$

The solution of these equations simultaneously gives $a = (\sqrt{2} - 1)L$, which locates the plastic hinge C. The same equations yield the limit load

$$p_{\text{ult}} = 2M_p/a^2 = 2M_p/[(\sqrt{2} - 1)L]^2$$

In problems with several concentrated forces applied to a beam, a search for the interior plastic hinge also has to be made. The least load for an assumed interior hinge under any one of the loads constitutes the solution of the problem. An equilibrium at any higher load requires moments greater than M_p and is therefore impossible. To determine this may require several trials.

Section 12-8
Elastic-plastic
bending of beams

EXAMPLE 12-20

A fixed beam of ductile material supports a uniformly distributed load, Fig. 12-27(a). Determine the limit load p_{ult}.

SOLUTION

According to the elastic analysis (see Example 12-12 and Fig. 12-15(c)) the maximum moments occur at the built-in ends and are equal to $p_o L^2/12$. Therefore

$$M_{yp} = p_{yp} L^2/12 \quad \text{or} \quad p_{yp} = 12 M_{yp}/L^2$$

On increasing the load, plastic hinges develop at the supports. The collapse mechanism is not formed, however, until a plastic hinge also occurs in the middle of the span, Figs. 12-27(b) and (c).

The maximum moment for a simply supported, uniformly loaded beam is $p_o L^2/8$. Therefore, as can be seen from Fig. 12-27(b), to

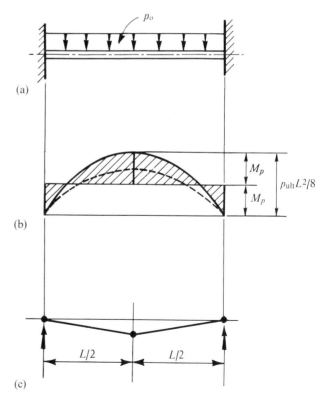

Fig. 12-27

Chapter 12
Statistically indeterminate problems

obtain the limit load in a clamped beam, this quantity must be equated to $2M_p$ with $p_o = p_{ult}$. Hence

$$p_{ult}L^2/8 = 2M_p \quad \text{or} \quad p_{ult} = 16M_p/L^2$$

Comparing this result with p_{yp}, one has

$$p_{ult} = \frac{4M_p}{3M_{yp}} p_{yp} = \tfrac{4}{3} k p_{yp}$$

As in Example 12-18, the increase of p_{ult} over p_{yp} depends on the shape factor k and the equalization of the maximum moments.

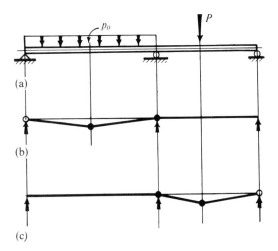

Fig. 12-28

The analysis of continuous beams proceeds in a manner analogous to the above. Ordinarily the collapse of such beams occurs locally in only one of the spans. For example, for the beam shown in Fig. 12-28(a), a mechanism can form as in Fig. 12-28(b) or as in Fig. 12-28(c) depending on the relative magnitudes of the loads and span lengths. Such problems revert to cases already considered. Local collapse mechanisms for the interior spans require the formation of three hinges resembling those in the last example. The mechanisms for frames can become quite complex; the treatment of such problems is beyond the scope of this text.*

12-9. CONCLUDING REMARKS

In practice, statically indeterminate members occur in numerous situations. Some of the methods for analyzing these members have been discussed in this chapter. Sometimes the stresses caused by indeterminacy, particularly those due to temperature, are undesirable. More often, however, members are deliberately arranged to be indeterminate as such members are stiffer, which is highly desirable in many cases. A reduction in stresses may also be accomplished. For example, the maximum bending moments in indeterminate beams are usually smaller than the maximum moments in similar determinate beams. This permits selection of smaller members and results in an economy of material.

There are also some disadvantages to using indeterminate members. Some uncertainty always exists as to whether the supports are capable of completely fixing the ends. Likewise, the supports may settle or move in relation to each other. Then the calculated elastic stresses or deflections may be seriously in error. These matters are of little concern in a statically determinate structure. Finally, the method of elastic analysis

---

* For further details see P. G. Hodge, *Plastic Analysis of Structures* (New York: McGraw-Hill Book Company, 1959).

becomes very involved when the degree of indeterminacy is high. However, this situation has been largely overcome by specialized methods and a wider use of digital computers.

For situations where the applied loads are static in character, and the materials employed are ductile, the plastic method of design offers advantages.

Section 12-9
Concluding remarks

PROBLEMS FOR SOLUTION

12-1. A $\frac{1}{2}$-in.² square bar fixed at both ends to immovable supports and carries two forces P_1 and P_2 as shown in the figure. The magnitude of P_2 is twice that of P_1. (a) Assuming linearly elastic behavior determine the reactions and the axial force distribution in the bar. Plot the axial force and the axial deformation diagrams. (b) If $\sigma_{yp} = 40$ ksi, determine the range of the contained plastic flow. Plot diagrams showing the variations in the magnitudes of the forces P_1 and P_2 as a function of their respective displacements. Let $E = 30 \times 10^6$ psi.

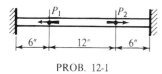

PROB. 12-1

12-2. A material possesses a nonlinear stress-strain relationship given as $\sigma = K\varepsilon^n$, where K and n are material constants. If a rod made of this material and of constant area A is initially fixed at both ends and is then loaded as shown in the figure, how much of the applied force P is carried by the left support? Ans. $P/[(a/b)^n + 1]$.

PROB. 12-2

12-3. A round bar of constant cross-sectional area is built in at both ends and is subjected to a torque T_1 as shown in the figure. (a) Assuming linearly elastic behavior of the material, determine the reactions. Plot the torque $T(x)$ and the angle of twist $\varphi(x)$ dia-

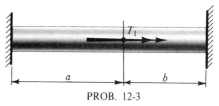

PROB. 12-3

grams. (b) If the bar is 2 in. in diameter, $a = 30$ in., and $b = 20$ in., determine and plot the relationship between the angle of twist φ at $x = 30$ in. and the applied torque T_1. Construct this diagram analogously to the one shown in Fig. 12-1(e). Assume the material to be elastic–perfectly plastic with $\tau_{yp} = 20$ ksi, and $G = 12 \times 10^6$ psi.

12-4. A solid brass, circular shaft is built in at both ends and two torques, $T_1 = 314$ lb-in. and $T_2 = 628$ lb-in., are applied to it as shown in the figure. Determine the torque at A, and plot the torque and the angle of twist diagrams. Assume that the material behaves linearly elastically with a $G = 5.6 \times 10^6$ psi. The diameter $d_1 = 2.83$ in. and $d_2 = 2.38$ in.

PROB. 12-4

12-5. A shaft built in at both ends is made of a 1-in. solid square bar and a $1\frac{1}{2}$-in. square tube welded at the juncture of the two sections to a plate which forms a pointer projecting in the horizontal direction, see figure. The square tube has a wall thickness of 0.040 in. Determine the vertical displacement of the tip of the pointer that would be caused by the

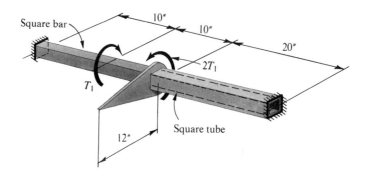

PROB. 12-5

application of the two torques T_1 and $2T_1$ when the maximum shearing stress in the shaft is 10 ksi. Neglect stress concentrations and let $G = 12 \times 10^6$ psi.

12-6. A circular stepped shaft built in at both ends has the dimensions shown in the figure. First this shaft is cut in two at section a-a. Then the right end of the left segment is twisted through an angle $\varphi = 0.060°$ around the x axis and welded back to the right segment. After the externally applied torque is removed, what is the residual torque in the shaft, and what is the final angle of twist φ of the joint from its original position? Assume $\tau_{yp} = 20{,}000$ psi, and $G = 5 \times 10^6$ psi.

12-7. A creep study of concrete was initiated at the University of California in 1930. One series of the experiments was completed in 1957. The typical arrangement used was as shown in the figure, and, initially a compressive stress of 800 psi was applied to the concrete cylinders by tightening the three steel rods. The spring constant of the large spring $k = 6{,}900$ lb per inch; the area of each rod $A = 0.20$ in.^{2}; the effective length of each rod $L = 24$ in.; the elastic modulus of the rods $E = 30 \times 10^6$ psi, and E of concrete was approximately 4×10^6 psi. If the change in deformation Δ due to shrinkage and creep in one of the concrete specimens after 27 years was found to be 0.0308 in., what change in stress occurred in the concrete? How well was the constant stress maintained? How does

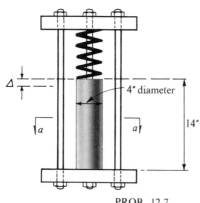

PROB. 12-6

Section a-a

PROB. 12-7

the total deformation in concrete compare with the elastic one? In your calculations include the change of stress in steel rods, but neglect the deformation of the end plates. *Ans.* $\Delta\sigma = 16.8$ psi.

12-8. A rigid platform rests on two aluminum bars ($E = 10^7$ psi) each 10.000 in. long. A third bar made of steel ($E = 30 \times 10^6$ psi) and standing in the middle is 9.995 in. long. (a) What will be the stress in the steel bar if a load P of 100 kips is applied on the platform? (b) How much do the aluminum bars shorten? *Ans.* (a) 15 ksi; (b) 0.01 in.

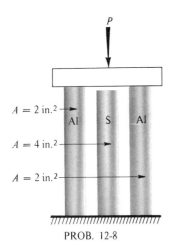

PROB. 12-8

12-9. Three wires support a load P hung from a rigid bar as shown in the figure. Establish a load-deflection diagram for this system if $L_1 = 100$ in., $A_1 = 0.10$ in.^{2}, $L_2 = 50$ in., and $A_2 = 0.20$ in.^{2} The material in the wires is linearly elastic–plastic material, $E = 10 \times 10^6$ psi and $\sigma_{yp} = 30,000$ psi. Clearly show

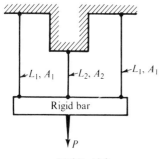

PROB. 12-9

the region of contained plastic flow on the diagram. *Ans.* $P_{yp} = 9$ kips, $P_{ult} = 12$ kips.

12-10. A load $P = 1,000$ lb is applied to a rigid bar suspended by three wires as shown in the figure. All wires are of equal size and the same material. For each wire $A = 0.10$ in.^{2}, $E = 30 \times 10^6$ psi, and $L = 10$ ft. If initially there were no slack in the wires, how will the applied load of 1,000 lb distribute between the wires? *Ans.* 83.3 lb, 333.3 lb, 583.3 lb.

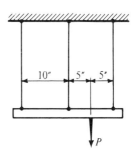

PROB. 12-10

12-11. A rigid bar is supported by a pin at A and two linearly elastic wires at B and C as shown in the figure. The area of the wire at B is 0.10 in.^{2}, and for the one at C is 0.20 in.^{2} Determine the reactions at A, B, and C caused by the applied load $P = 1.5$ kips. *Ans.* -1.2 k, 0.9 k, 1.8 k.

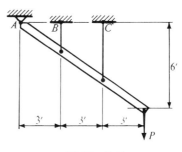

PROB. 12-11

12-12. If the bar in Prob. 4-34 is not rigid but has an $I = 0.222$ in.^{4} and, being of steel, an E of 30×10^6 psi, what forces will be developed in each wire? *Ans.* 428 lb in middle wire.

12-13. Suppose that in Example 12-4 the cross-sectional area of each bar is 2 in.^{2}, the

471

distance $L = 100$ in., and $\alpha = 30°$. The bars are made of steel with a well-defined yield stress of 40 ksi. Let the elastic modulus $E = 30 \times 10^6$ psi. During manufacture, by mistake, the middle bar was made 0.100 in. too short, i.e., before assembly, the three bars looked as shown in Fig. 12-10. (a) What residual stresses develop in the bars as a result of a forced assembly? Assume that no buckling of the bars can occur. (b) On the same graph show load-deflection diagrams analogous to Fig. 12-4(c) for the initially stress-free assembly, and the one with the residual stresses found above.

12-14. Five steel bars each having the cross-sectional area of 1 in.² are assembled in a symmetrical manner as shown in the figure. Assume that the steel behaves as linearly elastic–plastic material with $E = 30 \times 10^6$ psi, and $\sigma_{yp} = 36{,}000$ psi. Determine and plot the load deflection characteristics of joint A due to an application of a downward force P. Assume that initially all bars are free of stress.

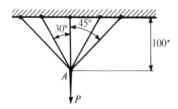

PROB. 12-14

12-15. Two vertical beams 60 in. long are connected at their midspans by a taut wire as shown in the figure. The EI for the beam on the left is 15×10^6 lb-in.² and for the beam on the right is 45×10^6 lb-in.² The cross-sectional area of the wire is 0.100 in.² and its $E = 10 \times 10^6$ psi. Find the stress in the wire after the two 600-lb forces are applied to the beams at midspan. Use the beam deflection formula given in Table 11 of the Appendix. *Ans*. 34.3 ksi.

12-16. The midpoint of a cantilever beam 18 ft long rests on the midspan of a simply supported beam 24 ft long. Determine the deflection of point A, where the beams meet, which results from the application of a 10-kip load at the end of the cantilever beam. The EI for both beams is the same and is constant. *Ans*. $3{,}290/EI$.

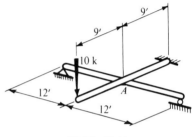

PROB. 12-16

12-17. A 30-in. cantilever beam of constant flexural rigidity, $EI = 10^7$ lb-in.², initially has a gap of 0.05 in. between its end and the spring. The spring constant $k = 10$ kips per inch. If a force of 100 lb, shown in the figure, is applied to the end of this cantilever, how much of this force will be carried by the spring? *Ans*. 40 lb.

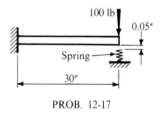

PROB. 12-17

12-18. One end of an 18 WF 50 beam is cast into concrete. It was intended to support the other end with a 1-in.² steel rod 12 ft long as shown in the figure. During the installation, however, the nut on the rod was poorly tightened and in the unloaded condition there is a ½-in. gap between the top of the nut and

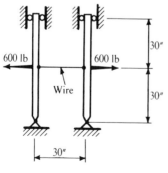

PROB. 12-15

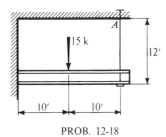

PROB. 12-18

the bottom of the beam. What tensile force will develop in the rod because a force of 15 kips is applied at the middle of the beam? Let $E = 30 \times 10^6$ psi. *Ans.* 2.03 kips.

12-19. The beam AB in an unloaded condition just touches a spring at midspan as shown in the figure. What is the spring stiffness k that will make the forces in all three supports equal for a uniformly distributed, vertical load p_0. Use Table 11 in the Appendix. *Ans.* $384\,EI/(7L^3)$.

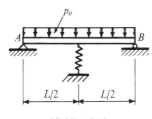

PROB. 12-19

12-20. At a given temperature an elastic cantilever just rests against the frictionless plane as shown in the figure. Calculate the maximum bending moment in the beam if the temperature of the beam is raised δT. Neglect the weight of the beam and the effect of axial force on bending deflection. The quantities A, E, I, and α are given. *Ans.* $\alpha L \delta T/[1/(AE) + L^2/(3EI)]$.

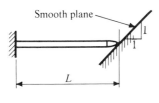

PROB. 12-20

12-21. A 1-in. stainless steel square bar 40 in. long lies between two parallel frictionless surfaces. If one side of this initially straight bar is maintained at a temperature 500°F higher than that of the opposite side, what is the deflection of the bar at the center from the chord through the ends? What moments applied at the ends would straighten out the bar? Assume that the temperature varies linearly across the thickness of the bar. Let $\alpha = 10 \times 10^{-6}$ per °F, and $E = 27 \times 10^6$ psi. For a useful relation in the solution of this problem see Prob. 11-2.

12-22. A $\frac{1}{16}$-in.-diameter copper wire is imbedded concentrically in a porcelain tube of $\frac{1}{4}$-in. diameter as shown in the figure. If the wire is bonded to the porcelain and if the assembly raises in temperature 100°F, what longitudinal tensile stress will be developed in the porcelain tube? For the copper, $A_{\text{cu}} = 0.00307$ in.², $\alpha_{\text{cu}} = 9.4 \times 10^{-6}$ per °F, $E_{\text{cu}} = 14.5 \times 10^6$ psi; and for the porcelain, $A_{\text{po}} = 0.0461$ in.², $\alpha_{\text{po}} = 4.7 \times 10^{-6}$ per °F, and $E_{\text{po}} = 8.0 \times 10^6$ psi. *Ans.* 404 psi.

PROB. 12-22

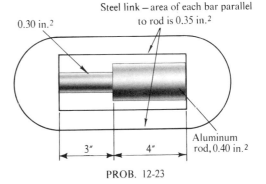

PROB. 12-23

12-23. An aluminum rod 7 in. long, having two different cross-sectional areas, is inserted into a steel link as shown in the figure. If at 60°F no axial force exists in the aluminum rod, what will be the magnitude of this force when the temperature rises to 160°F? $E_a = 10^7$ psi and $\alpha_a = 12.0 \times 10^{-6}$ per °F; $E_s = 30 \times 10^6$ psi and $\alpha_s = 6.5 \times 10^{-6}$. *Ans.* 1,650 lb.

12-24. Three steel wires attached to a rigid bar support a load of 300 lb. Initially this load is equally distributed between the three wires. What will each wire carry if the temperature of the right wire raises 84°F? Assume for all wires $E = 30 \times 10^6$ psi, $A = 0.011$ in.², and $\alpha = 6.5 \times 10^{-6}$ per °F. *Ans.* 70 lb, 160 lb, 70 lb.

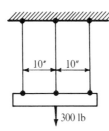

PROB. 12-24

12-25. A steel piano wire 30 in. long is stretched from the middle of an aluminum beam *AB* to a rigid support at *C* as shown in the figure. What is the increase in unit stress in the wire if the temperature drops 100°F? The cross-sectional area of the wire is 0.0001 in.²; $E = 30 \times 10^6$ psi. For the aluminum beam $EI = 1{,}040$ lb-in.² Let $\alpha_s = 6.5 \times 10^{-6}$ per °F, $\alpha_a = 12.9 \times 10^{-6}$ per °F. *Ans.* 6,500 psi.

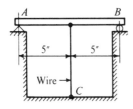

PROB. 12-25

12-26. Two steel cantilever beams *AB* and *CD* are connected by a taut steel wire *BC* having a length equal to 150 in. under initial no-load conditions, see figure. Determine the stress in the wire produced by a 2-kip load applied at *C* and a temperature drop, in the wire only, of 100°F. For beams *AB* and *CD*: $E = 30 \times 10^6$ psi, and $I = 24$ in.⁴ For wire *BC*: $E = 30 \times 10^6$ psi, $A = 0.1$ in.², and $\alpha = 6.0 \times 10^{-6}$ per °F. *Ans.* 1.16 k.

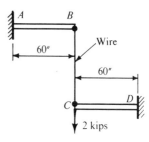

PROB. 12-26

12-27. A steel wire 200 in. in length with cross-sectional area equal to 0.25 in.² is stretched tightly between the midpoint of the simple beam and the free end of the cantilever as shown in the figure. Determine the deflection of the end of the cantilever as a result of a temperature drop of 50°F. For steel wire: $E = 30 \times 10^6$ psi, $\alpha = 6.5 \times 10^{-6}$ per °F. For both beams: $I = 21.3$ in.⁴ and $E = 1.5 \times 10^6$ psi. *Ans.* 4.08×10^{-2} in.

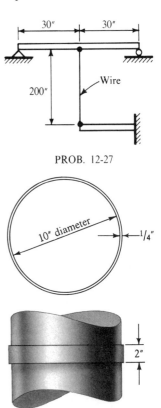

PROB. 12-27

PROB. 12-28

12-28. A thin ring is heated in oil 300°F above room temperature. In this condition the ring just slips on a solid cylinder as shown in the figure. Assuming the cylinder to be completely rigid, (a) determine the hoop stress which develops in the ring upon cooling, and (b) determine what bearing pressure develops between the ring and the cylinder. Let $\alpha = 10^{-5}$ per °F, and $E = 10^7$ psi.

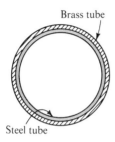

PROB. 12-29

12-29. A cylindrical pressure vessel is made by shrinking a brass shell over a mild steel shell. Both cylinders have a wall thickness of $\frac{1}{4}$ in. The nominal diameter of the vessel is 30 in. and is to be used in all calculations involving the diameter. When the brass cylinder is heated 100°F above room temperature, it exactly fits over the steel cylinder which is at room temperature. What is the stress in the brass cylinder when the composite vessels cool to room temperature? For brass $E_b = 16 \times 10^6$ psi, and $\alpha_b = 10.7 \times 10^{-6}$ per °F. For steel $E_s = 30 \times 10^6$ psi, and $\alpha_s = 6.7 \times 10^{-6}$ per °F. *Ans.* 11 ksi.

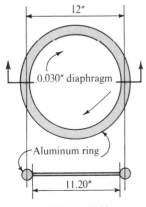

PROB. 12-30

12-30. A steel diaphragm 0.030 in. thick is stretched inside an aluminum-alloy ring as shown in the figure. The ring is made from a round rod of 0.80-in. diameter ($A = 0.50$ in.²). If the temperature is raised 100° F, what stresses are induced in the diaphragm? For the aluminum alloy $E_a = 10 \times 10^6$ psi, $G_a = 4 \times 10^6$ psi, and $\alpha_a = 13 \times 10^{-6}$ per °F. For the steel diaphragm $E_s = 30 \times 10^6$ psi, $G_s = 12 \times 10^6$ psi, and $\alpha_s = 6.5 \times 10^{-6}$ per °F. *Ans.* 10,700 psi.

12-31. Two metal strips of equal thickness but of different material are bonded together and are attached at one end to act as a cantilever as shown in the figure. Determine the tip deflection caused by a change in temperature of δT. Let the coefficient of expansion of one bar be α_1, and of the other bar be α_2. Assume that the elastic moduli for both materials are the same. Such bimetallic elements are widely used in temperature-control devices. (*Hint:* Consider that a longitudinal force and a moment exist in each strip. The strains at the joint are the same in the two materials. The deflection formula for a bar of constant curvature is given in Prob. 11-2.) *Ans.* $3(\alpha_2 - \alpha_1)(\delta T)L^2/(4h)$.

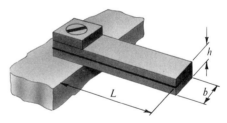

PROB. 12-31

12-32. If in Example 12-7, instead of a bolt, a rivet with no initial tension at 1,600°F is used in the assembly of the washers, what tensile stress will develop in the rivet when the temperature drops to 100°F? Let $E = 30 \times 10^6$ psi, $\sigma_{yp} = 40$ ksi, and $\alpha = 6.5 \times 10^{-6}$ per °F.

12-33. A small pressure vessel has been made for an industrial laboratory having the dimensions shown in the figure. The gasket arrangement is such that the internal pressure can act only on a projected surface of the head which is 23 in. in diameter. There are 20 bolts arranged on a 25-in. bolt circle diameter. The

475

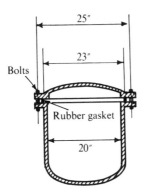

PROB. 12-33

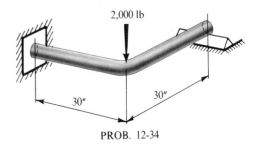

PROB. 12-34

bolts are $\frac{3}{4}$ in. in diameter. Controlled by the bolt strength, can this vessel operate at an internal pressure of 100 psi? The cross-sectional area at the root of the thread of each bolt is 0.302 in.² Before pressurization, the bolts will be tightened to develop in each one of them an initial force of 3 kips. The allowable stress for the bolts in tension is considered to be satisfactory at 18 ksi; however, at the root of the bolt threads a stress concentration factor of $2\frac{1}{2}$ is considered necessary.

12-34. An L-shaped steel shaft of 2.125-in. diameter is built in at one end to a rigid wall and is simply supported at the other end as shown in the figure. In plan the bend is 90°. What bending moment will be developed at the built-in end due to the application of a 2,000-lb force at the corner of the shaft? Assume $E = 30 \times 10^6$ psi, $G = 12 \times 10^6$ psi, and for simplicity let $I = 1.00$ in.⁴ and $J = 2.00$ in.⁴ *Ans.* 49,600 lb-in.

12-35. A steel wire 100 in. long is stretched from the end of the 1-in. standard steel pipe ABC to a rigid support D as shown in the figure. Compute the stresses acting on the element A, located at the support on top of the pipe, caused by a temperature drop in the wire of 100°F. Do not compute principal stresses. For the pipe $E = 30 \times 10^6$ psi, and $G = 12 \times 10^6$ psi; for the wire $A = 0.010$ in.², $E = 30 \times 10^6$ psi, and $\alpha = 6.5 \times 10^{-6}$ per °F. *Ans.* $\sigma_x = 2,970$ psi, $\tau_{xz} = 744$ psi.

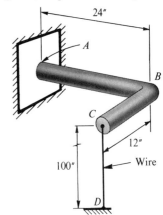

PROB. 12-35

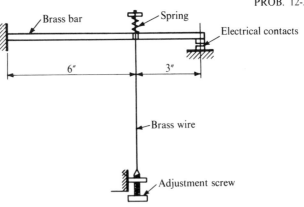

PROB. 12-36

12-36. A fire alarm is set up with a piece of brass wire as the sensitive element as schematically shown in the figure. If the gap is 0.084 in. when there is no tension in the wire, find the number of turns which should be taken on the screw to set the alarm so that the contacts will open at a temperature rise of 100° F. For the brass bar $EI = 6{,}000$ lb-in.^{2}, and the spring constant $k = 100$ lb per inch. The brass wire is 15 ft long, has an $A = 0.010$ in.^{2}, an $E = 15 \times 10^6$ psi, and an $\alpha = 10 \times 10^{-6}$ per °F. The adjusting screw has 20 threads per inch. *Ans.* 5.46 turns.

12-37. The temperature in a furnace is measured by means of a stainless steel wire place in it. The wire is fastened to the end of a cantilever beam outside the furnace. The strain measured by the strain gage glued to the outside of the beam is a measure of the temperature. Assuming that the full length of the wire is heated to the furnace temperature, what is the change in furnace temperature if the gage records a change in strain of -100×10^{-6} in. per inch. Assume that the wire has sufficient amount of initial tension to perform as intended. The mechanical properties of the materials are as follows: $\alpha_{ss} = 9.5 \times 10^{-6}$ per °F, $\alpha_a = 12 \times 10^{-6}$ per °F, $E_{ss} = 30 \times 10^6$ psi, $E_a = 10 \times 10^6$ psi, $A_{wire} = 5 \times 10^{-4}$ in.^{2}, $I_{beam} = 6.5 \times 10^{-4}$ in.^{4} The depth of the small beam is 0.25 in. *Ans.* 96.8°F.

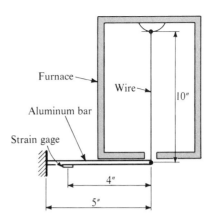

PROB. 12-37

12-38. Determine the total force which will develop in the bronze wire for the arrangement shown in the figure due to the application of 1,000-lb force F. Initially the wire is pre-stressed 38 lb. The steel shaft has an $I = 1.00$ in.^{4} and a $J = 2.00$ in.^{4} For the shaft, $E = 30 \times 10^6$ psi and $G = 12 \times 10^6$ psi. For the bronze wire, $A = 7.2 \times 10^{-2}$ in.^{2} and $E = 10 \times 10^6$ psi. The aluminum-alloy I beam is 5 in. deep and has the same cross-sectional dimensions as a 5-in., 14.75-lb steel I beam. For the beam, $E = 10 \times 10^6$ psi. *Ans.* 300 lb.

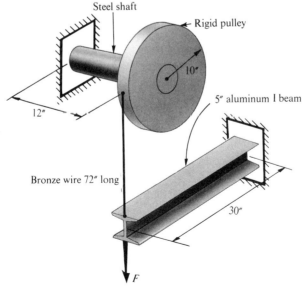

PROB. 12-38

12-39. Rework Example 12-10 by treating R_B and X_B as redundants. Employ the superposition equations, Eq. 12-11.

12-40. Using the superposition equations, Eq. 12-11, determine the reactions at the supports caused by the applied load for the beam shown in the figure. Treat moments at the supports as redundants, and determine the end rotations of the simply supported beam by any of the previously developed methods. (*Hint:* See Example 2-11, Fig. 2-30.) *Ans.* $M_a = -kL^3/30$, $M_b = -kL^3/20$.

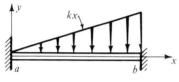

PROB. 12-40

12-41 and 12-42. For the beams loaded as shown in the figures, using the moment-area method, determine the redundant reactions and plot shear and moment diagrams. In all cases EI is constant. (*Hint:* In Prob. 12-42 treat the reaction on the right as the redundant.) *Ans.* M_A is given in parentheses by the figures.

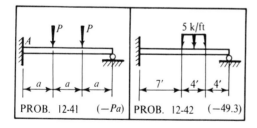

PROB. 12-41 $(-Pa)$ PROB. 12-42 (-49.3)

12-43. (a) Using the moment-area method, determine the redundant moment at the built-in end for the beam shown in the figure, and plot the shear and moment diagrams. Neglect the weight of the beam. (b) Select a WF beam using an allowable bending stress of 18,000 psi and a shearing stress of 12,000 psi. (c) Determine the maximum deflection of the beam between the supports and the maximum deflection of the overhang. Let $E = 29 \times 10^6$ psi. *Ans.* (b) 12 WF 27.

12-44 and 12-45. For the beams loaded as shown in the figures, using the moment-area method, (a) determine the fixed-end moments and plot shear and moment diagrams. Neglect the weight of the beams. (b) Express the maximum deflection in terms of the loads, distances, and EI. No adjustment for units need be made. *Ans.* For (a) by the figures.

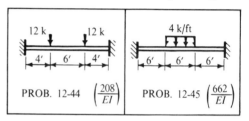

PROB. 12-44 $\left(\dfrac{208}{EI}\right)$ PROB. 12-45 $\left(\dfrac{662}{EI}\right)$

12-46. A 12 WF 36 beam is loaded as shown in the figure. Using the moment-area method and neglecting the weight of the beam, determine (a) the fixed-end moments, (b) the maximum bending stress, (c) the deflection at the midspan. *Ans.* (b) 13.9 ksi; (c) -0.133 in.

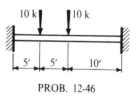

PROB. 12-46

12-47 and 12-48. For beams of constant flexural rigidity loaded as shown in the figures, using the moment-area method, determine the fixed-end moments.

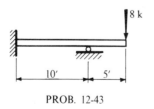

PROB. 12-43

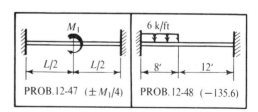

PROB. 12-47 $(\pm M_1/4)$ PROB. 12-48 (-135.6)

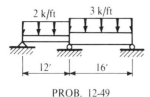

PROB. 12-49

12-49. For the continuous beam loaded as shown in the figure, using the moment-area method, determine the bending moment acting over the middle support, and plot shear and moment diagrams. The I of the beam in the right span is twice as large as that in the left span. *Ans.* $M_{max} = 68.3$ k-ft, $M_{min} = -60$ k-ft.

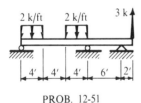

PROB. 12-50

12-50. A beam having a variable moment of inertia is loaded as shown in the figure. (a) Using the moment-area method determine the moment over the middle support. (b) Find all reactions.

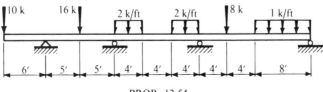

PROB. 12-51

12-51. For the continuous beam loaded as shown in the figure, using the moment-area method or the three-moment equation, determine the bending moment directly over the center support. The EI is constant. *Ans.* -13.44 k-ft.

12-52. A beam of constant flexural rigidity EI is continuous over four spans of equal length L. Plot shear and moment diagrams for this beam if throughout its length it is loaded with a uniformly distributed load of p_o lb per foot. Use the three-moment equations to determine the moments over the supports. (*Hint:* Take advantage of symmetry.) *Ans.* Moment over the center support is $p_o L^2/14$, the reactions at ends are $11 p_o L/28$ each.

12-53. Rework Example 12-14, Fig. 12-17, after assuming that both supports A and C are fixed. Use the three-moment equation. *Ans.* $M_A = 22.2$ k-ft, $M_B = -15.7$ k-ft.

12-54. Using the three moment equation, determine the moments over the supports for the beam loaded as shown in the figure. The EI is constant. *Ans.* -60 k-ft, -4.7 k-ft, -20.3 k-ft, and 0.

12-55. A restrained beam of ductile material is loaded with two concentrated forces P as shown in the figure. Determine the limit loads P_{ult}. Neglect the weight of the beam. (*Hint:* The possibility of a plastic hinge must be checked under each load.) *Ans.* $4M_p/L$.

12-56. Using limit analysis, calculate the value of P that would cause (flexural) collapse of the two-span beam shown. The beam has a rectangular cross section 4 in. wide and 10 in. deep. The yield stress is 36 ksi. Neglect the weight of the beam. *Ans.* 84.3 k.

12-57. A two-span, continuous, prismatic beam carries a concentrated force P in the middle of one span, and a uniformly distributed load p_o in the other span. Using the plastic method of analysis, determine the ratio of $p_o L$ to P necessary for the (flexural) collapse to occur in both spans simultaneously. Neglect the weight of the beam.

PROB. 12-54

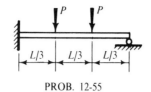

PROB. 12-55

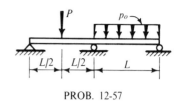

PROB. 12-57

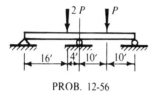

PROB. 12-56

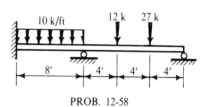

PROB. 12-58

12-58. Using the limit analysis approach, select a steel *WF* beam for the loading condition shown in the figure. Assume $\sigma_{yp} = 40$ ksi, a shape factor of 1.10, and use a load factor of 2. *Ans.* 14 WF 30.

12-59. A prismatic "weightless" beam is loaded as shown in the figure. What is the magnitude of the governing maximum moment? *Ans.* $p_o L^2/3$.

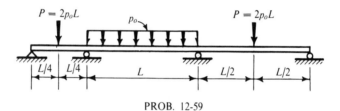

PROB. 12-59

Energy methods 13

13-1. INTRODUCTION

The material presented in the preceding chapters of the text was based on the Newtonian approach to mechanics, which relies on vectorial representations of equilibrium relations. Alternatively, mechanics can be treated through the Lagrangian approach by using scalar functions. It is the latter method based on energy concepts that is discussed in this chapter. By working with scalar functions, one finds the solution of many complex deformation problems is simpler than by working with the previously discussed methods.

Methods of computing the internal strain energy in members will be reviewed first. The expressions for the elastic strain energy in axially loaded bars and members subjected to torque will be supplemented by a relation for calculating the strain energy in beam bending. Then, by invoking the law of conservation of energy and equating the internal strain energy to the external work, the deflection of some members will be obtained.

The direct solution of problems by equating the external work to the internal strain energy turns out to be useful in cases where only one force is applied to a member. Therefore, next, a general theorem applicable to elastic systems subjected to any number of loads is derived for determining deflections and rotations of any element. This theorem is associated with the name of Castigliano.* This is followed by a still more general procedure based on the concept of virtual work. This method is applicable for determining deformations regardless of what causes them. Not only elastic deformation but also deformations due to temperature changes, plastic deformations, and those due to misfit of the fabricated elements can be determined. Solutions of deflection problems for beams, frames, trusses, and curved bars will be given as illustrations of the methods discussed.

13-2. ELASTIC STRAIN ENERGY

On substituting the strain energy density relation, Eq. 4-20, into Eq. 4-22, the general expression for the total *internal* strain energy in a *linearly elastic* body is obtained:

$$U = \frac{1}{2} \iiint_V (\sigma_x \varepsilon_x + \sigma_y \varepsilon_y + \sigma_z \varepsilon_z + \tau_{xy} \gamma_{xy} + \tau_{yz} \gamma_{yz} + \tau_{zx} \gamma_{zx}) \, dx \, dy \, dz$$

(13-1)

Integration extends over the volume of a body. Such a general expression is used in elasticity. In the technical mechanics of solids a less general class of problems is considered and Eq. 13-1 greatly simplifies. An expression

$$U = \frac{1}{2} \iiint_V (\sigma_x \varepsilon_x + \tau_{xy} \gamma_{xy}) \, dx \, dy \, dz \qquad (13\text{-}2)$$

is sufficient for determining the strain energy in axially loaded bars as well as in bent and sheared beams. Moreover, the last term of Eq. 13-2 written in the appropriate coordinates is all that is needed in the torsion problem of a circular shaft and for thin-walled tubes. These cases include the major types of problems treated in this text.

For linearly elastic material, for uniaxial stress, $\varepsilon_x = \sigma_x/E$, and for pure shear, $\gamma_{xy} = \tau_{xy}/G$. Thus Eq. 13-2 can be recast in the following form:

$$U = \underbrace{\iiint_V \frac{\sigma_x^2}{2E} \, dx \, dy \, dz}_{\text{for axial loading and bending of beams}} + \underbrace{\iiint_V \frac{\tau_{xy}^2}{2G} \, dx \, dy \, dz}_{\text{for shear in beams}} \qquad (13\text{-}3)$$

Several useful expressions of U as special cases can be developed from Eq. 13-3 by reducing the triple integrals to single ones.

---

* A. Castigliano, an Italian engineer (1847–84). The basic results were obtained from his thesis in 1873.

Strain energy for axially loaded bars

In such situations, $\sigma_x = P/A$, and at a given section through a beam $\iint dy\,dz = A$. Therefore, since P and A can be functions of x only, one has

$$U = \iiint_V \frac{\sigma_x^2}{2E}\,dV = \iiint_V \frac{P^2}{2A^2 E}\,dx\,dy\,dz$$

$$= \int_L \frac{P^2}{2A^2 E}\left[\iint_A dy\,dz\right]dx = \int_L \frac{P^2}{2AE}\,dx \qquad (13\text{-}4)$$

where a single integration along the bar length L gives the required quantity. If P, A, and E are constant, Eq. 13-4 becomes $U = P^2 L/(2AE)$.

Strain energy in bending

For this case, according to the elastic flexure formula for beams, $\sigma_x = -My/I$, Eq. 6-3. This relation must be substituted into the first right-hand term of Eq. 13-3. Then, noting that both M and I are functions of x only and that by definition $\iint y^2\,dy\,dz = I$, one has

$$U = \iiint_V \frac{\sigma_x^2}{2E}\,dV = \iiint_V \frac{1}{2E}\left(-\frac{My}{I}\right)^2 dx\,dy\,dz$$

$$= \int_L \frac{M^2}{2EI^2}\left[\iint_A y^2\,dy\,dz\right]dx = \int_L \frac{M^2}{2EI}\,dx \qquad (13\text{-}5)$$

Equation 13-5 reduces the volume integral for the elastic energy of a beam in bending to a single integral to be taken over the length L of a beam.

Strain energy for circular tubes in torsion

For this case the basic expression for the shearing strain energy is analogous to the last term of Eq. 13-3. Such an expression has been used previously in Example 5-4. Substituting into such an equation $\tau = T\rho/J$, Eq. 5-5, after some simplifications, one obtains

$$U = \iiint_V \frac{\tau^2}{2G}\,dV = \int_L \frac{T^2}{2GJ}\,dx \qquad (13\text{-}6)$$

The derivation of additional special cases for the strain energy will not be pursued further.

EXAMPLE 13-1

Find the energy absorbed by an elastic rectangular beam in pure bending in terms of the maximum stress and the volume of the material.

SOLUTION

The bending moment at every section of the beam is constant. Therefore, by direct application of Eq. 13-5, one has

$$U = \int_0^L \frac{M^2}{2EI}\,dx = \frac{M^2}{2EI}\int_0^L dx = \frac{M^2 L}{2EI}$$

And, since $\sigma_{\max} = Mc/I = 6M/bh^2$ and $I = bh^3/12$,

$$U = \frac{\sigma_{\max}^2}{2E}\left(\frac{bhL}{3}\right) = \frac{\sigma_{\max}^2}{2E}(\tfrac{1}{3}\,\text{vol})$$

For a given maximum stress, the volume of the material in this beam is only a third as effective for absorbing energy as it would be in a uniformly stressed bar where $U = (\sigma^2/2E)(\text{vol})$. This results from the existence of variable stresses in a beam. If the bending moment also varies along a prismatic beam, the volume of the material becomes even less effective. The same situation was observed earlier in axial loading and torsion (see Examples 4-3 and 5-4).

13-3. DISPLACEMENTS BY THE ENERGY METHOD

The principle of conservation of energy, which states that the energy can be neither created nor destroyed, can be adopted for determining the displacements of elastic systems due to the applied forces. The first law of thermodynamics expresses this principle as

$$\text{work done} = \text{change in energy} \tag{13-7}$$

For an adiabatic* process and when no heat is generated in the system, with the forces applied in a quasi-static manner,† the special form of this law for conservative systems reduces to

$$W_e = U \tag{13-8}$$

where W_e is the total work done by the externally applied forces during the loading process and U is the total strain energy stored in the system.

It is significant to note that alternatively

$$W_e + W_i = 0 \tag{13-9}$$

i.e., the external work W_e and the internal work W_i must equal zero. Therefore, from Eqs. 13-8 and 13-9, one has

---

* No heat is added or subtracted from the system.
† The forces are applied to the body so slowly that the kinetic energy can be neglected.

$$U = -W_i \tag{13-10}$$

where W_i has a negative sign because the deformations are opposed by the internal forces.

Several expressions for determining U were given in the preceding article. One can also note that, for the linearly elastic systems, if a force or a couple is applied to a body, it increases linearly from zero to its full value. Therefore, the external work W_e is equal to one-half the total force multiplied by the displacement in the direction of its action. The possibility of formulating both W_e and U provides the basis for utilizing Eq. 13-8 for determining displacements.

A few examples of the application of Eq. 13-8 follow.

EXAMPLE 13-2

Find the deflection of the free end of an elastic bar of constant cross-sectional area A and of length L due to an axial force P applied at the free end.

SOLUTION

As the force P is gradually applied to the bar, the external work $W_e = P\Delta/2$ where Δ is the deflection of the end of the bar. With P constant the expression for the internal strain energy, according to Eq. 13-4, is $U = P^2L/(2AE)$. Thus, from $W_e = U$,

$$\frac{P\Delta}{2} = \frac{P^2L}{2AE} \quad \text{and} \quad \Delta = \frac{PL}{AE}$$

which is the same as Eq. 4-33.

EXAMPLE 13-3

Find the rotation of the end of an elastic circular shaft with respect to the built-in end when a torque T is applied at the free end.

SOLUTION

As the torque T is gradually applied to the shaft, the external work $W_e = T\varphi/2$, where φ is the angular rotation of the free end in radians. The expression for the internal strain energy U for the circular shaft is given in Eq. 13-6. With T and J constant, this gives $U = T^2L/(2JG)$; then from $W_e = U$

$$\frac{T\varphi}{2} = \frac{T^2L}{2JG} \quad \text{and} \quad \varphi = \frac{TL}{JG}$$

which agrees with Eq. 5-13.

EXAMPLE 13-4

Find the maximum deflection due to a force P applied at the end of an elastic cantilever having a rectangular cross section, Fig. 13-1(a). Consider the effect of the flexural and shearing deformations.

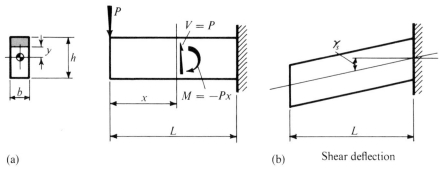

(a)   (b)   Shear deflection

Fig. 13-1. (For more accurate shearing distortion, see Fig. 7-12.)

SOLUTION

As the force P is applied to the beam, the external work $W_e = P\Delta/2$, where Δ is the total deflection of the end of the beam. The internal strain energy consists of two parts. One part is due to the bending stresses, the other is caused by the shearing stresses. According to Eq. 13-3 these strain energies may be directly superposed.

The strain energy in bending is obtained from Eq. 13-5, $U = \int M^2\,dx/(2EI)$, by noting that $M = -Px$. The strain energy in shear is found by using the second term of Eq. 13-3, $dU_{\text{shear}} = [\tau^2/(2G)]\,dV$. In this particular problem, the shear at every section is equal to the applied force P while the shearing stress τ, according to Example 7-3, is distributed parabolically as $\tau = [P/(2I)][(h/2)^2 - y^2]$. At any one level y, this shearing stress does not vary across either the breadth b or the length L of the beam. Therefore the infinitesimal volume dV in the shear energy expression is taken as $Lb\,dy$. By equating the sum of these two internal strain energies to the external work, the total deflection is obtained:

$$U_{\text{bending}} = \int_0^L \frac{M^2\,dx}{2EI} = \int_0^L \frac{(-Px)^2\,dx}{2EI} = \frac{P^2L^3}{6EI}$$

$$U_{\text{shear}} = \int_{\text{vol}} \frac{\tau^2}{2G}\,dV = \frac{1}{2G}\int_{-h/2}^{+h/2} \left\{\frac{P}{2I}\left[\left(\frac{h}{2}\right)^2 - y^2\right]\right\}^2 Lb\,dy$$

$$= \frac{P^2 Lb}{8GI^2}\frac{h^5}{30} = \frac{P^2 Lbh^5}{240G}\left(\frac{12}{bh^3}\right)^2 = \frac{3P^2L}{5AG}$$

where $A = bh$ is the cross section of the beam. Then

$$W_e = U = U_{\text{bending}} + U_{\text{shear}}$$

$$\frac{P\Delta}{2} = \frac{P^2L^3}{6EI} + \frac{3P^2L}{5AG} \quad \text{or} \quad \Delta = \frac{PL^3}{3EI} + \frac{6PL}{5AG}$$

Section 13-3
Displacements by
the energy method

The first term in the result, $PL^3/(3EI)$, is the ordinary deflection of the beam due to flexure. The second term is the deflection due to shear and can be interpreted as follows: The ratio $P/A = V/A$ is the average shearing stress τ_{av} across the section. This quantity divided by the shearing modulus G gives the shearing strain for a uniform stress. Since, however, the shearing stress varies across the section, a numerical correction factor, called here α, is necessary. In this problem $\alpha = 6/5$. On this basis (see Fig. 13-1(b)), for the constant shear occurring along the beam, the end deflection due to shear may be expressed in the following alternative forms:

$$\Delta_{\text{shear}} = \gamma_s L = \alpha \frac{\tau_{av}}{G} L = \alpha \frac{VL}{AG} = \frac{6PL}{5AG}$$

The factor α depends on the cross-sectional area of a member. In general, the shear V can vary across the span.

It is instructive to recast the expression for the total deflection Δ as

$$\Delta = \frac{PL^3}{3EI}\left(1 + \frac{3E}{10G}\frac{h^2}{L^2}\right)$$

where as before, the last term gives the deflection due to shear.

To gain further insight into this problem, replace in the last expression the ratio E/G by 2.5, a typical value for steels. Then

$$\Delta = (1 + 0.75 h^2/L^2)\Delta_{\text{bending}}$$

From this equation it can be seen that for a short beam—for example, one with $L = h$—the total deflection is 1.75 times that due to bending. Hence shear deflection is very important in comparable cases. On the other hand, if $L = 10h$, the deflection due to shear is less than 1 per cent. Small deflections due to shear are typical for ordinary, slender beams. This fact may be noted further from the original equation for Δ. There, whereas the deflection due to shear increases directly with the span length, the deflection due to bending increases as the cube of this distance. Therefore, as the length of a beam increases, the deflection due to bending quickly becomes dominant. For this reason it is usually possible to neglect the deflection due to shear. Of course, such a generalization is not always possible.

The approach illustrated in the above examples is limited to the determination of elastic deflections caused by one force at the point of its application. Otherwise intractable equations are obtained. For example, for two forces $\frac{1}{2} P_1 \Delta_1 + \frac{1}{2} P_2 \Delta_2 = U$, where Δ_1 and Δ_2 are the unknowns. An additional relationship between Δ_1 and Δ_2, except for cases of symmetry, is not available. This requires the development of more general methods. The remainder of the chapter discusses such methods based on the concepts of work or energy.

13-4. CASTIGLIANO'S DEFLECTION THEOREM

In calculating deflections of elastic systems, the following theorem may be often applied to advantage: *The partial derivative of the strain energy of a linearly elastic* system with respect to any selected force acting on the system gives the displacement of that force in the direction of its line of action.* The words *force* and *displacement* have a generalized sense and include, respectively, couple and angular rotation. This is Castigliano's (second) theorem. The derivation follows.

The expression for the internal strain energy for linearly elastic systems can be put essentially into the following form:

$$U = \iiint (\sigma^2/2E)\, dV$$

As was shown in Art. 13-2, however, since stresses depend on the applied forces, the strain energy of a given body can be also expressed as a quadratic function of the external forces $P_1, P_2, \ldots, P_k, \ldots, P_n, M_1, \ldots, M_p$, i.e.,

$$U = U(P_1, P_2 \cdots P_k \cdots P_n,\ M_1, M_2 \cdots M_p)$$

Let this energy correspond to a body such as shown in Fig. 13-2(a). The infinitesimal increase in this function δU for an infinitesimal increase in all of the applied forces δP_k's and δM_m's follows by applying the chain rule of differentiation. This gives

$$\delta U = \frac{\partial U}{\partial P_1}\delta P_1 + \frac{\partial U}{\partial P_2}\delta P_2 + \cdots + \frac{\partial U}{\partial P_k}\delta P_k \cdots + \frac{\partial U}{\partial M_p}\delta M_p$$

In this expression δP's, and δM's, are used instead of the ordinary differential notation to emphasize the linear independence of these quantities. From this point of view, if only the force δP_k were changed by an amount δP_k, Fig. 13-2(b), the strain energy increment would be

$$\delta U = \frac{\partial U}{\partial P_k}\delta P_k$$

Therefore, since the work of the reactions is zero, the total strain energy U' corresponding to the application of all external forces and δP_k, Fig. 13-2(c), is

$$U' = U + \delta U = U + \frac{\partial U}{\partial P_k}\delta P_k \tag{13-11}$$

---

* Generalization to nonlinear elastic systems is possible. See Art. 13-6.

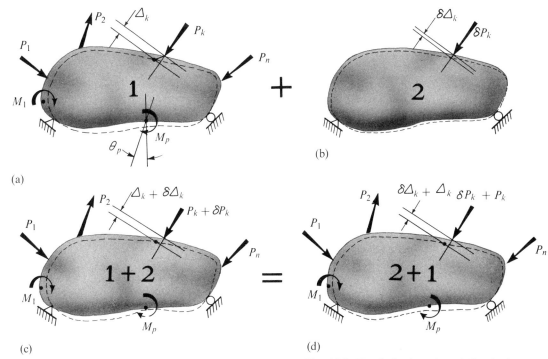

Fig. 13-2. For derivation of Castigliano's theorem.

An expression equal to this one will be formulated next by reversing the sequence of the load application, Figs. 13-2(a), (b), and (d). In applying δP_k first, an infinitesimal displacement $\delta \Delta_k$ is produced. For a linearly elastic body, the external work of $\delta P_k \, \delta \Delta_k /2$ corresponding to this can be neglected since it is of a second order. Further, the external work W_e done by the forces $P_1, P_2, \ldots, P_k, \ldots, M_p$ is unaffected by the presence of δP_k. On the other hand, during the application of these forces, the force δP_k does work in moving an amount Δ_k in the direction of P_k. This additional work equals $(\delta P_k) \Delta_k$. Therefore, the total work W'_e done by the external system of loading including the work done by δP_k, Fig. 13-2(d), is

$$W'_e = W_e + (\delta P_k)\Delta_k \qquad (13\text{-}12)$$

This relation can be set equal to Eq. 13-11 since the order of load application is immaterial, and the external work is equal to the internal strain energy:

$$W_e + (\delta P_k)\Delta_k = U + (\partial U/\partial P_k)\, \delta P_k$$

On simplifying

$$\Delta_k = \frac{\partial U}{\partial P_k} \qquad (13\text{-}13)$$

By retaining different terms in the derivation

$$\theta_p = \frac{\partial U}{\partial M_p} \tag{13-14}$$

which gives the rotation of a member at a point p in the direction of the applied moment M_p.

These versatile equations establish the proposition stated at the begining of this article. In applying them the solution of a problem is not limited by the number of the applied forces. The addition of a fictitious force at a location where no actual force is applied enables one to employ the above equations at any location on a body. On setting the fictitious force equal to zero, one determines the actual displacement in the end. This and other items in connection with the application of Castigliano's deflection theorem will be illustrated by the examples that follow.

EXAMPLE 13-5

Using Castigliano's theorem, verify the results of Examples 13-2, 13-3, and 13-4.

SOLUTION

In all these problems the expressions for the internal strain energy U have been formulated. Therefore, a direct application of Eq. 13-13 or 13-14 is all that is necessary to obtain the required results. In all cases the material obeys Hooke's law. Deflection of an axially loaded bar ($P = $ constant):

$$U = \frac{P^2 L}{2AE} \quad \text{hence} \quad \Delta = \frac{\partial U}{\partial P} = \frac{PL}{AE}$$

Angular rotation of a circular shaft ($T = $ constant):

$$U = \frac{T^2 L}{2JG} \quad \text{hence} \quad \varphi \equiv \theta = \frac{\partial U}{\partial T} = \frac{TL}{JG}$$

Deflection of a rectangular cantilever due to the end load P:

$$U = \frac{P^2 L^3}{6EI} + \frac{3P^2 L}{5AG} \quad \text{hence} \quad \Delta = \frac{\partial U}{\partial P} = \frac{PL^3}{3EI} + \frac{6PL}{5AG}$$

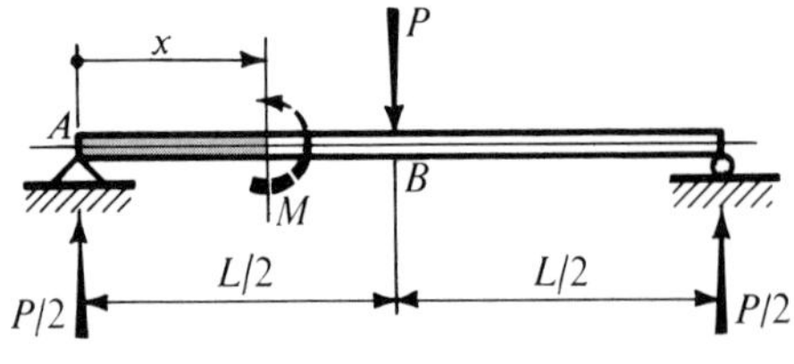

EXAMPLE 13-6

An elastic, prismatic beam is loaded as shown in Fig. 13-3. Using Castigliano's theorem, find the deflection due to bending caused by the applied force P at the center.

SOLUTION

The expression for the internal strain energy in bending is given by Eq. 13-5 as

Fig. 13-3

$$U = \int M^2/(2EI)\,dx$$

Section 13-4
Castigliano's
deflection theorem

Since according to Castigliano's theorem the required deflection is a derivative of this function it is advantageous to differentiate the expression for U before integrating. In problems where M is a complex function, this scheme is particularly useful. In this case the following relation becomes applicable:

$$\Delta = \frac{\partial U}{\partial P} = \int_0^L \frac{M}{EI}\frac{\partial M}{\partial P}\,dx \qquad (13\text{-}15)$$

Proceeding on this basis, one has from A to B:

$$M = +\frac{P}{2}x \quad \text{and} \quad \frac{\partial M}{\partial P} = \frac{x}{2}$$

On substituting these relations into Eq. 13-15 and observing the symmetry of the problem,

$$\Delta = 2\int_0^{L/2} \frac{Px^2}{4EI}\,dx = +\frac{PL^3}{48EI}$$

The positive sign indicates that the deflection takes place in the direction of the applied force P.

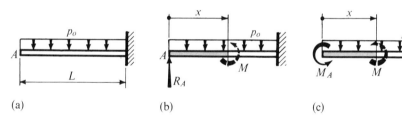

Fig. 13-4

EXAMPLE 13-7

Using Castigliano's theorem determine the deflection and the angular rotation of the end of a uniformly loaded cantilever, Fig. 13-4(a). The EI is constant.

SOLUTION

No forces are applied at the end of the cantilever where the displacements are to be found. Therefore, in order to be able to apply Castigliano's theorem, a fictitious force must be added corresponding to the displacement sought. Thus, as shown in Fig. 13-4(b), in addition to the specified loading, a force R_A has been introduced. This permits determining $\partial U/\partial R_A$, which with $R_A = 0$ gives the vertical deflection of point A. Applying Eq. 13-15 in this manner, one has

$$M = -\frac{p_o x^2}{2} + R_A x \quad \text{and} \quad \frac{\partial M}{\partial R_A} = +x$$

$$\Delta_A = \frac{\partial U}{\partial R_A} = \frac{1}{EI} \int_0^L \left(-\frac{p_o x^2}{2} + \cancel{R_A x}^{\,0}\right)(+x)\,dx = -\frac{p_o L^4}{8EI}$$

where the negative sign shows that the deflection is in the opposite direction to that assumed for force R_A. If R_A in the above integration were not set equal to zero, the end deflection due to p_o and R_A would be found.

The angular rotation of the beam at A can be found in a manner analogous to the above. A fictitious moment M_A is applied at the end, Fig. 13-4(c), and the calculations are made in much the same manner as before:

$$M = -\frac{p_o x^2}{2} - M_A \quad \text{and} \quad \frac{\partial M}{\partial M_A} = -1$$

$$\Delta_A = \frac{\partial U}{\partial M_A} = \frac{1}{EI} \int_0^L \left(-\frac{p_o x^2}{2} - \cancel{M_A}^{\,0}\right)(-1)\,dx = +\frac{p_o L^3}{6EI}$$

where the sign indicates that the sense of the rotation of the end coincides with the assumed sense of the fictitious moment M_A.

EXAMPLE 13-8

Determine the horizontal deflection for the simple elastic frame shown in Fig. 13-5(a). Consider only the deflection caused by bending. The flexural rigidity EI of both members is equal and constant.

SOLUTION

The strain energy function is a scalar. Therefore, the separate strain energies for the different elements of an elastic system can be added algebraically. After the total strain energy is determined, its partial derivative with respect to a force gives the displacement of that force. For the problems at hand, Eq. 13-15 is appropriate.

From A to B: $M = +Px$ and $\partial M/\partial P = +x$

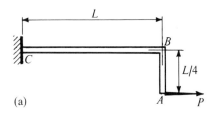
(a)

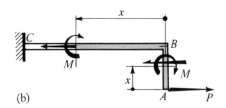
(b)

Fig. 13-5

From B to C: $\quad M = +\dfrac{PL}{4}\quad$ and $\quad \dfrac{\partial M}{\partial P} = +\dfrac{L}{4}$

$$\Delta_A = \frac{\partial U}{\partial P} = \frac{1}{EI}\int_0^{L/4} (+Px)(+x)\,dx + \frac{1}{EI}\int_0^L \left(+\frac{PL}{4}\right)\left(+\frac{L}{4}\right) dx$$

$$= +\frac{13PL^3}{192EI}$$

The reader should contrast the simplicity of this solution with the procedures discussed in Chapter 11 (see especially Fig. 11-15). Often it is possible to determine the displacements of the individual points in complex frames very simply by using an energy method. The complete freedom of choice in the sign convention of moments and in the location of the origin for each part or piece of a structure is an added attraction of the method.

In this problem in evaluating the elastic strain energy it would have been possible to include the axial energy for the member BC and the shearing energy for the member AB. If this were done, contributions to the deflection due to these causes could have been determined. The quantity determined gives the deflection of the frame due to bending only. Usually this is the dominant part of the total.

If the vertical deflection of point A were required, a fictitious vertical force F at A would have to be added. Then, as in the preceding example, $\partial U/\partial F$ with $F = 0$ would give the desired result.

In the analysis of statically indeterminate, linear systems, the superposition method discussed in the preceding chapter can always be followed (see Art. 12-4). In such an approach, Castigliano's theorem can be used for determining displacements in a structure reduced to statical determinancy. Alternatively, the redundants can be considered as unknown external forces and identified accordingly by algebraic symbols. Then the strain energy function will contain the redundant quantities as the unknowns. However, the kinematic conditions prescribed at each redundant provide the necessary conditions to solve a problem. This is best illustrated by an example.

EXAMPLE 13-9

Consider an elastic uniformly loaded beam clamped at one end and simply supported at the other, as represented in Fig. 13-4(b). Determine the reaction at A.

SOLUTION

The solution is analogous to that of Example 13-7 except that R_A must be treated as the unknown and not permitted to vanish. The key kinematic condition is

$$\Delta_A = \partial U/\partial R_A = 0 \tag{13-16}$$

which states that no deflection occurs at A due to the applied load p_o and R_A.

$$M = -\frac{p_o x^2}{2} + R_A x \quad \text{and} \quad \frac{\partial M}{\partial R_A} = +x$$

$$\Delta_A = \frac{\partial U}{\partial R_A} = \frac{1}{EI}\int_0^L \left(-\frac{p_o x^2}{2} + R_A x\right)(+x)\,dx = -\frac{p_o L^4}{8EI} + \frac{R_A L^3}{3EI} = 0$$

Therefore, $R_A = +3p_o L/8$, the result found in Example 12-9.

13-5. RECIPROCAL THEOREM

In linear structural systems the deflection Δ_i at point i due to forces P_i at i and P_j at j, according to Eq. 12-11, can be expressed as

$$\Delta_i = f_{ii} P_i + f_{ij} P_j$$

Similarly the deflection at j is

$$\Delta_j = f_{ji} P_i + f_{jj} P_j$$

where $f_{ii}, f_{ij}, f_{ji},$ and f_{jj} are the flexibility coefficients of a given system.

If the strain energy of the system due to the application of these forces is U, according to Castigliano's second theorem, Eq. 13-13, the same quantities are also given as

$$\Delta_i = \frac{\partial U}{\partial P_i} \quad \text{and} \quad \Delta_j = \frac{\partial U}{\partial P_j}$$

On equating the partial derivative of Δ_i with respect to P_j for the two expressions above, one has

$$\frac{\partial \Delta_i}{\partial P_j} = f_{ij} = \frac{\partial^2 U}{\partial P_j\, \partial P_i}$$

and similarly

$$\frac{\partial \Delta_j}{\partial P_i} = f_{ji} = \frac{\partial^2 U}{\partial P_i\, \partial P_j}$$

However, since the order of differentiation is immaterial,

$$f_{ij} = f_{ji} \tag{13-17}$$

This states that the deflection at any point i due to a unit force at any point j is equal to the deflection of j due to a unit force at i providing the directions of the forces and deflections in each of the two cases coincide. This is called the *reciprocal theorem* or *Maxwell's law of reciprocal deflections*.*

---

*This relationship was discovered by James Clerk Maxwell in 1864. The more general case was demonstrated by E. Betti in 1872.

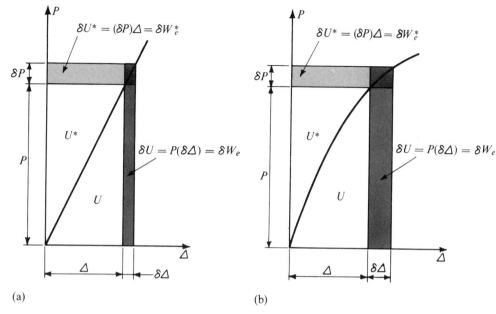

Fig. 13-6. Work and complementary work, and strain energy and complementary strain energy.

13-6. GENERALIZATION OF CASTIGLIANO'S THEOREMS

It is interesting to re-examine Castigliano's theorem of Art. 13-4 in relation to a force-deflection diagram for an axially loaded member. The linearly elastic response considered earlier is in Fig. 13-6(a). Note that for this case the elastic strain energy U is equal to the complementary strain energy U^* (see Art. 4-9). From this diagram it can be seen that for an increment of load δP, the complementary external work δW_e^* is $(\delta P)\Delta$. This in turn equals δU^*, hence $\delta U^* = (\delta P)\Delta$, from which in generalized form

$$\Delta_k = \frac{\partial U^*}{\partial P_k} \tag{13-18}$$

For the linearly elastic case where $U = U^*$ this expression reduces to Eq. 13-13, which is Castigliano's second theorem. An inspection of Fig. 13-6(b) shows however that this relation remains valid for elastically nonlinear material provided U^* is used instead of U.

On considering the external work δW_e done for a change in Δ and equating this to the strain energy δU, from either figure, one finds that $\delta U = P(d\Delta)$. Stated in a generalized form

$$P_k = \frac{\partial U}{\partial \Delta_k} \tag{13-19}$$

which is Castigliano's first theorem. Its meaning is analogous to that of the second theorem with the roles of deflection and displacement interchanged. This equation applies to both linear and nonlinear elastic materials.

*Chapter 13
Energy methods*

The application of Castigliano's first theorem and of the two theorems for nonlinear elastic materials will not be pursued in this text.*

13-7. VIRTUAL-WORK METHOD FOR DEFLECTIONS

It is possible to imagine that a real mechanical or a structural system in static equilibrium is arbitrarily displaced consistent with its boundary conditions or constraints. During this process the real forces acting on the system move through imaginary or virtual displacements. Alternatively, imaginary or virtual forces in equilibrium with the given system can be given real, kinematically admissible displacements. In either case one can formulate the imaginary or virtual work done. Here the discussion will be limited to the consideration of virtual forces undergoing real displacements.

For forces and displacements occurring in the above manner, the principle of conservation of energy remains valid. The change in the total work due to these disturbances must be zero. Therefore, for a conservative system, and according to Eq. 13-9,

$$\delta W_e + \delta W_i = 0 \quad \text{or} \quad \delta W_e = -\delta W_i \qquad (13\text{-}20)$$

where the notation δW_e and δW_i is used instead of dW_e and dW_i to emphasize that the change in work is virtual.

In the subsequent discussion it is more convenient to replace δW_i in the above equation by δW_{ei}, the external work on the internal elements of a body. This quantity is numerically equal to δW_i but has an opposite sign. In calculating δW_{ei}, in contrast to δW_i the internal deformations occur in the direction of the internal forces. Therefore

$$\delta W_e = \delta W_{ei} \qquad (13\text{-}21)$$

which expresses the virtual-work principle. For rigid body systems the term on the right side vanishes, whereas, for elastic systems, with the aid of Eq. 13-10 it can be seen that this term becomes δU. The restriction of the principle to the elastic response, however, is not implied in Eq. 13-21. It is the complete generality of the virtual-work equation that makes it a particularly valuable tool of analysis.

For determining the deflection of any point of a body due to any deformations occurring within a body, Eq. 13-21 can be put into a more suitable form. For example, consider a body such as shown in Fig. 13-7, for which the deflection of some point A in the direction A-B caused by deformation of the body is sought. For this, the virtual-work equation can be formulated by employing the following sequence of reasoning:

---

* Interested readers can consult T. Au, *Elementary Structural Mechanics* (Englewood Cliffs, N.J.: Prentice-Hall, Inc., 1963).

First, apply to the unloaded body an imaginary or virtual force δF acting in the direction *A-B*. This force causes internal forces throughout the body. These internal forces, designated as δf, Fig. 13-7(a), can be found in statically determinate systems.

Section 13-7
Virtual-work method for deflections

Next, with the virtual force remaining on the body, apply the actual or real forces, Fig. 13-7(b), or introduce the specified deformations, such as those due to a change in temperature. This causes real internal deformations ΔL, which may be computed. Owing to these deformations, the virtual-force system does work.

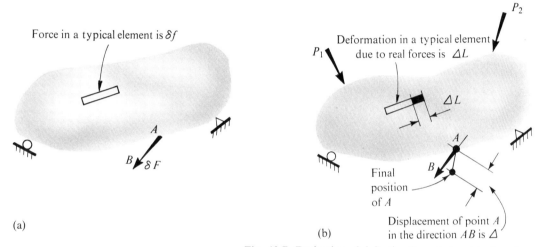

(a)

(b)

Fig. 13-7. Derivation of deflection formula by virtual work.

Therefore, since the external work done by the virtual force δF moving a real amount Δ in the direction of this force is equal to the total work done on the internal elements by the virtual forces δf's moving their respective real amounts ΔL, the special form of the virtual-work equation becomes*

$$\delta F \cdot \Delta = \sum \delta f \cdot \Delta L \tag{13-22}$$

Since all virtual forces attain their full values before real deformations are imposed, no factor of one-half appears anywhere in the equation. The summation, or, in general, an integration sign, is necessary on the right side of Eq. 13-22 to indicate that all internal work must be included.

Note that δF and δf need not be infinitesimal quantities. In Eq. 13-22 only their ratio is of consequence. Therefore, it is particularly convenient in applications to choose δF equal to unity, and to restate

---

* This equation represents the scalar (dot) product of the vectors.

Eq. 13-22 as

$$1 \cdot \Delta = \sum f \cdot \Delta L \tag{13-23}$$

where Δ = real deflection of a point in the direction of the applied virtual unit force

f = internal forces caused by the virtual unit force

ΔL = real internal deformations of a body

The real deformations can be due to any cause with the elastic ones being a special case. Tensile forces and elongations of members are taken positive. A positive result indicates that the deflection occurs in the same direction as the applied virtual force.

In determining the angular rotations of a member a unit couple is used instead of the unit force. In practice, the procedure of using the unit force or the unit couple in conjunction with virtual work is referred to as the *unit-dummy-load method*.

13-8. VIRTUAL-WORK EQUATIONS FOR ELASTIC SYSTEMS

For linearly elastic systems Eq. 13-23 can be specialized to facilitate the solution of problems. This is done here for axially loaded and for flexural members. Applications are illustrated by examples. In these examples it is interesting to note that if Castigliano's theorem is applied in the form given by Eq. 13-15, the solution procedure is nearly identical to that of virtual work.

Trusses

A virtual unit force must be applied at a point in the direction of the deflection to be determined.

If the real deformations are linearly elastic and are due only to axial deformations, $\Delta L = PL/(AE)$, and Eq. 13-23 becomes

$$1 \times \Delta = \sum_{i=1}^{n} \frac{p_i P_i L_i}{A_i E_i} \tag{13-24}$$

where p_i is the axial force in a member due to the virtual unit force and P_i is the force in the same member due to the real loads. The summation extends over all members of a truss.

Beams

If the deflection of a point on an elastic beam is wanted by the virtual-work method, a virtual unit force must be applied first in the direction in which the deflection is sought. This virtual force will set up internal bending moments at various sections of the beam designated by m, as is in Fig. 13-8(a). Next, as the real forces are applied to the beam, bending moments M rotate the "plane sections" of the beam $M\,dx/(EI)$ radians (Eq. 11-36). Hence the work done on an element of a beam by the virtual moments m is $mM\,dx/(EI)$. Integrating this over the length of the beam gives the external work on the internal elements. Hence the special form of Eq. 13-23 for beams becomes

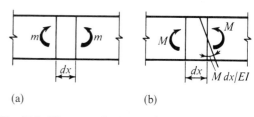

Fig. 13-8. Elements of a beam—(a) virtual bending moments m, (b) real bending moment M—and the rotation of sections they cause.

$$1 \times \Delta = \int_0^L \frac{mM\,dx}{EI} \qquad (13\text{-}25)$$

An analogous expression may be used to find the angular rotation of a particular section in a beam. For this case, instead of applying a virtual-unit force, a virtual unit couple is applied to the beam at the section being investigated. This virtual couple sets up internal moments m along the beam. Then, as the real forces are applied, they cause rotations $M\,dx/(EI)$ of the cross sections. Hence the same integral expression as in Eq. 13-25 applies here. The external work by the virtual unit couple is obtained by multiplying it by the real rotation θ of the beam at this couple. Hence

$$1 \times \theta = \int_0^L \frac{mM\,dx}{EI} \qquad (13\text{-}26)$$

In Eqs. 13-25 and 13-26, m is the bending moment due to the virtual loading, and M is the bending moment due to the real loads. Since both m and M usually vary along the length of the beam, both must be expressed by appropriate functions.

EXAMPLE 13-10

Find the vertical deflection of point B in the pin-jointed steel truss shown in Fig. 13-9(a) due to the following causes: (a) the elastic deformation of the members, (b) a shortening by 0.125 in. of the member AB by means of a turnbuckle, and (c) a drop in temperature of 120°F occurring in the member BC. The coefficient of thermal expansion of steel is 0.0000065 in. per inch per degree Fahrenheit. Neglect the possibility of lateral buckling of the compression member.

SOLUTION

Case (a). A virtual unit force is applied in the vertical direction as shown in Fig. 13-9(b), and the resulting forces p are determined and recorded on the same diagram (check these). Then the forces in each member due to the real force are also determined and recorded, Fig. 13-9(c). The solution follows by means of Eq. 13-24. The work is carried out in tabular form.

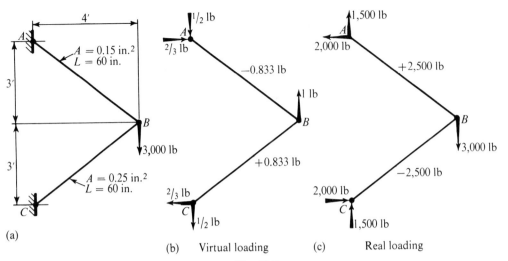

Fig. 13-9

Member	p, lb	P, lb	L, in.	A, in.²	pPL/A
AB	−0.833	+2,500	60	0.15	−833,000
BC	+0.833	−2,500	60	0.25	−500,000

From this table $\Sigma pPL/A = -1{,}333{,}000$. Hence

$$1 \times \Delta = \Sigma \frac{pPL}{AE} = \frac{-1{,}333{,}000}{(30)10^6} = -0.0444 \text{ lb-in.}$$

and $\Delta = -0.0444$ in.

The negative sign means that point B deflects down. In this case, "negative work" is done by the virtual force acting upward when it is displaced in a downward direction. Note particularly the units and the signs of all quantities. Tensile forces in members are taken positive, and vice versa.

Case (b). Equation 13-23 is used to find the vertical deflection of point B due to the shortening of the member AB by 0.125 in. The forces set up in the bars by the virtual force acting in the direction of the deflection sought are shown in Fig. 13-9(b). Then, since ΔL is −0.125 in. (shortening) for the member AB and is zero for the member BC,

$$1 \times \Delta = (-0.833)(-0.125) + (+0.833)(0) = +0.1042 \text{ lb-in.}$$

and $\Delta = +0.1042$ in. up.

Case (c). Again using Eq. 13-23 and noting that due to the drop in temperature, $\Delta L = -0.0000065(120)60 = -0.0468$ in. in the member BC,

$$1 \times \Delta = (+0.833)(-0.0468)$$
$$= -0.0390 \text{ lb-in.}$$
and $\Delta = -0.0390$ in. down

By superposition, the net deflection of point B due to all three causes is $-0.0444 + 0.1042 - 0.0390 = +0.0208$ in. up. To find this quantity, all three effects could have been considered simultaneously in the virtual-work equation.

EXAMPLE 13-11

Find the deflection at the midspan of a cantilever beam loaded as in Fig. 13-10(a). The EI of the beam is constant.

SOLUTION

The virtual force is applied at point A, whose deflection is sought, Fig. 13-10(b). The m diagram and the M diagram are shown in Figs. 13-10(c) and 13-10(d), respectively. For these functions, the same origin of x is taken at the free end of the cantilever. After these moments are determined, Eq. 13-25 is applied to find the deflection.

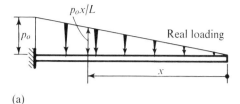
(a)

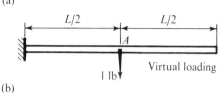
(b)

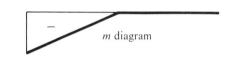
(c)

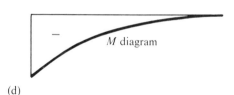
(d)

Fig. 13-10

$$M = -\frac{x}{2}\frac{p_o x}{L}\frac{x}{3} = -\frac{p_o x^3}{6L} \quad (0 \leq x \leq L)$$

$$m = 0 \quad (0 \leq x \leq L/2)$$

$$m = -1(x - L/2) \quad (L/2 \leq x \leq L)$$

$$1 \times \Delta = \int_0^L \frac{mM\,dx}{EI}$$

$$= \frac{1}{EI}\int_0^{L/2}(0)\left(-\frac{p_o x^3}{6L}\right)dx + \frac{1}{EI}\int_{L/2}^L\left(-x + \frac{L}{2}\right)\left(-\frac{p_o x^3}{6L}\right)dx$$

$$= \frac{49 p_o L^4}{3{,}480 EI} \text{ lb-in.}$$

The deflection of point A is numerically equal to this quantity. The deflection due to shear has been neglected.

Note that if Castigliano's theorem were used in this case a force P would have been applied at A. This together with p_o would give $M = -p_o x^3/(6L) - P(x - L/2)$ for $x \geq L/2$, whence

*Chapter 13
Energy methods*

$\partial M/\partial P = -(x - L/2)$,

which is precisely the m for $x \geq L/2$. Therefore, the application of Eq. 13-15 would result in an integral identical to that above.

EXAMPLE 13-12

Find the downward deflection of the end C caused by the applied force of 2 kips in the structure shown in Fig. 13-11(a). Neglect deflection caused by shear. Let $E = 10^7$ psi.

SOLUTION

A unit virtual force of 1 kip is applied vertically at C. This force causes an axial force in member DB and in the part AB of the beam, Fig. 13-11(b). Owing to this force, bending moments are also caused in the beam AC, Fig. 13-11(c). Similar computations are made and are shown in Figs. 13-11(d) and (e) for the applied real force. The deflection of

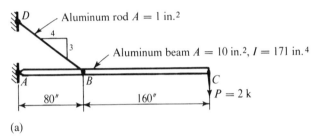

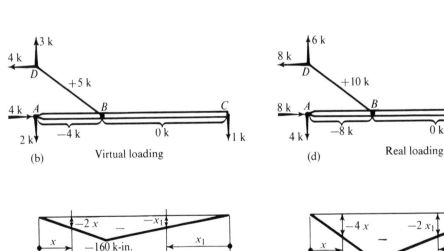

Fig. 13-11

point C depends on the deformations caused by the axial forces, as well as flexure, hence the virtual-work equation is

Section 13-8
Virtual-work equations for elastic systems

$$1 \times \Delta = \Sigma \frac{pPL}{AE} + \int_0^L \frac{mM\,dx}{EI}$$

The first term on the right side of this equation is computed in the table below. Then the integral for the internal virtual work due to bending is found. For the different parts of the beam, two origins of x's are used in writing the expressions for m and M, Figs. 13-11(c) and (e).

Member	p, kips	P, kips	L, in.	A, in.^{2}	pPL/A
DB	+5	+10	100	1.0	+5,000
AB	−4	−8	80	10.0	+256

From the table, $\Sigma pPL/A = +5{,}256$, or $\Sigma pPL/(AE) = +0.5256$ kip-in.

$$\int_0^L \frac{mM\,dx}{EI} = \int_0^{80} \frac{(-2x)(-4x)\,dx}{EI} + \int_0^{160} \frac{(-x_1)(-2x_1)\,dx_1}{EI}$$
$$= +2.39 \text{ kip-in.}$$

Therefore $\quad 1 \times \Delta = +0.5256 + 2.39 = 2.92$ kip-in.

and point C deflects 2.92 in. down.

Note that the work due to the two types of action was superposed. Also note that the origins for the coordinate system for moments may be chosen as convenient; however, the same origin must be used for the corresponding m and M.

EXAMPLE 13-13

Find the horizontal deflection, caused by the concentrated force P, of the end of the curved bar shown in Fig. 13-12(a). The flexural rigidity EI of the bar is constant. Neglect the effect of shear on the deflection.

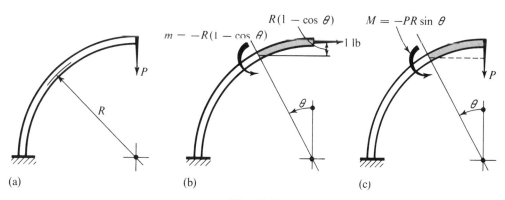

Fig. 13-12

Chapter 13
Energy methods

SOLUTION

If the radius of curvature of a bar is large in comparison with the cross-sectional dimensions (Art. 6-11), ordinary beam deflection formulas may be used replacing dx by ds. In this case, $ds = R\, d\theta$.

Applying a horizontal virtual force at the end in the direction of the deflection wanted, Fig. 13-12(b), it is seen that $m = -R(1 - \cos\theta)$. Similarly, for the real load, from Fig. 13-12(c), $M = -PR\sin\theta$. Therefore

$$1 \times \Delta = \int_0^L \frac{mM\, ds}{EI}$$

$$= \int_0^{\pi/2} \frac{[-R(1-\cos\theta)](-PR\sin\theta)R\, d\theta}{EI} = +\frac{PR^3}{2EI}\ \text{in-lb}$$

The deflection of the end to the right is numerically equal to this expression.

13-9. STATICALLY INDETERMINATE PROBLEMS

Statically indeterminate problems may be solved with the aid of the virtual-work method. For the linearly elastic systems the method of superposition discussed in Art. 12-4 may be used to particular advantage. In applying this approach the virtual work method merely provides the means of determining deflections of structures artificially reduced to statical determinacy. Much confusion is avoided if the above statement is clearly kept in mind.

EXAMPLE 13-14

Find the forces in the pin-jointed bars of the steel structure shown in Fig. 13-13(a) if a force of 3,000 lb is applied at B.

SOLUTION

The structure may be rendered statically determinate by cutting the bar DB at D. Then the forces in the members are as shown in Fig. 13-13(b). In this determinate structure, the movement of point D must be found. This can be done by applying a vertical virtual force at D, Fig. 13-13(c), and using the virtual-work method. However, since the $pPL/(AE)$ term for the member BD is zero, the vertical movement of point D is the same as that of B. In Example 13-10 the latter quantity was found to be 0.0444 in. down and is so shown in Fig. 13-13(b).

The movement of point D, shown in Fig. 13-13(b), violates the conditions of the problem, and a force must be applied to bring it back where it belongs. According to Eq. 12-11 this can be stated as

$$\Delta_D = f_{DD} X_D + \Delta_{DP} = 0$$

where the gap $\Delta_{DP} = -0.0444$ in.

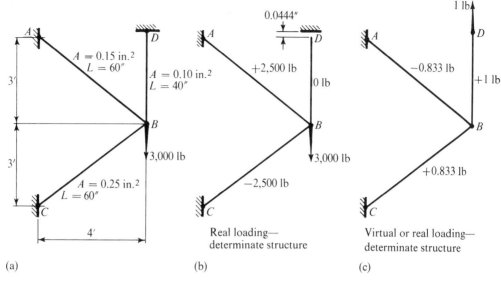

Fig. 13-13

To determine f_{DD} a 1-lb real force is applied at D and the virtual-work method is used to find the deflection due to this force. The forces set up in the determinate structure by the virtual and the real forces are numerically the same, Fig. 13-13(c). To differentiate between the two, forces in members caused by a real force are designated by p', by the virtual force by p. The solution is carried out in tabular form.

Member	p, lb	p', lb	L, in.	A, in.^{2}	$pp'L/A$
AB	−0.833	−0.833	60	0.15	+278
BC	+0.833	+0.833	60	0.25	+167
BD	+1.000	+1.000	40	0.10	+400

From the table, $\Sigma pp'L/A = +845$. Therefore, since

$$1 \times \Delta = \Sigma \frac{pp'L}{AE} = \frac{+845}{30(10)^6} = 0.0000281 \text{ lb-in.}$$

$$f_{DD} = 0.0000281 \text{ in.}, \quad \text{and} \quad 0.0000281 X_D - 0.0444 = 0$$

To close the gap of 0.0444 in., the 1-lb real force at D must be increased $X_D = 0.0444/0.0000281 = 1{,}580$ times. Therefore the actual force in the member DB is 1,580 lb. The forces in the other two members may now be determined from statics or by superposition of the forces shown in Fig. 13-13(b) with X_D times the p' forces shown in Fig. 13-13(c). By either method, the force in AB is found to be +1,180 lb (tension), and in BC, −1,180 lb (compression).

Chapter 13
Energy methods

In any given case to make certain that the elastic analysis, such as the one in the above example, is applicable, maximum stresses must be determined. For the solution to be correct, these must be in the linearly elastic range for the material used.

PROBLEMS FOR SOLUTION

13-1. A simple beam of rectangular cross section and span L is loaded with a concentrated force P at the middle of the span. Neglecting the weight of the beam and equating internal to external energy, (a) determine the maximum deflection caused by bending; (b) determine the maximum deflection caused by the shearing deformations. *Ans.* (a) $PL^3/(48EI)$.

13-2. Determine the span length for a simply supported 20-in. steel I beam weighing 75 lb per foot for supporting a concentrated load in the middle, so that the deflection due to shear is equal to that due to flexure. Let the ratio E/G be 2.5, and assume that in transmitting the shear only the area of the web is effective for which $\alpha = 1.0$. (See Example 13-4.) Make two separate sketches of the deflected beam due to the two causes. Neglect the weight of the beam.

13-3. A short, 3-in.-diameter steel shaft supporting a gear spans 6 in. between the centerlines of the bearings as shown in the figure. Idealizing the gear force as a concentrated force P and concentrating the reactions, determine the ratio of the deflection due to shear to that due to flexure. Assume the ratio E/G is 2.5 and note that for a circular cross section $\alpha = 10/9$. (*Hint:* See Example 13-4.) *Ans.* 0.52.

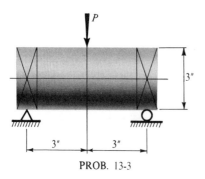

PROB. 13-3

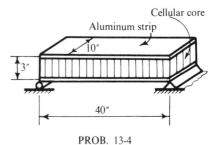

PROB. 13-4

13-4. A 10-in.-wide beam spanning 40 in. is of a sandwich construction, see figure. The two outer aluminum-alloy "skins" are each 0.050 in. thick ($E = 10^7$ psi); the interior light-weight core is 3 in. thick, having an

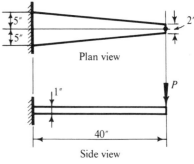

PROB. 13-5

effective $G = 10,000$ psi. (a) Determine the allowable uniformly distributed load so that the maximum bending stress will not exceed 8,000 psi. (b) Find the maximum total deflection caused by flexure and shear. Assume that the core carries only shear and no bending stress. (*Note:* In an actual problem bearing stresses at the supports have to be carefully investigated.)

13-5. Using an energy method, determine the vertical deflection of the free end of the cantilever beam shown in the figure due to the application of a force $P = 100$ lb. Consider only flexural effects, i.e., neglect shear deformations. Let $E = 30 \times 10^6$ psi. *Ans.* 0.112 in.

13-6. (a) In terms of P, L, and EI, calculate the amount of elastic strain energy stored in the beam shown in the figure, caused by the applied loads. (b) By equating the work done by the external forces to the change in the elastic strain energy, determine the deflection at the loads. (*Hint:* Due to symmetry, deflections at both loads are equal.) *Ans.* (b) $PL^3/(48EI)$.

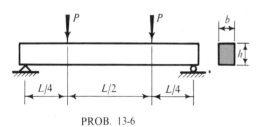

PROB. 13-6

13-7. For the beam shown in the figure, using an energy method, determine the deflection of the beam at the points of application of the loads. The moment of inertia of the cross section in the middle half of the beam is I_o. *Ans.* $0.029\, PL^2/(EI_o)$.

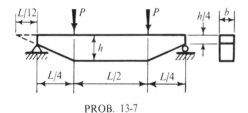

PROB. 13-7

13-8. Find the instantaneous maximum deflections and bending stresses for the rectangular steel beam shown in the figure when struck by a 30-lb weight falling from a height 3 in. above the top of the beam, if (a) the beam is on rigid supports, and (b) the beam is supported at each end on springs. The constant k for each spring is 1,667 lb per inch. (*Hint:* See Prob. 4-14, where it is shown that the impact factor by which a static force must be multiplied to obtain an equivalent dynamic

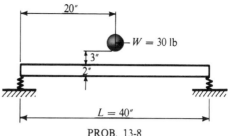

PROB. 13-8

force is $1 + \sqrt{1 + 2h/\Delta_{st}}$. Calculations made on this basis are not entirely reliable as local deformations and mass of the beam are neglected.) *Ans.* (a) 0.0785 in., 17,700 psi; (b) 0.255 in., 5,740 psi.

13-9. A man weighing 180 lb jumps onto a diving board from a height of 2 ft. If the board is of the dimensions shown in the figure, what is the maximum bending stress? Let $E = 1.6 \times 10^6$ psi. Use any method to establish the deflection characteristics of the board. (*Hint:* See the above problem and the answer to Prob. 13-15.) *Ans.* 8,560 psi.

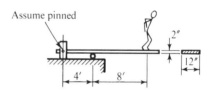

PROB. 13-9

13-10. Using Castigliano's theorem, find the deflection of the point of application of the force P on the beam of variable cross section shown in the figure. *Ans.* $13PL^3/(1,458EI)$.

13-11. For the beam shown in the figure, using Castigliano's theorem, determine (a) the

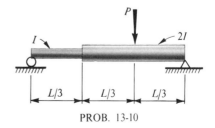

PROB. 13-10

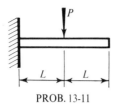

PROB. 13-11

deflection at the center of the beam, and (b) the tip deflection of the beam. Consider only flexural deformations. The EI is constant. Ans. (a) $PL^3/(3EI)$, (b) $5PL^3/(6EI)$.

13-12. Using Castigliano's theorem, find the deflection of the point of application of the force P. The flexural rigidity EI is constant over the entire length. Consider only bending deformations. Ans. $54P/(EI)$.

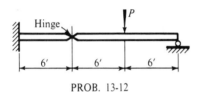

PROB. 13-12

13-13. Using Castigliano's theorem, determine the maximum deflection in terms of p_o, L, and EI for a uniformly loaded, simple beam having a constant EI. Ans. $5p_oL^4/(384EI)$.

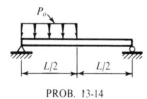

PROB. 13-14

13-14. Using Castigliano's theorem, determine the deflection at the center of the beam loaded as shown in the figure. The EI is constant. Ans. $5p_oL^4/(768EI)$.

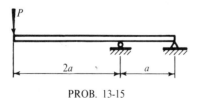

PROB. 13-15

13-15. An overhanging beam is loaded with a concentrated force P at the end as shown in the figure. Using Castigliano's theorem, find the deflection and rotation of the overhanging end caused by the force P. The EI is constant. Ans. $4Pa^3/(EI)$, $8Pa^2/(3EI)$.

13-16. (a) An overhanging beam is loaded with a couple M_o at the end as shown in the

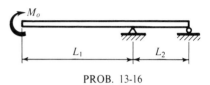

PROB. 13-16

figure. Using Castigliano's theorem, determine the deflection and rotation of the overhanging end due to M_o. (b) Let $L_1 = 2a$, and $L_2 = a$, and compare the end deflection for a unit moment M_o with the end rotation caused by a unit force P as found in the preceding problem. This is a special case of Maxwell's law of reciprocal deflections. Ans. (a) $M_oL_1(3L_1 + 2L_2)/(6EI)$, $M_o[L_1 + \tfrac{1}{3}L_2]/(EI)$.

13-17. Using an energy method, determine the horizontal deflection of point D for the structure shown in the figure due to the application of the force H. The EI is constant for the entire frame. Consider only flexural effects. Ans. $5HL^3/(3EI)$.

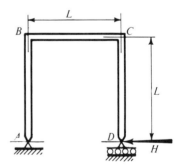

PROB. 13-17

13-18. By applying Castigliano's theorem determine the horizontal deflection of point A caused by the force F applied to the frame as shown in the figure. Consider only the de-

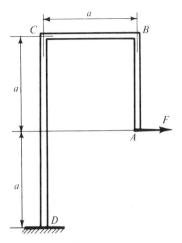

PROB. 13-18

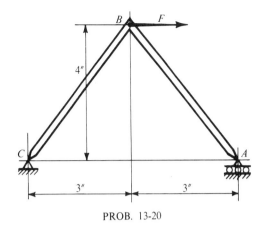

PROB. 13-20

flection due to bending. The EI is constant. Ans. $2Fa^3/(EI)$.

13-19. A rod of constant flexural rigidity EI is shaped into a planar member as shown in the figure. Find the horizontal deflection of the end A caused by the application of the vertical force F. Only flexural effects need be considered. The effect of the axial forces and shears on the deflection is to be neglected. Ans. $2Fa^3/(EI)$.

13-20. A planar elastic member is loaded as shown in the figure. Determine the horizontal movement of the end A caused by the application of the force F at B. Consider only flexural deformations. The EI is constant. Ans. $80F/(3EI)$.

13-21. A Z-shaped bar of constant EI is attached at one end as shown in the figure. Determine the rotation of the free end caused by the application of the vertical force P. Consider bending only. Ans. $66P/(EI)$.

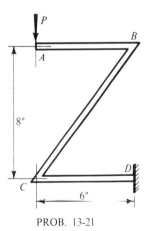

PROB. 13-21

13-22. A rod bent into the shape shown in the figure is rigidly attached at the top. (a) Find the vertical deflection of the bottom point due to the application of the force P. (b) Determine the horizontal deflection of the same point. Consider only flexural deflections. The bent rod lies in a plane. Ans. (a) $Pa^3(2\sqrt{2}/3 + 1)/(EI)$; (b) $3Pa^3(1 + \sqrt{2})/(2EI)$.

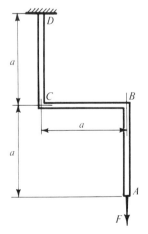

PROB. 13-19

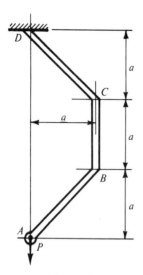

PROB. 13-22

13-23. A U-shaped member of constant EI has the dimensions shown in the figure. Determine the deflection of the applied forces away from each other. Consider only flexural effects. (*Hint:* Take advantage of symmetry.) *Ans.* $(2PL^3/3 + PL^2R\pi + PR^3\pi/2 + 4PLR^2)/(EI)$.

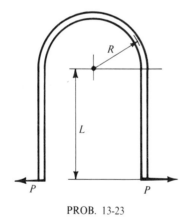

PROB. 13-23

13-24. In order to install a split ring used as a retainer on a machine shaft it is necessary to open up a gap of Δ by applying the forces P, as shown in the figure. If EI of the cross section of the ring is constant, determine the required magnitude of the forces P. *Ans.* $P = \Delta EI/(3\pi a^3)$.

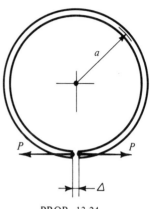

PROB. 13-24

13-25. A bent bar having a circular cross section is built in at one end and is loaded with a force F at the other end as shown in the figure. The force F acts normal to the plane of the bent bar. Using an energy method, determine (a) the translations of the free end along the coordinate axes, and (b) the rotation of the free end about the same axes. The constants E, G, I, and J are given.

13-26. A bar having a circular cross section is bent into a semicircle and is built in at one end as shown in the figure. Determine the

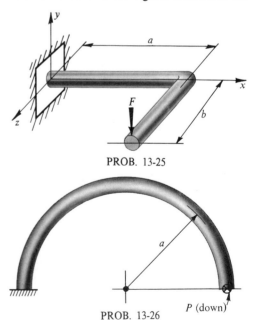

PROB. 13-25

PROB. 13-26

deflection of the free end caused by the application of force P acting normal to the plane of the semicircle. Neglect the contribution of shear deformation. *Ans.* $\Delta = \pi P a^3 [1/(EI) + 3/(GJ)]/2$.

13-27. Using an energy method, determine the horizontal deflection of the top joint of the truss shown in the figure due to the applied force $P = 25$ kips. All joints are pinned. The cross-sectional area of each bar $A = 1$ in.^{2} and $E = 30 \times 10^6$ psi.

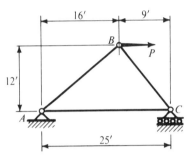

PROB. 13-27

13-28. Find the vertical deflection of point B caused by $F = 12$ kips for the structure shown in the figure. For member AB, the area $A = 0.50$ in.^{2}, and $E = 30 \times 10^3$ ksi. For member

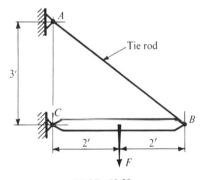

PROB. 13-28

BC, the area $A = 4$ in.^{2}, $EI = 10^6$ k-in.^{2}, and $E = 30 \times 10^3$ ksi. Use an energy method. *Ans.* 0.071 in.

13-29. Using an energy method, determine the vertical and the horizontal displacements of joint C for the truss shown in the figure due

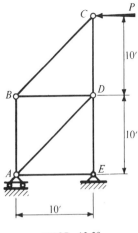

PROB. 13-29

to an applied force $P = 2$ kips. For simplicity, assume $AE = 1$ for all members.

13-30. Two steel wires BC and CD ($A = 0.10$ in.^{2} each) are arranged as shown in the figure. At D the wire CD is attached to a rigid support; at B the wire CB is attached to a vertical cantilever AB ($A = 2$ in.^{2}, $I = 6$ in.^{4}). Using an energy method, determine the vertical deflection at C caused by applying the force $P = 1{,}600$ lb. Neglect the contribution of shear deformation. *Ans.* 0.048 in.

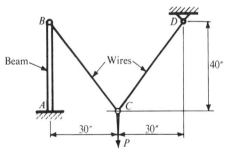

PROB. 13-30

13-31. A jib crane is arranged as shown in the figure. For member AC the cross-sectional area $A = 1$ in.^{2}; for member BD, $A = 5$ in.^{2}, and $EI = 10^6$ k-in.^{2} For both members $E = 30 \times 10^3$ ksi. Find the vertical deflection of point D due to the simultaneous application of the two loads. Use an energy method.

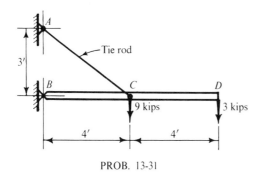

PROB. 13-31

Neglect the contribution of shear deformation. Ans. 0.405 in.

13-32. A pin-joined system of three bars, each having a cross-sectional area A, is loaded as shown in the figure. (a) Determine the vertical and horizontal displacements of the

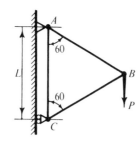

PROB. 13-32

joint B caused by the load P. (b) If by means of a turnbuckle the length of the member AC is shortened by $\frac{1}{2}$ in., what is the movement of the joint B? Ans. (a) $-9PL/(4AE)$, $-\sqrt{3}PL/(12AE)$; (b) $\frac{1}{4}$ in., $\sqrt{3}/12$ in.

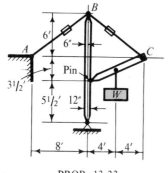

PROB. 13-33

13-33. For the mast and boom arrangement shown in the figure, (a) determine the vertical movement of the load W caused by lengthening the rod AB a distance of $\frac{1}{2}$ in. (b) By how much must the rod BC be shortened to bring the weight W to its original position? Ans. (a) 0.167 in., (b) 0.347 in.

13-34. For the beam shown in the figure determine the reaction at A. Use an energy method and treat the reaction as the redundant. Ans. $3p_oL/8$.

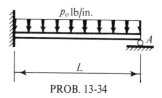

PROB. 13-34

13-35. For the beam shown in the figure, using an energy method, (a) determine the reaction at A treating it as the redundant. (b) Determine the moment at B treating it as a redundant. Ans. (a) $2p/3$; (b) $-PL/3$.

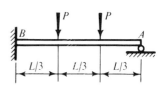

PROB. 13-35

13-36. A circular ring of a linearly elastic material is loaded by two equal and opposite forces P as shown in the figure. For this ring both A and I are constant. (a) Determine the largest bending moment caused by the applied

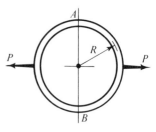

PROB. 13-36

forces, and plot the entire moment diagram. (b) Find the decrease in the diameter AB caused by the applied forces. Consider only flexural deformations. (*Hint:* Take advantage of symmetry and consider the moment at A as the redundant.) *Ans.* (a) PR/π, (b) $PR^3(2/\pi - \frac{1}{2})/(EI)$.

13-37. Three bars of a linearly elastic material are connected at the ends A, B, C, and D by pins as shown in the figure. Determine the force in the member AB caused by the applied load. Treat member AB as the redundant. The values of L/A are as follows: $\frac{7}{15}$ for AD, $\frac{7}{20}$ for AC, and 1 for AB. *Ans.* $+5.83$ kips.

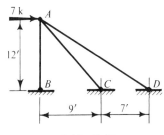

PROB. 13-37

13-38. Determine the reaction at A, treating it as a redundant, for the truss loaded as shown in the figure. The material of the truss is linearly elastic. The value of L/A for all members is 1. *Ans.* 18.75 kips.

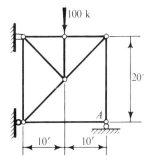

PROB. 13-38

13-39. A system of steel rods each having a cross-sectional area of 0.20 in.² is arranged as shown in the figure. At 50°F joint D is 0.10 in. away from its support. (a) At what temperature can the connection be made

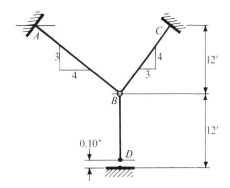

PROB. 13-39

without stressing any of the members? Let $E = 30 \times 10^6$ psi, and $\alpha = 6.5 \times 10^{-6}$ per °F. (b) What stresses will develop in the members if after making the connections at D the temperature drops to $-10°F$? *Ans.* (a) 85.7°F.

13-40. A continuous, linearly elastic beam is loaded by a concentrated force P as shown in the figure. (a) Set up the superposition equations, Eq. 12-11, treating the two unknown forces on the left as the redundants. Make use of the results in Probs. 13-15 and 13-16. (b) Solve the above equations simultaneously.

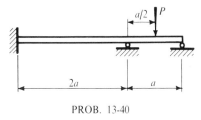

PROB. 13-40

13-41. A pipeline, such as shown in the figure, for pumping hot liquid from one pressure vessel into another also serves as an expansion loop. The pipe is a standard steel pipe having a nominal 4-in. diameter, weighing 10.79 lb per foot. ($E = 29 \times 10^6$ psi, and $\alpha = 6.5 \times 10^{-6}$ per °F.) In operation the temperature of this pipe is raised 400°F. If $L = R = 100$ in., make the following investigations: (a) Assuming that the pipe supports at A and B are capable of resisting only horizontal and vertical forces, find the bending moment at C

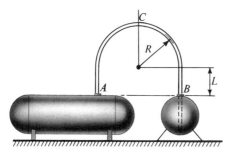

PROB. 13-41

caused by the raise in pipe temperature. Plot the bending moment diagram for the pipe. Make use of the results found in Prob. 13-23. (b) Set up the superposition equations for the above problem assuming that the pipe supports at A and B are so rigid that the pipe ends are completely fixed. Supplement the results of Prob. 13-23 by using an energy method to derive additional flexibility coefficients. (c) Solve the above equations simultaneously, and plot the bending moment diagram for the pipe caused by the change in temperature.

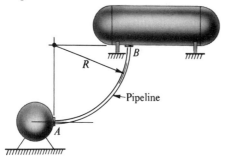

PROB. 13-42

13-42. A nominal 4-in.-diameter steel pipe weighing 10.79 lb per foot is bent into the form of one-quarter of a circle to a radius of $R = 100$ in. and is connected to two pressure vessels as shown in the figure. In operation this pipe can rise 500° F above its surroundings. For the material of the pipe, $E = 29 \times 10^6$ psi, and $\alpha = 6.5 \times 10^{-6}$ per °F. (a) Assuming that the pipe supports at A and B provide completely fixed conditions, set up superposition equations for the solution of this problem. Results found in Example 13-13, Fig. 13-12, are useful for this purpose. Derive additional flexibility coefficients using an energy method. Consider only flexural effects in establishing these quantities. (b) Solve the above equations simultaneously, and plot the bending moment diagram for the pipe caused by the change in temperature.

13-43. A pipeline connects two pressure vessels as shown in the figure. Assume that at A the support is very rigid and that this end of the pipe may be considered fixed. At B the support provides complete restraint against translation of the end, and provides good restraint for the moment (torque) around the y axis. At B the moment restraint around the x and the z axes is poor and is assumed to be zero. Set up a schematic set of superposition equations for the stress analysis of this problem. The analysis is to be made for the weight of the pipe and for a raise in its temperature. Outline the procedure you would follow to obtain the flexibility coefficients. Neglect the effect of axial force and shear on the flexibility of the system. State how the solution would be modified if the supports moved relative to each other.

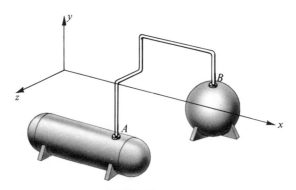

PROB. 13-43

Buckling of columns 14

14-1. INTRODUCTION

At the beginning of this text it was stated that the selection of structural and machine elements is based on three characteristics: strength, stiffness, and stability. The procedures of stress and deformation analyses were discussed in some detail in the preceding chapters. In this chapter, the question of the possible instability of structural systems will be considered. In such problems additional critical parameters must be found which determine whether a given configuration or deformation pattern is at all possible in a given system. This problem is unlike any encountered before.

As a simple intuitive example, consider a rod of diameter D subjected to an axial compressive force. If this rod, acting as a "column," were only of length D, no question of instability would arise, and a considerable force could be carried by this short member. On the other hand, if the same rod were made several diameters high, when subjected to an even smaller axial force than that which a short piece could carry, the rod could become laterally unstable through sidewise buckling and could collapse. An ordinary slender yardstick, if subjected to an axial compression, fails in this

Chapter 14
Buckling of columns

manner. The consideration of material strength alone is not sufficient to predict the behavior of such a member.

The same phenomenon occurs in numerous other situations where compressive stresses are present. Thin sheets, although fully capable of sustaining tensile loadings, are very poor in transmitting compression. Narrow beams, unbraced laterally, can snap sidewise and collapse under an applied load. Vacuum tanks, as well as submarine hulls, unless properly designed can severely distort under external pressure and can assume shapes which differ drastically from their original geometry. A thin-walled tube may wrinkle like tissue paper when subjected to a torque. (For example see Fig. 14-1*.) During some stages of firing, the thin casings of missiles are critically loaded in compression. These are crucially

---

* Figures taken from L. A. Harris, H. W. Suer, and W. T. Skene, "Model Investigations of Unstiffened and Stiffened Circular Shells," *Experimental Mechanics* (July 1961), pp. 3 and 5.

Fig. 14-1. (a) Typical buckle pattern for thin-walled cylinder in compression; (b) typical buckle pattern for pressurized cylinder in torsion. (Courtesy Dr. L. A. Harris of North American Aviation, Inc.)

important problems of engineering design. Moreover, usually the buckling or wrinkling phenomena observed in loaded members occurs rather suddenly. For this reason many of the structural instability failures are spectacular and very dangerous.

Section 14-2 Nature of the beam-column problem

The vast number of structural instability problems suggested by the above listing are beyond the scope of this text.* Here only the column problem will be considered. Using this problem as an example, however, brings out the essential features of the instability phenomenon and some of the basic analytical procedures of its analysis. This will be done by first investigating the behavior of slender, axially loaded bars which are simultaneously subjected to bending. Such members are referred to as *beam-columns*. The beam-column problems, in addition to having a significance of their own, enable one to determine the magnitudes of the critical axial loads at which buckling occurs.

The buckling of ideal, concentrically loaded columns will be considered next. This leads to discussion of the characteristic values or eigenvalues of the appropriate differential equations. The corresponding eigenfunctions give the buckled shapes of such columns. Both the elastic and the inelastic buckling of ideal columns will be discussed. Some information on eccentrically loaded columns also will be presented. An energy approach for determining buckling loads is introduced at the end of the chapter.

The chapter includes the design of columns together with illustrations of some typical formulas used in practice.

14-2. NATURE OF THE BEAM-COLUMN PROBLEM

The behavior of actual beam-columns can be best understood by considering first an idealized example, shown in Fig. 14-2(a). Here for simplicity a perfectly rigid bar of length L initially is held in a vertical position by a spring at A having a torsional stiffness k. Then a vertical force P and a horizontal force F are applied at the top of the bar. Unlike the procedure followed in all the previous problems, now the equilibrium equations must be written for the deformed state. Bearing in mind that $k\theta$ is the resisting moment developed by the spring at A, one obtains

$$\sum M_A = 0 \circlearrowright +, \qquad PL \sin \theta + FL \cos \theta - k\theta = 0$$

or

$$P = \frac{k\theta - FL \cos \theta}{L \sin \theta} \qquad (14\text{-}1)$$

---

* See for example, S. P. Timoshenko and J. M. Gere, *Theory of Elastic Stability*, (New York: McGraw-Hill Book Company (2nd ed.) 1961).

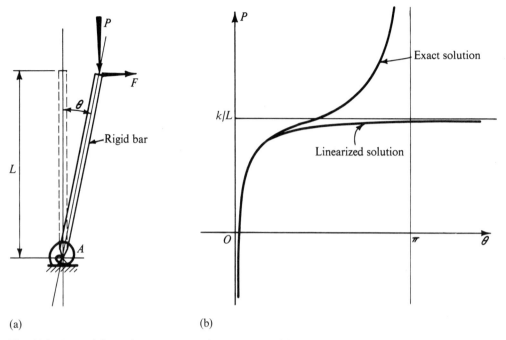

Fig. 14-2. Force-deformation response of a system with one degree of freedom.

The qualitative features of this result are shown in Fig. 14-2(b), and the corresponding curve is labeled as the exact solution. It is interesting to note that as $\theta \to \pi$, providing the spring continues to function, a very large force P can be supported by the system. For a force P applied in an upward direction, plotted downward in the figure, the angle θ decreases as P increases. In the analysis of problems in the preceding chapters, the term $PL \sin \theta$ would not have appeared at all.

The solution expressed by Eq. 14-1 is for arbitrarily large deformations. In complex problems it is very difficult to achieve solutions of such generality. Moreover, in the majority of applications large deformations cannot be tolerated. Therefore, it is usually possible to limit the investigation of the behavior of systems to small and moderately large deformations. In this problem this can be done by setting $\sin \theta \approx \theta$, and $\cos \theta \approx 1$. In this manner Eq. 14-1 simplifies to

$$P = \frac{k\theta - FL}{L\theta} \quad \text{or} \quad \theta = \frac{FL}{k - PL} \tag{14-2}$$

For small values of θ this solution is quite acceptable. On the other hand, as θ increases, the discrepancy between this linearized solution and the exact one becomes very large, Fig. 14-2(b).

For a certain critical combination of the parameters k, P, and L, the denominator $(k - PL)$ in the last term of Eq. 14-2 would become zero and presumably would give rise to an infinite angular rotation θ. This is completely unrealistic and is the result of a poor mathematical formulation of the problem. Nevertheless, such a solution provides a good guide to the magnitude of the axial force P at which the deflections become intolerably large. The asymptote to this solution, obtained by setting $(k - PL) = 0$, defines the critical force P_{cr} as

$$P_{cr} = k/L \qquad (14\text{-}3)$$

*Section 14-2
Nature of the
beam-column
problem*

It is significant to note that in real systems the large deformations associated with forces of the order of magnitude of P_{cr} usually cause stresses so high as to make the system unserviceable. On the other hand, the nonlinear analysis of structural systems, because of the change in geometry and because of the inelastic behavior of materials, is prohibitively complex. Therefore, in the buckling analysis of compression members, the determination of P_{cr} on a simplified basis, along the lines of the approach used in the above example, plays the key role.

Next, the above ideas will be employed in the solution of an elastic beam-column problem.

EXAMPLE 14-1

A beam-column is subjected to an axial force P and an upward transverse force F at its midspan, Fig. 14-3(a). Determine the equation of the elastic curve, and the critical axial force P_{cr}. Let $EI = $ constant.

SOLUTION

The free-body diagram for the deflected beam-column is shown in Fig. 14-3(b). This diagram permits the formulation of the total bending moment M which includes the effect of the axial force P multiplied by the deflection v. The total moment divided by EI could be set equal to the expression for the exact curvature, Eq. 11-8. However, as is customary, this curvature will be taken as d^2v/dx^2; i.e., the expression $M = EIv''$ will

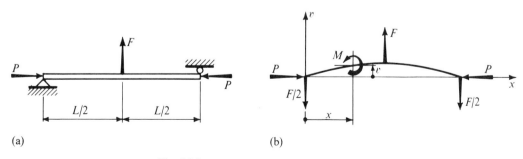

Fig. 14-3

be accepted. This yields accurate results only for small deflections and rotations. Just as in the preceding example, the acceptance of this approximation will lead to infinite deflections at the critical loads.

Thus, using the relation $M = EIv''$ and noting that for the left side of the span* $M = -(F/2)x - Pv$, one has

$$EIv'' = M = -Pv - (F/2)x \qquad (0 \leq x \leq L/2)$$

or

$$EIv'' + Pv = -(F/2)x$$

By dividing through by EI and letting

$$\lambda^2 = P/(EI) \qquad (14\text{-}4)$$

after some simplification, the governing differential equation becomes

$$\frac{d^2v}{dx^2} + \lambda^2 v = -\frac{\lambda^2 F}{2P} x \qquad (0 \leq x \leq L/2) \qquad (14\text{-}5)$$

The homogeneous solution for this differential equation has the well-known form of the one for simple harmonic motion; the particular solution equals the right-hand term divided by λ^2. Therefore, the complete solution is

$$v = C_1 \sin \lambda x + C_2 \cos \lambda x - (F/2P)x \qquad (14\text{-}6)$$

The constants C_1 and C_2 follow from the boundary condition $v(0) = 0$ and from a condition of symmetry $v'(L/2) = 0$. The first condition gives

$$v(0) = C_2 = 0$$

Since

$$v' = C_1 \lambda \cos \lambda x - C_2 \lambda \sin \lambda x - F/(2P)$$

with C_2 already known to be zero, the second condition gives

$$v'(L/2) = C_1 \lambda \cos \lambda L/2 - F/(2P) = 0$$

or

$$C_1 = F/[2P\lambda \cos (\lambda L/2)]$$

On substituting this constant into Eq. 14-6,

$$v = \frac{F}{2P\lambda} \frac{1}{\cos \lambda L/2} \sin \lambda x - \frac{F}{2P} x \qquad (14\text{-}7)$$

The maximum deflection occurs at $x = L/2$. Thus, after some simplifications,

$$v_{\max} = [F/(2P\lambda)](\tan \lambda L/2 - \lambda L/2) \qquad (14\text{-}8)$$

---

* Note especially that the deflection v in this case is shown as a positive quantity. If the beam were deflected downward, one would have $+P(-v) = -Pv$, which again is the same term as above. This is basic for preserving the invariance of the differential equation.

From this it may be concluded that the absolute maximum moment, occurring at the midspan, is

$$M_{\max} = \left| -\frac{FL}{4} - Pv_{\max} \right| = \frac{F}{2\lambda} \tan \frac{\lambda L}{2} \qquad (14\text{-}9)$$

Note that the expressions given by Eqs. 14-7, 14-8, and 14-9 become infinite if $\lambda L/2$ is a multiple of $\pi/2$ since this makes $\cos \lambda L/2$ equal to zero and $\tan \lambda L/2$ infinite. Stated algebraically this occurs when

$$\frac{\lambda L}{2} = \sqrt{\frac{P}{EI}} \frac{L}{2} = \frac{n\pi}{2} \qquad (14\text{-}10)$$

where n is an integer. Solving this equation for P, one obtains the magnitude of P causing either infinite deflections or bending moment. This corresponds to the condition of the critical axial force P_{cr} for this bar:

$$P_{\mathrm{cr}} = \frac{n^2 \pi^2 EI}{L^2} \qquad (14\text{-}11)$$

For the smallest critical force the integer $n = 1$. This result was first established by the great mathematician Leonhard Euler in 1757 and is often referred to as the *Euler buckling load*. Equation 14-11 will be discussed in greater detail in Art. 14-5.

It is important to note that the differential equation, Eq. 14-5, is of a different type than that used for beams loaded transversely only. For this reason, the singularity functions previously presented cannot be applied in these problems.

14-3. DIFFERENTIAL EQUATIONS FOR BEAM-COLUMNS

For a more complete understanding of the beam-column problem, it is instructive to derive several differential relations among the variables. For this purpose consider an element isolated from a beam column as shown in Fig. 14-4. Note especially that this element is shown in its deflected position. For ordinary, transversely loaded beams this is not necessary (see Fig. 2-25). The deflections considered in this treatment, however, are small in relation to the span of the beam-column, which permits the following approximations:

$$dv/dx = \tan \theta \approx \sin \theta \approx \theta, \quad \cos \theta \approx 1, \quad \text{and} \quad ds \approx dx$$

On this basis, the two equilibrium equations are

$$\sum F_y = 0 \uparrow +, \qquad p\,dx - V + (V + dV) = 0$$

$$\sum M_A = 0 \circlearrowleft +, \qquad M - P\,dv - V\,dx + p\,dx\frac{dx}{2} - (M + dM) = 0$$

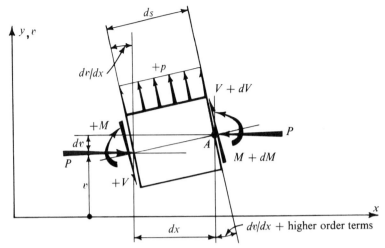

Fig. 14-4. An element of a beam column.

The first one of these equations yields

$$\frac{dV}{dx} = -p \tag{14-12}$$

which is identical to Eq. 2-4. The second, on neglecting the infinitesimals of higher order, gives

$$V = -\frac{dM}{dx} - P\frac{dv}{dx} \tag{14-13}$$

Therefore, for beam-columns, the shear V, in addition to depending on the rate of change in the moment M as in beams, now also depends on the magnitude of the axial force and the slope of the elastic curve. The latter term is the component of P along the inclined sections shown in Fig. 14-4.

In this development for curvature, the usual relation of the bending theory $d^2v/dx^2 = M/(EI)$ can be employed. On substituting Eq. 4-13 into Eq. 14-12 and making use of the above relation, one obtains two alternative differential equations for beam-columns:

$$\frac{d^2M}{dx^2} + \lambda^2 M = p \tag{14-14}$$

or

$$\frac{d^4v}{dx^4} + \lambda^2 \frac{d^2v}{dx^2} = \frac{p}{EI} \tag{14-15}$$

where for simplicity EI is assumed to be constant and as before $\lambda^2 = P/(EI)$. If $P = 0$, Eqs. 14-14 and 14-15 revert, respectively, to Eqs. 2-6 and 11-17 for transversely loaded beams. For the new equations the statement of the boundary conditions is the same as before (see Fig. 11-4) except that the shear is given by Eq. 14-13.

Section 14-3 Differential equations for beam-columns

For future reference the homogeneous solution of Eq. 14-15 and several of its derivatives are listed below:

$$v = C_1 \sin \lambda x + C_2 \cos \lambda x + C_3 x + C_4 \qquad (14\text{-}16a)$$

$$v' = C_1 \lambda \cos \lambda x - C_2 \lambda \sin \lambda x + C_3 \qquad (14\text{-}16b)$$

$$v'' = -C_1 \lambda^2 \sin \lambda x - C_2 \lambda^2 \cos \lambda x \qquad (14\text{-}16c)$$

$$v''' = -C_1 \lambda^3 \cos \lambda x + C_2 \lambda^3 \sin \lambda x \qquad (14\text{-}16d)$$

These relations are needed in some problems to express the boundary conditions for evaluating the constants C_1, C_2, C_3, and C_4.

EXAMPLE 14-2

A slender bar of constant EI is simultaneously subjected to the end moments M_o and an axial force P as shown in Fig. 14-5(a). Determine the maximum deflection and the largest bending moment.

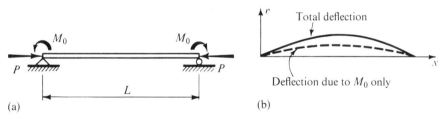

Fig. 14-5

SOLUTION

Within the span there is no transverse load. Therefore, the right-hand term of Eq. 14-15 is zero, and the homogeneous solution of this equation given by Eq. 14-16(a) is the complete solution. The boundary conditions are

$$v(0) = 0, \quad v(L) = 0, \quad M(0) = -M_o, \quad \text{and} \quad M(L) = -M_o$$

Since $M = EIv''$, with the aid of Eqs. 14-16(a) and 14-16(c) these conditions yield:

$$v(0) = \qquad\qquad\qquad + C_2 \qquad\qquad\qquad\qquad + C_4 = 0$$

$$v(L) = +C_1 \sin \lambda L \quad + C_2 \cos \lambda L \quad + C_3 L + C_4 = 0$$

$$M(0) = \qquad\qquad\qquad - C_2 EI\lambda^2 \qquad\qquad\qquad\qquad = -M_o$$

$$M(L) = -C_1 EI\lambda^2 \sin \lambda L - C_2 EI\lambda^2 \cos \lambda L \qquad\qquad\qquad = -M_o$$

Solving these four equations simultaneously

$$C_1 = \frac{M_o}{P}\left(\frac{1 - \cos \lambda L}{\sin \lambda L}\right), \quad C_2 = -C_4 = \frac{M_o}{P}, \quad \text{and} \quad C_3 = 0$$

Therefore, the equation of the elastic curve is

$$v = \frac{M_o}{P}\left(\frac{1 - \cos \lambda L}{\sin \lambda L} \sin \lambda x + \cos \lambda x - 1\right) \tag{14-17}$$

The maximum deflection occurs at $x = L/2$. After some simplifications, it is found to be

$$v_{\max} = \frac{M_o}{P}\left(\frac{\sin^2 \lambda L/2}{\cos \lambda L/2} + \cos \frac{\lambda L}{2} - 1\right) = \frac{M_o}{P}\left(\sec \frac{\lambda L}{2} - 1\right) \tag{14-18}$$

The largest bending moment also occurs at $x = L/2$. Its absolute maximum is

$$M_{\max} = |-M_o - Pv_{\max}| = M_o \sec \lambda L/2 \tag{14-19}$$

It is important to note that in slender members bending moments may be substantially increased by the presence of axial compressive forces. When such forces exist, the deflection caused by the transverse loading is magnified, Fig. 14-5(b). For tensile forces the deflections are reduced.

14-4. STABILITY OF EQUILIBRIUM

A perfectly straight needle balanced on its tip may be said to be in equilibrium. However, the least disturbance or imperfection in its manufacture would make this equilibrium impossible. This kind of equilibrium is said to be unstable, and it is imperative to avoid analogous situations in structural systems.

To clarify the problem further, again consider a vertical rigid bar with a torsional spring of stiffness k at the base as shown in Fig. 14-6(a). The behavior of such a bar subjected to a vertical force P and a horizontal force F was considered in Art. 14-2. The response of this system as P increases is shown in Fig. 14-6(b) for a large and a small F. The question then arises: How will this system behave if $F = 0$? This is the limiting case, and it corresponds to the investigation of pure buckling.

To answer this question analytically, the system must be deliberately displaced a small (infinitesimal) amount consistent with the boundary conditions. Then, if the restoring forces are greater than the forces tending to upset the system, the system is stable, and vice versa.

The rigid bar shown in Fig. 14-6(a) can experience only rotation as it cannot bend; i.e., the system has one degree of freedom. For an assumed rotation θ the restoring moment is $k\theta$, and, with $F = 0$, the upsetting

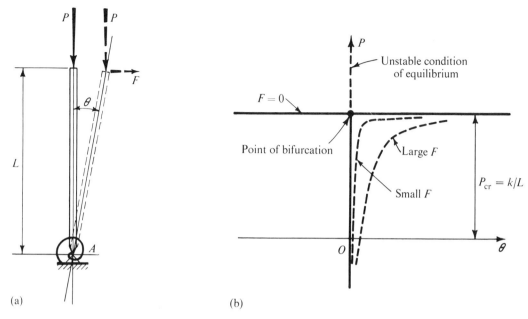

Fig. 14-6. Buckling behavior of a rigid bar.

moment is $PL \sin \theta \approx PL\theta$. Therefore, if

$$k\theta > PL\theta \quad \text{the system is stable}$$

and if

$$k\theta < PL\theta \quad \text{the system is unstable}$$

Right at the transition point $k\theta = PL\theta$, and the equilibrium is neither stable nor unstable but is neutral. The force associated with this condition is the critical or buckling load, which will be designated P_{cr}. For the bar system considered

$$P_{cr} = k/L \tag{14-20}$$

This condition establishes the inception of buckling. At this force two equilibrium positions are possible, the straight form and a deflected form infinitesimally near it. Since two branches of the solution are thus possible, this is called the *bifurcation point* of the equilibrium solution. For $P > k/L$ the system is unstable. As the solution has been linearized, there is no possibility of having θ become indefinitely large at P_{cr}. In the sense of large displacements, there is always a point of stable equilibrium at $\theta < \pi$.

The behavior of perfectly straight, concentrically loaded, elastic columns, i.e., of ideal columns, is highly analogous to the behavior described in the simple example above. From a linearized formulation of the problem, the critical buckling loads can be determined. Examples will be given in the following articles.

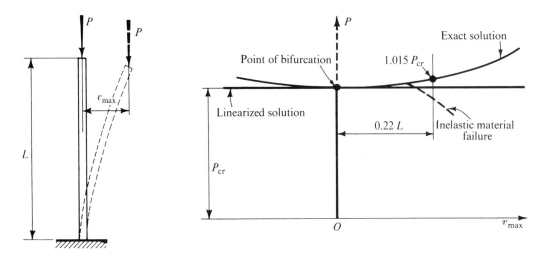

Fig. 14-7. Behavior of an ideal elastic column.

The critical loads do not describe the action of the buckling itself. By using an exact differential equation of the elastic curve for large deflections, it is possible* to find equilibrium positions higher than P_{cr} corresponding to the applied force P. The results of such an analysis are illustrated in Fig. 14-7. Note especially that increasing P_{cr} by a mere 1.5 per cent causes a maximum sidewise deflection of 22 per cent of the column length.† For practical reasons such enormous deflections can seldom be tolerated. Moreover, usually the material cannot resist the induced bending stresses. Therefore, real columns fail inelastically. In the vast majority of engineering applications, P_{cr} represents the ultimate capacity for a straight, concentrically loaded column.

14-5. EULER BUCKLING LOAD FOR PIN-ENDED COLUMNS

In order to formulate the governing relation for determining the buckling load of an ideal column, a small lateral deflection of the column axis must be permitted to occur. For the initially straight pin-ended column of Fig. 14-8(a), this is shown in Fig. 14-8(b).

For the slightly bent column of Fig. 14-8(b), the bending moment M at a general section is‡ $-Pv$, which on substituting into the differential

---

* See Timoshenko and Gere, *Theory of Elastic Stability*, p. 76.
† The fact that an elastic column continues to carry a load beyond the buckling stage can be demonstrated by applying a force in excess of the buckling load to a flexible bar or plate such as a carpenter's saw.
‡ If the end is clamped, the unknowns M_0 and $V_0 x$ appear in the expression. Also see footnote on p. 520.

equation for the elastic curve yields

$$\frac{d^2v}{dx^2} = \frac{M}{EI} = -\frac{P}{EI}v$$

Section 14-5
Euler buckling load
for pin-ended
columns

Then, as in Eq. 14-4, on letting $\lambda^2 = P/EI$, one has

$$\frac{d^2v}{dx^2} + \lambda^2 v = 0 \qquad (14\text{-}21)$$

This equation is seen to be the homogeneous part of Eq. 14-5 for a pin-ended beam-column. The solution of it is

$$v = C_1 \sin \lambda x + C_2 \cos \lambda x \qquad (14\text{-}22)$$

where the arbitrary constants C_1 and C_2 must be determined from the conditions at the boundary, which are

$$v(0) = 0 \quad \text{and} \quad v(L) = 0$$

Hence $v(0) = 0 = C_1 \sin 0 + C_2 \cos 0$, or $C_2 = 0$, and

$$v(L) = 0 = C_1 \sin \lambda L \qquad (14\text{-}23)$$

Equation 14-23 may be satisfied by taking $C_1 = 0$. As this corresponds to a condition of no buckling, this is a trivial solution. Alternatively, Eq. 14-23 also is satisfied if

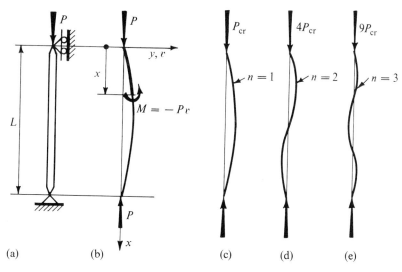

Fig. 14-8. Pin-ended column and its first three buckling modes.

$$\lambda L = \sqrt{P/EI}\, L = n\pi \tag{14-24}$$

where n is an integer. From this equation the characteristic values or eigenvalues for this differential equation, which make a buckled shape possible, require that

$$P_n = \frac{n^2 \pi^2 EI}{L^2} \tag{14-25}$$

Here n can assume all integral values. However, since interest centers on the least value at which a buckled shape can occur, n must be taken as unity. Therefore, the critical or the Euler buckling load for a column pinned at both ends is

$$P_{cr} = \frac{\pi^2 EI}{L^2} \tag{14-26}$$

where I must be the smallest or least moment of inertia of the cross-sectional area of a column and L is the column length. This case of a column pinned at both ends is often referred to as the *fundamental case*.

By substituting Eq. 14-24 into Eq. 14-22, with C_2 known to be zero, the buckled shape or mode of the column is obtained:

$$v = C_1 \sin n\pi x/L \tag{14-27}$$

This is the characteristic function or eigenfunction of this problem, and, since n can assume any integral value, there is an infinite number of such functions. In this linearized solution the amplitude C_1 of the buckling mode remains indeterminate. For $n = 1$, the elastic curve is a half-wave sine curve. This shape together with the modes corresponding to $n = 2$ and $n = 3$ are shown in Figs. 14-8(c), (d), and (e). The higher modes have no physical significance in buckling problems since the least critical load occurs at $n = 1$.

An alternative solution of the above problem may be obtained by using the differential equation of the fourth order for beam-columns with the transverse load set equal to zero. From Eq. 14-15 such an equation is

$$\frac{d^4 v}{dx^4} + \lambda^2 \frac{d^2 v}{dx^2} = 0 \tag{14-28}$$

For the case considered, the boundary conditions are:

$$v(0) = 0, \quad v(L) = 0, \quad M(0) = EIv''(0) = 0,$$

and

$$M(L) = EIv''(0) = 0$$

Using these conditions with the homogeneous solution of Eq. 14-28, together with its derivatives as given by Eqs. 14-16a and c, one obtains:

$$\begin{aligned}
&& + C_2 && &&+ C_4 = 0 & \\
C_1 \sin \lambda L && + C_2 \cos \lambda L && + C_3 L &&+ C_4 = 0 & \\
&& - C_2 \lambda^2 EI && && = 0 & \\
-C_1 \lambda^2 EI \sin \lambda L && - C_2 \lambda^2 EI \cos \lambda L && && = 0 &
\end{aligned}$$

Section 14-6
Elastic buckling of columns with different end restraints

To satisfy this set of equations, C_1, C_2, C_3, and C_4 could be set equal to zero, which gives a trivial solution. Alternatively, to obtain a nontrival solution, the determinant of the coefficients for a set of homogeneous algebraic equations must vanish.* Therefore, with $\lambda^2 EI = P$,

$$\begin{vmatrix} 0 & 1 & 0 & 1 \\ \sin \lambda L & \cos \lambda L & L & 1 \\ 0 & -P & 0 & 0 \\ -P \sin \lambda L & -P \cos \lambda L & 0 & 0 \end{vmatrix} = 0$$

The evaluation of this determinant leads to $\sin \lambda L = 0$, which is precisely the same condition as given by Eq. 14-23.

This approach is advantageous in problems with different boundary conditions where the axial force and EI remain constant throughout the length of the column. The method cannot be applied if the axial force extends over only a part of a member.

14-6. ELASTIC BUCKLING OF COLUMNS WITH DIFFERENT END RESTRAINTS

The same procedures as those discussed in the preceding article can be used to determine the elastic buckling loads of columns with different boundary conditions. The solutions of such problems are very sensitive to the end restraints. For example, the critical buckling load for a free-standing column,† Fig. 14-9(b), with a load at the top is

$$P_{cr} = \pi^2 EI/(4L^2) \tag{14-29}$$

In this extreme case the critical load is only one-fourth of that for the fundamental case, Eq. 14-26.

For a column fixed at one end and pinned at the other, Fig. 14-9(c):

$$P_{cr} = 2.05 \pi^2 EI/L^2 \tag{14-30}$$

---

* See, for example, F. B. Hildebrand, *Advanced Calculus for Applications*, (Englewood Cliffs, N.J.: Prentice-Hall, Inc., 1962), p. 188; or any text on linear algebra.
† A telephone pole having no external braces and with a heavy transformer at the top is an example.

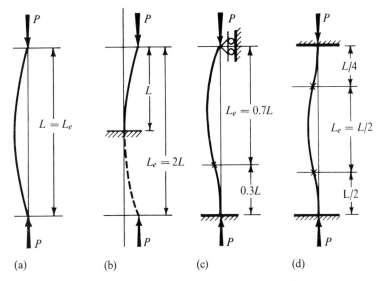

Fig. 14-9. Effective lengths of columns with different restraints.

Whereas, for a column fixed at both ends, Fig. 14-9(d),

$$P_{cr} = 4\pi^2 EI/L^2 \qquad (14\text{-}31)$$

The last two equations show that by restraining the ends the critical buckling loads are substantially raised above those in the fundamental case.

All the above formulas can be made to resemble the fundamental case providing that instead of the actual column length the effective column lengths are used. This length turns out to be the distance between the inflection points on the elastic curves or hinges if there are any. The effective column length L_e for the fundamental case is L, but for the above cases it is $2L$, $0.7L$, and $0.5L$, respectively. For a general case, $L_e = KL$, where K is the effective length factor, which depends on the end restraints.

In contrast to the classical cases shown in Fig. 14-9, actual compression members are seldom truly pinned or completely fixed against rotation at the ends. Because of the uncertainty regarding the fixity of the ends, columns are often assumed to be pin-ended. With the exception of the case shown in Fig. 14-8(b), where it cannot be used, this procedure is conservative.

The above equations become completely misleading in the inelastic range and must not be used in the form given (see Art. 14-8).

14-7. LIMITATION OF THE ELASTIC BUCKLING FORMULAS

In the above derivations of the buckling formulas for columns, it was tacitly assumed that the material behaves linearly elastically. To bring out this significant limitation, Eq. 14-26 can be written in a different form. By definition, $I = Ar^2$, where A is the cross-sectional area of a section and r is its radius of gyration. Substitution of this relation into Eq. 14-26 gives

$$P_{cr} = \frac{\pi^2 EI}{L^2} = \frac{\pi^2 E A r^2}{L^2}$$

or

$$\sigma_{cr} = \frac{P_{cr}}{A} = \frac{\pi^2 E}{(L/r)^2} \qquad (14\text{-}32)$$

Section 14-7 Limitation of the elastic buckling formulas

where the critical stress σ_{cr} for a column is defined as an average stress over the cross-sectional area A of a column at the critical load P_{cr}. The length* of the column is L, and r is the least radius of gyration of the cross-sectional area since the original Euler formula is in terms of the minimum I. The ratio L/r of the column length to the least radius of gyration is called the column *slenderness ratio*.

From Eq. 14-32 it can be concluded that the proportional limit of the material is the upper limit of the stress at which the column will buckle elastically. The necessary modification of the formula to include inelastic material response will be discussed in the next article.

EXAMPLE 14-3

Find the shortest length L for a pinned-ended steel column having a cross-sectional area of 2 in. by 3 in. for which the elastic Euler formula applies. Let $E = 30 \times 10^6$ psi and assume the proportional limit to be at 36 ksi.

SOLUTION

The minimum moment of inertia of the cross-sectional area $I_{min} = 3(2)^3/12 = 2$ in.^{4} Hence

$$r = r_{min} = \sqrt{\frac{I_{min}}{A}} = \sqrt{\frac{2}{2(3)}} = \frac{1}{\sqrt{3}} \text{ in.}$$

Then using Eq. 14-32 $\sigma_{cr} = \pi^2 E/(L/r)^2$ and solving it for the L/r ratio at the proportional limit

$$\left(\frac{L}{r}\right)^2 = \frac{\pi^2 E}{\sigma_{cr}} = \frac{\pi^2 (30) 10^6}{(36) 10^3} = 8{,}220$$

or

$$\frac{L}{r} = 90.7 \quad \text{and} \quad L = \frac{90.7}{\sqrt{3}} = 52.3 \text{ in.}$$

Therefore, if this column is 52.3 in. or more in length, it will buckle elastically as for such dimensions of the column the critical stress at buckling will not exceed the proportional limit for the material.

---

* Using the effective length L_e makes this expression general.

14-8. GENERALIZED EULER BUCKLING-LOAD FORMULAS

A typical compression stress-strain diagram for a specimen which is prevented from buckling may be represented as in Fig. 14-10(a). In the stress range from 0 to A the material behaves elastically. If the stress in a column at buckling does not exceed this range, the column buckles elastically. The hyperbola expressed by Eq. 14-32, $\sigma_{\text{cr}} = \pi^2 E/(L/r)^2$, is applicable in such a case. This portion of the curve is shown as ST in Fig. 14-10(b). It is important to recognize that this curve does not represent the behavior of one column but rather the behavior of an infinite number of the ideal columns of different lengths. The hyperbola beyond the useful range is shown in the figure by dashed lines.

A column with an L/r ratio corresponding to point S in Fig. 14-10(b) is the shortest column of a given material and size which will buckle elastically. A shorter column, having a still smaller L/r ratio, will not buckle at the proportional limit of the material. On the compression stress-strain diagram, Fig. 14-10(a), this means that the stress level in the column has passed point A and has reached some point B perhaps. At this higher stress level, it may be said that, in effect, a column of different material has been created since the stiffness of the material is no longer represented by the elastic modulus. At this point, the material stiffness is instantaneously given by the tangent to the stress-strain curve, i.e., by the tangent modulus E_t. The column remains stable if its new flexural rigidity $E_t I$ at B is sufficiently great, and it can carry a higher load. As the load is increased, the stress level rises, whereas the tangent modulus decreases. A

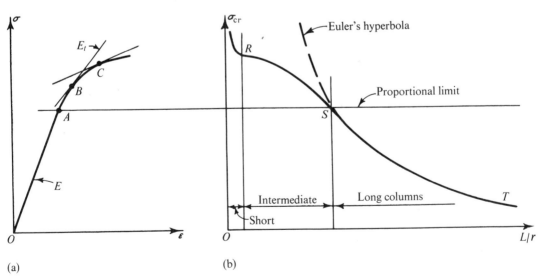

Fig. 14-10. (a) Compression stress-strain diagram; (b) critical stress in columns versus slenderness ratio.

column of ever "less stiff material" is acting under an increasing load. Substitution of the tangent modulus E_t for the elastic modulus E is then the only modification necessary to make the elastic buckling formulas applicable in the inelastic range. Hence the generalized Euler buckling-load formula, or the tangent modulus formula* becomes

Section 14-8 Generalized Euler buckling-load formulas

$$\sigma_{\text{cr}} = \frac{\pi^2 E_t}{(L/r)^2} \quad (14\text{-}33)$$

Since stresses corresponding to the tangent moduli can be obtained from the compression stress-strain diagram, the L/r ratio at which a column will buckle with these values can be obtained from Eq. 14-33. A plot representing this behavior for low and intermediate ratios of L/r is shown in Fig. 14-10(b) by the curve from R to S. Tests on individual columns verify this curve with remarkable accuracy.†

---

* The tangent modulus formula gives the carrying capacity of a column defined at the instant it tends to buckle. As a column deforms further, the fibers on the concave side continue to exhibit approximately the tangent modulus E_t. The fibers on the convex side, however, are relieved of some stress and rebound with the original elastic modulus E. These facts led to the establishment of the so-called *double-modulus theory* of load-carrying capacity of columns. The end results as obtained by this theory do not differ greatly from those obtained by the tangent modulus theory. For further details and significant refinements see F. R. Shanley, "Inelastic Column Theory," *Journal of Aeronautical Sciences*, **14**, no. 5 (May 1947); and F. Bleich, *Buckling Strength of Metal Structures* (New York: McGraw-Hill Book Company, 1952).
† See Bleich, *Buckling Strength of Metal Structures*, p. 20.

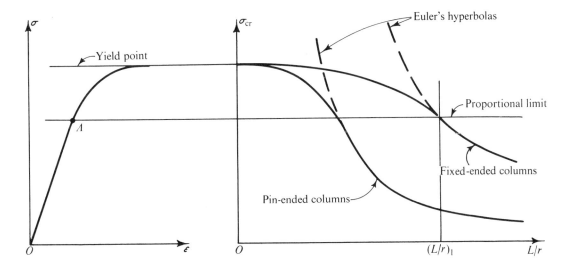

Fig. 14-11. Comparison of the behavior of columns with different end conditions.

Columns which buckle elastically sometimes are referred to as *long columns*. Columns having low L/r ratios, exhibiting essentially no buckling phenomena, are called *short columns*. At low L/r ratios, ductile materials "squash out" and can carry very large loads.

If the length L in Eq. 14-33 is treated as the effective length of a column, different end conditions can be analyzed. Following this procedure, for comparative purposes plots of critical stress σ_{cr} versus the slenderness ratio L/r for fixed-ended columns and pin-ended ones are in Fig. 14-11. It is important to note that the carrying capacity for the two cases is in a ratio of 4 to 1 only for columns having the slenderness ratio $(L/r)_1$ or greater. For smaller L/r ratios, progressively less benefit is derived from restraining the ends. At low L/r ratios the curves merge. It makes little difference whether a "short block" is pinned or fixed at the ends as strength rather than buckling determines the behavior.

14-9. ECCENTRICALLY LOADED COLUMNS

In the preceding discussion of column buckling the columns were assumed to be ideally straight. Since in reality all columns have some imperfections, the buckling loads obtained for ideal columns are the best possible. Such analyses only provide envelopes for the best possible performance of columns. It is not surprising, therefore, that the performance of columns has also been explored based on some statistically determined imperfections or possible misalignments of the applied loads. As an illustration of this approach, an eccentrically loaded column will be considered. This problem is important in itself.

An eccentrically loaded column is shown in Fig. 14-12(a). This loading is equivalent to a concentric axial force P and end moments $M_o = Pe$. Such a beam column has already been analyzed in Example 14-2, where it was found that because of the flexibility of the member the largest bending moment $M_{\max} = M_o \sec \lambda L/2$, Eq. 14-19. Therefore, the maximum compressive stress occurring at midheight on the concave side of the column may be computed as

$$|\sigma|_{\max} = \frac{P}{A} + \frac{Mc}{I} = \frac{P}{A} + \frac{M_{\max}c}{Ar^2} = \frac{P}{A}\left(1 + \frac{ec}{r^2}\sec\frac{\lambda L}{2}\right)$$

But $\lambda = \sqrt{P/(EI)} = \sqrt{P/(EAr^2)}$, hence

$$|\sigma|_{\max} = \frac{P}{A}\left(1 + \frac{ec}{r^2}\sec\frac{L}{r}\sqrt{\frac{P}{4EA}}\right) \tag{14-34}$$

This equation, since it contains a secant term, is often referred to as *the secant formula for columns*. For this equation to hold true, the maximum stress must remain within the elastic limit.

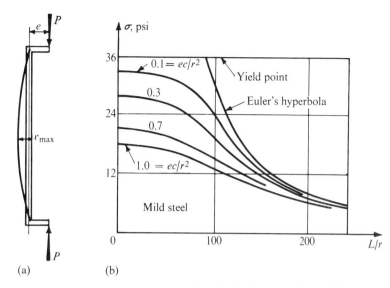

Fig. 14-12. Eccentrically loaded column.

Note that in Eq. 14-34 r may not be a minimum since it is obtained from the value of I associated with the axis around which bending occurs. In some cases a more critical condition for buckling may exist in the direction of no definite eccentricity. Also note that in Eq. 14-34 the relation between σ_{max} and P is not linear; σ_{max} increases faster than P. Therefore maximum stresses caused by axial forces cannot be superposed.

A plot of Eq. 14-34 for various assumed eccentricities* for mild steel columns of the same size is in Fig. 4-12(b). Note the close resemblance of the curves for small eccentricites and the column buckling curves established earlier. It should be noted, however, that these curves are made with the maximum stress as the criterion. This is not a true measure of the buckling capacity of a member. It can be shown that an additional axial load can be resisted beyond the point where the maximum stress at the critical section is reached. A discussion of this is beyond the scope of this text.†

14-10. DESIGN OF COLUMNS

For economy, the cross-sectional areas of columns, other than short blocks, should possess the largest possible least radius of gyration. This gives a smaller L/r ratio, which permits the use of a higher axial stress.

---

* This figure is taken from D. H. Young, "Rational Design of Steel Columns," *Trans. ASCE*, **101** (1936), 431.
† See Bleich, *Buckling Strength of Metal Structures*, chapter 1.

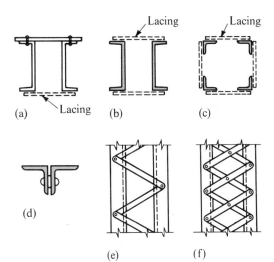

Fig. 14-13. Typical built-up cross sections of columns.

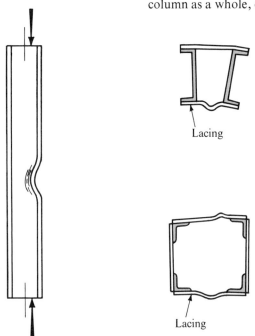

Fig. 14-14. Examples of local instability in columns.

Tubes form excellent columns. Wide-flange sections (which are also sometimes called H sections) are superior to I sections. In columns built up from rolled or extruded shapes, the individual pieces are spread out to obtain the desired effect. Cross sections for typical bridge compression members are shown in Figs. 14-3(a) and (b), for a boom in Fig. 14-13(c), and for an ordinary truss in Fig. 14-13(d). The angles in Fig. 14-13(d) are separated by spacers. The main shapes of Figs. 14-13(a), (b), and (c) are laced, or latticed, together by light bars, as shown in Figs. 14-13(e) and (f).

Obtaining a large r by placing a given amount of material away from the centroid of an area, as illustrated above, can reach a limit. The material can become so thin that it crumples locally. This behavior is termed *local instability*. When failure caused by local instability takes place in the flanges or the component plates of a member, the compression member becomes unserviceable. An illustration of local buckling is in Fig. 14-14. It is usually characterized by a change in the shape of a cross section. The equations derived earlier are for the instability of a column as a whole, or for primary instability. Discussion of the possibility of torsional instability, exemplified by the twisting of a whole section (which is a form of primary instability), is beyond the scope of this text.

After the chaotic situation which existed for many years with regard to the column-design formulas, now that the column-buckling phenomenon is more clearly understood only a few formulas are in common use. In most widely used specifications, a pair of formulas is given. One formula is used for small and intermediate values of L/r; the other for slender columns with large values of L/r. In the range of small and intermediate values of L/r, either a parabola or a straight line with a stipulated maximum is employed to define the critical stresses. For large values of L/r, Euler's hyperbola for the elastic response is used, Fig. 14-15. Sometimes the equations of the two complementary formulas have a common tangent at a selected value of L/r. A few specifications make use of the secant-formula approach with an assumed eccentricity based on the manufacturing tolerances.

In applying the design formulas it is

important to observe the following items:

1. The material for which the formula is written.
2. Whether the formula gives the working load (or stress) or whether it estimates the ultimate carrying capacity of a member. If the formula is of the latter type, a safety factor must be introduced.
3. The range of the applicability of the formula. Some empirical formulas if used beyond the specified range can lead to unsafe design. (See Fig. 14-15(b).)

It should be pointed out that the column formulas discussed above assume inviscid material, i.e., that the material possesses no viscoelastic or rheological properties. However, in some applications creep buckling of columns at sustained loads may be the primary consideration.*

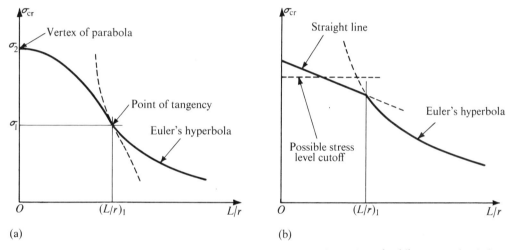

Fig. 14-15. Typical column-buckling curves for design.

14-11. COLUMN FORMULAS FOR CONCENTRIC LOADING

As examples of column-design formulas for nominally concentric loading, representative formulas for structural steel, aluminum alloy, and wood are given below. For the design of eccentrically loaded columns the reader is referred to more specialized books.

Column formulas for structural steel

The American Institute of Steel Construction† recommends the use of column

---

* See for example N. J. Hoff, "A survey of the Theories of Creep Buckling," *Proceedings, Third U.S. National Congress of Applied Mechanics*, 1958, p. 29 (published by ASME).

† See American Institute of Steel Constructing, *AISC Steel Construction Manual* (New York: AISC, Inc., 1963).

*Chapter 14
Buckling of columns*

formulas patterned on the scheme illustrated in Fig. 14-15(a). Since steels of many different yield strengths are manufactured, the formulas are stated in terms of σ_{yp}, which varies for different steels. The elastic modulus E for all steels is approximately the same. Euler's elastic buckling formula is specified for the slender columns beginning with $(L/r)_1 = C_c$ occurring at the slenderness ratio corresponding to one-half of the yield stress σ_{yp} of the steel. In order to fulfill this assumption, from Eq. 14-32, the slenderness ratio $C_c = (L/r)_1 = \sqrt{2\pi^2 E/\sigma_{yp}}$.

By using this equation, with $E = 29 \times 10^3$ ksi, the allowable stress for columns having a slenderness ratio larger than C_c becomes:

$$\sigma_{\text{allow}} = 149{,}000/(L_e/r)^2 \quad [\text{ksi}] \tag{14-35}$$

where L_e is the effective column length. A safety factor of 1.92 with respect to buckling is incorporated in Eq. 14-35. No columns are permitted to exceed an L_e/r of 200.

For an L_e/r ratio less than C_c, AISC specifies a parabolic formula:

$$\sigma_{\text{allow}} = \frac{[1 - (L_e/r)^2/(2C_c^2)]\sigma_{yp}}{\text{F.S.}} \tag{14-36}$$

where F.S., the factor of safety, is defined as

$$\text{F.S.} = 5/3 + 3(L_e/r)/(8C_c) - (L_e/r)^3/(8C_c^3)$$

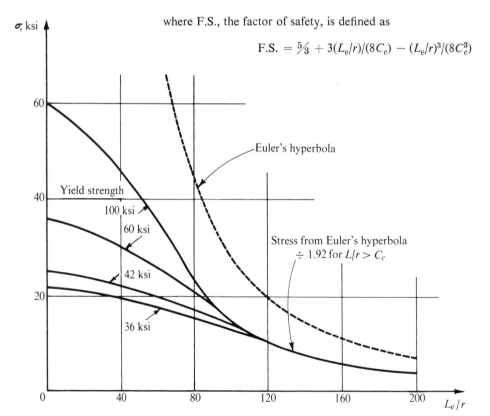

Fig. 14-16. Allowable stress for concentrically loaded columns per AISC specifications.

It is interesting to note that F.S. varies, being more conservative for the larger ratios of L_e/r. The equation chosen for F.S. approximates a quarter sine curve with the value of 1.67 at zero L_e/r and 1.92 at C_c. An allowable stress versus slenderness ratio for axially loaded columns of several kinds of structural steels is shown in Fig. 14-16.

Section 14-11 Column formulas for concentric loading

Column formulas for aluminum alloys

A large number of aluminum alloys are available for engineering applications. The yield and the ultimate strengths of such alloys vary over a considerable range. The elastic modulus for the alloys, however, is reasonably constant, and the Aluminum Company of America recommends the use of a factor of 102,000 ksi to represent the quantity $\pi^2 E$.* Hence, using Eq. 14-32, the ultimate strength formula for long columns with effective length L_e is

$$\sigma_{cr} = 102{,}000/(L_e/r)^2 \quad [\text{ksi}] \tag{14-37}$$

An inclined straight line, as in Fig. 14-15(b), is used for the column-strength curve for low and intermediate values of L_e/r. For some alloys this straight line is so chosen as to be tangent to Euler's hyperbola; for others, an angle is formed as in the figure. As an example, drawn from a very large number of cases given in the ALCOA handbook, for an extruded 2024-T4 alloy,

$$\sigma_{cr} = 44.8 - 0.313(L_e/r) \text{ ksi} \quad (0 \leq (L_e/r) \leq 64) \tag{14-38}$$

In applying this particular formula the L_e/r ratio must not exceed 64. Extrapolations of such formulas beyond their range of applicability is impermissible as they may give values of higher stress than that which is possible at Euler's buckling load. A factor of safety must be applied with Eqs. 14-37 and 14-38.

Column formulas for wood

The National Lumber Manufacturers Association† recommends the use of the Euler buckling load formula for solid wood columns. According to the recommendation the allowable stress is

$$\sigma_{\text{allow}} = \pi^2 E/[2.727(L/r)^2] = 3.619 E/(L/r)^2 \tag{14-39}$$

Here the allowable stress cannot exceed the value for compression of short blocks parallel to the grain for the particular species of wood. These stresses are increased for short-duration loading and are decreased for sustained loading. This formula is applicable for pin- and "square"-ended conditions.

For columns of square or rectangular cross section Eq. 14-39 becomes

$$\sigma_{\text{allow}} = 0.30 E/(L/d)^2 \tag{14-40}$$

where d is the smallest side dimension of a member.

---

* See *ALCOA Structural Handbook* (8th ed.) (Pittsburgh, Pa.: Aluminum Company of America 1960) p. 110.
† *NLMA National Design Specification* (Washington, D.C.: National Lumber Manufacturers Association, 1962).

EXAMPLE 14-4

Compare the allowable axial compressive loads for a 3-in.-by-2-in.-by-$\frac{1}{4}$-in. aluminum-alloy angle 43 in. long (a) it if acts as a pin-ended column, and (b) if it is so restrained that its effective length L_e is $0.9L$. Assume a factor of safety of 2.5 and that Eqs. 14-37 and 14-38 apply.

SOLUTION

The proportions of aluminum-alloy angles are the same as those of steel. Hence, for this size angle, from Table 7 of the Appendix, the least radius of gyration $r = 0.43$ in., and its area $A = 1.19$ in.^{2} The solution follows by applying Eq. 14-37 and using the specified factor of safety. Case (a): $L_e = L = 43$ in., $L_e/r = 43/(0.43) = 100$.

$$\sigma_{cr} = \frac{102,000,000}{(L_e/r)^2} = 10,200 \text{ psi}$$

$$P_{cr} = A\sigma_{cr} = (1.19)10,200 = 12,100 \text{ lb}$$

$$P_{allow} = \frac{P_{cr}}{F.S.} = \frac{12,100}{2.5} = 4,830 \text{ lb}$$

Case (b): $L_e = 0.9L = 38.7$ in., $L_e/r = (38.7)/(0.43) = 90$, and $\sigma_{cr} = 12,600$ psi.

$$P_{cr} = A\sigma_{cr} = (1.19)12,600 = 15,000 \text{ lb}$$

$$P_{allow} = \frac{P_{cr}}{F.S.} = \frac{15,000}{2.5} = 6,000 \text{ lb}$$

EXAMPLE 14-5

Using AISC column formulas select a 15-ft-long, pin-ended column to carry a concentric load of 200 kips. The structural steel is to be A441, having $\sigma_{yp} = 50$ ksi.

SOLUTION

The required size of the column may be found directly from the tables in the AISC Steel Construction Manual. However, this example provides an opportunity to demonstrate the trial-and-error procedure which is so often necessary in design, and the solution presented follows from using this method.

First try: Let $L/r = 0$ (a poor assumption for a column 15 ft long.) Then, from Eq. 14-36, since F.S. $= \frac{5}{3}$, $\sigma_{allow} = 50/(F.S.) = 30$ ksi and $A = P/\sigma_{allow} = 200/30 = 6.67$ in.^{2} From Table 4 in the Appendix, this requires an 8 WF 24 section, whose $r_{min} = 1.61$ in. Hence $L/r = 15(12)/(1.61) = 112$. With this L/r, the allowable stress is found using Eq. 14-35 or 14-36, whichever is applicable depending on C_c:

$$C_c = \sqrt{2\pi^2 E/\sigma_{yp}} = \sqrt{2\pi^2 \times 29 \times 10^3/50} = 107 < L/r = 112$$

hence $\quad \sigma_{allow} = 149,000/(112)^2 = 11.9$ ksi

This is much smaller than the initially assumed stress of 30 ksi, and another section must be selected.

Second try: Let $\sigma_{\text{allow}} = 11.9$ ksi as found above. Then $A = 200/11.9 = 16.8$ in.² requiring an 8 WF 58 section having $r_{\min} = 2.10$ in. Now $L/r = 15(12)/(2.10) = 85.7$, which is less than C_c found above. Therefore, Eq. 14-36 applies, and

$$\text{F.S.} = \tfrac{5}{3} + 3(85.7)/(8 \times 107) - (85.7)^3/(8 \times 107^3) = 1.90$$

and

$$\sigma_{\text{allow}} = [1 - (85.7)^2/(2 \times 107^2)]50/(1.90) = 17.9 \text{ ksi}$$

This stress requires $A = 200/17.9 = 11.2$ in.², which is met by an 8 WF 40 section with $r_{\min} = 2.04$ in. A calculation of the capacity for this section shows that the allowable axial load for it is 204 kips, which meets the requirements of the problem.

14-12. ON THE ENERGY APPROACH FOR DETERMINING BUCKLING LOADS

Section 14-12
On the energy approach for determining buckling loads

Stability problems can be treated in a very general manner using the energy methods. As an introduction to such methods, the basic criteria for determining the stability of equilibrium are derived in this article for conservative, linearly elastic systems.

To establish the stability criteria, a function Π called the *total potential* of the system must be formulated. This function is expressed as the sum of the internal potential energy U (strain energy) and the potential energy Ω (omega) of the external forces that act on a system, i.e.,

$$\Pi = U + \Omega \tag{14-41}$$

Disregarding a possible additive constant, $\Omega = -W_e$; i.e., the loss of potential energy during the application of the forces is equal to the work done on the system by the external forces. Hence, Eq. 14-41 can be rewritten as

$$\Pi = U - W_e \tag{14-42}$$

As is known from classical mechanics, for equilibrium the total potential Π must be stationary,* therefore its variation $\delta\Pi$ must equal zero, i.e.,

$$\delta\Pi = \delta U - \delta W_e = 0 \tag{14-43}$$

For conservative, elastic systems this relation agrees with Eq. 13-21, which states the virtual work principle. This condition can be used to determine the position of equilibrium. However, Eq. 14-43 cannot discern the type

---

* In terms of the ordinary functions this simply means that a condition exists where the derivative of a function with respect to an independent variable is zero and the function itself has a maximum, a minimum, a minimax, or a constant value.

of equilibrium and thereby establish the condition for the stability of equilibrium. Only by examining the higher order terms in the expression for the change $\Delta \Pi$ in the total potential Π can this be determined. Therefore, the more complete expression for the increment in Π as given by Taylor's expansion must be examined. Such an expression is

$$\Delta \Pi = \delta \Pi + \frac{1}{2!} \delta^2 \Pi + \frac{1}{3!} \delta^3 \Pi + \cdots \qquad (14\text{-}44)$$

Since for any type of equilibrium $\delta \Pi = 0$, it is the first nonvanishing term of this expansion that determines the type of equilibrium. For linear elastic systems the second term suffices. Thus, from Eq. 14-44, the stability criteria are

$\delta^2 \Pi > 0 \qquad$ for stable equilibrium

$\delta^2 \Pi < 0 \qquad$ for unstable equilibrium $\qquad (14\text{-}45)$

and $\quad \delta^2 \Pi = 0 \qquad$ for neutral equilibrium associated with the critical load

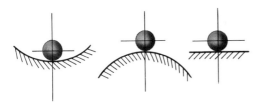

Fig. 14-17. Examples from rigid-body mechanics of stable, unstable, and neutral condition of equilibrium.

The meaning of these expressions may be clarified by examining the simple example shown in Fig. 14-17, where the shaded surfaces represent three different types of Π functions. In all three cases, the first derivative at the point of equilibrium of the ball is zero, and it is the second derivatives that determine the type of equilibrium.

For simple functions of Π the procedures for forming the derivatives, differentials, and variations are alike. If, however, the function of Π is expressed by integrals, the problem becomes mathematically much more complicated requiring the use of the calculus of variations. The treatment of such problems is beyond the scope of this text.*

EXAMPLE 14-6

Using the energy method, verify the critical load found before for a rigid bar with a torsional spring at the base, Fig. 14-6(a).

SOLUTION

For a displaced position of the bar, the strain energy in the spring is $k\theta^2/2$. For the same displacement the force P lowers an amount $L - L\cos\theta = L(1 - \cos\theta)$. Therefore

$$\Pi = U - W_e = \tfrac{1}{2} k\theta^2 - PL(1 - \cos\theta)$$

---

* For further details see for example H. L. Langhaar, *Energy Methods in Applied Mechanics* (New York: John Wiley & Sons, Inc., 1962).

If the study of the problem is confined to small (infinitesimal) displacements and since $\cos\theta = 1 - \theta^2/2! + \theta^4/4! + \ldots$, the total potential Π to a consistent order of accuracy simplifies to

$$\Pi = \frac{k\theta^2}{2} - \frac{PL\theta^2}{2}$$

Section 14-12
On the energy approach for determining buckling loads

Note especially that the ½ in the last term is due to the expansion of the cosine into the series. Full external force P acts on the bar as θ is permitted to change.

Having the expression for the total potential, one must solve two distinctly different problems. In the first problem a position of equilibrium is found. For this purpose Eq. 14-43 is applied:

$$\delta\Pi = \frac{d\Pi}{d\theta}\,\delta\theta = (k\theta - PL\theta)\,\delta\theta = 0 \quad \text{or} \quad (k - PL)\theta\,\delta\theta = 0$$

At this point of the solution k, P, and L must be considered constant, and $\delta\theta$ cannot be zero. Therefore an equilibrium position occurs at $\theta = 0$.

In the second, distinctly different, phase of the solution, according to the last part of Eq. 14-45, for neutral equilibrium,

$$\delta^2\Pi = \frac{d^2\Pi}{d\theta^2}\,\delta\theta^2 + \frac{d\Pi}{d\theta}\,\delta^2\theta = 0$$

or
$$(k - PL)(\delta\theta)^2 + (k - PL)\theta\,\delta^2\theta = 0$$

Since $\delta^2\theta$ also cannot be zero, for equilibrium at $\theta = 0$, the second term on the left side vanishes, whereas the first yields $P = k/L$, which is the critical buckling load.

PROBLEMS FOR SOLUTION

14-1. A rigid bar is held in a vertical position as shown in the figure by a spring having a spring constant k_1 lb per inch and a torsional spring having stiffness k_2 lb in. per radian. Find the critical load P_{cr} for this system.

14-2. Two rigid bars of equal length are connected by a frictionless hinge at A and are supported at B and D as shown in the figure. Determine the magnitude of the critical force P_{cr}. The spring attached to the lower bar at C is linearly elastic having a spring constant k. Ans. $ka/3$.

14-3. Two rigid bars each of length L are assembled into a vertical column by means of two torsional springs as shown in the figure.

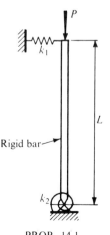

PROB. 14-1

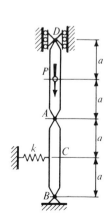

PROB. 14-2

The spring constant for each spring is k. (a) What are the two critical loads (eigenvalues) for this system? (*Hint:* Rotate the bars through small angles. Then, for the displaced position of the system, write two equations of equilibrium $\Sigma M_A = 0$ and $\Sigma M_B = 0$. Simultaneous solution of these two homogeneous algebraic equations yields the two eigenvalues of this system. The solution is achieved by setting the determinant of the coefficients for the unknown angles of rotation equal to zero. See an analogous solution for Eq. 14-28.) (b) Substitute the eigenvalues found above into either one of the equilibrium equations to find the ratio between the angles of rotation of the two bars. Sketch the deflected shape of the buckled column for the above ratios of angular rotations. These sketches represent the eigenfunctions of this system having two degrees of freedom. *Ans.* (a) $P_{cr} = P_{min} = (3 - \sqrt{5})k/(2L)$, $P_{max} = (3 + \sqrt{5})k/(2L)$; (b) 1.62, −0.62.

joint write an equation of equilibrium of moments. Then construct a matrix of the coefficients of bar rotations.) *Ans.* For example, $\Sigma M_3 = 0$ yields $Pa\,\delta\alpha_1 + Pa\,\delta\alpha_2 + (Pa - k)\,\delta\alpha_3 + k\,\delta\alpha_4 = 0$, where $\delta\alpha_i$ are small rotations of bars.

14-5. A beam AB of flexural rigidity EI is propped up at its center by a column CD. The column consists of two rigid bars connected at K by a hinge and a torsional spring with a spring constant k. Estimate the maximum vertical deflection of the beam caused by the application of the force $P = 4k/a$. *Ans.* $kL^3/(24aEI)$.

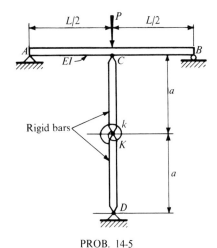

PROB. 14-5

14-6. Show that Eq. 14-16a is the solution of Eq. 14-15. How will the solution be modified if the axial force P is tensile?

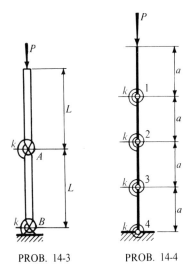

PROB. 14-3   PROB. 14-4

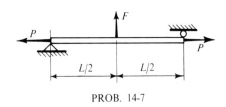

PROB. 14-7

14-4. A real column may be approximated by a series of rigid bars of length a each with an appropriate torsional spring at each joint as shown in the figure. In this problem all spring constants are assumed to be k each. Using this system as a model, set up a matrix equation for determining the critical load for a system having n degrees of freedom. (*Hint:* This is a generalization of the preceding problem. Deflect the column, and at each

14-7. Determine the equation of the elastic curve for the beam column loaded as shown in the figure. The EI is constant. *Ans.* $[F/(2P\lambda)](\text{sech}\,\lambda L/2)\sinh\lambda x - Fx/(2P)$.

14-8. A beam-column is loaded transversely

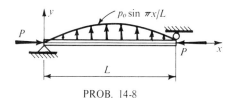

PROB. 14-8

by a sinusoidal load as shown in the figure. Show that the equation of the elastic curve is

$$v = \frac{p_o L^4}{\pi^4 EI}\left[\frac{1}{1 - P/P_{\mathrm{cr}}}\right]\sin\frac{\pi x}{L}$$

where $P_{\mathrm{cr}} = \pi^2 EI/L^2$ and the expression in the brackets is the magnification factor. Sketch a graph analogous to Fig. 14-1(b) showing the dependence of the deflection at the center on the magnitude of the axial force P.

14-9. A beam-column is subjected to a uniformly distributed transverse load p_o and an axial force P as shown in the figure. (a) By solving Eq. 14-14, subject to the prescribed boundary conditions, show that

$$M = -p_o[(\cos\lambda L - 1)\sin\lambda x/(\sin\lambda L) \\ - \cos\lambda x + 1]/\lambda^2$$

(b) How can one obtain the equation of the elastic curve from the above relation easily?

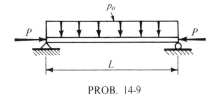

PROB. 14-9

14-10. A bar is initially curved so that its axis has the shape of a one-half sine wave, $v_o = a\sin\pi x/L$. If this bar is subjected to an axial

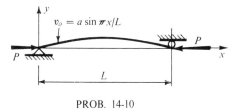

PROB. 14-10

compression as shown in the figure, show that the total deflection

$$v = v_o + v_1 = [1/(1 - P/P_{\mathrm{cr}})]a\sin\pi x/L$$

where $P_{\mathrm{cr}} = \pi^2 EI/L^2$ and the expression in the brackets is the magnification factor.

14-11. Rework Example 14-2 using Eq. 14-14. Also show that if $P = 0$, Eq. 14-18 reduces to $v_{\max} = M_o L^2/(8EI)$.

14-12. Derive Eq. 14-29, and show that a typical eigenfunction for an n-th mode in this case is $v_n = C_n(1 - \cos\lambda_n x)$ where $\lambda_n = (2n + 1)\pi/(2L)$ with n an integer. (For $n = 1$ see Fig. 14-9(b).)

14-13. Derive Eq. 14-30. (Here the transcendental equation for determining the critical roots is $\tan\lambda L = \lambda L$, which is satisfied when $\lambda L = 4.493$.)

14-14. Derive Eq. 14-31.

14-15. The cross section of a linearly elastic bar differs greatly in its two segments, so that the upper half has a flexural rigidity EI whereas the lower half can be considered as being infinitely rigid, see figure. Determine the critical load for this bar. (*Hint:* Divide the bar into two regions and make use of the continuity requirements at midheight. The use of second-order differential equations is recommended.) *Ans.* $1.67\pi^2 EI/L^2$.

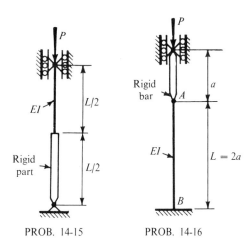

PROB. 14-15   PROB. 14-16

14-16. Determine the critical load for the column AB of constant EI for the system shown in the figure. (*Hint:* When point A is displaced horizontally an amount Δ, a horizontal force $P\Delta/a$ acts at A on the column AB. Also note that the moment at A must be zero.) *Ans.* $\tan \lambda L = 3\lambda L/2$, hence $\lambda L = 0.97$, and $P_{\text{cr}} = 0.94 EI/L^2$.

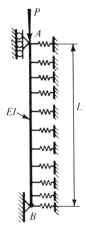

PROB. 14-17

14-17. A pin-ended bar of constant EI is supported along its length by an elastic foundation as shown in the figure. The foundation modulus is k lb per square inch and is such that when the bar deflects an amount v, a restoring force kv lb per inch is exerted by the foundation normal to the bar. First, satisfy yourself that the governing homogeneous differential equation for this problem is

$$EIv^{\text{iv}} + Pv'' + kv = 0$$

Then, show that the required eigenvalue of the above differential equation is

$$P_{\text{cr}} = \frac{\pi^2 EI}{L^2} \left[n^2 + \frac{\gamma}{n^2} \right]_{\min}$$

where
$$\gamma = \frac{kL^4}{\pi^4 EI}$$

Note that if $k = 0$, the minimum value of P_{cr} becomes the classical Euler load. (*Note:* See Prob. 11-28 for additional details on this type of problem.)

14-18. An allowable axial load for a 10-ft-long pin-ended column of a certain linearly elastic material is 4 kips. Five different columns made of the same material and having the same cross section have the supporting conditions shown in the figure. Using the column capacity for the 10-ft column as the criterion, what are the allowable loads for the five columns shown?

14-19. A piece of mechanical equipment is to be supported at the top of a 5-in. nominal diameter standard steel pipe as shown in the sketch. The equipment and its supporting platform weigh 5,500 lb. The base of the pipe will be anchored in a concrete pad, the top end will be unsupported. If the factor of safety required against buckling is 2.5, what is the maximum height of the column on which the equipment can be supported? Let $E = 30 \times 10^6$ psi. *Ans.* 23.8 ft.

14-20. A thin bar of stainless steel is axially precompressed 20 lb between two plates which are fixed at a constant distance of 6.28 in. apart, see figure. This assembly is

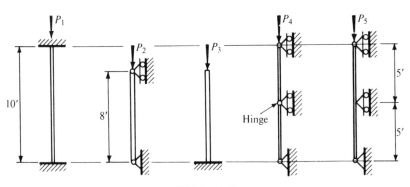

PROB. 14-18

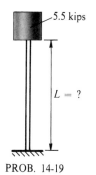

PROB. 14-19

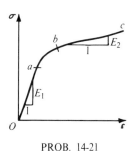

PROB. 14-20

made at 60° F. How high may the temperature of the bar rise, so as to have a factor of safety of 2 with respect to buckling? Assume $E = 28 \times 10^6$ psi, and $\alpha = 9.0 \times 10^{-6}$ per °F.

PROB. 14-21

14-21. Assume that a stress-strain diagram for a material is as shown in the figure. (a) Sketch a diagram showing the variation of E as a function of ε. (b) Sketch a graph of critical stress versus slenderness ratio, analogous to Fig. 14-10(b), for this material. Note that from 0 to a, and from b to c, the stress-strain relationship is linear with $E_1 > E_2$.

14-22. On the same diagram, reasonably close to scale, sketch the curves of the critical stress versus slenderness ratio for three different materials: an aluminum alloy, a low-carbon hot-rolled steel, and a Douglas Fir or a Southern Pine. Most of the required data are given in Table 1 of the Appendix. Also consult your text on materials science.* If some data are not readily available, make plausible assumptions.

14-23. The stress-strain curve in simple tension for an aluminum alloy is shown in the figure, where for convenience $\varepsilon \times 10^3 = e$. The alloy is linearly elastic for stresses up to 40 ksi; the ultimate stress is 50 ksi. (a) Idealize the stress-strain relation by fitting a parabola to the curve so that σ and $d\sigma/de = E_t$ is continuous at the proportional limit and so that the $\sigma = 50$ ksi line is tangent to the parabola. (b) Plot $E_t(\sigma)/E$ against $\sigma/\sigma_{\text{ult}}$ where E is the elastic modulus, σ_{ult} the ultimate stress, and E_t the tangent modulus at stress σ. (c) Plot in one graph σ_{cr} against L/r for fixed-fixed and pinned-pinned columns where σ_{cr} is based on E_t. Ans. (a) $\sigma = -2.5e^2 + 30e - 40$.

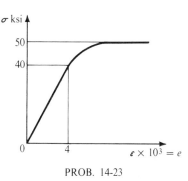

PROB. 14-23

14-24. The cross section of a compression member for a small bridge is made as shown in Fig. 14-13(a). The top cover plate is $\frac{1}{2}$ in. by 18 in. and the two 12 ⌷ 20.7 are placed 10 in. from back to back. If this member is 20 ft long, what is its slenderness

---

* A great deal of this kind of information is contained in *Metals Handbook* (8th ed.), Vol. 1, *Properties and Selection of Metals* (Metals Park, Novelty, Ohio: American Society for Metals, 1961).

ratio? (Check L/r in two directions.) *Ans.* 51.5.

14-25. A boom for an excavating machine is made of four $2\frac{1}{2}$-in.-by-$2\frac{1}{2}$-in.-by-$\frac{1}{4}$-in. steel angles arranged as shown in Fig. 14-13(c). Out-to-out dimensions of the square column (exclusive of the dimensions of the lacing bars) are 14 in. What axial load may be applied to this member if it is 52 ft long? Use the Euler formula for pinned columns and a factor of safety of 5. *Ans.* 28.8 kips.

14-26. If the capacity of the jib crane, the dimensions of which are shown in the figure, is to be 2 tons, what size standard steel pipe AB should be used? Use the Euler formula with a factor of safety of 3.5. Let $E = 30 \times 10^6$ psi. Neglect the weight of construction. *Ans.* $2\frac{1}{2}$ in.

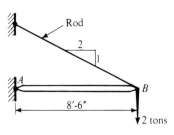

PROB. 14-26

14-27. A 1-in. round steel bar 4 ft long acts as a spreader bar in the arrangement shown in the figure. If cables and connections are properly designed, what pull P may be applied

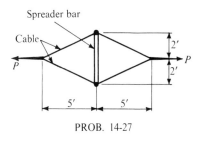

PROB. 14-27

to the assembly? Use Euler's formula and assume a factor of safety of 3. Let $E = 29 \times 10^6$ psi.

14-28. The pin-connected aluminum-alloy frame shown carries a concentrated load F. Assuming buckling can only occur in the

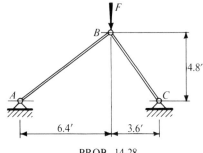

PROB. 14-28

plane of the frame, determine the value of F that will cause instability. Use the Euler formula as a criterion for member buckling. Take $E = 10 \times 10^6$ psi for the alloy. Both members have 2-in.-by-2-in. cross sections. *Ans.* 23,800 lb.

14-29. A machine bracket of steel alloy is to be made as shown in the figure. The compression member AB is so arranged that it can buckle as a pin-ended column in the plane ABC, but as a fixed-ended column in the direction perpendicular to this plane. (a) If the thickness of the member is $\frac{1}{2}$ in., what should its height h be to have equal probability of buckling in the two mutually perpendicular directions? (b) If $E = 28 \times 10^6$, and the factor of safety on instability is to be 2, what force F can be applied to the bracket? Assume that the bar designed in (a) controls the capacity of the assembly.

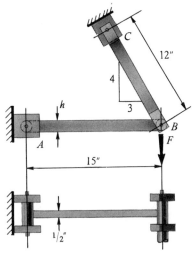

PROB. 14-29

14-30. The mast of a derrick is made of a standard rectangular 4-in.-by-2-in. steel tubing weighing 6.86 lb per foot. ($A = 2.02$ in.², $I_x = 1.29$ in.⁴, $I_z = 3.87$ in.⁴) If this derrick is assembled as indicated in the figure, what vertical force F, governed by the size of the mast, may be applied at A? Assume that all joints are pin-connected, and that the connection details are so made that the mast is loaded concentrically. The top of the mast is braced to prevent sidewise displacement. Use Euler's formula with a factor of safety of 3.5. Let $E = 29 \times 10^6$ psi. *Ans.* 9,800 lb.

Assume all joints are pin-connected and are capable of developing the full strength of a member. All members are 2½-in.-by-2-in.-by-¼-in. steel angles, whose weight is to be neglected. The allowable tensile stress is 18 ksi; the allowable compressive stress is to be determined by Euler's formula using a factor of safety of 3. Let $G = 12 \times 10^6$ psi, and $E = 30 \times 10^6$ psi. *Ans.* 1.48 kips.

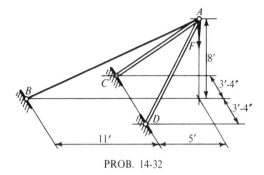

PROB. 14-32

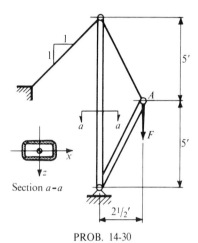

PROB. 14-30

14-31. One part of a structure is a light truss having the dimensions shown in the figure. The joints of this truss are braced transversely to maintain the truss as a planar structure. What force F may be applied to the truss?

14-32. A tripod made from 2-in. standard steel pipes is to be used for lifting loads vertically with a pulley at A as shown in the figure. What load rating may be assigned to this structure? Use the Euler formula and a factor of safety of 3. All joints may be considered pinned, and assume that the connection details, anchorages, and the tension rod AB are adequate. *Ans.* 5 kips.

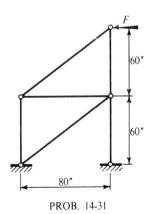

PROB. 14-31

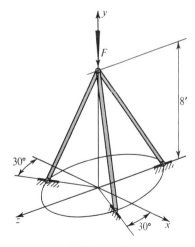

PROB. 14-33

14-33. A tripod is made from three 3-in.-by-3-in.-by-¼-in. steel angles each 10 ft long. In a plan view these angles are placed 120° apart, and the tripod is 8 ft high, see figure. Assuming that the ends of these angles are pin-ended, what is the magnitude of the allowable vertical downward force F that may be applied to the tripod? Use the Euler formula with a factor of safety equal to 3. Neglect the weight of the angles. Let $E = 30 \times 10^6$ psi. *Ans.* 8.25 kips.

14-34. What size WF column is required to carry a concentric load of 200 kips on a pin-ended column if it is 12 ft long? Use the AISC formulas and assume that A36 steel with $\sigma_{yp} = 36$ ksi will be used.

14-35. A compression member made of two 8 ⌋⌋ 11.5 is arranged as shown in Fig. 14-13(b). (a) Determine the distance back to back of the channels so that the moments of inertia about the two principal axes are equal. (b) If the member is pin-ended and is 32 ft long, what axial load may be applied according to the AISC code? Assume that A441 steel with $\sigma_{yp} = 50$ ksi will be used, and that the lacing is adequate.

14-36. A 14 WF 320 core section has two 24-in.-by-3-in. cover plates as shown in the figure. If this member is 20 ft long and is assumed to be pin-ended, what axial compressive force may be applied according to the AISC code? Assume that A242 steel with $\sigma_{yp} = 42$ ksi will be used.

PROB. 14-36

14-37. (a) Show how Eq. 14-40 is obtained from Eq. 14-39. (b) Apply Eq. 14-40 to obtain the allowable axial load on 4-in.-by-4-in. (nominal) Douglas Fir posts 8 ft and 14 ft long. Assume $E = 1.6 \times 10^6$ psi.

14-38. A narrow rectangular beam, such as shown in the figure, when loaded can collapse through lateral instability by twisting and

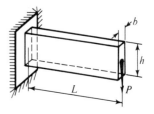

PROB. 14-38

displacing sidewise. It can be shown* that for the case shown the critical force which may be applied at the end is

$$P_{cr} = 4.013 \sqrt{B_1 C}/L^2$$

where $B_1 = hb^3E/12$ is the flexural stiffness of the beam around the vertical axis, and $C = \beta h b^3 G$ is the torsional stiffness. (For rectangular sections the coefficient β is given in a table in Art. 5-11.)

If a ½-in.-by-5-in. narrow rectangular cantilever is made of steel ($E = 30 \times 10^6$ psi, $E/G = 2.5$, and $\sigma_{yp} = 36$ ksi), and is loaded at the end, (a) find the maximum length L controlled by the strength of the material; (b) find the maximum length L controlled by lateral instability. (Note that for this case $\beta = 0.312$.)

14-39. Transversely loaded I beams which are laterally unsupported exhibit instability analogous to that described in the preceding problem. In practice, to guard against possible lateral instability, the allowable stress is reduced in comparison with the allowable stress for beams which are held laterally. A widely used formula, recommended by AISC, states that for A36 steel having a $\sigma_{yp} = 36$ ksi the allowable stress

$$\sigma_{\text{allow}} = \frac{12{,}000{,}000}{Ld/A_f} \leq 22{,}000 \text{ psi}$$

where L is the unbraced length of the compression flange, d is the depth of beam, and A_f is the area of the compression flange. (a) Using the above information, consider a laterally unsupported, 24 WF 100, simple beam 24 ft long, and determine what uniformly distributed load, including the weight of the

---

* See Timoshenko and Gere, *Theory of Elastic Stability*, p. 260.

beam, may be applied. (b) How much more load could be carried by the same beam if it were laterally braced so that the full allowable stress of 22,000 psi could be used? *Ans.* (a) 4,650 lb per foot.

14-40. It can be shown* that the critical value of the external pressure causing buckling of a thin-walled circular cylinder is

$$p_{cr} = \frac{E}{4(1-\nu^2)}\left(\frac{h}{R}\right)^3$$

---

*Timoshenko and Gere, *Theory of Elastic Stability*, p. 289.

where h is the wall thickness, and R is the radius of the cylinder. (a) Determine p_{cr} for a steel cylindrical tube having $h = 1$ in., and $R = 10$ in. Let $E = 30 \times 10^6$ psi, $\sigma_{yp} = 42,000$ psi, and $\nu = 0.25$. Compare this result with the maximum internal pressure which this cylinder could resist prior to yielding. (b) Repeat the problem for $h = 0.10$ in. and $R = 10$ in. (Note that if the cylinder is short, the end enclosures substantially modify the magnitude of the load at buckling.)

14-41. Using the energy approach rework Prob. 14-2.

Appendix tables

1. Typical Physical Properties of and Allowable Stresses for Some Common Materials.
2. Useful Properties of Areas.
3. American Standard Steel I-Beams, Properties for Designing.
4. Steel Wide-Flange Beams, Properties for Designing.
5. American Standard Steel Channels, Properties for Designing.
6. Steel Angles with Equal Legs, Properties for Designing.
7. Steel Angles with Unequal Legs, Properties for Designing.
8. Standard Steel Pipe.
9. Plastic Section Moduli.
10. American Standard Timber Sizes, Properties for Designing.
11. Deflections and Slopes of Elastic Curves for Variously Loaded Beams.

Acknowledgement: Data for Tables 3 through 10 are taken from *AISC Manual of Steel Construction* and are reproduced by permission of the American Institute of Steel Construction, Inc.

TABLE 1   TYPICAL PHYSICAL PROPERTIES OF AND ALLOWABLE STRESSES FOR SOME COMMON MATERIALS[a]

Material	Unit Weight, lb/in.³	Ultimate Strength, ksi			Yield Strength[g], ksi		Allow. Stresses[i], psi		Elastic Moduli × 10⁶ psi		Coef. of Thermal Expans. × 10⁻⁶ per °F
		Tens.	Comp.[c]	Shear	Tens.[h]	Shear	Tens. or Comp.	Shear	Tens. or Comp.	Shear	
Aluminum alloy {2024–T4 (extruded)	0.100	60	...	32	44	25			10.6	4.00	12.9
{(6061–T6)		38	...	24	35	20			10.0	3.75	13.0
Cast iron {Gray	0.276	30	120	...[e]	...	...			13	6	5.8
{Malleable		54	...	48	36	24			25	12	6.7
Concrete[b] {8 gal/sack	0.087	...	3	...[e]	...	...	−1,350[f]	66	3	...	6.0
{6 gal/sack		...	5	...	...	...	−2,250[f]	86	5	...	
Magnesium alloy, AM100A	0.065	40	...	21	22	...			6.5	2.4	14.0
Steel {0.2% Carbon (hot-rolled)	0.283	65	...	48	36	24	±24,000	14,500	30[k]	12	6.5
{0.6% Carbon (hot-rolled)		100	...	80	60	36					
{0.6% Carbon (quenched)		120	...	100	75	45					
{3½% Ni, 0.4% C		200	...	150	150	90					
Wood {Douglas Fir (coast)	0.018	...	7.4[d]	1.1[f]	...	...	±1,900[j]	120[j]	1.76	...	...
{Southern Pine (longleaf)	0.021	...	8.4[d]	1.5[f]	...	...	±2,250[j]	135[j]	1.76	...	...

[a] Mechanical properties of metals depend not only on composition but also on heat treatment, previous cold working, etc. Data for wood are for clear 2-in.-by-2-in. specimens at 12 per cent moisture content. True values vary.   [b] 8 gal/sack means 8 gallons of water per 94-lb sack of Portland cement. Values for 28-day-old concrete.   [c] For short blocks only. For ductile materials the ultimate strength in compression is indefinite; may be assumed to be the same as that in tension.   [d] Compression parallel to grain on short blocks. Compression perpendicular to grain at proportional limit 950 psi, 1,190 psi, respectively. Values from *Wood Handbook*, U.S. Dept. of Agriculture.   [e] Fails in diagonal tension.   [f] Parallel to grain.   [g] For most materials at 0.2 per cent set.   [h] For ductile materials compressive yield strength may be assumed the same.   [i] For static loads only. Much lower stresses required in machine design because of fatigue properties and dynamic loadings.   [j] In bending only. No tensile stress is allowed in concrete. Timber stresses are for select or dense grade.   [k] AISC recommends the value of 29 × 10⁶ psi.

TABLE 2
USEFUL PROPERTIES OF AREAS

AREAS AND MOMENTS OF INERTIA OF AREAS AROUND CENTROIDAL AXES

RECTANGLE

$A = bh$
$I_o = bh^3/12$

CIRCLE

$A = \pi R^2$
$I_o = J/2 = \pi R^4/4$

TRIANGLE

$A = bh/2$
$I_o = bh^3/36$

SEMICIRCLE

$A = \pi R^2/2$
$I_o = 0.110 R^4$

$2R$ $4R/(3\pi)$

THIN TUBE

$A = 2\pi R_{avg} t$
$I_o = J/2 \approx \pi R_{avg}^3 t$

HALF OF THIN TUBE

$A = \pi R_{avg} t$
$I_o \approx 0.095 \pi R_{avg}^3 t$

$2R_{avg}$ $(2/\pi) R_{avg}$

AREAS AND CENTROIDS OF AREAS

TRIANGLE

Centroid

$2b/3$ $b/3$

$A = bh/2$

TRIANGLE

$(a + L)/3$ $(b + L)/3$

$A = hL/2$

PARABOLA

Vertex

$3/8\, b$

$A = 2/3\, bh$

PARABOLA: $y = -ax^2$

Vertex

$3/4\, b$

$A = bh/3$

$y = -ax^n$

Vertex

$[(n + 1)/(n + 2)]\, b$

$A = bh/(n + 1)$

PARABOLA

Vertex

$l/2$ $l/2$

The area for any segment of a parabola is $A = 2/3\, hl$

TABLE 3

AMERICAN STANDARD STEEL *I* BEAMS

PROPERTIES FOR DESIGNING

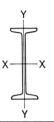

Nominal* Size	Weight per Foot	Area	Depth	Flange		Web Thickness	Axis x-x			Axis y-y		
				Width	Thickness		I	$\dfrac{I}{c}$	r	I	$\dfrac{I}{c}$	r
in.	lb	in.²	in.	in.	in.	in.	in.⁴	in.³	in.	in.⁴	in.³	in.
24 × 7⅞	120.0	35.13	24.00	8.048	1.102	.798	3010.8	250.9	9.26	84.9	21.1	1.56
	105.9	30.98	24.00	7.875	1.102	.625	2811.5	234.3	9.53	78.9	20.0	1.60
24 × 7	100.0	29.25	24.00	7.247	.871	.747	2371.8	197.6	9.05	48.4	13.4	1.29
	90.0	26.30	24.00	7.124	.871	.624	2230.1	185.8	9.21	45.5	12.8	1.32
	79.9	23.33	24.00	7.000	.871	.500	2087.2	173.9	9.46	42.9	12.2	1.36
20 × 7	95.0	27.74	20.00	7.200	.916	.800	1599.7	160.0	7.59	50.5	14.0	1.35
	85.0	24.80	20.00	7.053	.916	.653	1501.7	150.2	7.78	47.0	13.3	1.38
20 × 6¼	75.0	21.90	20.00	6.391	.789	.641	1263.5	126.3	7.60	30.1	9.4	1.17
	65.4	19.08	20.00	6.250	.789	.500	1169.5	116.9	7.83	27.9	8.9	1.21
18 × 6	70.0	20.46	18.00	6.251	.691	.711	917.5	101.9	6.70	24.5	7.8	1.09
	54.7	15.94	18.00	6.000	.691	.460	795.5	88.4	7.07	21.2	7.1	1.15
15 × 5½	50.0	14.59	15.00	5.640	.622	.550	481.1	64.2	5.74	16.0	5.7	1.05
	42.9	12.49	15.00	5.500	.622	.410	441.8	58.9	5.95	14.6	5.3	1.08
12 × 5¼	50.0	14.57	12.00	5.477	.659	.687	301.6	50.3	4.55	16.0	5.8	1.05
	40.8	11.84	12.00	5.250	.659	.460	268.9	44.8	4.77	13.8	5.3	1.08
12 × 5	35.0	10.20	12.00	5.078	.544	.428	227.0	37.8	4.72	10.0	3.9	.99
	31.8	9.26	12.00	5.000	.544	.350	215.8	36.0	4.83	9.5	3.8	1.01
10 × 4⅝	35.0	10.22	10.00	4.944	.491	.594	145.8	29.2	3.78	8.5	3.4	.91
	25.4	7.38	10.00	4.660	.491	.310	122.1	24.4	4.07	6.9	3.0	.97
8 × 4	23.0	6.71	8.00	4.171	.425	.441	64.2	16.0	3.09	4.4	2.1	.81
	18.4	5.34	8.00	4.000	.425	.270	56.9	14.2	3.26	3.8	1.9	.84
7 × 3⅝	20.0	5.83	7.00	3.860	.392	.450	41.9	12.0	2.68	3.1	1.6	.74
	15.3	4.43	7.00	3.660	.392	.250	36.2	10.4	2.86	2.7	1.5	.78
6 × 3⅜	17.25	5.02	6.00	3.565	.359	.465	26.0	8.7	2.28	2.3	1.3	.68
	12.5	3.61	6.00	3.330	.359	.230	21.8	7.3	2.46	1.8	1.1	.72
5 × 3	14.75	4.29	5.00	3.284	.326	.494	15.0	6.0	1.87	1.7	1.0	.63
	10.0	2.87	5.00	3.000	.326	.210	12.1	4.8	2.05	1.2	.82	.65
4 × 2⅝	9.5	2.76	4.00	2.796	.293	.326	6.7	3.3	1.56	.91	.65	.58
	7.7	2.21	4.00	2.660	.293	.190	6.0	3.0	1.64	.77	.58	.59
3 × 2⅜	7.5	2.17	3.00	2.509	.260	.349	2.9	1.9	1.15	.59	.47	.52
	5.7	1.64	3.00	2.330	.260	.170	2.5	1.7	1.23	.46	.40	.53

* Steel *I* beams are designated by giving their depth in inches first; then the letter I to designate an *I* beam; then the weight in pounds per linear foot. For example, 24 I 120.0.

TABLE 4

STEEL, WIDE-FLANGE BEAMS

PROPERTIES FOR DESIGNING

(ABRIDGED LIST)

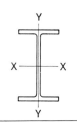

Nominal* Size	Weight per Foot	Area	Depth	Flange		Web Thickness	Axis x-x			Axis y-y		
				Width	Thickness		I	$\dfrac{I}{c}$	r	I	$\dfrac{I}{c}$	r
in.	lb	in.²	in.	in.	in.	in.	in.⁴	in.³	in.	in.⁴	in.³	in.
36 × 16½	230	67.73	35.88	16.475	1.260	.765	14988.4	835.5	14.88	870.9	105.7	3.59
36 × 12	150	44.16	35.84	11.972	.940	.625	9012.1	502.9	14.29	250.4	41.8	2.38
33 × 15¾	200	58.79	33.00	15.750	1.150	.715	11048.2	669.6	13.71	691.7	87.8	3.43
33 × 11½	130	38.26	33.10	11.510	.855	.580	6699.0	404.8	13.23	201.4	35.0	2.29
30 × 15	172	50.65	29.88	14.985	1.065	.655	7891.5	528.2	12.48	550.1	73.4	3.30
30 × 10½	108	31.77	29.82	10.484	.760	.548	4461.0	299.2	11.85	135.1	25.8	2.06
27 × 14	145	42.68	26.88	13.965	.975	.600	5414.3	402.9	11.26	406.9	58.3	3.09
27 × 10	94	27.65	26.91	9.990	.747	.490	3266.7	242.8	10.87	115.1	23.0	2.04
24 × 14	130	38.21	24.25	14.000	.900	.565	4009.5	330.7	10.24	375.2	53.6	3.13
24 × 12	100	29.43	24.00	12.000	.775	.468	2987.3	248.9	10.08	203.5	33.9	2.63
24 × 9	76	22.37	23.91	8.985	.682	.440	2096.4	175.4	9.68	76.5	17.0	1.85
21 × 13	112	32.93	21.00	13.000	.865	.527	2620.6	249.6	8.92	289.7	44.6	2.96
21 × 9	82	24.10	20.86	8.962	.795	.499	1752.4	168.0	8.53	89.6	20.0	1.93
21 × 8¼	62	18.23	20.99	8.240	.615	.400	1326.8	126.4	8.53	53.1	12.9	1.71
18 × 11¾	96	28.22	18.16	11.750	.831	.512	1674.7	184.4	7.70	206.8	35.2	2.71
18 × 8¾	64	18.80	17.87	8.715	.686	.403	1045.8	117.0	7.46	70.3	16.1	1.93
18 × 7½	50	14.71	18.00	7.500	.570	.358	800.6	89.0	7.38	37.2	9.9	1.59
16 × 11½	88	25.87	16.16	11.502	.795	.504	1222.6	151.3	6.87	185.2	32.2	2.67
16 × 8½	58	17.04	15.86	8.464	.645	.407	746.4	94.1	6.62	60.5	14.3	1.88
16 × 7	50	14.70	16.25	7.073	.628	.380	655.4	80.7	6.68	34.8	9.8	1.54
	36	10.59	15.85	6.992	.428	.299	446.3	56.3	6.49	22.1	6.3	1.45
14 × 16	142	41.85	14.75	15.500	1.063	.680	1672.2	226.7	6.32	660.1	85.2	3.97
	320†	94.12	16.81	16.710	2.093	1.890	4141.7	492.8	6.63	1635.1	195.7	4.17
14 × 14½	87	25.56	14.00	14.500	.688	.420	966.9	138.1	6.15	349.7	48.2	3.70
14 × 12	84	24.71	14.18	12.023	.778	.451	928.4	130.9	6.13	225.5	37.5	3.02
	78	22.94	14.06	12.000	.718	.428	851.2	121.1	6.09	206.9	34.5	3.00

* Steel *WF* beams are designated by giving their nominal depth in inches first; then the letters *WF* to designate a wide-flange beam; then the weight in pounds per linear foot. For example, 36 WF 230.
† Column core section.

(Table continued on next page.)

TABLE 4 (*Continued*)

Nominal* Size	Weight per Foot	Area	Depth	Flange Width	Flange Thickness	Web Thickness	Axis x-x I	Axis x-x $\frac{I}{c}$	Axis x-x r	Axis y-y I	Axis y-y $\frac{I}{c}$	Axis y-y r
in.	lb	in.²	in.	in.	in.	in.	in.⁴	in.³	in.	in.⁴	in.³	in.
14 × 10	74	21.76	14.19	10.072	.783	.450	796.8	112.3	6.05	133.5	26.5	2.48
	68	20.00	14.06	10.040	.718	.418	724.1	103.0	6.02	121.2	24.1	2.46
	61	17.94	13.91	10.000	.643	.378	641.5	92.2	5.98	107.3	21.5	2.45
14 × 8	53	15.59	13.94	8.062	.658	.370	542.1	77.8	5.90	57.5	14.3	1.92
	43	12.65	13.68	8.000	.528	.308	429.0	62.7	5.82	45.1	11.3	1.89
14 × 6¾	38	11.17	14.12	6.776	.513	.313	385.3	54.6	5.87	24.6	7.3	1.49
	34	10.00	14.00	6.750	.453	.287	339.2	48.5	5.83	21.3	6.3	1.46
	30	8.81	13.86	6.733	.383	.270	289.6	41.8	5.73	17.5	5.2	1.41
12 × 12	85	24.98	12.50	12.105	.796	.495	723.3	115.7	5.38	235.5	38.9	3.07
	65	19.11	12.12	12.000	.606	.390	533.4	88.0	5.28	174.6	29.1	3.02
12 × 10	53	15.59	12.06	10.000	.576	.345	426.2	70.7	5.23	96.1	19.2	2.48
12 × 8	40	11.77	11.94	8.000	.516	.294	310.1	51.9	5.13	44.1	11.0	1.94
12 × 6½	36	10.59	12.24	6.565	.540	.305	280.8	45.9	5.15	23.7	7.2	1.50
	31	9.12	12.09	6.525	.465	.265	238.4	39.4	5.11	19.8	6.1	1.47
	27	7.97	11.95	6.500	.400	.240	204.1	34.1	5.06	16.6	5.1	1.44
10 × 10	112	32.92	11.38	10.415	1.248	.755	718.7	126.3	4.67	235.4	45.2	2.67
	100	29.43	11.12	10.345	1.118	.685	625.0	112.4	4.61	206.6	39.9	2.65
	89	26.19	10.88	10.275	.998	.615	542.4	99.7	4.55	180.6	35.2	2.63
	77	22.67	10.62	10.195	.868	.535	457.2	86.1	4.49	153.4	30.1	2.60
	49	14.40	10.00	10.000	.558	.340	272.9	54.6	4.35	93.0	18.6	2.54
10 × 8	45	13.24	10.12	8.022	.618	.350	248.6	49.1	4.33	53.2	13.3	2.00
	39	11.48	9.94	7.990	.528	.318	209.7	42.2	4.27	44.9	11.2	1.98
	33	9.71	9.75	7.964	.433	.292	170.9	35.0	4.20	36.5	9.2	1.94
10 × 5¾	29	8.53	10.22	5.799	.500	.289	157.3	30.8	4.29	15.2	5.2	1.34
	21	6.19	9.90	5.750	.340	.240	106.3	21.5	4.14	9.7	3.4	1.25
8 × 8	67	19.70	9.00	8.287	.933	.575	271.8	60.4	3.71	88.6	21.4	2.12
	58	17.06	8.75	8.222	.808	.510	227.3	52.0	3.65	74.9	18.2	2.10
	48	14.11	8.50	8.117	.683	.405	183.7	43.2	3.61	60.9	15.0	2.08
	40	11.76	8.25	8.077	.558	.365	146.3	35.5	3.53	49.0	12.1	2.04
	35	10.30	8.12	8.027	.493	.315	126.5	31.1	3.50	42.5	10.6	2.03
	31	9.12	8.00	8.000	.433	.288	109.7	27.4	3.47	37.0	9.2	2.01
8 × 6½	28	8.23	8.06	6.540	.463	.285	97.8	24.3	3.45	21.6	6.6	1.62
	24	7.06	7.93	6.500	.398	.245	82.5	20.8	3.42	18.2	5.6	1.61
8 × 5¼	20	5.88	8.14	5.268	.378	.248	69.2	17.0	3.43	9.22	3.50	1.25
	17	5.00	8.00	5.250	.308	.230	56.4	14.1	3.36	7.44	2.83	1.22

TABLE 5

AMERICAN STANDARD STEEL CHANNELS

PROPERTIES FOR DESIGNING

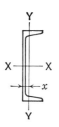

Nominal* Size	Weight per Foot	Area	Depth	Flange		Web Thickness	Axis x-x			Axis y-y			
				Width	Average Thickness		I	$\frac{I}{c}$	r	I	$\frac{I}{c}$	r	x
in.	lb	in.^{2}	in.	in.	in.	in.	in.^{4}	in.^{3}	in.	in.^{4}	in.^{3}	in.	in.
18 × 4†	58.0	16.98	18.00	4.200	.625	.700	670.7	74.5	6.29	18.5	5.6	1.04	.88
	51.9	15.18	18.00	4.100	.625	.600	622.1	69.1	6.40	17.1	5.3	1.06	.87
	45.8	13.38	18.00	4.000	.625	.500	573.5	63.7	6.55	15.8	5.1	1.09	.89
	42.7	12.48	18.00	3.950	.625	.450	549.2	61.0	6.64	15.0	4.9	1.10	.90
15 × 3⅜	50.0	14.64	15.00	3.716	.650	.716	401.4	53.6	5.24	11.2	3.8	.87	.80
	40.0	11.70	15.00	3.520	.650	.520	346.3	46.2	5.44	9.3	3.4	.89	.78
	33.9	9.90	15.00	3.400	.650	.400	312.6	41.7	5.62	8.2	3.2	.91	.79
12 × 3	30.0	8.79	12.00	3.170	.501	.510	161.2	26.9	4.28	5.2	2.1	.77	.68
	25.0	7.32	12.00	3.047	.501	.387	143.5	23.9	4.43	4.5	1.9	.79	.68
	20.7	6.03	12.00	2.940	.501	.280	128.1	21.4	4.61	3.9	1.7	.81	.70
10 × 2⅝	30.0	8.80	10.00	3.033	.436	.673	103.0	20.6	3.42	4.0	1.7	.67	.65
	25.0	7.33	10.00	2.886	.436	.526	90.7	18.1	3.52	3.4	1.5	.68	.62
	20.0	5.86	10.00	2.739	.436	.379	78.5	15.7	3.66	2.8	1.3	.70	.61
	15.3	4.47	10.00	2.600	.436	.240	66.9	13.4	3.87	2.3	1.2	.72	.64
9 × 2½	20.0	5.86	9.00	2.648	.413	.448	60.6	13.5	3.22	2.4	1.2	.65	.59
	15.0	4.39	9.00	2.485	.413	.285	50.7	11.3	3.40	1.9	1.0	.67	.59
	13.4	3.89	9.00	2.430	.413	.230	47.3	10.5	3.49	1.8	.97	.67	.61
8 × 2¼	18.75	5.49	8.00	2.527	.390	.487	43.7	10.9	2.82	2.0	1.0	.60	.57
	13.75	4.02	8.00	2.343	.390	.303	35.8	9.0	2.99	1.5	.86	.62	.56
	11.5	3.36	8.00	2.260	.390	.220	32.3	8.1	3.10	1.3	.79	.63	.58
7 × 2⅛	14.75	4.32	7.00	2.299	.366	.419	27.1	7.7	2.51	1.4	.79	.57	.53
	12.25	3.58	7.00	2.194	.366	.314	24.1	6.9	2.59	1.2	.71	.58	.53
	9.8	2.85	7.00	2.090	.366	.210	21.1	6.0	2.72	.98	.63	.59	.55
6 × 2	13.0	3.81	6.00	2.157	.343	.437	17.3	5.8	2.13	1.1	.65	.53	.52
	10.5	3.07	6.00	2.034	.343	.314	15.1	5.0	2.22	.87	.57	.53	.50
	8.2	2.39	6.00	1.920	.343	.200	13.0	4.3	2.34	.70	.50	.54	.52
5 × 1¾	9.0	2.63	5.00	1.885	.320	.325	8.8	3.5	1.83	.64	.45	.49	.48
	6.7	1.95	5.00	1.750	.320	.190	7.4	3.0	1.95	.48	.38	.50	.49
4 × 1⅝	7.25	2.12	4.00	1.720	.296	.320	4.5	2.3	1.47	.44	.35	.46	.46
	5.4	1.56	4.00	1.580	.296	.180	3.8	1.9	1.56	.32	.29	.45	.46
3 × 1½	6.0	1.75	3.00	1.596	.273	.356	2.1	1.4	1.08	.31	.27	.42	.46
	5.0	1.46	3.00	1.498	.273	.258	1.8	1.2	1.12	.25	.24	.41	.44
	4.1	1.19	3.00	1.410	.273	.170	1.6	1.1	1.17	.20	.21	.41	.44

* Steel channels are designated by giving their depth in inches first; then the symbol ⊔ to designate a channel; then the weight in pounds per linear foot. For example, 15 ⊔ 50.0.
† Car and shipbuilding channel; not an American standard.

TABLE 6

STEEL ANGLES
EQUAL LEGS

PROPERTIES FOR DESIGNING

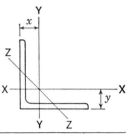

Size	Thickness	Weight per Foot	Area	Axis x-x and Axis y-y				Axis z-z
				I	$\frac{I}{c}$	r	x or y	r
in.	in.	lb	in.²	in.⁴	in.³	in.	in.	in.
8 × 8	1⅛	56.9	16.73	98.0	17.5	2.42	2.41	1.56
	1	51.0	15.00	89.0	15.8	2.44	2.37	1.56
	⅞	45.0	13.23	79.6	14.0	2.45	2.32	1.57
	¾	38.9	11.44	69.7	12.2	2.47	2.28	1.57
	⅝	32.7	9.61	59.4	10.3	2.49	2.23	1.58
	9/16	29.6	8.68	54.1	9.3	2.50	2.21	1.58
	½	26.4	7.75	48.6	8.4	2.50	2.19	1.59
6 × 6	1	37.4	11.00	35.5	8.6	1.80	1.86	1.17
	⅞	33.1	9.73	31.9	7.6	1.81	1.82	1.17
	¾	28.7	8.44	28.2	6.7	1.83	1.78	1.17
	⅝	24.2	7.11	24.2	5.7	1.84	1.73	1.18
	9/16	21.9	6.43	22.1	5.1	1.85	1.71	1.18
	½	19.6	5.75	19.9	4.6	1.86	1.68	1.18
	7/16	17.2	5.06	17.7	4.1	1.87	1.66	1.19
	⅜	14.9	4.36	15.4	3.5	1.88	1.64	1.19
	5/16	12.5	3.66	13.0	3.0	1.89	1.61	1.19
5 × 5	⅞	27.2	7.98	17.8	5.2	1.49	1.57	.97
	¾	23.6	6.94	15.7	4.5	1.51	1.52	.97
	⅝	20.0	5.86	13.6	3.9	1.52	1.48	.98
	½	16.2	4.75	11.3	3.2	1.54	1.43	.98
	7/16	14.3	4.18	10.0	2.8	1.55	1.41	.98
	⅜	12.3	3.61	8.7	2.4	1.56	1.39	.99
	5/16	10.3	3.03	7.4	2.0	1.57	1.37	.99
4 × 4	¾	18.5	5.44	7.7	2.8	1.19	1.27	.78
	⅝	15.7	4.61	6.7	2.4	1.20	1.23	.78
	½	12.8	3.75	5.6	2.0	1.22	1.18	.78
	7/16	11.3	3.31	5.0	1.8	1.23	1.16	.78
	⅜	9.8	2.86	4.4	1.5	1.23	1.14	.79
	5/16	8.2	2.40	3.7	1.3	1.24	1.12	.79
	¼	6.6	1.94	3.0	1.1	1.25	1.09	.80
3½ × 3½	½	11.1	3.25	3.6	1.5	1.06	1.06	.68
	7/16	9.8	2.87	3.3	1.3	1.07	1.04	.68
	⅜	8.5	2.48	2.9	1.2	1.07	1.01	.69
	5/16	7.2	2.09	2.5	.98	1.08	.99	.69
	¼	5.8	1.69	2.0	.79	1.09	.97	.69
3 × 3	½	9.4	2.75	2.2	1.1	.90	.93	.58
	7/16	8.3	2.43	2.0	.95	.91	.91	.58
	⅜	7.2	2.11	1.8	.83	.91	.89	.58
	5/16	6.1	1.78	1.5	.71	.92	.87	.59
	¼	4.9	1.44	1.2	.58	.93	.84	.59
	3/16	3.71	1.09	.96	.44	.94	.82	.59
2½ × 2½	½	7.7	2.25	1.2	.72	.74	.81	.49
	⅜	5.9	1.73	.98	.57	.75	.76	.49
	5/16	5.0	1.47	.85	.48	.76	.74	.49
	¼	4.1	1.19	.70	.39	.77	.72	.49
	3/16	3.07	.90	.55	.30	.78	.69	.49

TABLE 7

STEEL ANGLES
UNEQUAL LEGS

PROPERTIES FOR DESIGNING

(ABRIDGED LIST)

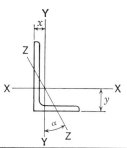

Size	Thickness	Weight per Foot	Area	Axis x-x				Axis y-y				Axis z-z	
				I	$\frac{I}{c}$	r	y	I	$\frac{I}{c}$	r	x	r	Tan α
in.	in.	lb	in.²	in.⁴	in.³	in.	in.	in.⁴	in.³	in.	in.	in.	
8 × 6	1	44.2	13.00	80.8	15.1	2.49	2.65	38.8	8.9	1.73	1.65	1.28	.543
	3/4	33.8	9.94	63.4	11.7	2.53	2.56	30.7	6.9	1.76	1.56	1.29	.551
	1/2	23.0	6.75	44.3	8.0	2.56	2.47	21.7	4.8	1.79	1.47	1.30	.558
8 × 4	1	37.4	11.00	69.6	14.1	2.52	3.05	11.6	3.9	1.03	1.05	.85	.247
	3/4	28.7	8.44	54.9	10.9	2.55	2.95	9.4	3.1	1.05	.95	.85	.258
	1/2	19.6	5.75	38.5	7.5	2.59	2.86	6.7	2.2	1.08	.86	.86	.267
6 × 4	3/4	23.6	6.94	24.5	6.3	1.88	2.08	8.7	3.0	1.12	1.08	.86	.428
	1/2	16.2	4.75	17.4	4.3	1.91	1.99	6.3	2.1	1.15	.99	.87	.440
5 × 3	1/2	12.8	3.75	9.5	2.9	1.59	1.75	2.6	1.1	.83	.75	.65	.357
	3/8	9.8	2.86	7.4	2.2	1.61	1.70	2.0	.89	.84	.70	.65	.364
	1/4	6.6	1.94	5.1	1.5	1.62	1.66	1.4	.61	.86	.66	.66	.371
4 × 3½	1/2	11.9	3.50	5.3	1.9	1.23	1.25	3.8	1.5	1.04	1.00	.72	.750
	3/8	9.1	2.67	4.2	1.5	1.25	1.21	3.0	1.2	1.06	.96	.73	.755
	1/4	6.2	1.81	2.9	1.0	1.27	1.16	2.1	.81	1.07	.91	.73	.759
4 × 3	1/2	11.1	3.25	5.1	1.9	1.25	1.33	2.4	1.1	.86	.83	.64	.543
	3/8	8.5	2.48	4.0	1.5	1.26	1.28	1.9	.87	.88	.78	.64	.551
	1/4	5.8	1.69	2.8	1.0	1.28	1.24	1.4	.60	.90	.74	.65	.558
3½ × 2½	1/2	9.4	2.75	3.2	1.4	1.09	1.20	1.4	.76	.70	.70	.53	.486
	7/16	8.3	2.43	2.9	1.3	1.09	1.18	1.2	.68	.71	.68	.54	.491
	3/8	7.2	2.11	2.6	1.1	1.10	1.16	1.1	.59	.72	.66	.54	.496
	5/16	6.1	1.78	2.2	.93	1.11	1.14	.94	.50	.73	.64	.54	.501
	1/4	4.9	1.44	1.8	.75	1.12	1.11	.78	.41	.74	.61	.54	.506
3 × 2½	1/2	8.5	2.50	2.1	1.0	.91	1.00	1.3	.74	.72	.75	.52	.667
	7/16	7.6	2.21	1.9	.93	.92	.98	1.2	.66	.73	.73	.52	.672
	3/8	6.6	1.92	1.7	.81	.93	.96	1.0	.58	.74	.71	.52	.676
	5/16	5.6	1.62	1.4	.69	.94	.93	.90	.49	.74	.68	.53	.680
	1/4	4.5	1.31	1.2	.56	.95	.91	.74	.40	.75	.66	.53	.684
3 × 2	1/2	7.7	2.25	1.9	1.0	.92	1.08	.67	.47	.55	.58	.43	.414
	7/16	6.8	2.00	1.7	.89	.93	1.06	.61	.42	.55	.56	.43	.421
	3/8	5.9	1.73	1.5	.78	.94	1.04	.54	.37	.56	.54	.43	.428
	5/16	5.0	1.47	1.3	.66	.95	1.02	.47	.32	.57	.52	.43	.435
	1/4	4.1	1.19	1.1	.54	.95	.99	.39	.26	.57	.49	.43	.440
	3/16	3.07	.90	.84	.41	.97	.97	.31	.20	.58	.47	.44	.446
2½ × 2	3/8	5.3	1.55	.91	.55	.77	.83	.51	.36	.58	.58	.42	.614
	5/16	4.5	1.31	.79	.47	.78	.81	.45	.31	.58	.56	.42	.620
	1/4	3.62	1.06	.65	.38	.78	.79	.37	.25	.59	.54	.42	.626
	3/16	2.75	.81	.51	.29	.79	.76	.29	.20	.60	.51	.43	.631

TABLE 8

STANDARD STEEL PIPE

	Dimensions					Properties		
Nom. Diam., In.	Outside Diam., In.	Inside Diam., In.	Thick- ness, In.	Weight per Foot, lb		I, in.^{4}	A, in.^{2}	r, in.
				Plain Ends	Thread & Cplg.			
1/8	.405	.269	.068	.24	.25	.001	.072	.12
1/4	.540	.364	.088	.42	.43	.003	.125	.16
3/8	.675	.493	.091	.57	.57	.007	.167	.21
1/2	.840	.622	.109	.85	.85	.017	.250	.26
3/4	1.050	.824	.113	1.13	1.13	.037	.333	.33
1	1.315	1.049	.133	1.68	1.68	.087	.494	.42
1 1/4	1.660	1.380	.140	2.27	2.28	.195	.669	.54
1 1/2	1.900	1.610	.145	2.72	2.73	.310	.799	.62
2	2.375	2.067	.154	3.65	3.68	.666	1.075	.79
2 1/2	2.875	2.469	.203	5.79	5.82	1.530	1.704	.95
3	3.500	3.068	.216	7.58	7.62	3.017	2.228	1.16
3 1/2	4.000	3.548	.226	9.11	9.20	4.788	2.680	1.34
4	4.500	4.026	.237	10.79	10.89	7.233	3.174	1.51
5	5.563	5.047	.258	14.62	14.81	15.16	4.300	1.88
6	6.625	6.065	.280	18.97	19.19	28.14	5.581	2.25
8	8.625	8.071	.277	24.70	25.00	63.35	7.265	2.95
8	8.625	7.981	.322	28.55	28.81	72.49	8.399	2.94
10	10.750	10.192	.279	31.20	32.00	125.9	9.178	3.70
10	10.750	10.136	.307	34.24	35.00	137.4	10.07	3.69
10	10.750	10.020	.365	40.48	41.13	160.7	11.91	3.67
12	12.750	12.090	.330	43.77	45.00	248.5	12.88	4.39
12	12.750	12.000	.375	49.56	50.71	279.3	14.58	4.38

TABLE 9

PLASTIC SECTION MODULI AROUND THE x-x AXIS

Shape	Plastic Modulus Z, in.^{3}	Shape	Plastic Modulus Z, in.^{3}
36 WF 230	942.7	15 I 42.9	68.6
33 WF 200	754.4	16 WF 36	63.9
30 WF 172	593.0	12 I 50	60.7
27 WF 145	452.0	10 WF 45	55.0
24 WF 130	369.2	14 WF 34	54.5
30 WF 108	345.5	12 I 40.8	52.5
24 I 120	298.0	12 WF 36	51.4
21 WF 112	278.0	8 WF 48	49.0
27 WF 94	277.7	14 WF 30	47.1
14 WF 142	254.8	10 WF 39	47.0
24 I 90	220.5	12 I 31.8	41.6
24 I 79.9	203.0	8 WF 40	39.9
24 WF 76	200.1	10 I 35	35.2
21 WF 62	144.1	10 I 25.4	28.0
20 I 65.4	137.3	8 WF 28	27.1
14 WF 78	134.0	8 WF 24	23.1
10 WF 100	130.1	8 I 23	19.2
14 WF 74	125.6	8 WF 20	19.1
16 WF 58	106.2	8 WF 17	15.8
10 WF 77	97.7	7 I 20	14.4

TABLE 10

TIMBER
AMERICAN STANDARD SIZES

PROPERTIES FOR DESIGNING

NATIONAL LUMBER MANUFACTURERS ASSOCIATION

Nominal Size	American Standard Dressed Size	Area of Section	Weight per Foot	Moment of Inertia	Section Modulus	Nominal Size	American Standard Dressed Size	Area of Section	Weight per Foot	Moment of Inertia	Section Modulus
in.	in.	in.²	lb	in.⁴	in.³	in.	in.	in.²	lb	in.⁴	in.³
2 × 4	1⅝ × 3⅝	5.89	1.64	6.45	3.56	10 × 10	9½ × 9½	90.3	25.0	679	143
6	5⅝	9.14	2.54	24.1	8.57	12	11½	109	30.3	1204	209
8	7½	12.2	3.39	57.1	15.3	14	13½	128	35.6	1948	289
10	9½	15.4	4.29	116	24.4	16	15½	147	40.9	2948	380
12	11½	18.7	5.19	206	35.8	18	17½	166	46.1	4243	485
14	13½	21.9	6.09	333	49.4	20	19½	185	51.4	5870	602
16	15½	25.2	6.99	504	65.1	22	21½	204	56.7	7868	732
18	17½	28.4	7.90	726	82.9	24	23½	223	62.0	10274	874
3 × 4	2⅝ × 3⅝	9.52	2.64	10.4	5.75	12 × 12	11½ × 11½	132	36.7	1458	253
6	5⅝	14.8	4.10	38.9	13.8	14	13½	155	43.1	2358	349
8	7½	19.7	5.47	92.3	24.6	16	15½	178	49.5	3569	460
10	9½	24.9	6.93	188	39.5	18	17½	201	55.9	5136	587
12	11½	30.2	8.39	333	57.9	20	19½	224	62.3	7106	729
14	13½	35.4	9.84	538	79.7	22	21½	247	68.7	9524	886
16	15½	40.7	11.3	815	105	24	23½	270	75.0	12437	1058
18	17½	45.9	12.8	1172	134						
						14 × 14	13½ × 13½	182	50.6	2768	410
						16	15½	209	58.1	4189	541
4 × 4	3⅝ × 3⅝	13.1	3.65	14.4	7.94	18	17½	236	65.6	6029	689
6	5⅝	20.4	5.66	53.8	19.1	20	19½	263	73.1	8342	856
8	7½	27.2	7.55	127	34.0	22	21½	290	80.6	11181	1040
10	9½	34.4	9.57	259	54.5	24	23½	317	88.1	14600	1243
12	11½	41.7	11.6	459	79.9						
14	13½	48.9	13.6	743	110	16 × 16	15½ × 15½	240	66.7	4810	621
16	15½	56.2	15.6	1125	145	18	17½	271	75.3	6923	791
18	17½	63.4	17.6	1619	185	20	19½	302	83.9	9578	982
						22	21½	333	92.5	12837	1194
6 × 6	5½ × 5½	30.3	8.40	76.3	27.7	24	23½	364	101	16763	1427
8	7½	41.3	11.4	193	51.6						
10	9½	52.3	14.5	393	82.7	18 × 18	17½ × 17½	306	85.0	7816	893
12	11½	63.3	17.5	697	121	20	19½	341	94.8	10813	1109
14	13½	74.3	20.6	1128	167	22	21½	376	105	14493	1348
16	15½	85.3	23.6	1707	220	24	23½	411	114	18926	1611
18	17½	96.3	26.7	2456	281	26	25½	446	124	24181	1897
20	19½	107.3	29.8	3398	349						
						20 × 20	19½ × 19½	380	106	12049	1236
						22	21½	419	116	16150	1502
8 × 8	7½ × 7½	56.3	15.6	264	70.3	24	23½	458	127	21089	1795
10	9½	71.3	19.8	536	113	26	25½	497	138	26945	2113
12	11½	86.3	23.9	951	165	28	27½	536	149	33795	2458
14	13½	101.3	28.0	1538	228						
16	15½	116.3	32.0	2327	300	24 × 24	23½ × 23½	552	153	25415	2163
18	17½	131.3	36.4	3350	383	26	25½	599	166	32472	2547
20	19½	146.3	40.6	4634	475	28	27½	646	180	40727	2962
22	21½	161.3	44.8	6211	578	30	29½	693	193	50275	3408

All properties and weights given are for dressed size only. The weights given above are based on assumed average weight of 40 lb per cubic foot.

TABLE 11

DEFLECTIONS AND SLOPES OF ELASTIC CURVES FOR VARIOUSLY LOADED BEAMS

Loading	Equation of Elastic Curve	
	Maximum Deflection	Slope at End
Cantilever with point load P at free end, length L	$v = \dfrac{P}{6EI}(2L^3 - 3L^2 x + x^3)$ $v_{max} = v(0) = \dfrac{PL^3}{3EI}$	$\theta(0) = -\dfrac{PL^2}{2EI}$
Cantilever with uniform load p_0, length L	$v = \dfrac{p_0}{24EI}(x^4 - 4L^3 x + 3L^4)$ $v_{max} = v(0) = \dfrac{p_0 L^4}{8EI}$	$\theta(0) = -\dfrac{p_0 L^3}{6EI}$
Simply supported beam with uniform load p_0	$v = \dfrac{p_0 x}{24EI}(L^3 - 2Lx^2 + x^3)$ $v_{max} = v(L/2) = \dfrac{5p_0 L^4}{384EI}$	$\theta(0) = -\theta(L) = \dfrac{p_0 L^3}{24EI}$
Simply supported beam with point load P at distance a from left	$v = \dfrac{Pb}{6EIL}\left[(L^2 - b^2)x - x^3 + \left(\dfrac{L}{b}\right)\langle x - a \rangle^3\right]$ See p. 398. When $a = b = \dfrac{L}{2}$, then $v = \dfrac{Px}{48EI}(3L^2 - 4x^2) \quad \left(0 \le x \le \dfrac{L}{2}\right)$ $v_{max} = v(L/2) = \dfrac{PL^3}{48EI}$	$\theta(0) = -\theta(L) = \dfrac{PL^2}{16EI}$
Simply supported beam with moment M_0 at left end	$v = -\dfrac{M_0 x}{6EIL}(L^2 - x^2)$ $v_{max} = v(L/\sqrt{3}) = -\dfrac{M_0 L^2}{9\sqrt{3}EI}$	$\theta(0) = -\dfrac{\theta(L)}{2} = -\dfrac{M_0 L}{6EI}$
Simply supported beam with two symmetric point loads P at distance a from ends	$v_a = v(a) = \dfrac{Pa^2}{6EI}(3L - 4a)$ $v_{max} = v(L/2) = \dfrac{Pa}{24EI}(3L^2 - 4a^2)$	$\theta(0) = \dfrac{Pa}{2EI}(L - a)$

Index

Abbreviations and symbols, xvii
Allowable stress:
    definition, 82
    table of, 554
Angle of twist:
    circular shaft, 153
    hollow shafts, 171
    rectangular shaft, 167
Angle sections, properties for designing, 560, 561
Anisotropy, 100
Area-moment method (*see* Moment-area method)
Areas, useful properties of, 555
Axes:
    neutral, 180, 262
    principal, 185, 186
Axial force, 23
Axial loads (*see also* Columns)
    definition, 23
    deformation due to, 123
    stresses due to, 70

Beam-columns, 517, 521
Beams:
    bending moments in, 23
    bending stresses in, 183, 193
    built-in (*see* Beams, fixed)
    cantilever, 15
    center of twist (*see* Shear center)

Beams (*cont.*):
    classification, 15
    constant strength, 361
    continuous, 15
    cover plated, 229
    crippling of, 359
    curved, 208, 503
    deflection (*see* Deflection)
    design of, 189, 356, 360
    elastic curve for, 46, 382, 389, 414
    elastic section modulus of, 189
    elastic strain energy in, 483
    fixed, 15
    flexure formula for, 183
    inelastic bending, 192
    lateral instability, 178, 550
    maximum bending stresses in, 189, 354
    neutral axis in, 180
    of two materials, 202
    of variable cross section, 360
    overhanging, 15
    plastic analysis of, 198, 462
    plastic section modulus of, 198
    prismatic, 225, 356
    radius of curvature, 180, 379
    reactions, 12, 16
    reinforced concrete, 205

565

Beams (cont.):
  restrained, 15
  shear in, 21
  shearing stresses in, 219, 229, 235, 354
  simple, 15
  simply supported, 15
  skew bending of, 260
  statically indeterminate, 9, 400, 450
  unsymmetrical, 185
Bearing stress, 72
Bending:
  of beams (see Beams)
  combined with axial loads, 255, 265, 521
  deflections due to, 378
  inelastic, 192
  pure, 177
  skew, 261
  strain energy in, 483
  stresses due to, 183
Bending moment:
  definition, 23
  diagrams, 25
  diagrams, sign convention for, 10, 24
  diagrams by summation method, 36
  and elastic curve, relation between, 46, 382
  and shear, relation between, 34
Biaxial stress, 67
Bifurcation point, 525
Body forces, 4, 68
Boltzmann's principle, 121
Boundary conditions:
  for bars in torsion, 156
  for beams in bending, 385
Bredt's formula for tubes in torsion, 171
Buckling:
  of beams, 178, 550
  of columns, 515, 526, 529
  of vacuum chambers, 352, 551
Bulk modulus, 315

Cantilever, 16
Castigliano's theorems, 488, 495
Center of twist (see Shear center)
Centroids of areas, 555
Channel sections, properties for designing, 559
Clapeyron's equation, 458
Coefficient:
  of thermal expansion, 104, 441
  of viscosity, 116

Columns:
  critical load, 528, 529
  critical stress, 531
  design, 535
  double-modulus theory for, 533
  eccentrically loaded, 534
  Euler's formula, 526
  formulas for concentrically loaded, 537
  secant formula, 534
  slenderness ratio, 531
  tangent modulus formula, 533
Combined stresses, equations for, 255, 262, 265, 273
Compatibility, 97, 135
Compound stresses, 252
Concentration of stress (see Stress, concentration factor)
Constant strength beams, 361
Constitutive relations, 93, 99, 109, 284, 316
Continuous beams:
  analysis, 402, 405, 447, 456
  definition, 16
Contraflexure, 46
Creep, 83
  of axially loaded bars, 120
  of bars in torsion, 164
  of beams, 387
Creep compliance, 121
Crippling of web, 359
Critical sections, 145, 352, 354
Curvature, radius of, 180, 379
Curved beams:
  deflection of, 503
  stresses in, 208

d'Alembert's principle, 5, 88
Deflection:
  of axially loaded rods, 123
  of beams:
    by Castigliano's theorem, 488
    due to impact, 507
    due to shear, 487
    integration methods for, 389
    moment-area method for, 411, 450
    by singularity functions, 404
    in skew bending, 408
    statically indeterminate, 400, 504
    strain energy method for, 484
    superposition method for, 406
    table of, 564
    virtual work method for, 496, 499
  of frameworks, 140, 438, 498

Deflection (*cont.*):
   of helical springs, 274
Design:
   of axially loaded members, 85
   of beams, 189, 198
   of columns, 535
   of complex members, 362
   of torsion members, 152
Deviation, tangential, 413
Diagrams:
   axial-force, 25
   bending-moment, 25
   shear, 25
Differential equation of elastic curve, 382
Differential equations of equilibrium, 32, 67
Dilatation, 315
Dirac delta function, 50
Doublet, 50
Dynamic loading, 136, 137, 507

Eccentric loading:
   of columns, 534
   of members, 265
   of short blocks, 267
Effective column length, 530
Efficiency of a joint, 352
Elastic curve, 46, 379, 414
Elasticity:
   definition, 111
   modulus of, 101
Elastic limit, 112
Elastic modulus, 101
Elastic strain energy:
   in bending, 483
   for multiaxial stress, 108
   for shearing stress, 107
   in torsion, 151
   for uniaxial stress, 105
Endurance limit, 84
Equilibrium:
   equations of static, 9
   stability of, 524, 542
Euler's formula, 521, 528

Factor of safety, 85
Factors of stress-concentration:
   in bending, 200, 213
   for helical springs, 274
   in tension or compression, 130
   in torsion, 162
Failure, theories of (*see* Yield and fracture criteria)

Fatigue, 84
Fiber stress, definition, 84
Flange, definition, 201
Flexibility coefficients, 126, 447
Flexural rigidity, 385
Flexure formula:
   for curved beams, 208
   for straight beams, 181
Formulas:
   for centroids and moments of inertia of areas, 555
   for deflection of beams, 564
Fracture criteria, 316
Frameworks, deflection of, 140, 438, 498
Free body, definition of, 4
Fringe, 342

Helical springs, 272
Hinge, plastic, 463
Hooke's law:
   for anisotropic material, 99
   generalized, 100
   for isotropic materials, 100
Hooks, stresses in, 208
Hoop stress, 346
Horsepower and torque relation, 145
Hysteresis, 114

I beams:
   crippling in, 359
   shearing stresses in, 236
   table of properties for designing, 556
Impact:
   deflection due to, 136, 507
   loading, 136
Indeterminate (*see* Statically indeterminate members)
Inelastic behavior:
   of beams, 192, 408, 462
   of torsion members, 158
Inertia, moment of, 183, 555
Inflection point, 46, 414
Interaction curve, 259
Internal work, 484, 496
Invariant:
   of strain, 310, 315
   of stress, 297, 315
Isoclinic, 340
Isotropy, definition, 100

Joints, 20, 351

Kelvin-Voigt solid (see Voigt-Kelvin solid)
Kern, 269
Keyways, 163

Lateral instability of beams, 178, 550
Limit analysis, 435, 462
Linear strain, definition of, 94, 96
Line of zero stress or strain, 257, 267
Loads:
  axial, 62
  concentrated, 13
  dead, 360, 374
  distributed, 14
  impact, 136, 507
  live, 360, 374
Localized stress (see Factors of stress-concentration)
Longitudinal stress in cylinder, 349

Margin of safety, 85
Materials:
  anelastic, 118
  homogeneous, 100
  isotropic, 100
  orthotropic, 100
  table of physical properties, 554
  viscoelastic, 116
Maximum Distortion Energy Theory, 319
Maximum Normal Stress Theory, 323
Maximum shearing stress, 290
Maximum Shearing Stress Theory, 316
Maxwell solid, 118
Members of two materials, 202
Membrane analogy for torsion, 169
Method of sections:
  for axially loaded members, 69
  for beams, 182, 190
  definition, 3
  for torsion members, 144
Mises' yield condition, 322
Modulus:
  bulk, 315
  of elasticity, 101
  of resilience, 113
  of rigidity, 101
  of rupture in bending, 196
  of rupture in torsion, 159
Modulus of elasticity related to modulus of rigidity, 102, 314
Mohr's circle:
  for strain, 308

Mohr's circle (cont.):
  for stress, 295, 302
Moment-area method:
  for determinate beams, 411
  for indeterminate beams, 450
  and superposition, 450
Moment diagrams (see also Bending moment):
  definition, 25
  method of constructing, 25, 36, 41
Moment of inertia:
  parallel-axis theorem for, 186
  of plane areas, 183, 186, 555
  polar, 147
  principal axes of, 185, 186
  table of, 555

Necking, 110
Neutral axis, 180
Neutral surface, 180, 185
Normal stress:
  in axially loaded members, 70, 334
  in bending, 183, 338
  definition, 63
  maximum and minimum, 290

Octohedral shearing stress, 322

Parallel-axis theorem, 186
Photoelastic method of stress analysis, 340
Pipe, standard steel, 562
Plane strain, 98
Plane stress, 67, 287
Plastic analysis:
  of beams, 198, 462
  of torsional members, 162
Point of inflection, 46
Poisson's ratio, 101, 102
Polar moment of inertia, 147
Potential, total, 541
Pressure vessels, thin-walled, 351
Principal axes of inertia, 186
Principal planes of bending, 185, 408
Principal strain, 309
Principal stress, 289
Properties:
  of angles, 560, 561
  of channels, 559
  of I beams, 556, 562
  of pipe, 562
  of rectangular timber, 563
  of WF beams, 557, 562

Proportional limit, 111

Radius:
  of curvature, 180, 379
  of gyration, 531
Ramberg-Osgood formula, 115
Reactions, calculation of, 16
Reinforced concrete beams, 205
Relation between E, G, and v, 102, 314
Relation between shear and bending moment, 34
Relations between stress, curvature, and bending moment, 381
Relaxation modulus, 121
Repeated loading, 83
Residual stress, 84, 130, 161, 198
Resilience, modulus of, 113
Restrained beams, 15, 450
Rigidity, flexural, 385
Rigidity, modulus of, 161
Rosettes, strain, 311
Rupture, modulus of, 159, 196

St. Venant's principle, 132
Secant formula for columns, 534
Section modulus:
  elastic, 189
  plastic, 198
  tables, 556–63
Shaft (*see* Torsion)
Shape factor, 198
Shear:
  and bending moment, relation between, 34
  definition, 21
  diagrams, 25
  diagrams by summation method, 34
  double, 74
  sign convention for, 10, 22
Shear center, 243
Shear diagonal, 295
Shear diagrams (*see* Shear)
Shear flow, 170, 223
Shearing deflections of beams, 487
Shearing force in beams (*see* Shear)
Shearing modulus of elasticity, 101
Shearing strain, 95, 96, 307
Shearing stress:
  in circular shafts, 148
  definition, 64
  formula for beams, 230
  maximum, 290
  in noncircular shafts, 166, 169

Shells of revolution, 344
Sign convention:
  for moment, 10, 24
  for shear, 10, 22
  for stresses, 64
Simple beam, definition, 15
Singularity functions, 47, 404
Skew bending, 261
Slenderness ratio, 531
S–N diagrams, 84
Spherical pressure vessels, 349
Spring constant, 126, 276
Springs, helical:
  deflection of, 274
  stresses in, 272
Stability of equilibrium, 524, 542
Statical moment of area, 224
Statically indeterminate beams:
  by Castigliano's theorem, 493
  definition, 9, 401
  by integration, 400
  by moment-area, 450
  by superposition, 406
  by three-moment equation, 456
  by virtual work, 504
Statically indeterminate members:
  axially loaded, 434
  in frames, 438
  in torsion, 436
Strain:
  definition, 95
  due to temperature, 104, 441
  irrotational, 98
  linear, 94, 96
  maximum, 309
  Mohr's circle for, 308
  natural, 94
  plane, 98
  principal, 309
  pure, 98
  residual, 120
  shearing, 95, 96
  tensor, 97
  thermal, 104
  transformation of, 304
Strain energy (*see also* Elastic strain energy):
  complementary, 106, 495
Strain rosettes, 311
Strength, ultimate, 82
Stress:
  allowable, 82, 554
  bearing, 72, 74

Stress (cont.):
  bending, 183
  biaxial, 67
  combined (see Combined stresses)
  compound, 252
  concentration factor:
    for axially loaded members, 130
    in bending, 200, 213
    definition, 132
    in springs, 274
    in torsion, 163
  conditions for uniform, 71
  critical, 531
  in curved bars, 208, 258
  definition of, 62
  fiber, 84
  flexure, 183
  hoop, 346
  impact, 136
  on inclined planes, 284, 288
  maximum and minimum normal, 290
  maximum shearing, 290
  Mohr's circle for, 295
  normal, 63, 290
  plane, 67
  at a point, 284
  principal, 289
  residual, 84, 130, 161, 198
  shearing, 64, 72
  state of, 283
  tangential, 346
  tensile, 64
  tensor, 64, 65, 320
  three-dimensional, 66, 320
  torsional shearing, 147, 158, 166, 169
  transformation of, 283, 284
  triaxial, 67
  true, 110
  two-dimensional, 67
  uniaxial, 67
  working, 86
  yield point, 111
Stress-strain:
  diagrams, 110, 114
  idealizations, 114
Stress trajectories, 340
Suddenly applied loads, 136, 507
Superposition:
  of deflections, 406, 408, 444
  principle of, 99
  statically indeterminate problems solved by, 448, 450
  of stresses, 160, 198, 255

Supports, diagrammatic conventions for, 11

Tables, index, 553
Tangential deviation, 413
Tangent modulus, definition, 115
Temperature or thermal stresses, 441
Tensile stress, 64
Tensor:
  deviatoric stress, 320
  dilatational stress, 320
  distortional stress, 320
  spherical stress, 320
  strain, 97
  stress, 64, 65
Theories of failure (see Yield and fracture criteria)
Thermal stresses, 441
Thin-walled pressure vessels, 349
Three-moment equation, 456
Thrust, 23
Timber, sizes, properties for designing, 563
Torque, internal, 145
Torsion:
  angle of twist due to, 155
  assumptions of theory, 145
  of circular shafts, 147, 152
  elastic energy in, 151
  formula, elastic, 147
  of hollow members, 148, 169
  inelastic, 158
  of noncircular solid bars, 166
  of rectangular bars, 167
  of viscoelastic circular bars, 164
Toughness, 113
Trajectories, stress, 340
Transformation:
  of strain, 304
  of stress, 67, 284
Transformed sections, 202
Tresca yield condition, 319
Triaxial stress, 67
True stress, 110
Twist, angle of (see Torsion)

Ultimate strength, 82
Uniaxial stress, 67, 70
Uniform strength beams (see Constant strength beams)
Uniform stress, conditions for, 71
Unit impulse (or step) function, 50
Unsymmetrical bending (see Skew bending)

Unsymmetrical sections subjected to bending, 185

Variable cross-section, beams of, 361
Vertical shear (*see* Shear)
Virtual load or force, 497
Virtual work method for deflections, 496
Viscoelastic behavior, 116, 164
Voigt-Kelvin solid, 116

Wahl correction factor for helical springs, 274
Web, definition, 201
Wide flange beams, properties for designing, 557
Working stress, 86

Yield and fracture criteria, 283, 316, 324
Yield point or strength, 111
Young's modulus, 101